Biochemistry and Molecular Biology of Parasites

Biochemistry and Molecular Biology of Parasites

Edited by

J. JOSEPH MARR

Ribozyme Pharmaceuticals Inc., 2950 Wilderness Place, Boulder, CO 80301, USA

and

MIKLÓS MÜLLER

The Rockefeller University, 1230 York Avenue, New York, NY 10021, USA

ACADEMIC PRESS

Harcourt Brace & Company, Publishers

London   San Diego   New York
Boston   Sydney   Tokyo   Toronto

ACADEMIC PRESS LIMITED
24–28 Oval Road
LONDON NW1 7DX

US Edition Published by
ACADEMIC PRESS INC.
San Diego, CA 92101

This book is printed on acid free paper

A catalogue record for this book is available from the British Library

ISBN 0-12-473345-X

Cover illustration
The editors are indebted to Dr James McSwiggen of Ribozyme Pharmaceuticals, Inc., for the computer re-creation of the ribozyme pictured on the front cover.

Typeset by Doyle Graphics Ltd, Tullamore, Co. Offaly, Republic of Ireland

Printed and bound in the United Kingdom
Transfered to Digital Printing, 2011

Contents

Contributors

C. J. Bacchi Haskins Laboratories and Department of Biology, 41 Park Row, Pace University, New York, NY 10038-1502, USA

R. L. Berens Division of Infectious Diseases, University of Colorado Health Sciences Center, 4200 East Ninth Avenue, Denver, CO 80262, USA

G. H. Coombs Parasitology Laboratory, Institute of Biomedical and Life Sciences, Joseph Black Building, University of Glasgow, Glasgow G12 8QQ, Scotland, UK

R. E. Davis Department of Zoology, University of Wisconsin-Madison, Madison, Wisconsin 53706, USA

R. Docampo Department of Veterinary Pathobiology, University of Illinois, Urbana, IL 61801, USA

J. F. Dubremetz Unité 42 INSERM, 369 rue J Guesde, 59650 Villeneuve d'Ascq Cedex-France

R. H. Fetterer United States Department of Agriculture, Agricultural Research Service, Livestock and Poultry Sciences Institute, Parasite Biology and Epidemiology Laboratory, Beltsville, MD 20705, USA

W. R. Fish Pediatric Endocrinology, State University of New York — Health Science Center at Syracuse, 750 East Adams Street, Syracuse, NY 13210, USA

H. R. Gamble United States Department of Agriculture, Agricultural Research Service, Livestock and Poultry Sciences Institute, Parasite Biology and Epidemiology Laboratory, Beltsville, MD 20705, USA

T. G. Geary Animal Health Discovery Research, Upjohn Laboratories, Kalamazoo, MI 49001, USA

B. G. Harris Department of Biochemistry and Molecular Biology, The University at North Texas, Health Science Center, 3516 Camp Bowie Boulevard, Fort Worth, TX 76107-2699, USA

R. Komuniecki Department of Biochemistry, University of Toledo, Toledo, OH 43605, USA

E. C. Krug Division of Infectious Diseases, University of Colorado Health Sciences Center, 4200 East Ninth Avenue, Denver, CO 80262, USA

†B. T C. Lockwood Laboratory for Biochemical Parasitology, Department of Zoology, University of Glasgow, Glasgow G12 8QQ, Scotland, UK

†Died 8 October 1993.

J. H. McKerrow   Departments of Pathology, Medicine and Pharmaceutical Chemistry, University of California, San Francisco, CA 94143 and Department of Veterans Affairs Medical Center, San Francisco, CA 94121, USA

J. J. Marr   Ribozyme Pharmaceuticals Inc., 2950 Wilderness Place, Boulder, CO 80301, USA

M. Müller   The Rockefeller University, 1230 York Avenue, New York, NY 10021, USA

T. Nilsen   Department of Molecular Biology and Microbiology, Case Western Reserve University School of Medicine, 10900 Euclid Avenue, Cleveland, OH 44106, USA

M. J. North   Department of Biological and Molecular Sciences, University of Stirling, Stirling FK9 4LA, Scotland, UK

F. R. Opperdoes   International Institute of Cellular and Molecular Pathology and University of Louvain, ICP-TROP/7439, avenue Hippocrate. 74, B-1200 Brussels, Belgium

M. Parsons   Seattle Biomedical Research Institute, 4 Nickerson Street, Seattle, WA 98109, and Department of Pathobiology, University of Washington, Seattle, WA 98195, USA

A. O. W. Stretton   Department of Zoology, University of Wisconsin-Madison, Madison, WI 53706, USA

D. P. Thompson   Animal Health Discovery Research, Upjohn Laboratories, Kalamazoo, MI 49001, USA

J. W. Tracy   Departments of Comparative Biosciences and Pharmacology, and Environmental Toxicology Center, University of Wisconsin-Madison, Madison, WI 53706-1102, USA

S. J. Turco   Department of Biochemistry, University of Kentucky Medical Center, Lexington, KT 40536, USA

B. Ullman   Department of Biochemistry and Molecular Biology, Oregon Health Sciences University, 3181 SW Sam Jackson Park Road, Portland, OR 97201, USA

E. Ullu   Department of Internal Medicine and Cell Biology, Yale University School of Medicine, 333 Cedar Street/LCI 801, New Haven, CT 06520-8022, USA

J. F. Urban   United States Department of Agriculture, Agricultural Research Service, Livestock and Poultry Sciences Institute, Parasite Immunobiology Laboratory, Beltsville, MD 20705, USA

E. A. Vande Waa   Department of Comparative Biosciences, University of Wisconsin-Madison, 2015 Linden Drive West, Madison, WI 53706-1102, USA

N. Yarlett   Haskins Laboratories and Department of Biology, 41 Park Row, Pace University, New York, NY 10038-1502, USA

Preface

Parasitology developed as a part of medicine practised in the tropical and subtropical areas of the globe. Later, it also became a major veterinary discipline. Although it has remained in these roles, it has taken on the characteristics of a science in its own right. The organisms which comprise this discipline are varied in their morphologies and astoundingly complex in their life cycles. In the course of these lives they enter and leave ecosystems and thereby undergo significant metabolic and genetic alterations. For these reasons parasitic organisms have become model systems for the study of biochemistry and molecular biology. The phenomena of aerobic fermentation, compartmentalization of enzyme systems, metabolic shifts to accompany morphologic changes, rapid alterations in membrane chemistry, and genetic changes associated with adaptation to the host have recommended parasites to the attention of scientists of many disciplines.

The objective of this volume is to present the modern scientific disciplines which have parasitology as a common resource in a setting that will allow and encourage the reader to place the biochemistry and molecular biology of these organisms in their biological context. The format is of a multiauthor volume, in keeping with the many branches of science which have made their places in parasitology. The traditional separations within these have been eliminated in so far as possible to facilitate correlation. The chapters are cross-referenced and grouped in a manner which should be self-reinforcing. For this reason, for example, helminth intermediary metabolism is placed with that of protozoa in order to draw the appropriate parallels and contrasts, rather than placing it in a section devoted to helminths. We believe that this organizational arrangement will permit the reader to create a composite of the biochemistry of all these organisms and not be distracted by their morphological and taxonomic differences. We believe these correlations are important since this discipline has traditionally been taught by artificially subdividing it according to the number of cells in the organism, the residence of the parasite within the intestinal lumen or tissues of the host, or morphology. Science has demonstrated the fundamental unity of biochemistry and molecular biology and parasitology is ripe for unifying concepts.

The volume begins with molecular biology which is presented as it relates to the cell biology of these organisms. Although protozoa have been exploited to great advantage in the understanding of molecular biology, the knowledge derived is conceptual to molecular biology and not restricted to protozoa. The reader is referred to molecular biology texts for information pertinent to that discipline and to standard biochemistry texts for corresponding information. Where the molecular biology is relevant to parasitic organism function as we understand it, it has been included and is, therefore, also distributed throughout the text where appropriate.

Carbohydrate metabolism and energetics are given first consideration in the biochemistry discussions. More is known of these than other areas of biochemistry of these organisms. The aerobic and anaerobic metabolism of protozoa is contrasted with the fundamentally anaerobic metabolism of helminths. The theme of aerobic fermentation can be found among all of these organisms. This is followed by amino acid and protein metabolism; much new information on proteases is included in this section. In the chapter on purine and pyrimidine metabolism the common theme of *de novo* purine synthesis in protozoa is the framework upon which the species' variations are presented. Pyrimidine synthesis is more varied since effective salvage pathways exist in virtually all organisms. Polyamine metabolism, a basic discipline which, like purine metabolism, has emerged as a promising source of chemotherapeutic possibilities, is presented as a separate section. Lipid metabolism is understood best in relatively few species and, for this reason, we have elected to concentrate on these organisms rather than attempt to provide a compendium of information which cannot yet be understood in its biological context. Nucleic acid and protein synthesis are the only anabolic functions which have been kept separate. The others are captured within the foregoing chapters since they are integral to the catabolic activities of those pathways. Nucleic acid and protein synthesis provide compounds which are relatively stable and represent end-products in the organism. Antioxidant mechanisms and the metabolism of xenobiotics represent metabolism directed to the defense of the organism against exogenous chemicals. They are placed at the end of the sections of intermediary metabolism since they draw upon that knowledge. Antioxidant mechanisms have been given separate treatment because they have been well studied for a longer time, are more focused and, therefore, better understood.

Cell surfaces are considered next in the transition from biochemistry into physiology. The membrane glycoproteins of protozoans are presented both with respect to their biochemistry and the dynamics of their insertion into this cellular structure. This is correlated with the section on lipid metabolism. Surfaces of helminths are next discussed in order to contrast these multicellular organisms with their single-celled counterparts. Cyst structures, although originally considered for inclusion, were reluctantly eliminated since the state of the art does not yet permit the correlations with biology and physiology that can be made for cell membranes. Cell organelles make an appearance as the volume progresses toward consideration of correlative multicellular physiology. These organelles provide many of the mechanisms which allow intracellular functions to occur and, thereby, permit intercellular activities to supervene. Some of the best understood aspects of physiology are in the neuromuscular systems of helminths and in their reproduction and development. Invasion mechanisms of protozoans are presented as an aspect of physiology of these cells since they require a co-opting of the biochemistry of the receiving organism and, therefore, an intercellular interaction. These conclude the integrative sections of the volume. The final chapter is a review of the foregoing information with respect to current chemotherapy and an attempt to predict where these basic sciences may be applied to medicine in the future.

Immunological aspects of parasitology, a field which is both broad and detailed, has not been included. This was done in order to contain the size and maintain the focus of the volume. This has required that certain other important aspects of cell biology also be excluded, such as the antigenic shifts of trypanosomes. These are regrettable but necessary omissions and are well described in other texts of molecular biology and immunology. The biochemistry and cell biology of the variable surface glycoprotein have been included, however.

We are proud of the group of collaborating authors in this volume. The knowledgeable scientist will recognize their contributions. We believe they have written clearly, comprehensively and well. Presentations by these seasoned investigators should be of interest to the experienced investigator, the graduate student and the newcomer.

We must list first among the acknowledgements our authors. Each has provided contributions but also has reviewed the contributions of others. This has given an internal perspective which was of great assistance to the editors. External reviewers have been thoughtful and generous with their criticism and the volume has benefited accordingly. The patience of Academic Press and Dr Tessa Picknet in particular are gratefully acknowledged. Particular thanks are owed to Jean Smith, in Boulder, Colorado and to Karrie Polowetzky in New York who were of enormous assistance in organizing and preparing the manuscript.

J. JOSEPH MARR and
MIKLÓS MÜLLER

1 Molecular Biology of Protozoan and Helminth Parasites

ELISABETTA ULLU[1] and TIMOTHY NILSEN[2]

[1]Department of Internal Medicine and Cell Biology, Yale University School of Medicine, New Haven, CT and [2]Department of Molecular Biology and Microbiology, Case Western Reserve University School of Medicine, Cleveland, OH, USA

SUMMARY

The intent of this chapter is to offer an overview of selected aspects of the molecular biology of parasitic protozoans and helminths. Important topics, i.e. DNA rearrangements associated with antigenic variation in protozoans and chromosome diminution in nematodes have been omitted. Nevertheless, it should be apparent that molecular analysis of parastic organisms has been remarkably productive in revealing unusual and unexpected pathways of gene expression. We emphasize that, although such phenomena as *trans*-splicing and RNA editing were discovered in parasites, they are not restricted to parasites, and thus cannot be considered adaptations to parasitism. This in no way diminishes the importance of the discoveries and there is every reason to suspect that further investigations in parasitic organisms will continue to provide novel insights into mechanisms of eukaryotic gene expression in general while simultaneously suggesting targets for chemotherapeutic intervention.

1. INTRODUCTION

Within the last ten years, the application of modern molecular biological approaches has provided a wealth of knowledge regarding gene structure, organization and expression in parasitic organisms. It is difficult to provide broad coverage of all topics

Biochemistry and Molecular Biology of Parasites
ISBN 0-12-473345-X

of interest in protozoan and helminth molecular biology. This discussion is restricted to trypanosomatids as 'representative' protozoans and nematodes as 'representative' helminths. This choice is dictated by the fact that these organisms, because of their experimental tractability, have been most fruitful in yielding biochemical insight into parasite gene expression. The three most important areas, transcription, *trans*-splicing and RNA editing were selected because, in addition to providing important insight into parasite gene expression, they directly impinge upon our understanding of the molecular biology of eukaryotic cells in general. In this regard, basic principles of molecular biology will not be discussed. The reader is referred to any of several excellent textbooks for appropriate background.

2. TRANSCRIPTION IN TRYPANOSOMES AND NEMATODES

2.1. Trypanosome Gene Expression (an Overview)

Compared to other eukaryotic organisms, trypanosomes are unusual in that they contain a number of tandem arrays of genes coding for housekeeping and developmentally regulated proteins, i.e. tubulins, calmodulin, ubiquitin, surface antigens, glycolytic enzymes. Although most protein coding genes appear to be repetitive, some single copy genes have been identified. The spacing between protein coding regions is quite short and ranges from less that one hundred to a few hundred nucleotides. This appears to be the case between repeating units within a gene cluster and also between genes encoding different proteins. This degree of packaging of genetic information is somewhat higher than in other unicellular eukaryotic organisms of similar genetic complexity like the yeast *Saccharomyces cerevisiae*.

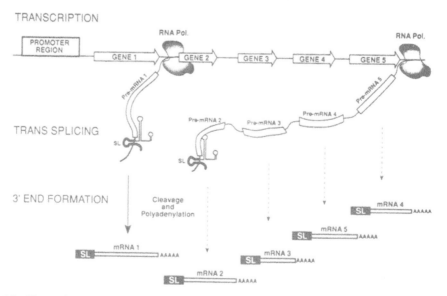

FIG. 1.1   Biogenesis of trypanosomatid mRNA. A schematic view of trypanosome gene expression. Mature mRNAs are generated from polycistronic pre-mRNAs via two RNA processing reactions, *trans*-splicing and polyadenylation. For details, see text.

Messenger RNA molecules of trypanosomatids are chimaeras of two different gene transcripts, the SL (spliced leader) RNA, which contributes the 5' end of the mature mRNA, and the pre-mRNA, which provides the body of the messenger RNA coding for the various proteins; joining of these molecules occurs by *trans*-splicing (Fig. 1.1). Thus, in trypanosomatids, the 5' end of mRNA is formed by an RNA processing reaction rather than by transcription initiation as is the case in most higher eukaryotes. In this context *trans*-splicing can be viewed as an alternate and unusual pathway of mRNA 5' end formation.

The formation of trypanosome mRNA 3' ends is thought to occur, as it does in higher eukaryotes, through an RNA processing reaction which involves cleavage at the polyadenylation site with the concomitant addition of a poly(A) tail to the newly formed mRNA 3' end. However, no consensus sequence similar to the AAUAAA polyadenylation signal has been identified in trypanosomatid mRNAs and it is at present unclear whether specific pre-mRNA signals are required for mRNA 3' end formation. In summary, the biogenesis of mature trypanosome mRNA takes place exclusively *via* RNA processing and requires two cleavage reactions of the pre-mRNA, one at the splice acceptor site (discussed in section 3 below) and the other at the site of addition of the poly(A) tail.

2.2. Trypanosomatid Genes Are Organized as Polycistronic Transcription Units

Soon after it was realized that the 5' end of mature mRNA is produced by RNA processing, the structure of the mRNA primary transcripts which are the substrates for *trans*-splicing became a major point of interest. In particular, because of the tandem reiteration of many genes and the small size of intergenic sequences, it was suggested that trypanosomatid transcription units might encompass more than one gene and that the corresponding primary RNA transcripts might be polycistronic. Indeed a considerable amount of evidence accumulated suggesting that this was the case (1–3). In particular: (i) nuclear run-on assays and nuclease protection assays of newly synthesized RNA *in vivo* showed that intergenic regions are continuously transcribed together with the protein coding regions; (ii) the kinetics of UV inactivation were consistent with the presence of a promoter sequence located upstream of large chromosomal regions which included several genes; (iii) dicistronic and polycistronic transcripts were identified in steady-state RNA and their accumulation was increased by treatments which inhibited *trans*-splicing. However, all the above evidence was rather indirect. Convincing proof that polycistronic transcription units could be functional came when efficient DNA transfection and expression systems were established for a number of trypanosomatids. Using synthetic constructs under controlled conditions, it has been demonstrated that intergenic regions from a number of gene clusters do not possess detectable promoter activity. However, the same intergenic regions, when sandwiched between two reporter genes in the form of dicistronic transcription units under the control of the PARP promoter (see below), provide RNA processing signals of polyadenylation and *trans*-splicing of the upstream and downstream mRNA, respectively.

Although dicistronic and polycistronic mRNAs can be identified in steady-state RNA, two lines of evidence suggest that these molecules probably are not authentic precursors to mature mRNAs. First, in *Trypanosoma brucei* cells, *trans*-splicing at three different splice acceptor sites (the α and β tubulin, and the actin RNA coding regions), occurs within less than one minute from the time the polymerase has traversed the splice site (4). Considering the average rate of elongation of RNA polymerase II, it is almost certain that *trans*-splicing of pre-mRNA is accomplished co-transcriptionally. Second, in isolated tryanosome nuclei it appears that cleavage at the polyadenylation site of a heat shock protein (HSP 70) pre-mRNA occurs on nascent RNA chains (5). Thus, it is unlikely that polycistronic pre-mRNAs are synthesized *in toto*. In light of these findings the dicistronic and polycistronic mRNAs detected in steady-state conditions most likely represent aberrant products which have failed to undergo complete processing by the *trans*-splicing and polyadenylation machineries.

Polycistronic transcription units are typical of prokaryotic organisms and of some eukaryotic viruses. Trypanosomes (and nematodes which also *trans*-splice, see below) represent the only examples of eukaryotes which transcribe their genes in this way. At present, there is no evidence that the various genes encoded in polycistronic transcription units in trypanosomes are related in their function as are operons in bacteria. Likewise, we do not know what the average size of a transcription unit is in trypanosomatids and whether there is any attenuation of transcription in regions that are promoter-distal relative to those that are promoter-proximal. This latter consideration has obvious relevance for the control of gene expression in these organisms.

2.3. Promoter Structure of Trypanosome Protein Coding Genes

Transcription of most protein coding genes in eukaryotic cells, including trypanosomes, is carried out by RNA polymerase II in concert with a variety of general and specific transcription factors. However, in African trypanosomes there are two known exceptions to this rule: the gene families coding for the variant surface glycoproteins (VSG) of bloodstream forms and the *Procyclic Acidic Repetitive Protein* (PARP). Whereas RNA polymerase II-mediated transcription is sensitive to low concentrations of the elongation inhibitor α-amanitin, transcription of both VSG and PARP genes is resistant to high concentrations of the inhibitor. This initial finding, corroborated by other lines of evidence, supports the notion that RNA polymerase I, the same RNA polymerase that transcribes ribosomal RNA genes, is responsible for transcription of these two gene families (6). However, the lack of an *in vitro* transcription system precludes at present the unequivocal identification of RNA polymerase I as the relevant polymerase of VSG and PARP gene transcription. Notwithstanding this ambiguity, the early analysis of the mode of transcription of these two gene families combined with recent advances at introducing foreign DNA into trypanosomatids, have provided the basis for the conclusive identification of promoters in *T. brucei*. *Cis*-acting DNA sequences which are necessary and sufficient for transcription initation of these gene families (promoters) have been identified. For the VSG locus the promoter sequences lie approximately 45–40 kb upstream from the VSG gene (7) whereas the PAPP promoter is within a few hundred nucleotides upstream of its mRNA body. A detailed mutational analysis of the PARP promoter sequence has revealed that the promoter structure is complex, composed of at least three sequence elements located within 250 bp 5′ to the transcription start site (8, 9). The topological arrangement of both of these sequence elements (whose integrity is essential for promoter activity) in relation to each other and to the transcription start site is reminiscent of typical promoters for RNA polymerase I.

Mutational analysis of the PARP promoter showed that promoter sequences and signals required for *trans*-splicing of the pre-mRNA do not overlap. These observations stimulated a series of elegant experiments which demonstrated that a chimaeric transcription unit consisting of the *T. brucei* ribosomal RNA promoter linked to a 3' splice site acceptor region and followed by a reporter gene is efficiently transcribed and *trans*-spliced in *T. brucei* cells (6). These are the only known examples of RNA pol I-mediated expression of eukaryotic protein coding genes and they underscore the possibility that some endogenous trypanosome genes, like the VSG and the PARP genes, might indeed be transcribed by RNA polymerase I.

Although in the past years, much has been learned about the structure of VSG and PARP gene promoters, at present essentially nothing is known about the promoter structure of the majority of protein coding genes in trypanosomatids. These are the genes which are transcribed by an RNA polymerase with the α-amanitin sensitivity typical of other eukaryotic RNA polymerase IIs. Thus far, experiments aimed at measuring the density of RNA polymerase molecules along chromosomal regions transcribed by RNA polymerase II have failed to detect discontinuities in the transcription maps which would be consistent with the presence of promoter sequences. Furthermore, attempts to demonstrate promoter activity in trypanosome DNA sequences located immediately upstream or in between genes of a tandem array have met with no success. The available evidence, although of a negative nature, points to the possibility that promoters for RNA polymerase II might be quite sparse in the trypanosomatid genome. A more provocative, but perhaps less likely, interpretation is that typical RNA polymerase II promoter sequences could be absent from trypanosomatid DNA. However, if this were true, how would RNA polymerase II initiate transcription? Possibly, the transcription complex has little, if any, sequence specificity and could initiate transcription at many different locations along chromosomal DNA. Such an explanation could account for the homogeneous rates of transcription observed at several chromosomal loci.

2.4. Regulation of Gene Expression in Trypanosomes

Although it is not known for sure how many RNA polymerase II promoters exist in the genome of trypanosomatids, it is now generally accepted that the general mode of transcription of both housekeeping and developmentally regulated protein coding genes is polycistronic. Since for most protein coding genes there is no evidence of transcriptional control, how is gene expression regulated in trypanosomatids? Relevant to this is the observation that in steady-state conditions the cellular representation of different mature mRNAs (even those derived from the same transcription unit) can vary widely. The best studied example comes from the VSG transcription unit, which consists of the VSG gene and several upstream genes. Here mRNAs encoded from the upstream genes are much less abundant than the VSG mRNA itself. Since nuclear run-on experiments indicate that the rate of transcription along the VSG transcription unit is quite homogeneous, it has been inferred that mRNA abundance is primarily regulated by post-transcriptional mechanisms (10). These could include pre-mRNA stability, *trans*-splicing or mRNA 3' end maturation (see Fig. 1.1). Each of these steps could conceivably be regulated and could modulate the output of mature mRNA from a given gene. Indeed, differential mRNA stability has been proposed to be a major factor in the developmental regulation of the VSG and PARP gene products during differentiation of African trypanosomes. At present there is no direct evidence that *trans*-splicing is a regulating process nor that the rate of *trans*-splicing varies between different splice acceptor sites. The same is true for the process of mRNA 3′ end formation. However, some experiments support the notion that these post-transcriptional mechanisms may play a pivotal role in regulating gene expression in trypanosomatids. For instance, it has been demonstrated that when *trans*-splicing is inhibited by destroying either U2, U4 or U6 snRNAs (see below), unspliced tubulin pre-mRNA is rapidly degraded, possibly because the 5′ end of newly synthesized pre-mRNA is not capped (4). These findings suggest that the output of mature mRNA from the corresponding pre-mRNA could be determined by the balance between the rate of pre-mRNA degradation and the rate of *trans*-splicing at a given splice acceptor site in the pre-mRNA. Further support for this notion comes from the observation that inhibition of *trans*-splicing results in almost complete inhibition of the formation of the 3′ end of mature mRNA molecules (11). This finding has led to the hypothesis that in trypansomatids there might be a hierarchical order in the pathway of pre-mRNA processing reactions. In this model *trans*-splicing, or perhaps assembly of a *trans*-splicing complex on the pre-mRNA, is a prerequisite for subsequent recognition of the mRNA 3′ end formation and polyadenylation signals. An alternative and intriguing possibility is that in the pre-mRNA the signals required for *trans*-splicing partially or perhaps completely overlap with those determining the choice of the polyadenylation sites.

2.5. Transcription of Nematode Protein Coding Genes

The discussion above should make it clear that there are significant gaps in our understanding of transcription in trypanosomatids. Unfortunately, even less is known regarding transcriptional control elements and transcription units encoding mRNAs in nematodes. Only a few promoters have been tentatively mapped and this work was carried out exclusively in the free-living nematode, *Caenorhabditis elegans*. Our knowledge of promoters and transcription units in other nematodes is non-existent. However, there are intriguing parallels between trypanosome gene expression and nematode gene expression which suggest that lessons learned in trypanosomes may be applicable to nematodes. Nematodes also carry out *trans*-splicing and recent evidence from work in *C. elegans* suggests that *trans*-splicing may serve the same function as it does in trypanosomes [i.e. processing polycistronic transcription units (12)].

In *C. elegans* two distinct SL RNAs are used for *trans*-splicing (SL1 and 2). Investigations designed to determine the specificity of SL1 or SL2 addition revealed that SL2-accepting pre-mRNAs were encoded by genes that were located only a short distance (~ 100 nt) 3′ of upstream genes transcribed in the same orientation. Transgenic analysis using a specific gene pair demonstrated that promoter elements driving expression of the SL2 accepting gene were not present in either the intergenic region or in the upstream gene itself. Instead expression of both genes was driven by elements well 5′ of the upstream gene. In additional experiments, a heat shock promoter was inserted immediately in front of the gene pair. In this case, mature message for the downstream gene was made, it contained the SL2 leader sequence, and its appearance was dependent on heat shock. This experiment demonstrated that when a polycistronic RNA was created artificially, it was capable of yielding mature, correctly *trans*-spliced mRNA. From these and other analyses, it seems clear that polycistronic transcription units exist in *C. elegans*. Furthermore, it appears that a major (if not the only) determinant of SL2 addition in *C. elegans* is that the accepting pre-mRNAs be encoded by genes internal in such transcription units.

These studies may have a broader significance to our understanding of the biological role of nematode *trans*-splicing in general. In this regard, it seems that the demonstration of polycistronic transcription units in nematodes is of fundamental importance. If such transcription units are common in nematodes, it suggests that the major function of *trans*-splicing in these organisms (as in trypanosomes) is in the maturation of 5' ends of mRNAs located within long poly-pre-mRNAs. In this view, SL2 *trans*-splicing in *C. elegans* would reflect a specialized form of *trans*-splicing which is used only when adjacent genes are in close proximity to each other. SL1 *trans*-splicing in *C. elegans*, and *trans*-splicing in other nematodes [which apparently lack alternative SL RNAs (see section 3)] would be used when adjacent pre-mRNAs are more widely spaced in the primary transcript. Although this notion has intrinsic appeal in that it would provide a functional link between *trans*-splicing in nematodes and trypanosomes, it remains largely speculative. Clearly, systematic transcription unit mapping is necessary before any conclusions can be drawn regarding the role (if any) of nematode *trans*-splicing in processing polycistronic mRNAs.

2.6. Transcription of the Nematode SL RNA

In contrast to our dearth of knowledge regarding promoters for mRNAs in nematodes, we have a detailed understanding of transcription of nematode *trans*-spliced leader RNAs. These RNAs resemble the U snRNAs (see section 3) which are required for *cis*-splicing and are transcribed by RNA polymerase II. Dissection of the SL RNA gene promoter became possible when it was shown that the *Ascaris* SL RNA gene was accurately and efficiently transcribed by RNA polymerase II from cloned templates in *Ascaris* embryo extracts (13, 14). This cell-free system permitted the identification of SL RNA transcriptional control elements through mutational analysis. These experiments revealed that the SL RNA gene represents an unusual RNA polymerase II transcription unit.

Efficient initation of SL RNA synthesis required two sequence elements, one of which was centered approximately 50 nucleotides upstream from the transcriptional start site. Remarkably, the second sequence element was the 22nt SL sequence itself; mutations in the SL sequence abolished transcription *in vitro*. DNase I footprinting showed that the SL sequence bound a protein factor; the boundaries of the footprint exactly coincided with the SL sequence. Competition experiments indicated that binding of the factor was directly correlated with transcription (13). Further analysis has indicated that the 22nt binding factor is ∼ 60 kDa protein. It will be of considerable interest to see if this factor shares any homology with known transcription factors involved in RNA polymerase II transcription.

Additional experiments characterizing SL RNA transcription showed that 3' end formation (termination) of SL RNA synthesis *in vitro* was also unusual in that it depended exclusively upon gene-internal sequences. This contrasts to the situation in vertebrate U snRNA genes where the primary determinant of 3′ end formation is the so called 3′ end formation box located a short distance downstream of mature 3′ ends (15, 16).

The organization and expression of the SL RNA gene has some interesting implications. The nematode SL sequence has been perfectly conserved in widely diverged nematodes. This conservation was commonly interpreted to mean that the 22nt SL sequence present on *trans*-spliced mRNAs must have some important post-transcriptional function. The finding that the SL sequence is an essential promoter element for its own synthesis suggests an alternative explanation for sequence conservation, i.e. the conservation could be dictated by constraints imposed by the binding specificity of a transcription factor.

3. TRANS-SPLICING

3.1. Overview

Intermolecular (*trans*)-splicing is, by definition, an RNA processing reaction which precisely joins exons derived from separately transcribed RNAs. Two types of *trans*-splicing are known to occur in nature. The first type is directly analogous to the *cis*-splicing of group II introns and is confined to 'split' introns in organelles of plants and fungi. The second type is analogous to the snRNP-mediated removal of introns from nuclear pre-mRNAs and is known as spliced-leader addition *trans*-splicing (17–20). This type of *trans*-splicing has been demonstrated in a variety of lower eukaryotes including trypanosomatids, *Euglena*, trematodes and nematodes (21–23). In the trypanosomes all pre-mRNAs acquire their 5′ terminal exon (the spliced leader, SL) from a small RNA termed the SL RNA. The evidence suggests that there are no conventional (*cis*) introns in trypanosomes (17). In the other organisms that carry out spliced leader addition *trans*-splicing, only a subset of pre-mRNAs receive the SL. In further contrast to trypanosomes, *trans*-spliced pre-mRNAs in these organisms contain conventional introns which are processed by *cis*-splicing.

The removal of intervening sequences from pre-mRNAs proceeds through two consecutive transesterification reactions. In the first step, the 3′-5′ phosphodiester bond at the splice donor site is cleaved, releasing the 5′ exon. Concomitant with this cleavage is the formation of a 2′-5′ phosphodiester bond between the 5′ end of the intron and an adenosine residue about 30 nt upstream of the splice acceptor site. The resultant intermediate is known as the lariat or 2/3 intermediate. The second step of splicing also involves the cleavage and concomitant ligation of phosphodiester linkages. The 3′ end of the 5′ exon generated in step 1 attacks the phosphodiester bond at the splice acceptor site resulting in two ligated exons and the release of the intron in the form of a lariat. Subsequent to intron release, the 2′-5′ bond is cleaved by an enzyme known as a debranching enzyme and the resultant linear intron is degraded in the nucleus.

Trans-splicing follows the same two steps (Fig. 1.2). In the first step the SL (5′ exon) is cleaved from the rest of the SL RNA. Concomitant with this cleavage is formation of a 2′-5′ phosphodiester bond between the 'intron' portion of the SL RNA and the

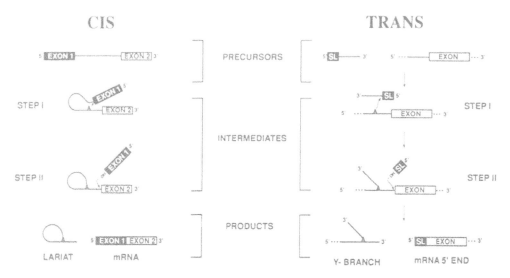

FIG. 1.2 *Cis* and *trans*-splicing proceed through analogous two-step reaction pathways. A schematic illustration of the similarities between *cis* and *trans*-splicing. See text for details.

branch site. Because the SL RNA and pre-mRNA are encoded as separate molecules, the intermediate generated by the first step of *trans*-splicing is a Y branched structure instead of a lariat (24, 25). In the second step of *trans*-splicing the 3' OH on the SL attacks the acceptor site, yielding ligated exons and releasing the Y branch structure which is subsequently debranched and degraded.

3.2. Mechanism of *Trans*-Splicing: the Spliced-Leader RNAs

*U*ridine-rich *S*mall *N*uclear RNAs (U snRNAs) are abundant short RNAs in eukaryotic cells. A subset of these RNAs participates in the catalysis of splicing (see below). The spliced leader RNAs (SL-RNAs) of both trypanosomes and nematodes bear striking similarities to the U snRNAs required for *cis*-splicing (Fig. 1.3.) This is particularly clear in nematodes where the SL RNA possesses both a trimethylguanosine cap structure and an Sm binding site (18,20). The Sm binding site of snRNAs conforms to the consensus sequence RR U_nRR (where R is a purine) and is located in a single-stranded region between two stem loops. In U snRNAs the Sm binding site is necessary for assembly of snRNAs into snRNPs (small nuclear ribonucleoproteins) because it promotes the binding of core Sm proteins (26, 27).

Several years ago a number of groups demonstrated that the Sm binding sequence of the *C. elegans* SL RNA was functional since the *C. elegans* SL RNP was precipitable from *C. elegans* extract with human Sm antisera. Furthermore, the SL RNA became precipitable with Sm antisera when incubated in HeLa cell extracts (28). It was subsequently established that assembly of the *Ascaris* SL RNA into an Sm snRNP was an absolute prerequisite for its participation in *trans*-splicing *in vitro* (see below) (29).

The relationship between the trypanosome SL RNA and the trypanosome U snRNAs was less obvious, since proteins associated with U snRNPs in trypanosomes do not contain epitopes recognized by Sm antisera. Furthermore, the trypanosome SL

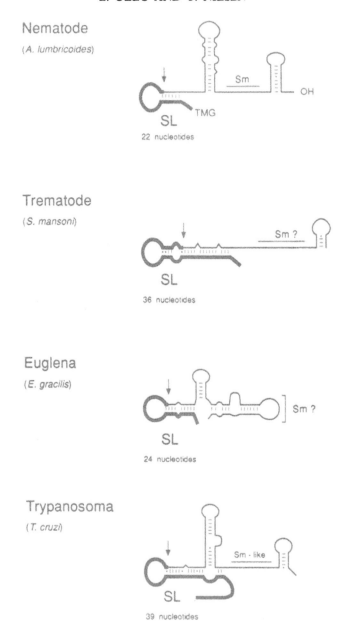

FIG. 1.3 SL RNAs have similar structures and resemble U snRNAs. Schematic representation of secondary structures of SL RNAs from organisms known to carry out *trans*-splicing. See text for details.

RNA does not have the TMG cap characteristic of U snRNAs but instead has a monomethylated guanosine as the capping nucleotide. This RNA also has a highly unusual array of base modifications in its first few nucleotides, the structural details of which have only recently emerged (30). However, the relationship between the trypanosome SL RNP and U snRNPs has been clarified recently with the successful affinity purification of SL RNPs and snRNPs from *T. brucei* (31). These analyses have shown convincingly that the SL RNA and snRNAs in trypanosomes share a common set of core proteins resembling the core proteins of other eukaryotic Sm snRNPs.

Collectively, these observations have reinforced the hypothesis originally proposed (32) that SL RNAs represent a special type of U snRNP where an exon (the SL) is fused to an snRNA-like sequence. Unlike the snRNAs required for *cis*-splicing, the SL RNA is consumed during *trans*-splicing. An intriguing question that remains the topic of speculation is whether SL RNAs predated the snRNAs as we know them today or alternatively arose from the union of a pre-existing snRNA and a free exon (33).

3.3. Features of the Nematode SL RNA Essential for Function in Trans-Splicing

An homologous cell free system which utilizes synthetic SL RNA (generated by *in vitro* transcription) in *trans*-splicing has permitted a detailed dissection of the nematode SL RNA (34). Site-directed mutagenesis and chemical modification interference studies have revealed that critical functional elements in the SL RNA are confined to remarkably short regions of the molecule, all of which reside in the snRNA-like domain (35). Surprisingly, exon sequences (the 22 nt SL) are irrelevant for SL RNA function in *trans*-splicing *in vitro* (36). The fact that the 22 nt SL was not important for SL RNA participation in *trans*-splicing was not anticipated since this sequence has been stringently conserved in evolutionarily distant nematodes. However, this sequence conservation can be explained in part by the fact that at the DNA level, the SL sequence itself is a promoter element essential for SL RNA synthesis by RNA polymerase II (see above).

Within the snRNA-like domain, essential sequences are confined to the single stranded region between the second and third stem loops. This sequence contains the Sm binding site and adjacent nucleotides. Mutational analyses have shown that there is an absolute correlation between the capacity of the SL RNA to assemble into an Sm snRNP and its ability to function in *trans*-splicing. Thus, it is clear that the SL RNA participates in *trans*-splicing as an Sm snRNP. The Sm binding site is critical for SL RNA function; cap structure, which constitutes the other similarity between SL RNAs and U snRNAs, is not. Although the trimethylguanosine cap structure is clearly not required for SL RNA function in *trans*-splicing *in vitro*, it may be required for proper subcellular localization of the SL RNP *in vivo*. Studies in other systems have demonstrated that the TMG cap on U snRNAs is part of bipartite nuclear localization signal (27).

The lack of SL RNA cap recognition in nematode *trans*-splicing stands in apparent contrast to recent results obtained in trypanosomes. Two groups have shown that trypanosome SL RNAs, undermethylated in the cap structure, do no function in *trans*-splicing. These results suggest that the unique trypanosome SL RNA cap may be required at some point in the assembly of the *trans*-spliceosome. However, this interpretation is not clear, since it is possible that, as with other snRNAs, the cap could serve to localize the trypanosome SL RNA to the nucleus. Until the pathway of SL RNP assembly is worked out or a trypanosome cell free system is established this will remain an open question. Nevertheless, the enzymes that catalyze cap modification in

trypanosomes present themselves as attractive candidates for chemotherapeutic intervention since such modification is clearly required for productive use of the
trypanosomatid SL RNA.

3.4. Significance of Essential Sequences Within the Nematode SL RNA

Functionally relevant sequence elements within the nematode SL RNA are confined to
the splice donor site and the Sm binding region. Although it is clear that the Sm binding
site is required for assembly of the SL RNA into an Sm snRNP, recent experiments
have shown that this single-stranded region may have an additional role in *trans*-
splicing. Functional Sm binding sequences, derived from other U snRNAs, fail to
support SL RNA function when substituted into the SL RNA: this indicates that the
primary sequence of the Sm binding region of the SL RNA is important. Such sequence
constraints can be imposed for a variety of reasons, one of which is the necessity to
interact by base-pairing with another RNA. Indeed, crosslinking experiments have
shown that the SL RNA interacts with U6 snRNA via a base-pairing interaction. In
the SL RNA the base-paired region spans the sequences critical for its function (35). In
U6 snRNA, the region of base pairing adjoins and overlaps the region of U6 known to
interact with U2 snRNA in higher eukaryotes and in trypanosomes (37,38). While the
functional significance of the SL/U6 interaction remains to be established experimentally, it may be that this RNA–RNA interaction facilitates entry of the SL RNA into
the *trans*-spliceosome. If true, it would provide some insight into one of the fundamental questions in *trans*-splicing; i.e., how do the two substrates (the SL RNA and
pre-mRNA) efficiently associate within the nucleus in the absence of significant
sequence complementarity?

3.5. Mechanism of *Trans*-Splicing: Required snRNP Cofactors

Conventional *cis*-splicing requires the participation of five U snRNPs, U1, U2, U4, U6
and U5 (39). U1 snRNP is involved in 5′ splice site identification and recent evidence
suggests a role for U1 snRNP in acceptor site recognition as well. U5 snRNP has also
recently been shown to be involved in splice site recognition whereas U2 snRNP
identifies the branch point on the pre-mRNA. U4 and U6 join the spliceosome together
as a base-paired double snRNP which does not appear to recognize specific features of
the pre-mRNA. U6 may be a catalytic RNA which promotes the actual splicing of
exons. See reference (40) for recent discussion and illustration of RNA–RNA interactions in *cis*-splicing.

   Trypanosomes contain homologs of U2, U4 and U6 snRNAs (41, 42). RNase H
degradation studies showed that these snRNAs are required for *trans*-splicing (4). To
date no homologs of U1 or U5 snRNAs have been found in trypanosomes, leading to
the notion that neither of these snRNAs is required for *trans*-splicing. If true (and
evidence against a role for U1 and U5 in *trans*-splicing in trypanosomes is exclusively
negative) it would suggest that *trans*-splicing is significantly simpler than *cis*-splicing,
but would leave the problem of how the 5′ and 3′ splice sites are juxtaposed. A detailed
discussion of the role of U snRNAs in *trans*-splicing is contained in ref (20).

   Since nematodes carry out both *cis*- and *trans*-splicing, it is not surprising that they
contain a complete complement of U snRNAs including U1 and U5. *In vitro* analysis,

using targeted degradation has shown that, as in trypanosomes, U2, U4 and U6 are required for *trans*-splicing. To date, it has not been possible to unambiguously determine the role or lack thereof of U1 or U5 snRNAs in nematode *trans*-splicing. This is clearly an important question since the answer will provide significant insight into the similarities or differences between *trans*-splicing in nematodes and trypanosomes.

3.6. Biological Roles of 5′ Leader Addition by *Trans*-Splicing

As discussed above, *trans*-splicing serves to mature 5′ ends of mRNAs embedded with polycistronic transcription units in trypanosomes and may serve the same function in nematodes. However, *trans*-splicing may be important in other aspects of mRNA metabolism.

In addition to providing a discrete 5′ end for mRNAs, *trans*-splicing also provides each mRNA with the cap structure derived from the SL RNA. Although it remains to be established whether translation in trypanosomes or nematodes is cap dependent, it seems likely that the cap, as it is in other eukaryotes, will be essential for mRNA recognition by ribosomes. It remains to be determined whether spliced leader or cap addition is necessary for transport of mRNAs from the nucleus or influences mRNA stability.

Because not all nematode mRNAs acquire the SL (43), *trans*-splicing might be a regulatory mechanism in these organisms. However, work in several laboratories has failed to reveal that *trans*-splicing of any pre-mRNA is regulated in a developmental, tissue-specific or stess-induced fashion. Furthermore, there is no suggestion that *trans*-splicing is restricted to a group of mRNAs which encode functionally related proteins. Much of the impetus for seeking a regulatory function for *trans*-splicing arose from the perception that only a minority (10–15%) of nematode messages was *trans*-spliced. However, this may have been an underestimate. The percentage of *trans*-spliced mRNAs was initially arrived at by interpretation of hybrid-arrest translation experiments carried out in rabbit reticulocyte lysates programmed with *C. elegans* mRNAs. Subsequent experiments demonstrated that *trans*-spliced nematode messages retained the SL RNA's trimethylguanosine cap structure (see above). Since it has been established that reticulocyte lysates translate trimethylguanosine-capped messages very poorly it seems probable that, in the early experiments proteins encoded by *trans*-spliced mRNAs could have been underrepresented in total translation products. Hybrid-arrest translation experiments indicate that 80–90% of *Ascaris* mRNAs are *trans*-spliced. If these findings can be generalized to other nematode species, they would indicate that *trans*-splicing plays a much more central role in nematode gene expression than was previously suspected.

4. RNA EDITING

4.1. Editing, the Phenomenon

In trypanosomatids the mitochondrial DNA [kinetoplast DNA (kDNA)] consists of two sets of circular molecules, namely 20–50 maxicircles (20–39 kb in size) and 5000–10 000 interlocked minicircles (0.5–2.5 kb in size). The maxicircles are analogous

```
EDITED COIII mRNA:    5'... AuCCAGuAuuGuuuuuuAuGGuuuuuuAuGuAGU GAGuuuGuuuuAuuuAuGGCG ...3'
                            |•||•••|||||•••|||•|||•••|||•||||||||||||| |||
        gRNA Tb1:     3'...ₙUUUAUUAUAGUGAAAAAUGUCAAGAAUGUAUCGCUCAAACAAAAUAUAUA ...5'
                                                                    Anchor
```

FIG. 1.4 An example of kinetoplastid pan-editing. A portion of the mature *T. brucei* cytochrome oxidase subunit III is shown above a guide RNA. Uridines inserted via editing are shown in lower case. Insertion of the uridines within the box is directed by a different guide RNA not shown. For details see text and ref. (45).

to the mitochondrial DNA of other eukaryotes and code for mitochondrial rRNAs and for some of the protein components of the respiratory chain. Sequencing of maxicircle DNA from various species brought about the remarkable discovery of kinetoplast RNA editing, one of the most astonishing findings in the field of RNA processing (44). Initial analysis revealed that the gene coding for cytochrome oxidase subunit II was interrupted by a frameshift; however, the frameshift was eliminated by insertion of four uridine residues in the corresponding mRNA. Since this early report, a wealth of information has been accumulated on the phenomenon of kinetoplast RNA editing which involves the non-encoded addition and, to a lesser extent, deletion of uridine residues in kinetoplast mRNAs. RNA editing not only eliminates internal frameshifts but also provides otherwise untranslatable open reading frames with the initiator codon AUG. Most startling was the discovery that editing can be so extensive as to remodel more that 50% of the mRNA sequence (pan-editing). Examples of the pan-editing occur in the cytochrome oxidase subunit III, NADH dehydrogenase and ATPase 6 mRNAs of *T. brucei* (Fig. 1.4).

The process of editing is post-transcriptional and seems to occur in a 3' to 5' direction. Directionality of editing is suggested by the sequence analysis of many partially edited mRNAs (perhaps editing intermediates) which contain edited 3' sequences and unedited 5' sequences. Editing is quite specific since it occurs only in certain regions of a transcript and is also remarkably precise since translatable open reading frames are generated.

4.2. Mechanism of RNA Editing

The existence of editing raises important fundamental questions: for example, where is the information for the edited sequences stored and how is the information transferred to the mRNA. A possible answer to the first question came from the discovery of small kinetoplast RNAs (60–70 nt in size), termed guide RNAs (gRNAs) (Fig. 1.4), that are encoded in both minicircle and maxicircle DNA. The main features of gRNAs are as follows. gRNAs contain at their 5' end a short anchor sequence complementary to a region immediately 3' of an edited site. This is followed by a sequence which is complementary to the edited version of the mRNA sequence. At the 3' end of the gRNAs there is a polyU tail which is added post-transcriptionally. Thus, gRNAs appear well suited to store the information for the edited sequence and must be important in the editing pathway. How the information contained within the guides is transferred from gRNAs to mRNAs is still an open question. Two models have been proposed to account for the editing pathway, the so called enzyme cascade and the transesterification models. At present, experimental evidence to distinguish between these hypotheses is lacking. To account for the structure of the edited sequences, primary RNA transcripts (pre-edited RNAs) must undergo a number of cleavage and ligation reactions at the editing sites with concomitant addition or deletion of U residues. The transesterification model is at present favored because of its intrinsic simplicity and its analogy to the chemistry of RNA splicing. It is proposed that editing is achieved through multiple rounds of two-step transesterification reactions. In the first step the gRNA base pairs to the pre-edited site positioning the 3′ OH group of the terminal uridine for attack on the phosphodiester bond at the editing site. Such a reaction would produce two intermediates, a free 5′ fragment of the pre-edited RNA cleaved at the edited site and a chimaeric molecule consisting of the gRNA covalently linked to the 3′ portion of the pre-edited RNA. In the second step, the 3′ OH of the pre-mRNA 5′ fragment would attack the chimaeric molecule, with concomitant ligation of the 5′ and 3′ end fragments of the pre-mRNA. Depending on the precise location of this second nucleophilic attack, the pre-mRNA would have one or more Us inserted or deleted at the editing site and the gRNA would be shortened or lengthened accordingly. This process would presumably be repeated until the edited site was fully base-paired to the complementary sequences in the gRNA. Experimental support for this model comes from the finding that chimera-forming activities are indeed present in kinetoplast extracts and that chimeric molecules can be detected *in vivo*. However, this evidence can not be considered definitive since chimera formation could conceivably be accomplished by a battery of enzymatic activities. Such enzymes would include an endonuclease specific for the edited sites, an RNA ligase and a terminal uridylyl-transferase. Each of these activities has been demonstrated in kinetoplast extracts. Whether or not enzyme activities are required for the editing pathway will become clear once an *in vitro* system capable of carrying out a complete editing cycle is available.

It should be stressed that all the available evidence about putative intermediates in the editing pathway has been obtained from inspection of cDNA molecules copied from cellular RNAs in steady-state condition. Although it is commonly assumed that these molecules are *bona fide* intermediates in the editing pathway, the possibility exists that at least some of them could be aberrant processing products. From all these studies, it has become apparent that the process of editing is incredibly complex. This is especially true for the cases of pan-editing, like that observed in the cytochrome oxidase subunit III mRNA of *T. brucei*. Many different gRNAs are required for this process and they must be utilized sequentially presumably in a 3′ to 5′ direction. How these different gRNA molecules are specifically recruited and in which order they act is the subject of intense investigation.

In addition to mechanistic questions, the existence of RNA editing adds an unanticipated layer of complexity to our understanding of gene expression. Editing clearly violates the 'central dogma' of molecular biology and examples of RNA editing have now been found in diverse species from humans to slime molds. Time will tell how widespread the phenomenon is and in what contexts it is used. It will be of great interest to determine what advantage is conferred by evolving (or retaining) this baroque mechanism of the mRNA maturation.

5. REFERENCES

1. Johnson, P. J., Kooter, J. M. and Borst, P. (1987) Inactivation of transcription by UV irradiation of *T. brucei* provides evidence for a multicistronic transcription unit including a VSG gene. *Cell* **51**: 273–281.
2. Muhich, M. L. and Boothroyd, J. C. (1988) Polycistronic transcripts in trypanosomes and their accumulation during heat shock: evidence for a precursor role in mRNA synthesis. Mol. *Cell Biol.* **8**: 3837–3846.
3. Tschudi, C. and Ullu, E. (1988) Polygene transcripts are precursors to calmodulin mRNAs in trypanosomes. *EMBO J.* **7**: 455–463.
4. Tschudi, C. and Ullu, E. (1990) Destruction of U2, U4 or U6 small nuclear RNAs blocks *trans*-splicing in trypanosome cells. *Cell* **61**: 459–466.
5. Huang, J. and van der Ploeg, L. H. T. (1991) Maturation of polycistronic pre-mRNA in *Trypanosoma brucei*: analysis of *trans*-splicing and poly (A) addition at nascent RNA transcripts from the hsp 70 locus. *Mol. Cell Biol.* **11**: 3180–3190.
6. Chung, H. M., Lee, M. G-S. and van der Ploeg, L.H.T. (1993) RNA polymerase I-mediated protein coding gene expression in *Trypanosoma brucei*. *Parasit. Today* **8**: 414–418.
7. Zomerdijk, J. C. B. M., Ouellette, M., ten Asbroek, A. L. M. A. *et al.* (1990) The promoter for a variant surface glycoprotein gene expression site in *Trypanosoma brucei*. *EMBO J.* **9**: 2791–2801.
8. Brown, S. D., Huang, J. and van der Ploeg, L. H. T. (1992) The promoter for the procyclic acidic repetitive protein (PARP) genes of *Trypanosoma brucei* shares features with RNA polymerase I promoters. *Mol. Cell Biol.* **12**: 2644–2652.
9. Sherman, D. R., Janz, L., Hung, M. and Clayton, C. (1991) Anatomy of the parp gene promoter of *Trypanosoma brucei*. *EMBO J.* **10**: 3379–3386.
10. Pays, E., Coquelet, H., Tebabi, P. *et al.* (1990) *Trypanosoma brucei*: constitutive activity of the VSG and procyclin promoters. *EMBO J.* **10**: 3145–3151.
11. Ullu, E., Matthews, K. R. and Tschudi, C. (1993) Temporal order of RNA process reactions in trypanosomes: rapid *trans*-splicing precedes polyadenylation of newly-synthesized tubulin transcripts. *Mol. Cell Biol.* **13**: 720–725.
12. Spieth, J., Lea, K., Brooke, B. and Blumenthal, T. (1993) Operons in *C. elegans*: polycistronic mRNA precursors are processed by *trans*-splicing of SL2 to downstream coding regions. *Cell* **73**: 521–532.
13. Hannon, G. J., Maroney, P. A., Ayers, D. G., Shambaugh, J. D. and Nilsen, T. W. (1990). Transcription of a nematode *trans*-spliced leader RNA requires internal elements for both initiation and 3′ end formation. *EMBO J.* **9**: 1915–1921.
14. Maroney, P. A., Hannon, G. J. and Nilsen, T. W. (1990) Transcription and cap trimethylation of a nematode spliced leader RNA in a cell free system. *Proc. Natl. Acad. Sci. USA* **87**: 709–713.
15. Ach, R. A. and Weiner, A. M. (1987) The highly conserved U small nuclear RNA 3′-end formation signal is quite tolerant to mutation. *Mol. Cell Biol.* **7**: 2070–2079.
16. Parry, H. D., Scherly, D. and Mattaj, I. W. (1989) 'Snurpogenesis': the transcription and assembly of U snRNP components. *Trends Biochem. Sci.* **14**: 15–19.
17. Agabian, N. (1990) *Trans*-splicing of nuclear pre-mRNAs. *Cell* **61**: 1157–1160.
18. Blumenthal, T. and Thomas, J. (1988) *Cis* and *trans*-mRNA splicing in *C. elegans*. *Trends Genet.* **4**: 305–308.
19. Borst, P. (1986) Discontinuous transcription and antigenic variation in trypanosomes. *Annu. Rev. Biochem.* **55**: 701–732.
20. Nilsen, T. W. (1993) *Trans*-splicing of nematode pre-messenger RNA *Annu. Rev. Microbiol.* **47**: 413–440.
21. Krause, M. and Hirsh, D. (1987) A *trans*-spliced leader sequence on actin mRNA in *C. elegans*. *Cell* **49**: 753–761.
22. Rajkovic, A., Davis, R. E., Simonsen, J. N. and Rottman, F. M. (1990) A spliced leader is present on a subset of mRNAs from the human parasite *Schistosoma mansoni*. *Proc. Natl. Acad. Sci. USA* **87**: 8879–8883.

23. Tessier, L.-H., Keller, M., Chan, R., Fournier, R., Weil, J. H. and Imbault, P. (1991) Short leader sequences may be transferred from small RNAs to pre-mature mRNAs by *trans*-splicing in *Euglena EMBO J.* **10**: 2621–2625.
24. Murphy, W. J., Watkins, K. P. and Agabian, N. (1986) Identification of a novel Y branch structure as an intermediate in trypanosome mRNA processing: evidence for *trans*-splicing. *Cell* **47**: 517–525.
25. Sutton, R. and Boothroyd, J. C. (1986) Evidence for *trans*-splicing in trypanosomes. *Cell* **47**: 527–535.
26. Luhrmann, R. (1988) snRNP proteins. In: *Small Nuclear Ribonucleoprotein Particles* (ed. Birnstiel, M. L.) Springer-Verlag, pp. 71–99.
27. Mattaj, I. W. (1988) U snRNP assembly and transport. In: *Small Nuclear Ribonucleoprotein Particles* (ed. Birnstiel, M. L.) Springer-Verlag, pp. 100–114.
28. Bruzik, J. P., van Doren, K., Hirsh, D. and Steitz, J. A. (1988) *Trans*-splicing involves a novel form of small ribonucleoprotein particles. *Nature* **335**: 559–562.
29. Maroney, P. A., Hannon, G. J. Denker, J. A. and Nilsen, T. W. (1990) The nematode spliced leader RNA participates in *trans*-splicing as an Sm snRNP. *EMBO J.* **9**: 3667–3673.
30. Bangs, J. D., Crain, P. F., Hashizume, T., McCloskey, J. A. and Boothroyd, J. C. (1992) Mass spectrometry of mRNA cap 4 from trypanosomatids reveals two novel nucleosides. *J. Biol. Chem.* **267**: 9805–9815.
31. Palfi, Z., Günzl, A., Cross, M. and Bindereif, A. (1991) Affinity purification of *Trypanosoma brucei* snRNPs reveals common and specific protein components. *Proc. Natl. Acad. Sci. USA* **88**: 9097–9101.
32. Sharp, P. A. (1987) *Trans*-splicing: variation on a familiar theme? *Cell* **50**: 147–148.
33. Laird, P. (1989) *Trans*-splicing in trypanosomes — archaism or adaptation? *Trends Genet.* **5**: 204–208.
34. Hannon, G. J., Maroney, P. A., Denker, J. A. and Nilsen, T. W. (1990) *Trans*-splicing of nematode pre-messenger RNA *in vitro*. *Cell* **61**: 1247–1255.
35. Hannon, G. J., Maroney, P. A., Yu, Y.-T., Hannon, G. E. and Nilsen, T. W. (1992) Interaction of U6 snRNA with a sequence required for function of the nematode SL RNA in *trans*-splicing interacts with U6 snRNA. *Science* **258**: 1775–1780.
36. Maroney, P. A., Hannon, G. J., Shambaugh, J. D. and Nilsen, T. W. (1991) Intramolecular base pairing between the nematode spliced leader and its 5′ splice site is not essential for *trans*-splicing *in vitro*. *EMBO J.* **10**: 3869–3875.
37. Hausner, T.-P., Giglio, L. M. and Weiner, A. M. (1990) Evidence for base-pairing between mammalian U2 and U6 small nuclear ribonucleoprotein particles. *Genes Dev.* **4**: 2146–2156.
38. Watkins, K. P. and Agabian, N. (1991) *In vivo* UV cross-linking of U snRNAs that participate in trypanosome *trans*-splicing. *Genes Dev.* **5**: 1859–1869.
39. Moore, M. J., Query, C. C. and Sharp, P. A. (1993) Splicing of precursors to messenger RNAs by the spiceosome. In: *The RNA World* (eds Gesterland, R. and Adkins, S.), Cold Spring Harbor Press: Cold Spring Harbor, NY, pp. 303–358.
40. Nilsen, T. W. (1994) RNA–RNA interactions in the spliceosome: Unraveling the ties that bind. *Cell* **78**: 1–4.
41. Mottram, J., Perry, K. L., Lizardi, P. M., Luhrmann, R., Agabian, N. and Nelson, R. G. (1989) Isolation and sequence of four small nuclear U RNA genes of *Trypanosoma brucei* subsp. *brucei*: identification of the U2, U4 and U6 RNA analogs. *Mol. Cell. Biol.* **9**: 1212–1223.
42. Tschudi, C., Richards, F. F. and Ullu, E. (1986) The U2 RNA analogue of *Trypanosoma brucei gambiense*: implications for splicing mechanisms in trypanosomes. *Nucleic Acids Res.* **14**: 8893–8903.
43. Bektesh, S. L., van Doren, K. V. and Hirsh, D. (1988) Presence of the *Caenorhabditis elegans* spliced leader on different mRNAs and in different genera of nematodes. *Genes Dev.* **2**: 1277–1283.
44. Stuart, K. (1993) RNA editing in trypanosomatid mitochondria. *Annu. Rev. Microbiol.* **45**: 327–344.
45. Pollard, V. W., Rohrer, S. P. Michelotti, E. F. *et al.* (1990) Organization of minicircle genes for guide RNAs in *Trypanosoma brucei*. *Cell* **63**: 783–790.

2 Carbohydrate and Energy Metabolism in Aerobic Protozoa

FRED R. OPPERDOES

International Institute of Cellular and Molecular Pathology and University of Louvain, Brussels, Belgium

SUMMARY

The Trypanosomatidae are characterized by a high flexibility of their energy metabolism. They contain a single mitochondrion at all developmental stages of which the contribution to the overall ATP generation varies significantly from one member of the trypanosomatid family to another and from one life-cycle stage to another. As a consequence the contribution of carbohydrate catabolism to the overall energy generation is also highly variable. The extreme is encountered in the bloodstream forms of some African trypanosomes where cytochromes and tricarboxylic acid (TCA) cycle enzymes are absent and glycolysis is the sole source of metabolic ATP; these organisms have a very active glycolytic pathway. In all the Trypanosomatidae studied, glycolysis takes place via the Embden–Meyerhoff pathway of which the early enzymes are sequestered inside a subcellular organelle: the glycosome. At present, the order Kinetoplastida, to which the Trypanosomatidae belong, is the only known group of living organisms where this kind of compartmentation has been found. The glycosome is a member of the family of microbodies to which also belong the peroxisomes, present in almost all other eukaryotic cells, and the glyoxysomes, typical of germinating plants.

The first part of this chapter discusses the African trypanosome *Trypanosoma brucei* and, where appropriate, other species of the *Trypanosoma*, and *Leishmania* genera will be dealt with. Data about other members of the Trypanosomatidae, such as *Crithidia* sp., *Herpetomonas* sp., *Leptomonas* sp. and *Phytomonas* sp. is highly fragmented and incomplete and therefore only dealt with where appropriate.

Biochemistry and Molecular Biology of Parasites
ISBN 0-12-473345-X

The second part of this chapter deals with the carbohydrate metabolism of the malaria parasite *Plasmodium*. Virtually nothing is known about the carbohydrate metabolism of the stages of the life cycle other than the one that infects the red blood cell. Even for this stage information on glycolysis and mitochondrial metabolism is still highly fragmented. Energy generation occurs predominantly or exclusively through fermentation of glucose with the production of lactate as the major end-product. Little is known about the role and properties of the individual enzymes involved in carbohydrate metabolism, mainly because of difficulties in obtaining sufficient parasite material free of host cell contamination. Rapid progress is now being made through the molecular cloning of several of the glycolytic enzymes.

1. THE TRYPANOSOMATIDAE

1.1. Substrates and End-products of Metabolism

1.1.1. African trypanosomes

The African trypanosomes go through a complex life cycle comprising, in the vertebrate host, long slender, intermediate and short stumpy bloodstream forms, in the midgut of the tsetse fly the procyclic trypomastigote stage and in the insect's salivary glands and proboscis the epimastigote and metacyclic stages. No comprehensive study of carbohydrate metabolism through each stage of the life cycle of any one species has been made so far and little is known about the metabolism of the epimastigote and metacyclic stages. Also little is known about the metabolism of African trypanosomes other than the extensively studied *T. brucei*.

1.1.1.1. *Vertebrate stages.* All bloodstream stages (long slender, intermediate and short stumpy) of *T. brucei* actively catabolize glucose, fructose, mannose and glycerol. Due to the absence of a mitochondrial TCA cycle and a functional respiratory chain, the long slender forms are incapable of oxidizing amino acids or fatty acids. Short-stumpy stages are able to utilize α-ketoglutarate in addition to glucose, fructose and glycerol (1). Bloodstream forms contain neither carbohydrate stores nor other energy reserves. Thus, depletion of exogenous substrate leads to a rapid drop of ATP levels and loss of motility.

Under aerobic conditions the long slender forms metabolize glucose quantitatively to pyruvate. Trace amounts of CO_2 and sometimes glycerol have been reported as end products. However, CO_2 may result from a spontaneous decarboxylation of pyruvate, whereas the glycerol most likely results from a partial anaerobiosis of the cells during culture or incubation. Pyruvate cannot be converted into lactate because of the absence of lactate dehydrogenase; enzymatic decarboxylation of pyruvate does not occur in these forms due to the absence of pyruvate decarboxylase.

In the short stumpy form pyruvate is further metabolized because a mitochondrial pyruvate decarboxylase is derepressed. In pleomorphic *T. rhodesiense*, comprising a high percentage of stumpy forms, the major end-products of glucose metabolism are pyruvate (75%), glycerol (5%), acetate (9%), succinate (1%) and CO_2 (3%).

Reoxidation of glycolytically produced NADH in the glycosome is mediated through a glycerol-3-phosphate–dihydroxyacetone-phosphate cycle comprising the glycosomal NAD-linked glycerol-3-phosphate dehydrogenase and the mitochondrial FAD-linked glycerol-3-phosphate dehydrogenase–oxidase complex. The terminal oxidase reduces molecular oxygen to water without H_2O_2 as intermediate. This reaction does not involve cytochromes and is insensitive to the classical inhibitors of mitochondrial respiration, such as cyanide and antimycin (Fig. 2.1).

Bloodstream-form trypanosomes do not exhibit a Pasteur effect. Under anaerobic conditions, or when oxygen consumption via the mitochondrial glycerol-3-phosphate oxidase is inhibited with salicyl hydroxamic acid (SHAM), long slender bloodstream forms continue to utilize glucose at about the same rate as under aerobic conditions. This is due to the fact that the glycerol-3-phosphate–dihydroxyacetone phosphate cycle is now inoperative, preventing the oxidation of glycerol-3-phosphate. The latter compound accumulates inside the glycosome, while glycosomal ATP is trapped by the phosphorylation of glucose and fructose-6-phosphate in the hexokinase (HK) and phosphofructokinase (PFK) reactions, respectively. Thus, anaerobiosis leads to high glycerol-3-phosphate and ADP concentrations and a low ATP concentration, and glycerol that rapidly equilibrates over biological membranes diffuses out of the glycosome. This creates in the glycosome the right conditions for a reversal of an otherwise irreversible glycerol kinase reaction by mass action, resulting in the utilization of glycerol-3-phosphate and ADP and the formation of glycerol and ATP. As a consequence, under anaerobic conditions, or with SHAM to inhibit respiration, glucose is dismutated into equimolar amounts of pyruvate and glycerol, with net synthesis of 1 molecule of ATP, per molecule of glucose consumed (2–4). Cells survive and remain motile under anaerobic conditions, although cellular ATP levels drop significantly (5). Glycerol, under anaerobic conditions, cannot serve as a substrate, because glycerol-3-phosphate cannot be oxidized to dihydroxyacetone phosphate (DHAP) without molecular oxygen. At concentrations above several millimolar, glycerol becomes toxic. At these glycerol concentrations the reversal of the glycerol kinase reaction is no longer possible and glycolysis comes to a halt, resulting in a total disappearance of cellular ATP (6). This aspect of trypanosome glycolysis has been exploited in the treatment of experimental animals infected with either T. brucei or T. rhodesiense. The administration of a combination of SHAM and glycerol has been shown to lead to an almost immediate lysis and disappearance of the parasites from the circulation (7). Permanent cures, however, were only obtained at concentrations of the compounds that were toxic to the animals.

1.1.1.2. *Insect stages*. Procyclic culture forms of T. rhodesiense actively metabolize glucose, fructose, mannose or glycerol, but do not oxidize other mono- or disaccharides (8). The major end-product of glucose metabolism is CO_2 (55% of glucose carbon), together with small amounts of acetate (3%) and succinate (4%). For T. brucei procyclics 17% of the glucose carbon was recovered as succinate and 8% as alanine (9). The production of CO_2 was not measured. Proline and the tricarboxylic acid cycle intermediates α-ketoglutarate and succinate also support respiration in T. rhodesiense.

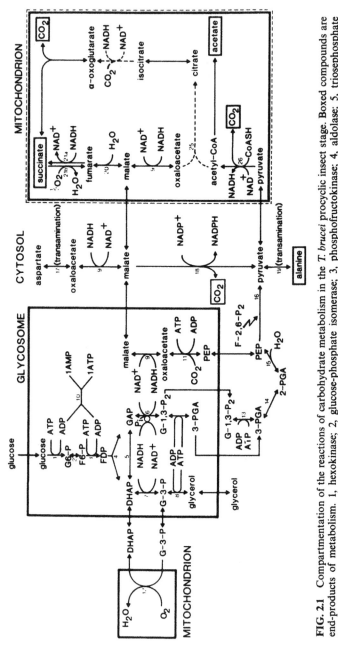

FIG. 2.1 Compartmentation of the reactions of carbohydrate metabolism in the *T. brucei* procyclic insect stage. Boxed compounds are end-products of metabolism. 1, hexokinase; 2, glucose-phosphate isomerase; 3, phosphofructokinase; 4, aldolase; 5, triosephosphate isomerase; 6, glyceraldehyde-phosphate dehydrogenase; 7, glycerol-3-phosphate dehydrogenase; 8, glycerol kinase; 9, malate dehydrogenase; 10, adenylate kinase; 11, PEP carboxykinase; 12, glycerol-3-phosphate oxidase; 13, phosphoglycerate kinase; 14, phosphoglycerate mutase; 15, enolase; 16, pyruvate kinase; 17, aspartate aminotransferase; 18, malic enzyme; 19, alanine aminotransferase; 20, fumarate hydratase; 21a, fumarate reductase; 21b, succinate dehydrogenase; 22, α-oxoglutarate decarboxylase; 23, isocitrate dehydrogenase; 24, aconitase; 25, citrate synthetase; 26, pyruvate dehydrogenase. Modified from reference 3.

22

Respiration with proline as substrate is up to two times higher than with the other substrates (Chapter 5).

Under anaerobic conditions procyclic stages of *T. rhodesiense* are capable of utilizing glucose and glycerol, provided CO_2 is present (8). Most of the carbon is recovered as succinate (75 and 63%, respectively) and a smaller amount of acetate (25 and 4%, respectively). Anaerobic data for *T. brucei* are not available. The glycosomal phosphoenolpyruvate carboxykinase (PEPCK) and malate dehydrogenase, virtually absent from bloodstream forms, but fully expressed in the procyclics, are believed to fix CO_2 and produce succinate as an anaerobic end-product (4, 10).

Virtually no information is available concerning the metabolism of the other two insect stages, i.e. epimastigotes and metacyclic trypomastigotes.

1.1.2. Trypanosoma cruzi *and* Leishmania *spp.*

1.1.2.1. *Insect stages.* The insect stages of *T. cruzi* (epimastigotes), and *Leishmania* spp. (promastigotes) have similar carbohydrate metabolism. Study of the glucose metabolism of these two organisms (11) has shown that *T. cruzi* preferred glucose over amino acids for its growth, whereas the reverse was true for *L. mexicana*. The first organism used glucose completely during the log phase of growth, whereas the latter used glucose only at the end of the log phase and at the beginning of the stationary phase (11). Both organisms produced succinate and much smaller amounts of acetate. *L. mexicana* produced small amounts of pyruvate. None of the cells studied produced any L-lactate or malate. *Leishmania* produced ammonia throughout the culture, whereas *T. cruzi* produced ammonia only after glucose had been consumed. Promastigotes of *Leishmania b. panamensis* with glucose as the sole carbon source produced glycerol, succinate, acetate, pyruvate, alanine and D-lactate, but no L-lactate, in addition to CO_2. Cells incubated with glycerol as the sole carbon source released acetate, succinate, D-lactate and CO_2 (12, 13). Both alanine and glutamate were oxidized via the TCA cycle at rates comparable or greater than the rate of oxidation of glucose (14). Under anaerobic conditions with glucose as substrate, more D-lactate, glycerol, pyruvate, succinate and alanine, but less acetate, were produced as the major end-products (15), whereas glucose consumption in the absence of CO_2 decreased (14). With CO_2 it was restored to aerobic levels, indicating that CO_2 fixation under anaerobic conditions is essential to maintain a high glycolytic flux (12, 13, 15). D-Lactate is produced by at least four species of *Leishmania*. *T. brucei* procyclics only produced L-lactate (16). D-Lactate is formed from glyceraldehyde-3-phosphate via the methylglyoxal bypass (Fig. 2.2) and the enzyme methylglyoxal reductase in *L. donovani* was shown to be the most active of all catabolic enzymes (17).

When the amount of glucose present in the growth medium was such that it was not rate-limiting glycolysis when *L. donovani* was grown in the chemostat, the label in glucose rapidly accumulated intracellularly as a compound distinct from glucose to a concentration of 30 mM glucose equivalents (18). Although the nature of the accumulated product was not identified, this is most likely the mannose-containing polysaccharide that has been shown to accumulate in stationary-phase cells of the same organism (19). No such accumulation was observed in cells that were grown under glucose starvation.

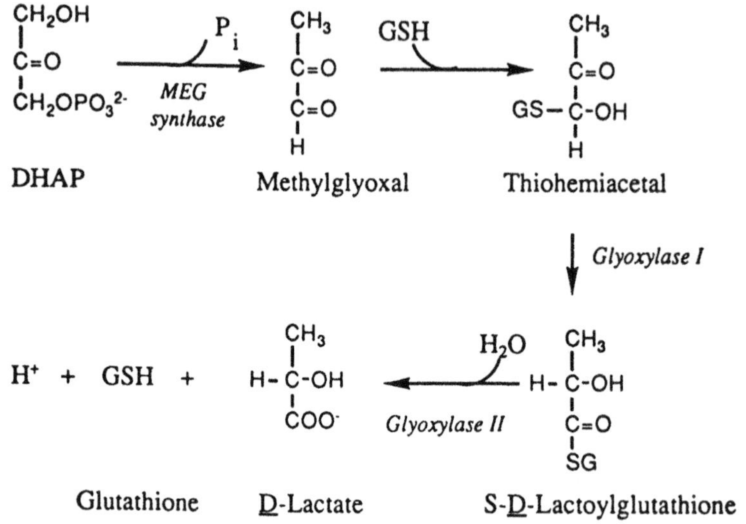

CH_2OH
|
$C=O$ $\xrightarrow[\substack{MEG \\ synthase}]{P_i}$ CH_3
|
$CH_2OPO_3^{2-}$

CH_3
|
$C=O$
|
$C=O$
|
H

    GSH

CH_3
|
$C=O$
|
$GS-C-OH$
|
H

DHAP   Methylglyoxal   Thiohemiacetal

$\downarrow$ *Glyoxylase I*

CH_3
|
$H-C-OH$
|
COO^-

$\xleftarrow[\text{Glyoxylase II}]{H_2O}$

CH_3
|
$H-C-OH$
|
$C=O$
|
SG

H^+ + GSH +

Glutathione   D-Lactate   S-D-Lactoylglutathione

FIG. 2.2   The methylglyoxal pathway.

Osmotic conditions influence metabolism drastically. *L. major* promastigotes respond to such variations by regulating their intermediary metabolism. Increased osmolality interferes with mitochondrial function, resulting in an inhibition of the oxidation of glucose, alanine, glutamate, glycerol and fatty acids (20). Hyperosmotic stress also alters the fluxes through the pathways of intermediary metabolism by increasing pyruvate, alanine and D-lactate production and by decreasing the production of acetate and succinate. The net result is an increase in the intracellular pool size of alanine, which counteracts the loss of water and reduction in cell volume that would otherwise occur (15).

Studies of the carbohydrate metabolism of *T. cruzi* (21) have shown that phosphoenolpyruvate serves as the acceptor of the primary CO_2-fixation reaction. This resulted in the formation of oxaloacetate and malate and the excretion of succinate. The central role of PEPCK in energy metabolism in insect-stage trypanosomatids has been illustrated in the case of *T. cruzi* epimastigotes, using 3-mercaptopicolinic acid, a powerful inhibitor of this enzyme (22). Inhibition led to a twofold reduction in the anaerobic production of succinate and a similar decrease in glucose consumption, while the production of alanine, via the transamination of pyruvate, increased threefold.

1.1.2.2. *Amastigote-like stages.* Knowledge about energy metabolism of the intracellular amastigote stages of *T. cruzi* and *Leishmania* spp. is limited. The fragmentary information available suggests that amastigotes have a significantly increased β-oxidation of fatty acids and reduced needs for the consumption of proline and glucose (23, 24).

Axenically cultured amastigote-like cells of *T. cruzi* have an essentially glycolytic metabolism. They ferment glucose to succinate and acetate and do not seem to excrete ammonia. Only after they reach stationary phase and transform to epimastigotes do they acquire the ability to oxidize substrates such as amino acids (25). Whether these amastigote-like cells resemble in their carbohydrate metabolism the intracellular stages in the mammalian host remains an open question.

1.2. Enzymes of Carbohydrate Metabolism

Most of the enzymes involved in the Embden–Meyerhoff pathway of *T. brucei* have been identified (26; see ref. 4 for a review). Many enzymes have been isolated and characterized and their kinetic properties studied in detail (26–37). Several of their genes, glucose phosphate isomerase (GPI) (33), aldolase (38, 39), triosephosphate isomerase (TIM) (40), glycosomal and cytosolic glyceraldehyde-3-phosphate dehydrogenase (GAPDH) (41, 42), glycosomal and cytosolic phosphoglycerate kinase (PGK) (43, 44), pyruvate kinase (PK) (45), PEPCK (46, 47) have been sequenced and some enzymes have also been expressed as recombinant proteins (PGK, ref. 48; aldolase, TIM, GAPDH, PK, Michels, P.A.M. *et al.* unpublished). Apart from the glycosomal HK and PFK, which significantly differ from their counterparts from other organisms (26, 27, 30), the other glycosomal enzymes strongly resemble their homologs from other organisms with respect to subunit mass, subunit composition and kinetic properties.

The glycosomal enzymes from *T. brucei* carry a high net positive charge (isoelectric points of 8.8–10.2, ref. 26) at neutral pH, whereas the same enzymes from other Trypanosomatidae such as *L. mexicana* (42) and *T. cruzi* (49) as well as the cytosolic isoenzymes from *T. brucei* have a more neutral or even a negative charge. Some correlation exists between the overall positive charge and the rate of the glycolytic flux catalyzed by these enzymes. Since these enzymes function inside the glycosomal compartment, where the concentrations of negatively charged glycolytic intermediates are relatively high (26), a positively charged surface would facilitate the interaction of the enzymes with the negatively charged phosphorylated intermediates and thus increase glycolytic flux.

1.3. Pentose Phosphate Shunt

The pentose phosphate pathway in *T. rhodesiense* represents only a minor part (0.6%) of the overall glucose utilization (50), in keeping with the high glycolytic flux. This contrasts with other trypanosomatids, such as *T. cruzi* (51) and *Leishmania b. panamensis* (14), where the contribution of the pathway may represent a considerable portion of the total glucose consumption. Both procyclic and bloodstream form *T. brucei* are capable of metabolizing glucose via the oxidative segment of the pentose phosphate pathway to produce D-ribose-5-phosphate for the synthesis of nucleic acids and the reduction of NADP for biosynthetic reactions. However, only procyclic forms are able to use the non-oxidative segment of the pathway to cycle carbon between pentose and hexose in order to use D-glyceraldehyde-3-phosphate as a net product of the pathway. Ribulose-5-phosphate-3-epimerase and ketolase were only detected in procyclics and not in bloodstream forms, whereas all the other enzymes of the pathway were present in both forms (52).

1.4. Regulation of Carbohydrate Metabolism

The maximal activities of the individual glycolytic enzymes of the bloodstream form of *T. brucei* are all in large excess of the overall glucose consumption rate of $85 \, \text{nmol} \, \text{min}^{-1} \, (\text{mg protein})^{-1}$. It is not possible to identify the rate-limiting step in glycolysis. The enzymes HK and PFK fulfil an important regulatory role in most eukaryotic cells, but apparently do not have such a function in *T. brucei*. All the enzymes of the first part of the pathway are sequestered inside the glycosome and this is probably the explanation for the fact that neither of the two enzymes regulate the pathway. The suggestion that the glycolytic pathway is not regulated is supported by the fact that in *T. brucei* bloodstream forms the aerobic to anaerobic transition is characterized by the absence of a Pasteur effect, and analysis of the levels of glycolytic intermediates before and after such a transition does not reveal a cross-over point between glucose-6-phosphate (Glc-6-P) and phosphoenol pyruvate. The levels of all metabolites and ATP decreased upon anaerobiosis (53). This indicates that the rate-limiting step in the pathway must be located before or at the level of the formation of Glc-6-P, which is in agreement with the observation that at glucose concentrations below 5 mM the rate-limiting step in the pathway is the transport of glucose over the plasma membrane (54) (Chapter 12).

Pyruvate kinase (PK) from *T. brucei* and the other Trypanosomatidae, which is located in the cytosol, is modulated by adenine nucleotides and inorganic phosphate (35,36) and the enzyme is also activated by nanomolar concentrations of fructose-2,6-bisphosphate (Fru-2,6-P_2) (55,56). The fact that in these organisms Fru-2,6-P_2 modulates the activity of PK, rather than that of PFK, as in most eukaryotic cells, is most likely related to the fact that PFK in the Kinetoplastida is sequestered inside the glycosome. In the vertebrate stage the cellular concentrations of Fru-2,6-P_2 and phosphoenolpyruvate are inversely related (56), which indicates that modulation of PK activity occurs indeed.

The regulation of the glycolytic flux in the insect-stage of *T. brucei* is entirely different from that of the vertebrate stage, and probably does not differ too much from the insect stages of other Trypanosomatidae, such as the *Leishmania* promastigote and the *T. cruzi* epimastigote stage. The procyclic insect stage of *T. brucei* has a poor ability to convert glucose into pyruvate. This is mainly due to the relatively low amounts of several of the glycolytic enzymes, particularly HK and PK (29,57). Due to the absence of Fru-2,6-P_2 the relatively low activity of PK is now mainly regulated by the cytosolic phosphate potential ([ATP]/[ADP].[P_i]), the availability of the TCA cycle intermediates oxaloacetate and acetyl-CoA and the cytosolic concentration of phosphoenolpyruvate, for which the $S_{0.5}$ in the absence of Fru-2,6-P_2 has increased by more than 10-fold (35,55).

1.5. Why a Glycosome?

It is not clear why only the Kinetoplastida have part of their glycolytic pathway sequestered inside an organelle. Glycosomes have not been detected in any other protists. Compartmentation may increase the efficiency of the pathway (Table 2.1). *T. brucei* bloodstream form is among the most actively glycolizing cells ever described, though it devotes only 9% of its total cellular protein to the pathway (26). Compartmentation of glycolysis allows the concentrations of glycolytic enzymes, intermediates and cofactors to be sufficiently high and allows the enzymes to be completely saturated (26,58). The reversal of the glycerol kinase reaction is an example of an otherwise

TABLE 2.1 Comparison of the glycolytic efficiency of *T. brucei* bloodstream form and glucose-grown yeast

Characteristic	*T. brucei*	Yeast
Glycolytic rate (nmol min^{-1} mg^{-1})	85	50
Glycolytic protein (% of total)	9	60
GAPDH (% of total protein)	0.5	5
Aldolase (% of total protein)	1.2	10
Glycolytic compartment (% of total)	4	> 50
Metabolites involved in the flux (%)	25	100

impossible reaction that can occur because of its compartmentation inside the glycosome (see above).

It is difficult to understand how the sequestration, one after another, of the glycolytic enzymes into a microbody could have led to an evolutionary advantage to the cell. It seems more likely that all glycosomal enzymes entered the ancestral kinetoplastid together probably as part of a prokaryotic endosymbiont. Eventually the endosymbiont lost all of its DNA and most of its metabolic functions, but retained its glycolytic enzymes. All microbodies (peroxisomes, glyoxysomes and glycosomes) share some common enzymes and pathways, such as catalase, β-oxidation of fatty acids and ether lipid biosynthesis and they all use the same type of topogenic signals for the import of their proteins. Therefore, microbodies probably have a monophyletic origin (59). The reason why the Trypanosomatidae maintained a sequestration of the glycolytic pathway, while all other eukaryotes could afford to lose this probably advantageous property, can only be guessed at. Its loss could be related to the fact that the other eukaryotes succeeded in integrating important regulatory properties in the enzymes HK and PFK, allowing the strict regulation of glycolysis and gluconeogenesis within one and the same compartment, thus preventing futile cycling and hydrolysis of cytosolic ATP (59). Trypanosomatid HK and PFK are not regulated, as in other eukaryotes. Nevertheless, the insect stages are capable of synthesizing polysaccharides and the complex carbohydrates of their surface structures through gluconeogenesis (60), a pathway probably only active while living on an amino acid diet in the insect gut. It is probably to prevent this futile cycling that the Kinetoplastida have maintained their glycolytic activity sequestered in microbodies (59).

2. THE PLASMODIA

Erythrocytic stages of the malaria parasite store no reserve carbohydrates. They must be constantly supplied with glucose. In infected erythrocytes, glucose metabolism undergoes a significant increase, which may be as much as 50–100-fold. This has been observed in several species of malaria parasites (61, 62), but the exact amount of sugar consumed depends on the stage of the parasite, the degree of parasitemia, and the species.

In red cells infected with *Plasmodium falciparum* almost all the glucose utilized passes through anaerobic glycolysis to lactic acid (63, 64). Both the parasite and the host cell lack a complete TCA cycle and the host cell lacks functional mitochondria (65). There is evidence for a TCA cycle in the avian malarial parasites, where the enzymes isocitrate dehydrogenase and succinate dehydrogenase have been detected. Malate dehydrogenase has been found in both avian and mammalian parasites, but appears to be cytosolic. In *P. knowlesi* lactate production accounted for only 50% of the glucose consumed; this parasite is capable of metabolizing pyruvate and lactate. Respiration of erythrocytes infected with *P. knowlesi* was maximally stimulated by lactate (66). Although intraerythrocytic stages depend mainly on glycolysis for their energy production, some aspects of mitochondrial function must be crucial to their survival, since inhibitors of electron transport and mitochondrial protein synthesis have significant antimalarial effects. The contribution of the mitochondria to the cellular ATP pool is relatively small and can be quickly compensated for by other sources (67, 68). Mitochondria of the rodent parasite *P. yoelii* are cristate, whereas those of *P. falciparum* are essentially acristate (69). They are capable of oxidizing NADH, glycerol-3-phosphate, and succinate. Proline and dihydro-orotate are also oxidized but to a lesser extent. The mitochondria of *P. falciparum*, but not those of *P. yoelii*, oxidize glutamate. In mitochondria the cytochromes aa_3, b,c and c_1 have been identified and the antimalarial hydroxynaphthoquinone 566C80 has been shown to interfere with the function of the bc_1 complex (70, 71). The respiratory chain 'Site I' NADH ubiquinone oxidoreductase seems to be inoperational in *Plasmodium* (69), which may mean that NADH-fumarate reductase is involved in the reoxidation of mitochondrial NADH.

The biochemical basis for the drastic increase in glycolytic rate of the infected host cell is not yet well understood. Enzymes of glycolysis have been described in almost all malaria parasites studied, but only in the species *P. falciparum* has a complete set of glycolytic enzymes been identified in parasite extracts (72). Some enzymes, such as HK, enolase, PK (72), glucose phosphate isomerase (GPI) (73) and PGK (74) vastly increased over the corresponding levels in uninfected cells, whereas others hardly increased with parasitemia (GAPDH and TIM). Several non-glycolytic enzymes such as glucose-6-phosphate dehydrogenase, diphosphoglycerate mutase and adenylate kinase even decreased in activity. Most enzymes, except for glucose-6-phosphate dehydrogenase, could be recovered from the parasites themselves after lysis of the host cells with saponin (72) and in most cases electrophoretically distinct bands of enzyme activity were also seen. The gene coding for the *P. falciparum* glucose-6-phosphate dehydrogenase has recently been cloned (75). So far the enzymes HK, GPI, PFK, aldolase, PGK and lactate dehydrogenase have been studied in some detail and some enzymes have been (partially) purified. The increase of glycolytic activity is the result of the expression of the parasite-derived enzymes of the glycolytic pathway in the infected cell. The pathway for the synthesis of 2,3-diphosphoglycerate is absent (76). The amount of information about the constituent enzymes of the glycolytic pathway of *Plasmodium* will be dramatically increased by the application of gene cloning techniques. Already the genes for HK, GPI, aldolase, TIM, PGK, enolase and LDH have been cloned and analysed and some of the gene products have now been overexpressed and produced in quantities sufficient for more detailed biochemical studies.

3. REFERENCES

1. Bowman, I. B. R. and Flynn, I. W. (1976) Oxidative metabolism of trypanosomes. In: *Biology of the Kinetoplastida* (eds Lumsden, W. H. R. and Evans, D. A.), vol. 1, Academic Press, New York, pp. 435–476.
2. Opperdoes, F. R. and Borst, P. (1977) Localization of nine glycolytic enzymes in a microbody-like organelle in *Trypanosoma brucei*: the glycosome. *FEBS Lett.* **80**: 360–364.
3. Fairlamb, A. H. and Opperdoes, F. R. (1986) Carbohydrate metabolism in African trypanosomes, with special reference to the glycosome. In: *Carbohydrate Metabolism of Cultured Cells* (ed. Morgan, M. J.), Plenum Press, New York, pp. 183–224.
4. Opperdoes, F. R. (1987) Compartmentation of carbohydrate metabolism in trypanosomes. *Ann. Rev. Microbiol.* **41**: 127–151.
5. Opperdoes, F. R., Borst, P. and Fonck, K. (1976) The potential use of inhibitors of glycerol-3-phosphate oxidase for chemotherapy of African trypanosomes. *FEBS Lett.* **62**: 169–172.
6. Fairlamb, A. H., Opperdoes, F. R. and Borst, P. (1976) New approach to screening drugs for activity against African trypanosomes. *Nature* **265**: 270–271.
7. Clarkson, Jr, A. B. and Brown, F. H. (1976) Trypanosomiasis: an approach to chemotherapy by the inhibition of carbohydrate metabolism. *Science* **194**: 204–206.
8. Ryley, J. F. (1962) Studies on the metabolism of protozoa. 9. Comparative metabolism of the culture and bloodstream forms of *Trypanosoma rhodesiense*. *Biochem. J.* **85**: 211–223.
9. Cross, G. A. M., Klein, R. A. and Linstead, D. J. (1975) Utilization of amino acids by *Trypanosoma brucei* in culture: L-threonine as a precursor for acetate. *Parasitology* **71**: 311–326.
10. Opperdoes, F. R. and Cottem, D. (1982) Involvement of the glycosome of *Trypanosoma brucei* in carbon dioxide fixation. *FEBS Lett.* **143**: 60–64.
11. Cazzulo, J. J., Franke de Cazzulo, B. M., Engel, J. C. and Cannata, J. J. (1985) End products and enzyme levels of aerobic glucose fermentation in trypanosomatids. *Mol. Biochem. Parasitol.* **16**: 329–343.
12. Darling, T. N., Davis, D. G., London, R. E. and Blum, J. J. (1987) Products of *Leishmania braziliensis* glucose metabolism: release of D-lactate and, under anaerobic conditions, glycerol. *Proc. Natl. Acad. Sci. USA* **84**: 7129–7133.
13. Darling, T. N., Davis, D. G., London, R. E. and Blum, J. J. (1989) Carbon dioxide abolishes the reverse pasteur effect in *Leishmania major* promastigotes. *Mol. Biochem. Parasitol.* **33**: 191–202.
14. Keegan, F. P., Sansone, L. and Blum, J. J. (1987) Oxidation of glucose, ribose, alanine and glutamate by *Leishmania braziliensis panamensis*. *J. Protozool.* **34**: 174–179.
15. Walsh, M. J. and Blum, J. J. (1992) Effects of hypoxia and acute osmotic stress on intermediate metabolism in *Leishmania* promastigotes. *Mol. Biochem. Parasitol.* **50**: 205–214.
16. Darling, T. N., Balber, A. E. and Blum, J. J. (1988) A comparative study of D-lactate and L-lactate formation by four species of *Leishmania* and of *Trypanosoma lewisi* and *Trypanosoma brucei gambiense*. *Mol. Biochem. Parasitol.* **30**: 253–258.
17. Ghoshal, K., Banerjee, A. B. and Ray, S. (1989) Methylglyoxal-metabolizing enzymes of *Leishmania donovani* promastigotes. *Mol. Biochem. Parasitol.* **35**: 21–29.
18. Ter Kuile, B. H. and Opperdoes, F. R. (1993) Uptake and turnover of glucose in *Leishmania donovani*. *Mol. Biochem. Parasitol.* **60**: 313–322.
19. Keegan, F. P. and Blum, J. J. (1992) Utilization of a carbohydrate reserve comprised primarily of mannose by *Leishmania donovani*. *Mol. Biochem. Parasitol.* **53**: 193–200.
20. Blum, J. J. (1991) Effect of osmotic pressure on the oxidative metabolism of *Leishmania major* promastigotes. *J. Protozool.* **38**: 229–233.
21. Frydman, B., de los Santos, C., Cannata, J. J. and Cazzulo, J. J. (1990) Carbon-13 nuclear magnetic resonance analysis of [1-^{13}C]glucose metabolism in *Trypanosoma cruzi*. Evidence for the presence of two alanine pools and of two CO_2 fixation reactions. *Eur. J. Biochem.* **192**: 363–368.

22. Urbina, J. A., Osorno, C. E. and Rojas, A. (1990) Inhibition of phosphoenol pyruvate carboxykinase from *Trypanosoma (Schizotrypanum) cruzi* epimastigotes by 3-mercaptopicolinic acid: in vitro and in vivo studies. *Arch. Biochem. Biophys.* **282**: 91–99.

23. Hart, D. T., Vickerman, K. and Coombs, G. H. (1981) Respiration of *Leishmania mexicana* amastigotes and promastigotes. *Mol. Biochem. Parasitol.* **4**: 39–51.

24. Rainey, P. M., Spithill, T. W., McMahon-Pratt, D. and Pan, A. A. (1990) Biochemical and molecular characterization of *Leishmania pifanoi* amastigotes in continuous axeneous culture. *Mol. Biochem. Parasitol.* **49**: 111–118.

25. Engel, J. C., Franke de Cazzulo, B. M., Stoppani, A. O., Cannata, J. J. and Cazzulo, J. J. (1987) Aerobic glucose fermentation by *Trypanosoma cruzi* axenic culture amastigote-like forms during growth and differentiation to epimastigotes. *Mol. Biochem. Parasitol.* **26**: 1–10.

26. Misset, O., Bos, O. J. M. and Opperdoes, F. R. (1986) Glycolytic enzymes of *Trypanosoma brucei*. Simultaneous purification, intraglycosomal concentrations and physical properties. *Eur. J. Biochem.* **157**: 441–453.

27. Misset, O. and Opperdoes, F. R. (1984) Simultaneous purification of hexokinase, class-I fructose-bisphosphate aldolase, triosephosphate isomerase and phosphoglycerate kinase from *Trypanosoma brucei*. *Eur. J. Biochem.* **144**: 475–483.

28. Misset, O. and Opperdoes, F. R. (1987) The phosphoglycerate kinases from *Trypanosoma brucei*: a comparison of the glycosomal and cytosolic isoenzymes and their sensitivity towards suramin. *Eur. J. Biochem.* **162**: 493–500.

29. Misset, O., Van Beeumen, J., Lambeir, A.-M., Van der Meer, R. and Opperdoes, F. R. (1987) Glyceraldehyde-phosphate dehydrogenase from *Trypanosoma brucei*. Comparison of the glycosomal and cytosolic isoenzymes. *Eur. J. Biochem.* **162**: 501–507.

30. Cronin, C. N. and Tipton, K. F. (1985) Purification and regulatory properties of phosphofructokinase from *Trypanosoma brucei*. *Biochem. J.* **227**: 113–124.

31. Lambeir, A.-M., Opperdoes, F. R. and Wierenga, R. K. (1987) Kinetic properties of triosephosphate isomerase from *Trypanosoma brucei*: a comparison with the rabbit muscle and yeast enzymes. *Eur. J. Biochem.* **168**: 69–74.

32. Lambeir, A.-M., Loiseau, A., Kuntz, D. A., Vellieux, F. M. D., Michels, P. A. M. and Opperdoes, F. R. (1991) The cytosolic and glycosomal glyceraldehyde-phosphate dehydrogenase from *Trypanosoma brucei*. Kinetic properties and comparison with homologous enzymes. *Eur. J. Biochem.* **198**: 429–435.

33. Marchand, M., Kooystra, U. and Wierenga, R. K. *et al.* (1989) Glucose-phosphate isomerase from *Trypanosoma brucei*. Cloning and characterization of the gene and analysis of the enzyme. *Eur. J. Biochem.* **184**: 455–464.

34. Callens, M., Kuntz, D. A. and Opperdoes, F. R. (1991) Kinetic properties of fructose bisphosphate aldolase from *Trypanosoma brucei* compared to aldolases from rabbit muscle and *Staphylococcus aureus*. *Mol. Biochem. Parasitol.* **47**: 1–10.

35. Callens, M., Kuntz, D. A. and Opperdoes, F. R. (1991) Characterization of pyruvate kinase of *Trypanosoma brucei* and its role in the regulation of carbohydrate metabolism. *Mol. Biochem. Parasitol.* **47**: 19–30.

36. Callens, M. and Opperdoes, F. R. (1992) Some kinetic properties of pyruvate kinase from *Trypanosoma brucei*. *Mol. Biochem. Parasitol.* **50**: 235–244.

37. Krakow, J. L. and Wang, C. C. (1990) Purification and characterization of glycerol kinase from *Trypanosoma brucei*. *Mol. Biochem. Parasitol.* **43**: 17–25.

38. Clayton, C. (1986) Structure and regulated expression of genes encoding fructose bisphosphate aldolase in *Trypanosoma brucei*. *EMBO J.* **4**: 2997–3003.

39. Marchand, M., Poliszczak, A., Gibson, W. C., Wierenga, R. K., Opperdoes, F. R. and Michels, P. A. M. (1988) Characterization of the genes for fructose-bisphosphate aldolase in *Trypanosoma brucei*. *Mol. Biochem. Parasitol.* **29**: 65–76.

40. Swinkels, B. W., Gibson, W. C., Osinga, K. A., Kramer, R., Veeneman, G. H. and Borst, P. (1986) Characterization of the gene for the microbody (glycosomal) triosephosphate isomerase of *Trypanosoma brucei*. *EMBO J.* **5**: 1291–1298.

41. Michels, P. A. M., Poliszczak, A. and Osinga, K. A. *et al.* (1986) Two tandemly linked identical genes code for the glycosomal glyceraldehyde-phosphate dehydrogenase in *Trypanosoma brucei*. *EMBO J.* **5**: 1049–1056.

42. Hannaert, V., Blaauw, M., Kohl, L., Allert, S., Opperdoes, F. R. and Michels, P. A. M. (1992) Molecular analysis of the cytosolic and glycosomal glyceraldehyde-3-phosphate dehydrogenase in *Leishmania mexicana*. *Mol. Biochem. Parasitol.* **55**: 115–126.

43. Osinga, K. A., Swinkels, B. W. and Gibson, W. C. *et al.* (1985) Topogenesis of microbody enzymes: a sequence comparison of the genes for the glycosomal (microbody) phosphoglycerate kinases. *EMBO J.* **4**: 3811–3817.

44. Swinkels, B. W., Evers, R. and Borst, P. (1988) The topogenic signal of the glycosomal (microbody) phosphoglycerate kinase of *Crithidia fasciculata* resides in a carboxy-terminal extension. *EMBO J.* **7**: 1159–1165.

45. Allert, S., Ernst, I., Poliszczak, A., Opperdoes, F. R. and Michels, P. A. M. (1991) Molecular cloning and analysis of two tandemly linked genes for pyruvate kinase of *Trypanosoma brucei*. *Eur. J. Biochem.* **200**: 19–27.

46. Kueng, V., Schlaeppi, K., Schneider, A. and Seebeck, T. (1989) A glycosomal protein (P60) which is predominantly expressed in procyclic *Trypanosoma brucei*. Characterization and DNA sequence. *J. Biol. Chem.* **264**: 5203–5209.

47. Parsons, M. and Smith, J. M. (1989) Trypanosome glycosomal protein P60 is homologous to phosphoenolpyruvate carboxykinase. *Nucleic Acids Res.* **17**, 6411.

48. Alexander, K., Hill, T., Schilling, J. and Parsons, M. (1990) Microbody phosphoglycerate kinase of *Trypanosoma brucei*: expression and complementation in *Escherichia coli*. *Gene* **90**: 215–220.

49. Kendall, G., Wilderspin, A. W. F., Ashall, F., Miles, M. A. and Kelly, J. M. (1990) *Trypanosoma cruzi* glycosomal glyceraldehyde-3-phosphate dehydrogenase does not conform to the 'hotspot' model of topogenesis. *EMBO J.* **9**: 2751–2758.

50. Grant, P. T. and Fulton, J. D. (1975) Catabolism of glucose by strains of *Trypanosoma rhodesiense*. *Biochem. J.* **66**: 242–250.

51. Bowman, I. B. R., Tobie, E. J. and Von Brandt, T. (1963) CO_2 fixation studies with the culture form of *Trypanosoma cruzi*. *Comp. Biochem. Physiol.* **9**: 105–114.

52. Cronin, C. N., Nolan, D. P. and Voorheis, H. P. (1989) The enzymes of the classical pentose-phosphate pathway display differential activities in the procyclic and bloodstream forms of *Trypanosoma brucei*. *FEBS Lett.* **244**: 26–30.

53. Visser, N. and Opperdoes, F. R. (1980) Glycolysis in *Trypanosoma brucei*. *Eur. J. Biochem.* **103**: 623–632.

54. Ter Kuile, B. H. and Opperdoes, F. R. (1991) Glucose uptake by *Trypanosoma brucei*. *J. Biol. Chem.* **266**: 857–862.

55. Van Schaftingen, E., Opperdoes, F. R. and Hers, G.-H. (1985) Stimulation of *Trypanosoma brucei* pyruvate kinase by fructose 2,6-bisphosphate. *Eur. J. Biochem.* **153**: 403–406.

56. Van Schaftingen, E., Opperdoes, F. R. and Hers, H.-G. (1987) Effects of various metabolic conditions of the trivalent arsenical melarsen oxide on the intracellular levels of fructose 2,6-bisphosphate and of glycolytic intermediates in *Trypanosoma brucei*. *Eur. J. Biochem.* **166**: 653–661.

57. Hart, D. T., Misset, O., Edwards, S. W. and Opperdoes, F. R. (1984) A comparison of the glycosomes (microbodies) isolated from *Trypanosoma brucei* bloodstream form and cultured procyclic trypomastigotes. *Mol. Biochem. Parasitol.* **12**: 25–35.

58. Visser, N., Opperdoes, F. R. and Borst, P. (1981) Subcellular compartmentation of the glycolytic intermediates in *Trypanosoma brucei*. *Eur. J. Biochem.* **118**: 521–526.

59. Michels, P. A. M. and Hannaert, V. (1994) The evolution of Kinetoplastid glycosomes. *J. Bioenerg. Biomembr.* **26**: 213–219.

60. Keegan, F. P. and Blum, J. J. (1993) Incorporation of label from acetate and laurate into the mannan of *Leishmania donovani* via the glyoxylate cycle. *J. Euk. Microbiol.* **40**: 730–732.

61. Sherman, I. W. (1979) Biochemistry of *Plasmodium* (malarial parasites). *Microbial Rev.* **43**: 453–495.

62. Homewood, C. A. and Neame, K. D. (1980) Biochemistry of malarial parasites. In: *Malaria* (ed. Kreir, J. P.), Vol. 1, Academic Press, New York, pp. 345–405.

63. McKee, R. W. (1951) Biochemistry of *Plasmodium* and the influence of antimalarials. In: *Biochemistry and Physiology of Protozoa* (ed. Hutner, S. H. and Lwoff, A.). Academic Press, New York, pp. 251–322.

64. Von Brandt, T. (1973) *Biochemistry of Parasites.* Academic Press, New York.
65. Blum, J. J. and Ginsburg, H. (1984) Absence of α-ketoglutarate dehydrogenase activity and presence of CO_2-fixing activity in *Plasmodium falciparum* grown in vitro in human erythrocytes. *J. Protozool.* **31**: 167–169.
66. Ali, S. N. and Fletcher, K. A. (1985) Carbohydrate metabolism of malarial parasites 1. Metabolism of lactate in *P. knowlesi*-infected monkey erythrocytes. *Comp. Biochem. Physiol.* **80B**: 725–729.
67. Fry, M. (1991) Mitochondria of *Plasmodium*. In: *Biochemical Protozoology* (eds Coombs, G. and North, M.) Taylor and Francis, London and Washington, pp. 154–167.
68. Fry, M., Webb, E. and Pudney, M. (1990) Effect of mitochondrial inhibitors on adenosine-triphosphate levels in *Plasmodium falciparum*. *Comp. Biochem. Physiol. B* **96**: 775–782.
69. Fry, M. and Beesley, J. E. (1991) Mitochondria of mammalian *Plasmodium* spp. *Parasitology* **102**: 17–26.
70. Hudson, A. T., Randall, A. W. and Fry, M. *et al.* (1985) Novel anti-malarial hydroxynaphthoquinones with potent broad spectrum anti-protozoal activity. *Parasitology* **90**: 45–55.
71. Fry, M. and Pudney, M. (1992) Site of action of the antimalarial hydroxynaphthoquinone, 2-[trans-4-(4'-chlorophenyl)cyclohexyl]-3-hydroxy1,4-naphthoquinone (566C80). *Biochem. Pharmacol.* **43**: 1545–1553.
72. Roth, E. F., Calvin, M. C., Max-Audit, I., Rosa, J. and Rosa, R. (1988) The enzymes of the glycolytic pathway in erythrocytes infected with *Plasmodium falciparum* malaria parasites. *Blood* **72**: 1922–1925.
73. Shrivastava, I. K., Schmidt, M., Grall, M., Certa, U., Garcia, A. M. and Perrin, L. H. (1990) Identification and purification of glucose phosphate isomerase of *Plasmodium falciparum*. *Mol. Biochem. Parasitol.* **54**: 153–164.
74. Grall, M., Shrivastava, I. K., Schmidt, M., Garcia, A.-M., Mauel, J. and Perrin, L. H. (1992) *Plasmodium falciparum*: identification and purification of the phosphoglycerate kinase of the malaria parasite. *Exp. Parasitol.* **75**: 10–18.
75. O'Brien, E., Kurdi-Haidar, B. and Wanachiwanawin, W. *et al.* (1994) Cloning of the glucose-6-phosphate dehydrogenase gene of *Plasmodium falciparum*. *Mol. Biochem. Parasitol.* **64**: 177–358.
76. Roth, E. F., Joulin, V. and Miwa, S. *et al.* (1988) The use of enzymopathic red cells in the study of malarial parasite glucose metabolism. *Blood* **71**: 1408–1413.

3 Energy Metabolism in Anaerobic Protozoa

GRAHAM H. COOMBS[1] and MIKLÓS MÜLLER[2]

[1]Institute of Biomedical and Life Sciences, University of Glasgow, Glasgow, UK and [2]The Rockefeller University, New York, NY, USA

SUMMARY

Giardia, Entamoeba and trichomonads are parasitic protozoa which can be described as anaerobic or microaerophilic because: (i) they do not require oxygen for survival and multiplication; (ii) they do not utilize oxygen as a major terminal electron acceptor in their energy metabolism and (iii) they can tolerate some oxygen but are damaged by higher concentrations. These organisms do not contain mitochondria and rely on fermentative processes via an extended glycolytic pathway for ATP generation. Fermentative metabolism persists in all these species, even when oxygen is present. Inorganic pyrophosphate rather than ATP is used in one or more of the glycolytic reactions. The glycolytic pathway is either not regulated or differs in its regulation from that operating in mitochondrion-containing eukaryotic cells. The central catabolic step, oxidative decarboxylation of pyruvate, is catalyzed by an iron–sulfur enzyme, pyruvate: ferredoxin oxidoreductase, and not by the pyruvate dehydrogenase complex characteristic of mitochondria. The reoxidation of ferredoxin reduced in this process occurs through the reduction of organic intermediates of metabolism (in *Giardia* and *Entamoeba*) or in the formation of hydrogen (in trichomonads). The energy metabolism of *Giardia* and *Entamoeba* is not compartmentalized, with all reactions occurring in the cytosol or on the cytosolic surfaces of membranes. Glycolysis in trichomonads is also cytosolic, but pyruvate oxidation and hydrogen formation occur in a separate organelle — the hydrogenosome. The enzymes of energy metabolism in these anaerobic parasites are more closely related to

Biochemistry and Molecular Biology of Parasites
ISBN 0-12-473345-X

those found in eubacteria and eukaryotes than to those present in archaea bacteria. The characteristic metabolic properties of these parasites account for their
selective susceptibility to the anti-anaerobe 5-nitroimidazole drugs.

An anaerobic or microaerophilic mode of life characterizes a number of parasitic
protozoa. Most of these organisms are not closely related and there are many major
differences among them in cell structure and their underlying molecular organizations.
They are considered together in this chapter, however, because they share a number of
metabolic peculiarities that can be regarded as adaptations for life in environments
where oxygen is limited. The evolutionary origin of these features remains obscure,
though they may represent ancestral traits. Certainly many anaerobic parasitic protozoa belong to lineages which represent early branches of the eukaryotic evolutionary
tree.

Anaerobic might not be an entirely accurate description of these parasites. It is,
however, a useful term to designate organisms which (a) do not require oxygen for their
survival and multiplication and (b) do not utilize oxygen as a major terminal electron
acceptor in their energy metabolism. It also appropriately reflects a major cytological
hallmark of the organisms discussed in detail in this chapter — none of them contain
respiratory organelles of a mitochondrial nature, that is with cytochrome-mediated
electron transport and electron transport-linked oxidative phosphorylation (1). Thus
their energy conservation processes are significantly less efficient than, and very
different from, those of aerobic eukaryotic cells which contain typical mitochondria.

This chapter focuses primarily on four parasites, all lumen dwellers, which have been
chosen for two main reasons. First, they are among the few anaerobes which have been
studied in sufficient detail to permit a detailed description. Secondly, they represent two
rather different types of subcellular organization (Fig. 3.1). These organisms are *Giardia
lamblia*, an inhabitant of the small intestine of mammals, *Entamoeba histolytica*, which
lives in the large intestine of humans, and two members of the family Trichomonadidae,
Trichomonas vaginalis and *Tritrichomonas foetus*, which are parasites of human and
bovine genitourinary tracts, respectively. The first two parasites lack any apparent
subcellular compartmentation of their energy metabolism, whereas the trichomonads
have a cytosol/hydrogenosome compartmentation (Fig. 3.1). Several other amitochondrial organisms, and indeed a few parasitic protozoa that possess mitochondria, will
also be mentioned, albeit in brief, in this chapter, but those parasitic protozoa such as
microsporidia, which lack mitochondria but have not been explored with biochemical
methods, are not included. Parasites such as *Trypanosoma brucei* which can survive
anaerobic conditions but only multiply in the presence of oxygen and as such could be
considered as facultative anaerobes are described in Chapter 2.

It is generally thought that oxygen is at a very low concentration in the natural
habitats of the anaerobic parasites. This is probably too simplistic a view, however, for
although oxygen is rapidly consumed by microbial inhabitants and is frequently not
detectable at the center of the lumen (at least in the mammalian large intestine, which
has been studied), it is likely that oxygen will be more available near to the mucosal
linings of the gut and vagina, and it is these regions which the four parasites under
discussion mainly inhabit. It is also likely that there will be significant fluctuations of
oxygen concentrations at these sites, depending on the physiological state of the organ.

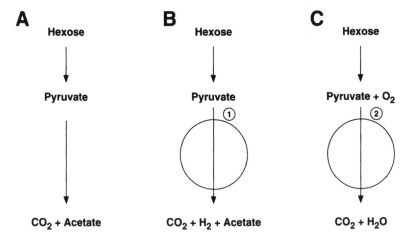

FIG. 3.1 Main types of subcellular organization of carbohydrate catabolism in eukaryotes. (A) No compartmentation (*Giardia* sp., *Entamoeba* sp.); (B) cytosolic/hydrogenosomal compartmentation (*Trichomonas vaginalis*, other trichomonads); (C) cytosolic/mitochondrial compartmentation (most eukaryotic cells). 1, hydrogenosome; 2, mitochondrion.

Indeed, data suggest that oxygen tensions in rat and sheep intestines may reach 60 μM (2), and that concentrations of 40 μM can occur near to the vaginal epithelium (3). In addition, a most important aspect of *E. histolytica* infections is invasion of tissues such as liver, whereas *T. foetus* invades the placenta. These parasites must be exposed to relatively high oxygen concentrations at least on some occasions while in their host. The cysts of *Entamoeba* and *Giardia* also clearly experience aerobic conditions when they are outside a host and they must have adaptations that enable them to remain viable for extended periods under such conditions. Nevertheless, the parasites have been considered to be anaerobes because the evidence available suggested that oxygen provides no significant benefit to the organisms and that atmospheric concentrations of oxygen adversely affect growth of the organisms *in vitro*. As low concentrations of oxygen do not kill the parasites, they have been said to be aerotolerant anaerobes. This contrasts with protozoa which live in the gut of termites, in that they are thought to be unable to withstand exposure to even low oxygen tensions and thus appear to be obligate anaerobes.

Recent evidence, however, suggests that *T. vaginalis* grows optimally in the presence of small amounts of oxygen (4), higher concentrations being inhibitory, which would mean that it should be classified as a microaerophile. Previous studies resulted in similar conclusions for *E. histolytica* (5). *Giardia lamblia* has not been rigorously investigated in this way. One problem raised by these results is to understand how such low levels of oxygen could be beneficial and attempts to unravel the paradox have been hampered by the difficulties encountered in studying the use of such small amounts of oxygen and so elucidating the mechanisms involved. All four parasites unquestionably do take up oxygen at a high rate when it is present (as discussed below), although this has generally been considered to be a means of detoxifying the potentially hazardous material and not to be involved in energy conservation. This is one aspect of the anaerobic nature of the parasites that requires further study.

Current data show that the parasites have several features consistent with their metabolism being of an anaerobic type. Oxygen, when used as an electron acceptor, is not involved in energy conservation. The organisms catabolize carbohydrates in a basically fermentative fashion (irrespective of whether oxygen is present or not). Glycolysis is central to this process, which yields organic acids and alcohols as major end products (1). The pathways of carbohydrate catabolism in *E. histolytica*, *G. lamblia* and *T. vaginalis* are depicted in Figs. 3.2 and 3.3. The glycolytic sequences themselves are classical in the sense that glucose is catabolized to pyruvate, but there are several features that distinguish the pathways from that operating in most eukaryotes. Some seem to be adaptations to an anaerobic existence and the roles of inorganic pyrophosphate and iron–sulfur proteins will be discussed in some detail below, as will the unusual regulatory mechanisms. Recent years have heralded the first sequence information on some of the glycolytic enzymes of these parasites. All enzyme sequences reported so far show that the anaerobes' enzymes are homologs of isofunctional enzymes in other eukaryotes and eubacteria and only more distantly related to the corresponding archaebacterial enzymes [for example, glyceraldehyde 3-phosphate dehydrogenase (6)]. As yet, however, the evolutionary relationships of the enzymes of the core metabolism of these amitochondrial protozoa are still obscure.

A most notable characteristic of these anaerobic parasites is the presence of enzymes specific for inorganic pyrophosphate (PP_i) rather than ATP (1). Such an observation was made for *E. histolytica* during the pioneering work of Reeves and his co-workers in the 1960s and 1970s and more recently similar findings have been reported for trichomonads and *Giardia* (7–10). In the case of *E. histolytica*, PP_i was reported to be involved in four reactions (5). In contrast, with *T. vaginalis* only phosphofructokinase (PFK) has been shown to be PP_i-specific (8) whereas in *Giardia* both phosphofructokinase and pyruvate,phosphate dikinase are PP_i-specific (9, 10). A perceived advantage to the cell of PP_i-dependent PFK is that the net ATP production in glycolysis is three molecules per glucose rather than the more common two. Such a difference in energy yield would be highly significant for a cell with fermentative metabolism (such as an anaerobe) which is unable to fully oxidize pyruvate to yield further ATP (the theoretical yield being 36 ATP per glucose for cells with mitochondrial respiration).

Such a strategy could not be successful, however, in the absence of appropriately high concentrations of PP_i, which in mammalian cells is normally kept at low concentrations through the action of cytosolic inorganic pyrophosphatase. Trichomonads (11) and *E. histolytica* (12), however, are devoid of such an activity. The accepted dogma is that PP_i is rapidly hydrolyzed by a cytosolic inorganic pyrophosphatase in mammalian cells in order to avoid inhibition of nucleic acid biosynthesis. An unanswered question is how trichomonads are able to avoid negative consequences of high PP_i concentrations. Another implication of PP_i-PFK is that this step in glycolysis can not be regulated as tightly as it is through ATP-PFK in mammalian cells (7). Indeed it appears that PP_i-PFK is not a point of control for glycolysis in these anaerobic organisms and is not affected by concentrations of ATP/ADP/AMP which are key controllers of mammalian PFK. Nor is the main effector molecule of mammalian PFK, fructose-2,6-bisphosphate, active in this way in these anaerobic organisms.

So how is glycolysis controlled in these anaerobes? A search for regulated enzymes of glycolysis in *E. histolytica* did not detect any (5). No information is available for *G. lamblia*. In *T. vaginalis*, however, pyruvate kinase is modulated by ribose-5-phosphate

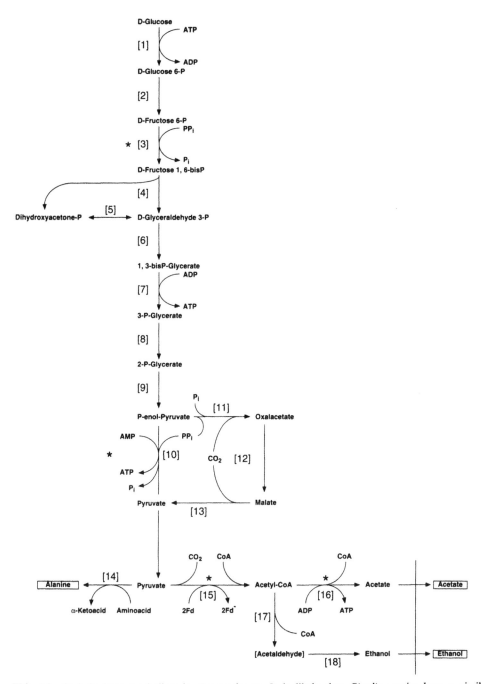

FIG. 3.2 Carbohydrate catabolism in *Entamoeba* sp. It is likely that *Giardia* species have a similar metabolism, as indicated by the detection of the same metabolic end-products and certain characteristic enzymes (marked with an asterisk *). Enzymes: glucokinase [1], phosphoglucose isomerase [2], phosphofructokinase (PP$_i$)* [3], aldolase [4], triosephosphate isomerase [5], glyceraldehyde-3-phosphate dehydrogenase [6], phosphoglycerate kinase [7], phosphoglyceromutase [8], enolase [9], pyruvate phosphate dikinase* [10], PEP carboxyphosphotransferase [11], malate dehydrogenase [12], malate dehydrogenase (decarboxylating) [13], alanine aminotransferase [14], pyruvate:ferredoxin oxidoreductase* [15]. Thiokinase* [16], bifunctional acetyl-CoA reductase-alcohol dehydrogenase [17–18], alanine amino Fd = ferredoxin. The ferredoxin reduced in reaction [15] is probably coupled to reaction [17] by an, as yet, unknown mechanism. Vertical line (lower right) represents the cell membrane. Metabolic end-products are boxed.

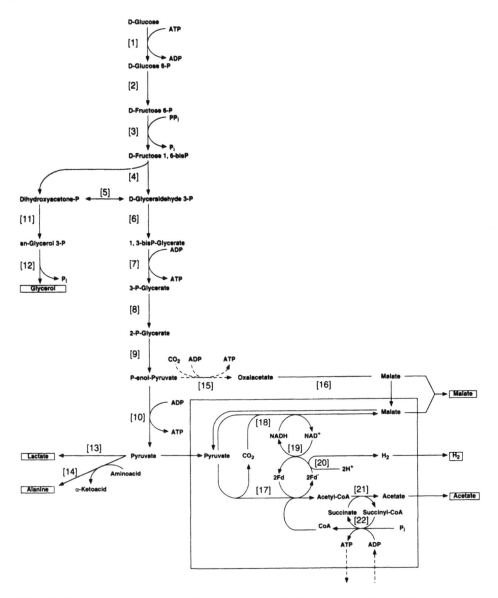

FIG. 3.3 Carbohydrate metabolism in *Trichomonas vaginalis*. Enzymes: glucokinase [1], phosphoglucose isomerase [2], phosphofructokinase (PP_i) [3], aldolase [4], triose phosphate isomerase [5], glyceraldehyde-3-phosphate dehydrogenase [6], phosphoglycerate kinase [7], phosphoglyceromutase [8], enolase [9], pyruvate kinase [10], glycerol-3-phosphate dehydrogenase [11], glycerol-3-phosphatase [12], lactate dehydrogenase [13], alanine aminotransferase [14], phosphoenolpyruvate carboxykinase [15], malate dehydrogenase [16], pyruvate:ferredoxin oxidoreductase [17], malate dehydrogenase (decarboxylating) [18], NAD(P):ferredoxin oxidoreductase [19], hydrogenase [20], acetate:succinate CoA-transferase [21], succinate thiokinase [22]. Fd = ferredoxin. Arrows indicate the assumed physiologic directions of the reactions. Dashed arrows indicate postulated reactions. The box designates the hydrogenosomal location of reactions [17] to [22], but does not reflect the relationship of the enzymes to the organelle's envelope. Modified with permission from ref. 40. Metabolic end-products are boxed.

and glycerate-3-phosphate (13). Ribose-5-phosphate has not been observed as a glycolytic regulator in other organisms, whereas glycerate-3-phosphate is known as a negative effector of *Escherichia coli* pyruvate kinase.

Other parasitic protozoa have also been reported to possess PP_i-PFK but not ATP-PFK. This is the case with the amoeba *Naegleria fowleri* (normally a free-living organism, but able to invade human brain with fatal consequences) and the coccidia *Toxoplasma gondii* and *Eimeria tenella* (7, 14). These organisms were not generally considered to be anaerobes, and are thought to have functional mitochondria. *Eimeria*, however, does have a phase within the gut lumen of its host and it is conceivable that one adaptation of this developmental stage of the life cycle for survival in this low oxygen environment is a fermentative metabolism with PP_i-PFK. Similarly, *T. gondii* has a largely fermentative metabolism, at least at some stages in the life cycle (15). Interestingly, *E. tenella* is also similar to *T. vaginalis* in having pyruvate kinase regulated with unusual effectors (14). Further studies are required to discover the extent to which there is stage variation in the energy metabolism of these coccidian parasites, but current data suggest that they may represent an unusual type of anaerobe—one which is adapted for anaerobiosis but encounters an environment low in oxygen for only a short phase in its life cycle.

Energy metabolism is fermentative in each of *Entamoeba*, *Giardia* and the trichomonads, but the end-products differ between species. None of the organisms are homolactic fermenters but instead produce a number of reduced compounds (acids and alcohols) (boxed on Figs. 3.2 and 3.3). The perceived advantage of this transglycolytic metabolism is that it results in the generation of additional energy. Whereas this is undoubtedly true in the case of acetate production, the advantage to the parasites of producing some of the compounds, for example lactate production by *T. vaginalis*, has not been resolved. There is a need for production of reduced compounds in order to maintain redox balance (as detailed later), but this does not appear to account fully for the variety of end-products that are released.

The four parasites under discussion all have complex enzymatic machinery involved in connecting glycolysis itself to the further metabolism of its products (Figs. 3.2 and 3.3). The pathways depicted in the figures have been proposed on the basis of enzyme activities detected and the overall carbon balances reported, but the relative carbon flow through the various branches still remains to elucidated. Both C_3 and C_4 compounds play a significant role in this further metabolism, showing that carboxylation of the glycolytically formed phosphoenolpyruvate (PEP) and pyruvate is an important part of the energy metabolism. It is clear, however, that there are significant differences between the pathways that operate in *G. lamblia* and *E. histolytica*, which in this respect are similar, and the trichomonads.

The better studied of the first pair is *E. histolytica*, in which several additional PP_i-linked enzymes are involved (Fig. 3.2). The direct conversion of PEP into pyruvate is not catabolized by pyruvate kinase but by pyruvate,phosphate dikinase. The same overall result is also produced through the sequential action of three enzymes, PEP carboxykinase (PP_i-forming), malate dehydrogenase and malate dehydrogenase (decarboxylating). These form a bypass to the direct conversion. What is the advantage to the parasite of having these two routes for converting PEP into pyruvate? By contributing equally to the carbon flow, the two branches could functionally complement each other. One produces the PP_i consumed by the other. In addition, the coenzyme specificity of the two malate dehydrogenases means that the bypass functions as a virtual NADH/NADPH transhydrogenase (5). *G. lamblia* has been studied in less detail, but may well be similar to *E. histolytica*. It definitely has a pyruvate,phosphate dikinase instead of pyruvate kinase (10).

Carboxylation of C_3 compounds is also important in trichomonads (Fig. 3.3). Both *T. vaginalis* and *T. foetus* are thought to contain an ADP-linked PEP carboxykinase which produces oxaloacetate. They also have malate dehydrogenase of high activity, and malate dehydrogenase (decarboxylating) activity (in both the cytosol and hydrogenosomes, see below). Pyruvate kinase of *T. vaginalis* has been characterized (13), but this enzyme could not be detected in *T. foetus*. The latter species also lacks pyruvate,phosphate dikinase, suggesting that formation of pyruvate in this parasite occurs primarily via malate, involving a carboxylation and decarboxylation step.

Most oxidoreductases of these parasites are similar to those in other eukaryotes in using pyridine nucleotides (NAD and NADP) as redox partners. In addition, however, their core metabolism is characterized by an important involvement of iron–sulfur proteins, which is also a characteristic of anaerobic eubacteria. The key step in pyruvate metabolism, its oxidative decarboxylation to acetyl-CoA, is catalyzed by an iron–sulfur enzyme, pyruvate:ferredoxin oxidoreductase. The enzyme has been characterized from *T. vaginalis* (16), *T. foetus* (17) and *E. histolytica* (18), and detected in *G. lamblia* (19). It is similar in properties and has sequence homology to isofunctional enzymes in anaerobic eubacteria. The enzyme is, however, fundamentally different from the mitochondrial pyruvate dehydrogenase complex, which catalyzes the same overall reaction but involves a different electron acceptor. The electron acceptor used by the enzyme of these anaerobic parasites is known to be ferredoxin, also an iron–sulfur protein (1). Interestingly, however, the trichomonad protein, a [2Fe-2S] ferredoxin with some similarities to mitochondrial proteins, belongs to a different subfamily to the protein in *E. histolytica*, which is a 2[4Fe-4S] ferredoxin like those in anaerobic eubacteria. The *G. lamblia* ferredoxin is probably similar to the latter, although it is yet to be fully characterized.

Reduced ferredoxin, generated in the parasites during the conversion of pyruvate into acetyl-CoA, can be reoxidized in a number of ways, In *E. histolytica* and probably also in *Giardia* the electrons are utilized in the production of ethanol by the bifunctional enzyme, acetyl-CoA reductase–alcohol dehydrogenase. This enzyme differs significantly from the alcohol dehydrogenase that occurs in most organisms (20). The intermediate steps of the electron transfer await elucidation. In trichomonads under anaerobic conditions, the electrons are removed from reduced ferredoxin in the formation of hydrogen gas. This reaction is catalyzed by hydrogenase, another iron–sulfur enzyme. Such an enzyme is a characteristic constituent of various groups of anaerobic eubacteria, but among eukaryotes it is known to occur only in anaerobic protozoa. Ferredoxin can also be reoxidized directly by oxygen (the significance of this in the living parasite is unclear, but it probably plays an important role in mediating resistance of trichomonads to 5-nitroimidazoles, see Chapter 9) or NAD/NADP (see Fig. 3.3).

Acetyl-CoA, formed from pyruvate, can also be converted into acetate, another end-product of these organisms. The mechanism of its formation differs between species, but is always coupled to ATP production. In *Entamoeba* and *Giardia* it is catalyzed by a single enzyme, acetate thiokinase (18, 19). This enzyme is known to occur in just a few prokaryotes. In the trichomonads, acetate formation is mediated by acetate: succinate-CoA-transferase, which transfers the CoA moiety of acetyl-CoA to succinate (21). The resulting succinyl-CoA serves as substrate to succinyl thiokinase, an enzyme typical of mitochondria and many prokaryotes.

The mechanism of pyruvate oxidation in these anaerobic parasitic protozoa, and in particular its difference from those of mammalian enzymes, has a significant pharmacological implication. Metronidazole and other 5-nitroimidazole derivatives are effective in treating infections caused by these parasites. The selective activity of these compounds is related to their reductive activation to short-lived toxic compounds in the target organisms (Chapter 10). The reduction is achieved by reduced ferredoxin, the main (and perhaps sole) source of electrons to reduce ferredoxin being pyruvate: ferredoxin oxidoreductase. It is noteworthy that, except in the case of acquired drug resistance, the antianaerobic activity of 5-nitroimidazoles persists even when oxygen is present, although the efficacy is often less under aerobic conditions. This shows that there is a considerable metabolic flow through pyruvate:ferredoxin oxidoreductase even in the presence of oxygen, and that not all of the reduced ferredoxin is reoxidized by the oxygen present. The presence of hydrogenosomes and cytosolic oxidases in trichomonads are thought to aid in protecting the parasite against oxygen, as described below.

The subcellular organization of carbohydrate catabolism in *Giardia* and *E. histolytica* is apparently very simple, all of the reactions discussed above occur in the cytosol or at the cytosolic face of the membranous components of the cell (for instance, pyruvate:ferredoxin oxdoreductase is membrane-associated). Subcellular compartmentation does not seem to play an important part in the energy metabolism of these organisms (Fig. 3.1).

One of the most interesting peculiarities of certain other anaerobic protozoa, however, is that enzymes involved in the further oxidation of pyruvate are located in membrane-bounded organelles. These were first discovered in trichomonads and have become known as hydrogenosomes, the name reflecting the role of the organelle as the site of hydrogen production (21) (Chapter 13). Although these organelles utilize pyruvate, just as mitochondria do, they differ markedly from mitchondria in their structure and metabolic capabilities. The hydrogenosomes of *T. vaginalis* are spherical to ovoid, number some 30 to 40, and are situated close to the axostyle and costa. This location means that they are relatively distant from the cell surface, and so from the oxygen in the parasite's environment. This is thought to be an important factor that helps the organelles retain an anaerobic microenvironment, any oxygen entering the parasite being consumed by cytosolic enzymes (NAD/NADP oxidases) before permeating as far as the hydrogenosomes. The maintenance of an anaerobic environment within the organelles is necessary for the continued functioning of the contained catabolic machinery, some enzymes of which are highly oxygen sensitive.

How does the presence of hydrogenosomes affect energy metabolism? Current knowledge suggests that they have relatively little effect in overall terms, although they may play roles that are yet to be realized. The current view, however, is that their presence enables trichomonads to live in relatively high oxygen tensions, higher than those that *Giardia* and *Entamoeba* can withstand.

The explanation for the difference in metabolic compartmentation between trichomonads and *Giardia/Entamoeba* is unknown but a variety of evolutionary pathways can be envisaged. A possibility that fits well with current knowledge is that the ability to catabolize pyruvate to acetate via pyruvate:ferredoxin oxidoreductase was acquired independently by trichomonads, *Entamoeba* and *Giardia* and that this occurred because of the selective advantage that it provides for cells living anaerobically. A similar ability is possessed by some anaerobic prokaryotes and also other anaerobic protozoa including rumen ciliates and fungi and some free-living flagellates and ciliates (1). Thus the similarities could be accounted for by convergent evolution. Such diverse origins would also explain the distinct differences between these organisms. Not only are there obvious structural differences but also significant differences at the molecular level. For instance, only those organisms with hydrogenosomes possess hydrogenase, the conversion between acetyl CoA and acetate is catabolized by a single enzyme in *Entamoeba* but two in trichomonads (see Figs. 3.2 and 3.3), and the ferredoxins of trichomonads, *Entamoeba* and *Giardia* differ considerably in structure (that of trichomonads being somewhat similar to mitochondrial iron–sulfur proteins).

The evolutionary origin of hydrogenosomes, the differences between hydrogenosomes of distantly related species, and their relationship (if any) to mitochondria are topics arousing considerable interest at present. The similarity between trichomonad ferredoxin and mitochondrial proteins and the finding that hydrogenosomal proteins have amino-terminal leader sequences apparently responsible for targeting them to the organelle together provide some support for the recent suggestion that hydrogenosomes and mitochondria have a similar origin or may have evolved from a common ancestor. Present evidence suggests that hydrogenosomes must have been invented on several independent occasions.

The current consensus is that *Giardia* branched from the main evolutionary trunk of eukaryotes before the advent of mitochondria (22), which is consistent with them lacking any metabolic compartmentation. *Entamoeba*, however, represents a later branch, emerging after several lineages of mitochondrion-containing organisms. This suggests that even the impressive metabolic similarity of *Giardia* and *Entamoeba* might be the result of convergence.

Acetate production from pyruvate is advantageous to the parasites in terms of energy yield. If a molecule of glucose was broken down entirely via this pathway, to carbon dioxide and acetate, then the total energy yield would be five ATPs per glucose. Another consequence, however, is that the NAD reduced to NADH during glycolysis is not reoxidized. Such a lack of redox balance is not consistent with the continued welfare of the cells and clearly cannot be the case. Other compounds are produced by the organisms as a means of ensuring redox balance.

The strategies used differ between cells. Trichomonads produce glycerol in roughly equal amounts to acetate (23), whereas *Entamoeba* and *Giardia* produce ethanol. Although these reduced products could account for redox balance, they are not the only ones released by the parasites. The selective advantage to *T. vaginalis* of producing lactate is unclear, as is the reason for succinate production by *T. foetus*. Succinate is produced by a number of other parasites, including aerobic protozoa such as *Leishmania* (Chapter 2) and anaerobic parasitic worms including *Ascaris* (Chapter 4). The final enzyme in the pathway of *Ascaris* is known as fumarate reductase and is linked to energy production (see Chapter 4), whereas a similar activity is a key mitochondrial enzyme leading into the respiratory chain in trypanosomatids and malaria parasites (see Chapter 2). The enzyme of *T. foetus* is cytosolic and acts in neither of these ways (1).

The explanations for some of these anomalies may be resolved by future studies. It is far from certain that all end-products of energy metabolism have been identified. It was as recently as 1985 that glycerol production by trichomonads was discovered using nuclear magnetic resonance (NMR) techniques (24). The use of NMR technology provides an objective assessment of the range of end-products formed. Subsequently alanine production by *G. lamblia* was also found by applying this technology (25), although the functional significance of this end-product is yet to be resolved. However, even NMR is limited in scope in that only products resulting from metabolism of the substrate provided will be detected. HPLC techniques can be applied to detect all products of a class, for instance acids. Application of such methodology has revealed that propionate is released by *T. vaginalis*, presumably as a breakdown product of a substrate other than glucose (26). Current evidence suggests that it is a product of the use of the amino acids methionine and threonine as energy substrates (see Chapter 5).

Most experimental studies with trichomonads, *Giardia* and *Entamoeba* have involved glucose or a glucose oligomer such as maltose as the supplied and abundant energy substrate. It is very possible that other substrates are also used by the parasites naturally. Recent data suggests that a variety of amino acids can be used by each of the four anaerobic parasitic protists and the role of amino acids as energy substrates is discussed in Chapter 5. Fatty acids, however, have not been found to serve as substrates for energy metabolism in the anaerobes.

All four parasites can use carbohydrates other than glucose as a source of energy. It has been reported that maltose is the preferred substrate for *T. vaginalis*, although glycogen and starch can also be used. It appears, however, that this parasite only has a glucose transporter in its cell membrane (27) and cannot take up the other carbohydrates. Thus the utilization of these oligomers and polymers depends on their extracellular breakdown on or close to the cell surface (28). All four parasites contain large amounts of glycogen. This is considered to be an energy store, but the role it plays is obscure. It seems likely (although it does not seem to have been investigated) that the cyst stages of *Giardia* and *Entamoeba* could rely on such a store for survival. *T. vaginalis* and *T. foetus*, however, have no cyst stage. Transmission of the trichomonads that parasitize the urinogenital tract of mammals is thought to be direct during sexual contact with there being no other developmental forms of the parasite. Glycogen stores could be useful to an organism that experiences fluctuating availability of food, but does *T. vaginalis*? This question does not appear to have been addressed, is it possible that availability of nutrients changes during the menstrual cycle?

Trichomonas vaginalis and *E. histolytica* are both phagocytic and able to ingest microorganisms and, in the case of *E. histolytica* at least, mammalian cells. These are digested via lysosomal systems of the parasites which contain a variety of hydrolases, a topic dealt with in Chapter 4. This process presumably makes a range of nutrients available to the parasites. Determining which are used preferentially and those that comprise the usual diet of the parasite is a difficult task and one yet to be carried out.

This review has so far concentrated on the multiplicative, feeding stages (trophozoites) of the three main groups of parasites, the stages that are parasitic and responsible for the diseases. Transmission of both *Giardia* and *Entamoeba*, however, involves a cyst stage. Cysts are resistant, non-feeding forms able to survive an unfavorable environment for an extended period of time. One aid in long-term survival is the low metabolic rate of cysts. Thus study of the energy metabolism of these stages is not an enticing prospect. It is not surprising, therefore, that they have been little studied, although in recent years there has been renewed interest in the cysts of *Giardia* (29).

Most of our knowledge of the energy metabolism of these anaerobic parasitic protozoa has come from studies on trophozoite stages cultured axenically *in vitro* and in many cases involved parasite lines that had been grown *in vitro* through very many cycles of division over a great number of years since their initial isolation. It is very likely that such lines had changed considerably in response to the culture conditions to which they had been exposed *in vitro*. In addition, probably only some lines were initially grown up, those lines most adapted for the culture conditions. How similar is the energy metabolism of these *in vitro* lines to those growing in a host? This is an important but unanswered question. That environmental conditions can affect the parasite is clear. *T. vaginalis* growing in agar plates have an amoeboid appearance, with no flagella apparent. Over the years there have been reports of similar lines *in vivo*. Is the energy metabolism of these forms the same as the free-living phenotype? Recent data suggest that at least in terms of protease activity the two forms are similar (30). High concentrations of exogenous iron are beneficial to the growth of *T. vaginalis*, perhaps because the iron concentration can be the limiting factor for the synthesis of iron–sulfur proteins such as ferredoxin (31). *T. vaginalis* also grows more rapidly and to a higher density in cultures with a gas phase containing 5% carbon dioxide than in lower carbon dioxide tensions (4). The reason for this is not clear, but it seems likely that it is a reflection of an improved ability to generate energy. It has been proposed that carbon dioxide is an important controller of hydrogenosomal metabolism with high carbon dioxide concentration resulting in a higher flux from pyruvate to malate, although the consequences of this have not been elucidated (32).

Oxygen is present in the environments of the four parasites at various and changing concentrations. The organisms indeed consume oxygen, at rates that are comparable to rates observed for aerobic protists. The affinity for oxygen (K_m of respiration) in trichomonads was found to be as high as that of organisms with mitochondria (around 100 nM oxygen) (33). Respiration of *E. histolytica* exhibits lower affinity (34). The nature of the terminal oxidases remains somewhat an enigma, as discussed below.

The effects of oxygen on the overall metabolism of the organisms have been studied in some detail. Most results concern the behavior of organisms in air-saturated media, whereas recent work noted that low concentrations of oxygen also have significant effects (4). In general, the overall oxidation state of the end-products increases under aerobic conditions, although the physiological significance of this is in most cases not clear.

Alanine production by *Giardia* ceases in the presence of oxygen (35). *T. vaginalis* produces more acetate and less lactate in the presence of oxygen, whereas *T. foetus* similarly switches from succinate to acetate. It has been suggested that as the acetate pathway is energy-linked, such a switch could be beneficial to the parasite. But the concentration of oxygen necessary to affect the end-products has a deleterious effect on growth. An explanation for the enhanced growth in the presence of small amounts oxygen (4) has not been reported.

Oxygen entering *T. vaginalis* is mainly consumed by reaction in the cytosol with NADH or NADPH, reactions catalyzed by the appropriate oxidases (36). These appear to be cytosolic enzymes that are not linked to energy production or any other substrate and, as suggested above, may function in preventing oxygen permeating through to the hydrogenosomes. In this way they protect the cell against oxygen damage. The consequence of increased oxygen entering the cell is that more NAD(P)H is consumed and so less is available for reductive reactions such as the production of lactate and succinate. Thus the observed switch in metabolism resulting from the introduction of oxygen is probably a reflection of oxygen detoxification rather than an energetically advantageous change.

A large body of evidence has been accumulated on the capabilities of these four anaerobic parasites and their energy metabolism when in axenic culture. A remaining need, however, is to find out more about their natural environments and then mimic them *in vitro* in order to determine just how the cells produce energy when parasitizing a mammal. Some studies with these aims are in progress, but lack of good data on the natural environments is a major obstacle.

This review has concentrated on a few major parasites, which are the most studied of the anaerobic amitochondriate protozoa. Other groups of protozoa also have amitochondriate species, all living in habitats with low oxygen concentrations. One of the most important of such habitats is the rumen of ruminants, an anaerobic fermentor where cellulose is converted into utilizable material. A number of ciliates and fungi which live in this niche contain hydrogenosomes and have an anaerobic fermentative metabolism (37, 38). Their energy metabolism is quite similar but not identical to that which occurs in trichomonads. It is likely, however, that the hydrogenosomes in these organisms arose independently from the trichomonad organelles, as discussed earlier. The question as to whether the rumen ciliates are beneficial symbionts or harmful parasites is still a contentious issue. A full discussion of rumen organisms is beyond the scope of this chapter, for further information see reference 39.

This overview of the energy metabolism of anaerobic parasitic protozoa has detailed how the metabolic strategies of these atypical eukaryotic organisms deviate markedly from those of typical aerobic organisms which comprise the majority of eukaryotes, including, of course, the anaerobic parasites' hosts. The energy metabolism of the anaerobes is clearly well adapted to their luminal habitats which contain little or no oxygen. Its evolutionary history remains to be elucidated, but there are already clear indications that the strategy has appeared several times during eukaryotic evolution.

REFERENCES

1. Müller, M. (1988) Energy metabolism of protozoa without mitochondria. *Annu. Rev. Microbiol.* **42**: 465–488.
2. Atkinson, H. J. (1976) The respiratory physiology of nematodes. In: *The Organisation of Nematodes* (ed. Croll, N. A.), Academic Press, London, pp. 243–272.
3. Wagner, G. and Levin, R. (1984) Human vaginal pH and sexual arousal. *Fertil. Steril.* **41**: 389–394.
4. Paget, T. A. and Lloyd, D. (1990) *Trichomonas vaginalis* requires traces of oxygen and high concentrations of carbon dioxide for optimal growth. *Mol. Biochem. Parasitol.* **41**: 65–72.

5. Reeves, R. E. (1984) Metabolism of *Entamoeba histolytica* Schaudinn, 1903. *Adv. Parasitol.* **23**: 105–142.
6. Markoš, A., Miretsky, A. and Müller, M. (1993) A glyceraldehyde 3-phosphate dehydrogenase with eubacterial features in the amitochondriate eukaryote, *Trichomonas vaginalis*. *J. Mol. Evol.* **37**: 631–643.
7. Mertens, E. (1993) ATP versus pyrophosphate: glycolysis revisited in parasitic protists. *Parasitol. Today* **9**: 122–126.
8. Mertens, E., Van Schaftingen, E. and Müller, M. (1989) Presence of a fructose-2,6-bisphosphate-insensitive pyrophosphate:fructose-6-phosphate phosphotransferase in the anaerobic protozoa *Tritrichomonas foetus*, *Trichomonas vaginalis* and *Isotricha prostoma*. *Mol. Biochem. Parasitol.* **37**: 183–190.
9. Mertens, E. (1990) Occurrence of pyrophosphate:fructose-6-phosphate 1-phosphotransferase in *Giardia lamblia* trophozoites. *Mol. Biochem. Parasitol.* **40**: 147–150.
10. Hrdý, L, Mertens, E. and Nohynková, E. (1993) *Giardia intestinalis:* detection and characterization of a pyruvate phosphate dikinase. *Exp. Parasitol.* **76**: 438–441.
11. Searle, S. M. J. and Müller, M. (1991) Inorganic pyrophosphatase of *Trichomonas vaginalis*. *Mol. Biochem. Parasitol.* **44**: 91–96.
12. Reeves, R. E. (1976) How useful is the energy in inorganic pyrophosphate? *Trends Biochem. Sci.* **1**: 53–55.
13. Mertens, E., Van Schaftingen, E. and Müller, M. (1992) Pyruvate kinase from *Trichomonas vaginalis*, an allosteric enzyme stimulated by ribose 5-phosphate and glycerate 3-phosphate. *Mol. Biochem. Parasitol.* **54**: 13–20.
14. Denton, H., Thong, K.-W. and Coombs, G. H. (1994) *Eimeria tenella* contains a pyrophosphate-dependent phosphofructokinase and a pyruvate kinase with unusual allosteric regulators. *FEMS Microbiol. Lett.* **115**: 87–92.
15. Wang, C. C. (1982) Biochemistry and physiology of coccidia. In: *Biology of Coccidia* (ed. Long, P. L.), Edward Arnold, London, pp. 168–227.
16. Williams, K., Lowe, P. N. and Leadlay, P. F. (1987) Purification and characterization of pyruvate:ferredoxin oxidoreductase from the anaerobic protozoon *Trichomonas vaginalis*. *Biochem. J.* **246**: 529–536.
17. Docampo, R., Moreno, S. N. J. and Mason, R. P. (1987) Free radical intermediates in the reaction of pyruvate:ferredoxin oxidoreductase in *Tritrichomonas foetus* hydrogenosomes. *J. Biol. Chem.* **262**, 12417–12420.
18. Reeves, R. E., Warren, L. G., Susskind, B. and Lo, H.-S. (1977) An energy-conserving pyruvate-to-acetate pathway in *Entamoeba histolytica*. Pyruvate synthase and a new acetate thiokinase. *J. Biol. Chem.* **252**, 726–731.
19. Lindmark, D. G. (1980) Energy metabolism of the anaerobic protozoon *Giardia lamblia*. *Mol. Biochem. Parasitol.* **1**: 1–12.
20. Yang, W., Li, E., Kairong, T. and Stanley, Jr, S. L. (1994) *Entamoeba histolytica* has an alcohol dehydrogenase homologous to the multifunctional *adhE* gene product of *Escherichia coli*. *Mol. Biochem. Parasitol.* **64**: 253–260.
21. Müller, M. (1993) The hydrogenosome. *J. Gen. Microbiol.* **139**: 2879–2889.
22. Sogin, M. L. (1991) Early evolution and the origin of eukaryotes. *Curr. Opin. Gen. Dev.* **1**: 457–463.
23. Steinbüchel, A. and Müller, M. (1986) Glycerol, a metabolic end product of *Trichomonas vaginalis* and *Tritrichomonas foetus*. *Mol. Biochem. Parasitol.* **20**: 45–55.
24. Chapman, A., Linstead, D. J., Lloyd, D. and Williams, J. (1985) ^{13}C-NMR reveals glycerol as an unexpected major metabolite of the protozoan parasite *Trichomonas vaginalis*. *FEBS Lett.* **191**: 287–292.
25. Edwards, M. R., Gilroy, F. V., Jimenez, B. M. and O'Sullivan, W. J. (1989) Alanine is a major end product of metabolism by *Giardia lamblia*: a proton nuclear magnetic resonance study. *Mol. Biochem. Parasitol.* **37**: 19–26.
26. Lockwood, B. C. and Coombs, G. H. (1991) Amino acids catabolism in anaerobic protists. In: *Biochemical Protozoology* (eds Coombs, G. H. and North, M. J.), Taylor & Francis, London, pp. 113–122.

27. Ter Kuile, B. H. and Müller, M. (1993) Interaction between facilitated diffusion of glucose across the plasma membrane and its metabolism in *Trichomonas vaginalis*. *FEMS Microbiol. Lett.* **110**: 27–32.
28. Ter Kuile, B. H. and Müller, M. (1995) Maltose utilization by extracellular hydrolysis followed by glucose transport in the amitochondriate eukaryote, *Trichomonas vaginalis*. *Parasitology* **110**: 37–44.
29. Jarroll, E. L. (1991) *Giardia* cysts: their biochemistry and metabolism. In: *Biochemical Protozoology* (eds Coombs, G. H. and North, M. J.), Taylor & Francis, London, pp. 52–60.
30. Scott, D. A., North, M. J. and Coombs, G. H. (1995) *Trichomonas vaginalis*: amoeboid and flagellar forms synthesise similar proteinases. *Exp. Parasitol.* **80**: 345–348.
31. Petersen, K. M. and Alderete, J. F. (1984) Iron uptake and increased intracellular enzyme activity follow host lactoferrin binding by *Trichomonas vaginalis* receptors. *J. Exp. Med.* **160**: 398–410.
32. Steinbüchel, A. and Müller, M. (1986) Anaerobic pyruvate metabolism of *Tritrichomonas foetus* and *Trichomonas vaginalis* hydrogenosomes. *Mol. Biochem. Parasitol.* **20**: 57–65.
33. Lloyd, D., Williams, J., Yarlett, N. and Williams, A. G. (1982) Oxygen affinities of the hydrogenosome-containing protozoa *Tritrichomonas foetus* and *Dasytricha ruminantium*, and two aerobic protozoa, determined by bacterial bioluminescence. *J. Gen. Microbiol.* **128**: 1019–1022.
34. Weinbach, E. C. and Diamond, L. S. (1974) *Entamoeba histolytica*: I. Aerobic metabolism. *Exp. Parasitol.* **35**: 232–243.
35. Paget, T. A., Raynor, M. H., Shipp, D. W. E. and Lloyd, D. (1990) *Giardia lamblia* produces alanine anaerobically but not in the presence of oxygen. *Mol. Biochem. Parasitol.* **42**: 63–68.
36. Linstead, D. J. and Bradley, S. (1988) The purification and properties of two soluble reduced nicotinamide:acceptor oxidoreductases from *Trichomonas vaginalis*. *Mol. Biochem. Parasitol.* **27**: 125–133.
37. Yarlett, N., Hann, A. C., Lloyd, D. and Williams, A. (1981) Hydrogenosomes in the rumen protozoon *Dasytricha ruminantium* Schuberg. *Biochem. J.* **200**: 365–372.
38. Yarlett, N., Orpin, C. G., Munn, E. A., Yarlett, N. C. and Greenwood, C. A. (1986) Hydrogenosomes in the rumen fungus *Neocallimastix patriciarum*. *Biochem. J.* **236**: 729–739.
39. Williams, A. G. and Coleman, G. S. (1991) *The Rumen Protozoa*. Springer-Verlag, New York.
40. Müller, M. (1989) Biochemistry of *Trichomonas vaginalis*. In: *Trichomonads Parasitic in Humans* (ed. Honigberg, B. M.), Springer-Verlag, New York, pp. 53–83.

4 Carbohydrate and Energy Metabolism in Helminths

RICHARD KOMUNIECKI[1] and BEN G. HARRIS[2]

[1]Department of Biology, University of Toledo, Toledo, OH, and [2]Department of Biochemistry and Molecular Biology, The University of North Texas, Health Science Center, Fort Worth, TX, USA.

SUMMARY

Carbohydrate is an essential energy source in all adult parasitic helminths and its metabolism is often predominantly anaerobic, even in the presence of oxygen. Similarities in habitat are often reflected in similarities in energy metabolism, regardless of the phylogenetic position of helminth. Blood and tissue-dwelling helminths, such as schistosomes and filaria, which are continuously bathed in glucose, excrete primarily lactate as an end-product of carbohydrate metabolism, although aerobic pathways also have the capacity to contribute significantly to energy generation. In contrast, lumen-dwelling helminths, such as the cestode, *Hymenolepis diminuta*, and the nematode, *Ascaris suum*, which reside where glucose availability is episodic and oxygen tensions are low, rely on stored glycogen for energy and possess unique mitochondrial pathways designed to increase energy generation in the absence of oxygen. These pathways use unsaturated organic acids as terminal electron acceptors, instead of oxygen, and enzymes whose properties often differ significantly from those of their aerobic counterparts.

1. GLYCOGEN METABOLISM

Stored glycogen is the primary source of energy for many parasitic helminths. In *Ascaris suum* muscle or *Hymenolepis diminuta*, glycogen may be stored at concentrations that are 10–30 times that of human muscle and 2–4 times that of human liver.

Biochemistry and Molecular Biology of Parasites
ISBN 0-12-473345-X

The disaccharide, trehalose, also is an important storage form and in a few organisms, such as *Haemonchus contortus*, may be more abundant than glycogen (1). Little is known about the regulation of trehalose metabolism but, in the filaria *Brugia pahangi* and *Acanthoceilonema viteae*, trehalose metabolism appears to be altered by glucose availability (2).

Glycogen phosphorylase has been purified from adult *A. suum* muscle (3). The dephosphorylated enzyme, which requires AMP for activity, can be phosphorylated and activated by rabbit muscle phosphorylase kinase. Kinetic data and the observation that phosphorylase is not at equilibrium with its substrates and products, indicate that phosphorylase catalyzes the rate-limiting step of glycogenolysis in *A. suum* muscle. Phosphorylase activity correlates well with the rate of glycogenolysis observed *in vivo* (4). Glycogen synthase has also been purified from *A. suum* muscle and is about 3% of the soluble protein (5). It is converted into an inactive D-form when phosphorylated by ox heart cAMP-dependent protein kinase and binds about 1 mol of phosphate/mol of subunit. A second glycogen synthase activity has also been described and purified from *A. suum* muscle (6). Designated GSII, it occurs as a protein–carbohydrate complex, which contains proteins with apparent molcular weights of 140 000 and 66 000, and a glycoprotein with a carbohydrate/protein mass ratio of 3 : 1. GSII is totally dependent on glucose-6-phosphate for activity. GSII is not phosphorylated by the ox heart cAMP-dependent protein kinase. The physiological role of these two synthase activities is not clear but GSII appears to represent the bulk of the synthase activity. In *H. diminuta* and *Vampirolepis microstoma*, glycogen synthase and phosphorylase activities appear to be regulated by phosphorylation/dephosphorylation in a manner similar to that described for *A. suum*.

In *A. suum*, the glycogen content of adult muscle correlates well with glycogen synthase and phosphorylase activities (4, 7). Surprisingly, perfusion of ascarid muscle/cuticle preparations with glucose alone activates glycogen synthase and inactivates glycogen phosphorylase. Additional exogenous signals are not necessary. However, perfusion with serotonin (5-hydroxytryptamine), acetylcholine or γ-aminobutyric acid (GABA) all markedly stimulate glycogen metabolism (8–10). Serotonin stimulation is mediated by cAMP. Acetylcholine and GABA have no effect on cAMP levels but acetylcholine stimulates, and GABA inhibits, muscle contraction. Thus, the interconversion of the *A. suum* glycogen synthase and glycogen phosphorylase is regulated by phosphorylation/dephosphorylation and is mediated by both cAMP-dependent and cAMP-independent pathways, as has been observed for vertebrate skeletal muscle.

Glycogen metabolism has been studied also in schistosomes (11). As described above for isolated *A. suum* muscle, glucose is the sole substrate for glycogen synthesis in *Schistosoma mansoni*. Lactate or other trioses are not incorporated into glycogen, although all of the enzymes necessary for glyconeogenesis are present (12). Both glycogen synthesis and degradation appear to occur concurrently (13). Similarly, in *A. suum* muscle-cuticle preparations, the level of phosphorylation of glycogen phosphorylase never drops below 50%, even during periods of substantial glycogen synthesis (4). Therefore, there may be spatial separation of the glycogen being synthesized and that being degraded.

Serotonin appears to be an important regulator of glycogen metabolism in most parasitic helminths. In flatworms, such as *Fasciota hepatica* or *Schistosoma mansoni*, serotonin stimulates both glycogen breakdown and muscle contraction; in nematodes, such as *A. suum*, only glycogen breakdown is affected. In *F. hepatica*, the serotonin stimulation of glycogen breakdown is mediated by adenyl cyclase and a cAMP-dependent protein kinase, as in epinephrine stimulation in mammalian tissues (14). A similar serotonin-stimulated second message system also operates in *S. mansoni* (15). Serotonin receptors have been identified in *H. diminuta* (16) and *A. suum* (17), but cannot easily be classified using criteria described for mammalian serotonin receptor subtypes.

2. GLYCOLYSIS

All helminths examined to date use at least a portion of the glycolytic pathway for energy generation (Fig. 4.1). In most parasitic helminths, hexokinase, phosphofructokinase (PFK) and pyruvate kinase appear to be rate-limiting enzymes. The *A. suum* hexokinase has an elevated apparent K_m for glucose (5 mM) and is much less sensitive to inhibition by glucose-6-phosphate (18). Similar observations have been made for the *H. diminuta* and *Dirofilaria immitis* hexokinases. The high levels of glucose-6-phosphate

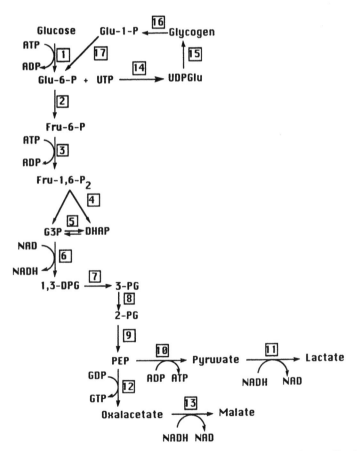

FIG. 4.1 Glycogenolysis and glycolysis in parasitic helminths. (1) hexokinase; (2) glucosephosphate isomerase; (3) phosphofructokinase; (4) aldolase; (5) triosephosphate isomerase; (6) glyceraldehyde-3-P dehydrogenase; (7) phosphoglycerate kinase; (8) phosphoglyceromutase; (9) enolase; (10) pyruvate kinase; (11) lactate dehydrogenase; (12) phosphoenolpyruvate carboxykinase; (13) malate dehydrogenase; (14) UDP glucosyl transferase; (15) glycogen synthase; (16) glycogen phosphorylase; (17) phosphoglucomutase.

required for hexokinase inhibition may permit glucose phosphorylation to proceed while still maintaining substantial glucose-6-phosphate for glycogen synthesis. Glucose levels must exceed 5 mM during the perfusion of *A. suum* muscle-cuticle preparations before substantial glycogen synthesis can be demonstrated (4). A similar relationship between elevated glucose levels and glycogen synthesis has been described for *S. mansoni* (13).

Phosphofructokinase (PFK) purified from *A. suum* muscle is inhibited by ATP and stimulated by AMP, K^+, NH_4^+, phosphate, and hexose bisphosphate (19). It has been estimated that the *A. suum* PFK functions at 1–2% of its maximal velocity *in vivo*. Both phosphorylated and dephosphorylated forms of the *A. suum* PFK exhibit similar specific activities, but the dephosphorylated enzyme has a 2–2.5-fold lower activity in a physiological assay system. Phosphorylation of the dephosphorylated PFK with the mammalian cAMP-dependent protein kinase results in the incorporation of about 1 mol of phosphate/mol of subunit and completely restores activity in the physiological assay system (20). Therefore, it appears that PFK activity in *A. suum* muscle can be regulated, not only by metabolites, but also by covalent modification catalyzed by a cAMP-dependent protein kinase. This contrasts with mammalian muscle, where the effect of PFK phosphorylation is poorly defined and ranges from inhibition to mild stimulation (20).

The activity of the *F. hepatica* PFK is also modified by phosphorylation and the phosphorylated enzyme is still stimulated by AMP and fructose-2,6-bisphosphate (Fru-2,6-P_2) (21). Phosphorylation may also play a physiological role in the regulation of the *F. hepatica* PFK, since serotonin exhibits a marked stimulatory effect on PFK activity in intact *F. hepatica* (22). The *D. immitis* PFK is also regulated by phosphorylation, Fru-2,6-P_2 and AMP (23).

Phosphorylation has complex effects on PFK activity and, in characterizing its regulation, it is important to appreciate the difficulty in assessing the physiological relevance of allosteric effectors identified *in vitro*. Phosphorylation of the *A. suum* PFK clearly lowers its $S_{0.5}$ value for fructose-6-phosphate, but this effect alone does not stimulate the enzyme to the extent required for activity under physiological conditions. PFK has both hysteretic and cooperative properties that can be correlated with the pK_a of a critical histidine residue (24). These properties can be partially explained by the inhibition of the enzyme by ATP and alkaline pH. However, raising the pH alone does not completely abolish the enzyme's allosteric properties. Allosteric effectors and phosphorylation have additional effects on PFK activity, after ATP inhibition of the enzyme is overcome. Phosphorylation, AMP and Fru-2,6-P_2 may all act in concert and affect different aspects of PFK regulation. Thus, all three factors must be considered to adequately understand the regulation of PFK activity in parasitic helminths.

Some of the regulatory enzymes involved in the phosphorylation/dephosphorylation of helminth glycolytic enzymes have been characterized. Two distinct protein kinases have been purified from *A. suum* muscle (25, 26). One is cAMP-dependent and appears to be similar to the corresponding mammalian enzyme. Its apparent K_m for PFK is similar to the intracellular concentration of the PFK, suggesting that PFK may be a substrate *in vivo*. The second kinase activity is cAMP-independent and not affected by the *A. suum* cAMP binding protein, Walsh inhibitor, ox heart R_{II} subunit or antiserum to the ox heart cAMP-dependent protein kinase (25). This kinase phosphorylates histone, phosvitin, and the *A. suum* PFK and phosphorylase *b*. It is responsible for about 75% of the PFK phosphorylating ability in *A. suum* muscle homogenates.

Additional work is required to establish the relative importance of these protein kinases *in vivo*. A cAMP-dependent protein kinase also phosphorylates *F. hepatica* PFK (27).

A novel phosphorylated peptide has been isolated from the *A. suum* PFK, after phosphorylation with the mammalian cAMP-dependent protein kinase (28). Its sequence, Ala-Lys-Gly-Arg-Ser-Asp-SerP-Ile-Val-Pro-Thr, is unlike the phosphorylation sequence found in rabbit muscle PFK. The predicted amino acid sequence of *H. contortus* PFK exhibits about 70% similarity with mammalian PFKs but no sequence is present at the carboxy-terminus which corresponds to the consensus cAMP-dependent phosphorylation site identified in mammalian PFKs (29). However, although not noted by the authors, *H. contortus* PFK does have a sequence near its amino-terminus that contains 7 of the 11 amino acids in the *A. suum* phosphorylated peptide. The role of this phosphorylation site in the regulation of PFK activity requires further study.

It has been hypothesized that mammalian PFK arose as a result of the gene duplication of an ancient prokaryotic gene; the amino-terminal half shows 32% sequence identity to the carboxy-terminal half. When the mammalian PFK is treated with trypsin, it is cleaved into two peptides of similar size. Treatment of *A. suum* PFK with trypsin yields similar results, suggesting a structure similar to that of the mammalian enzyme. In fact, many helminth enzymes, especially those not catalyzing rate-limiting reactions, such as aldolase and glyceraldehyde-3-phosphate dehydrogenase, have structures similar to the corresponding enzymes of the host (30).

Glycolytically-derived phosphoenolpyruvate (PEP) is metabolized through two distinct pathways in parasitic helminths (Fig. 4.1). In schistosomes and most filaria, lactate is the major glycolytic end-product and it is formed by the action of pyruvate kinase and lactate dehydrogenase. In other helminths, such *A. suum* and *F. hepatica*, which reside in microaerobic habitats, malate is the major cytoplasmic product. Malate is formed by the action of PEP carboxykinase and malate dehydrogenase, as a prelude to further anaerobic metabolism in the mitochondria, as described in Section 3. The pyruvate kinase/PEP carboxykinase branch-point has been studied intensively and it has been suggested that the ratio of the two activities or their differential regulation may be important in dictating whether lactate or malate is the predominant product. In addition, the high P_{CO_2} of the host gut has been identified as a potential factor driving CO_2-fixation and malate formation. Unfortunately, pyruvate kinase has never been purified to homogeneity from any parasitic helminth. PEP carboxykinase has been more extensively characterized. It has been isolated from *H. diminuta* (31) and *A. suum* (32). The purified *A. suum* enzyme functions efficiently in the direction of CO_2 fixation; it has an apparent K_m for CO_2 that is three times lower than the mammalian enzyme (32).

3. HELMINTH MITOCHONDRIAL METABOLISM AND ENERGY GENERATION

All adult helminths examined to date contain mitochondria and the role played by oxygen in their metabolism has been a source of debate for a number of years. It is now clear that adult helminths exhibit a wide variation in their ability to use oxygen as a terminal electron acceptor. In general, two basic metabolic plans have emerged. Blood and tissue-dwelling helminths, such as schistosomes and filaria, convert most of their abundant supplies of environmental glucose into lactate and survive well, at least in the short term, in the absence of oxygen (33, 34). However, these organisms still appear to contain cristate mitochondria capable of aerobic respiration (35, 36). Recent work suggests that these mitochondria may preferentially use carbon derived from amino acids, such as glutamine, as substrates (37, 38). The contribution of aerobic metabolism to these lactate-forming organisms is difficult to quantify, although they do not appear to exhibit a marked Pasteur effect. However, as pointed out by others, the complete oxidation of glucose to CO_2 is energetically much more efficient than lactate formation, so that even if only a small percentage of the glucose were oxidized completely, it could still contribute significantly to the overall energy balance to the organism (39). In contrast to blood and tissue dwelling helminths, lumen-dwelling helminths, such as *A. suum* and *F. hepatica*, contain unusual, less cristate, mitochondria which generate a significant portion of their energy anaerobically. These organelles lack a functional tricarboxylic acid cycle and are capable of using unsaturated organic acids, such as fumarate, as terminal electron-acceptors (40).

Exceptions to these generalizations exist. For example, some filaria, such as *Litomosoides carinii*, have a greater requirement for oxygen and accumulate products other than lactate as a consequence of glucose dissimilation (41). Alternatively, some small lumen-dwelling nematodes, such as *Nippostrongylus braziliensis*, which reside close to the mucosa where oxygen tensions are higher than in the lumen, have a functional tricarboxylic acid cycle and rely on oxygen as a terminal electron acceptor (42).

3.1. Anaerobic Mitochondrial Metabolism in Lumen-Dwelling Helminths

Lumen-dwelling helminths exhibit a wide range of potential end-products of mitochondrial metabolism, but their mitochondrial pathways are surprisingly similar given their diverse phylogenetic origins. In all of these organisms, malate appears to be the primary mitochondrial substrate and intramitochondrial reducing power is generated by oxidative decarboxylation of malate and pyruvate. However, these organisms differ in the pathways used to reoxidize intramitochondrial NADH. In the cestode *H. diminuta* redox balance is maintained by the efficient reduction of fumarate to succinate and *H diminuta* forms succinate and acetate as end-products of its anaerobic malate dismuta-tion (43) (Fig. 4.2). In the trematode *F. hepatica*, succinate is further decarboxylated to propionate, and *F. hepatica* forms primarily propionate and acetate as end-products (44, 45). In the nematode *A. suum*, acetate and propionate are further metabolized to the branched-chain fatty acids, 2-methylbutanoate and 2-methylpentanoate, providing additional avenues for the oxidation of excess reducing power (46, 47). *Ascaris suum* forms a complex mixture of acetate, propionate, succinate, 2-methylbutanoate, and 2-methylpentanoate, as end-products of its malate dismutation. Finally, some organisms, such as *H. contortus* and *Trichostrongylus colubriformis*, also form neutral volatile compounds, such as propanol (48, 49). However, the mechanism and site of propanol formation are not well understood. Summaries of end-product formation in different helminths can be found in a number of reviews (50–52). The factors dictating the type and ratios of end-products in these organisms are complex. However, it is clear that habitat and worm size, as they affect the availability of oxygen and glucose, play significant role in the pathways used to generate energy and maintain redox balance In addition, it has been noted that the pK_a values for the volatile acids are much lower

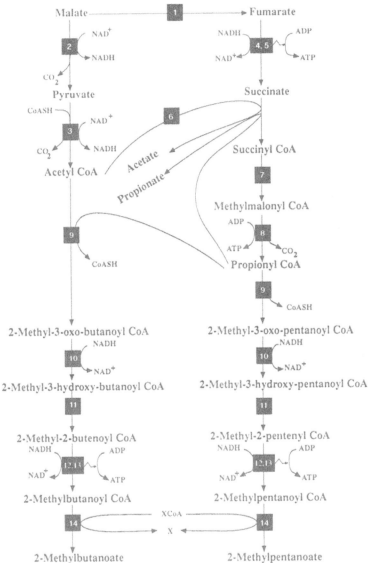

FIG. 4.2 Malate metabolism in mitochondria from body wall muscle of adult *Ascaris suum*. (1) Fumarase; (2) 'malic' enzyme; (3) pyruvate dehydrogenase complex; (4) complex I; (5) succinate-coenzyme Q reductase (complex II, fumarate reductase); (6) acyl CoA transferase; (7) methylmalonyl CoA mutase; (8) methyl-malonyl CoA decarboxylase; (9) propionyl CoA condensing enzyme; (10) 2-methyl acetoacetyl CoA reductase; (11) 2-methyl-3-oxo-acyl CoA hydratase; (12) electron-transfer flavoprotein; (13) 2-methyl branched-chain enoyl CoA reductase; (14) acyl CoA transferase.

than those for succinate or lactate, which may facilitate their excretion or minimize their effects on tissue acidification (53).

The anaerobic mitochondrial metabolism of *A. suum* has been used as a model for other invertebrates. *A. suum* is one of the few unusually large helminths from which substantial amounts of homogeneous tissue can be dissected for use in the isolation of mitochondria or protein purification. Most of the enzymes involved the anaerobic dismutation of malate in mitochondria from adult *A. suum* body wall muscle have been purified to homogeneity and at least partially characterized (Fig. 4.2).

In aerobic mitochondria, cytoplasmically generated pyruvate is oxidatively decarboxylated by the pyruvate dehydrogenase complex as a prelude to further tricarboxylic acid cycle oxidation. In contrast, in anaerobic muscle mitochondria of adult *A. suum*, tricarboxylic acid cycle activity is not significant and the levels of citrate synthase, aconitase, isocitrate dehydrogenase, and α-ketoglutarate dehydrogenase are low or barely detectable. In these anaerobic mitochondria, cytoplasmically generated malate, not pyruvate, is the primary mitochondrial substrate. Malate enters the *A. suum* mitochondrion through a phosphate-dependent porter system and a portion of it is decarboxylated to pyruvate by the action of 'malic' enzyme, generating reducing power in the form of NADH. The NADH then reduces the remaining malate, via fumarate, to succinate (42). Fumarate reduction is mediated by a membrane-bound electron-transport system, involving NADH dehydrogenase, rhodoquinone-9 and fumarate reductase (54, 55). In *H. diminuta*, 'malic' enzyme is $NADP^+$-linked, and this helminth contains an active, energy-linked, $NADPH:NAD^+$ transhydrogenase activity which converts NADPH into NADH (56).

The ability of ascarid mitochondria to catalyze the efficient, energy-linked, NADH-dependent reduction of fumarate to succinate has been recognized for a number of years. However, only recently have the major respiratory complexes of the inner ascarid mitochondrial membrane been isolated and characterized. Both complex I (NADH: rhodoquinone oxidoreductase) and complex II (succinate:rhodoquinone oxidoreductase) are involved in NADH-dependent fumarate reduction (54, 57, 58). Although the polypeptide composition of both respiratory complexes is superficially similar to those of their aerobic counterparts, the ratio of fumarate reductase to succinate dehydrogenase activity in adult *A. suum* mitochondrial membranes is about 400 times greater (59). Aerobic mitochondria catalyze the NADH-dependent reduction of fumarate slowly and the factors dictating the reversed direction of this reaction in helminth mitochondria have only recently begun to be elucidated. First, the presence of a novel quinone, rhodoquinone-9, appears to be critical for significant rates of fumarate reduction (60). It has a redox potential ($E_0 = -63\,\mathrm{mV}$) which is much more negative than that of ubiquinone ($E_0 = +110\,\mathrm{mV}$), the quinone commonly found in aerobic mitochondria. The redox potential of rhodoquinone is much closer to the NADH/NAD^+ couple ($E_0 = -320\,\mathrm{mV}$), and it appears that rhodoquinone is more efficient at shuttling electrons from NADH to fumarate ($E_0 = +30\,\mathrm{mV}$). In fact, in reconstitution studies using pentane-extracted ascarid mitochondria or isolated respiratory complexes, ubiquinone is unable to replace rhodoquinone in supporting NADH-dependent fumarate reduction (60). Second, complex II, which comprises about 8% of the total membrane protein in adult *A. suum* mitochondria, contains a low potential cytochrome, b_{558}, whose properties differ significantly from those of cytochrome b_{560}, commonly found in complex IIs isolated from aerobic mitochondria (54, 55, 59). Cytochrome b_{558} is reduced by NADH or succinate and reoxidized by fumarate in the presence of the appropriate *A. suum* respiratory complexes. These data suggest that cytochrome b_{558} plays an important role in determining the direction of electron-flow in the ascarid organelle.

Additional intramitochondrial NADH is generated from the oxidative decarboxylation of pyruvate to acetyl CoA catalyzed by the pyruvate dehydrogenase complex (PDC). This may be coupled to fumarate reduction or drive the synthesis of branched-chain fatty acids. In *A. suum*, no other reactions have been identified which provide reducing power in sufficient quantity to account for the amounts of reduced products formed. It was unexpected to find substantial PDC activity in these mitochondria. Intramitochondrial $NADH/NAD^+$ and acyl CoA/CoA ratios are elevated dramatically in these anaerobic organelles (61, 62) to ratios which potently inhibit the activity of PDCs isolated from aerobic mitochondria (63, 64). Inhibition results from both end-product inhibition and the stimulation of PDH_a kinase activity which catalyzes the phosphorylation and inactivation of the complex. In most anaerobic organisms, the PDC is either absent or not regulated by covalent modification. The subunit composition of the PDC from adult *A. suum* muscle differs substantially from PDCs isolated from both yeast and mammals and its regulation is modified to function under the reducing conditions present in the ascarid organelle (64–66). Both *A. suum* PDC and its associated PDH_a kinase and PDH_b phosphatase are much less sensitive to end-product inhibition (63, 64, 67). Recently, distinct, physiologically significant aerobic and anaerobic isozymes of the α-pyruvate dehydrogenase subunit have been identified. The anaerobic subunit contains only two phosphorylation sites and cannot be fixed in an inactive state by phosphorylation as has been reported for the mammalian enzyme (68, 69). The study of PDC activity in other helminths is warranted, especially those with a more aerobic component to their metabolism.

The branched-chain fatty acids, 2-methylbutanoate and 2-methylpentanoate are the ultimate products of glucose degradation in *A. suum* muscle and these volatile acids accumulate (> 50 mM total) in the perienteric fluid. Branched-chain fatty acids rival Cl^- as the most abundant anions in the perienteric fluid. They appear to exit the worm by diffusion through the cuticle, leaving open the possibility of their further metabolism as they pass through potentially aerobic mitochondria in the hypodermis. 2-Methylbutanoate and 2-methylpentanoate are formed by the condensation of an acetyl CoA and a propionyl CoA or two propionyl CoAs, respectively, with the subsequent reduction of the condensation products. Enzymes in the pathway differ significantly from the corresponding enzymes of β-oxidation found in mammalian mitochondria. Differences might be anticipated since the ascarid enzymes function physiologically in the direction of acyl CoA synthesis, not oxidation. The final reaction in the pathway, the NADH-dependent reduction of 2-methyl branched-chain enoyl CoAs, requires both mitochondrial membrane-bound and soluble components and is rotenone-sensitive (70–72). Recent work indicates that complex I, rhodoquinone, and electron-transport flavoprotein reductase comprise the membrane-bound components (72). Two flavoproteins, ETF and 2-methyl branched-chain enoyl CoA reductase, are the soluble components (70–71) (Fig. 4.3).

Similar straight- and branched-chain fatty acids have been reported from other helminths, such as *Echinostoma leie* and *Paragonimus westermani*, and will surely be detected in other lumen-dwelling helminths with improved assay techniques. However, volatile acids, such a 2-methylbutanoate, can also be formed from the degradation of amino acids, so that their presence alone does not prove that the ascarid pathway is operative. This appears to be the case in *F. hepatica*.

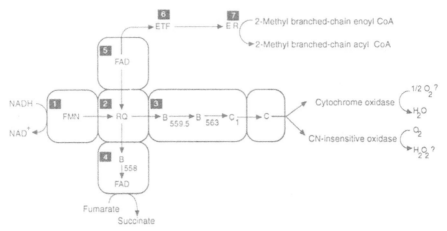

FIG. 4.3 Electron-transport in mitochondria from body wall muscle of adult *Ascaris suum*. (1) NADH: rhodoquinone reductase (complex I); (2) rhodoquinone; (3) reduced rhodoquinone cytochrome C reductase (complex III); (4) reduced rhodoquinone:fumarate reductase (complex II); (5) electron-transfer flavoprotein reductase; (6) electron-transfer flavoprotein; (7) 2-methyl branched-chain enoyl CoA reductase.

3.2. Anaerobic Energy Generation

Our understanding of the processes regulating anaerobic mitochondrial energy meta-bolism is still quite limited. At least four potential sites for energy-generation have been identified: two substrate-level phosphorylations associated with methylmalonyl CoA decarboxylation and acyl CoA hydrolysis, and two site I, electron-transport associated phosphorylations coupled to the NADH-dependent reductions of fumarate and 2-methyl branched-chain enoyl CoAs. Energy-conservation associated with methyl-malonyl CoA decarboxylation during propionate formation and fumarate reduction during succinate formation is well documented (40, 73, 74). The NADH-dependent reduction of fumarate involves the generation of a proton-gradient across the mitochondrial membrane and is coupled to a rotenone-sensitive, site I, phosphoryla-tion. Little is known about the nature of the proton-gradient or ionic fluxes. The membrane potential generated with malate in adult *A. suum* muscle mitochondria is insensitive to cyanide, dissipated with uncouplers, such as FCCP, not inhibited by anaerobiosis, and similar in magnitude to that generated in aerobic mitochondria (75). An oligomycin-sensitive ATPase activity has been identified in isolated *A. suum* muscle mitochondria which appears to differ from the corresponding activity in mammalian mitochondria in its sensitivity to uncouplers. Interestingly, the mitochondrial gene coding for subunit 8 of the ATP synthase does not appear to be present in the *A. suum* mitochondrial genome (76).

The importance of energy generation from acyl CoA hydrolysis, although feasible, is less well documented, although thiokinase activities have been measured in helminth extracts. In adult *A. suum* muscle, the intramitochondrial levels of free CoASH are very low and the transfer of the CoA moiety appears to be mediated by a number of distinct CoA transferases (62, 77). Low free CoASH levels may be critical for the formation of branched-chain fatty acids, since the initial reaction in this sequence, catalyzed by propionyl CoA condensing enzyme, is potently inhibited by free CoASH (78). Low free

CoASH levels are maintained by the PDC whose apparent K_m for CoASH is much lower than values reported for complexes isolated from aerobic mitochondria (64). Whether similar constraints on CoA metabolism apply to organisms that do not form branched-chain fatty acids remains to be determined.

Based on energetic considerations and the rotenone-sensitivity of the pathway, it is likely that the NADH-dependent reduction of 2-methyl branched-chain enoyl CoAs is coupled to the generation of a mitochondrial proton gradient and associated phosphorylation (46, 70). However, to date, this hypothesis has not been verified experimentally. When isolated *A. suum* mitochondria are incubated with malate *in vitro*, they form succinate and pyruvate, not the volatile acids characteristic of carbohydrate fermentation in intact muscle strips. The esterification of $^{32}P_i$ measured in these experiments is associated with succinate formation and gives no indication of the energy-generating potential of pathways leading to volatile acid formation. Complete carbon and energetic balances have never been measured experimentally for any physiologically functional helminth mitochondrial preparation. Based on theoretical calculations, the energy-generating efficiency of branched-chain fatty acid synthesis is inferior to that of acetate and propionate, but branched-chain fatty acid synthesis may provide increased flexibility in regulating intramitochondrial $NADH/NAD^+$ ratios under the reducing conditions in the host gut, by serving as an important sink for excess reducing power (46).

3.3. Oxygen as a Terminal Electron-Acceptor

All helminths and isolated helminth mitochondria will take up oxygen if it is available, but the terminal reactions involved in this process have remained elusive. Helminth mitochondria have varying degrees of aerobic capacity, with muscle mitochondria from large organisms like *A. suum*, representing the anaerobic extreme. Branched electron-transport chains have been postulated for a number of helminth mitochondria (Fig. 4.3), with one branch leading to fumarate as a terminal electron acceptor and the other leading to oxygen through either a classical, CN-sensitive, cytochrome *c* oxidase or a CN-insensitive, H_2O_2 forming, terminal oxidase. None of these potential oxidases have ever been isolated and purified from helminth mitochondria. Their presence has been inferred from studies using inhibitors of oxygen uptake, such as KCN or SHAM. These studies are limited by our understanding of the effects of these various inhibitors in helminth systems and have often been conducted using either fumarate or oxygen as a terminal acceptor without measuring the relative rates of energy generation. Therefore, their physiological meaning is difficult to interpret.

In *A. suum* muscle mitochondria, perhaps the most anaerobic of all helminth mitochondria, reduced cytochrome *c* oxidase activity is barely detectable and oxygen utilization results exclusively in the formation of hydrogen peroxide (73). In fact, the ratio of complex II to complex III is about 100 times greater than in rat liver mitochondria, attesting to the importance of NADH-dependent fumarate reduction in the ascarid organelle (54, 58). Substantial peroxide formation also has been observed in mitochondria isolated from other helminths, such as *H. diminuta* and *Ascaridia galli* (79, 81). In isolated *A. suum* mitochondria, no additional energy-generation appears to be associated with oxygen uptake, but succinate formation is slightly reduced (73). Incubation in air often alters the ratios of fermentative products excreted by intact helminths, with acetate formation usually increased and products of the reductive arm of the pathway, such as succinate, propionate and branched-chain fatty acids, reduced. The shift in end-product formation is probably the result of a decrease in intra-mitochondrial NADH/NAD$^+$ ratios. A full understanding of the role of oxygen in helminth mitochondria must await a more careful characterization of their terminal oxidases.

4. THE AEROBIC/ANAEROBIC TRANSITION DURING HELMINTH DEVELOPMENT

Two major generalizations may be made about the development aspects of energy generation in parasitic helminths. First, all adult helminths use fermentative pathways to some extent and excrete reduced organic acids as end-products of carbohydrate metabolism, even in the presence of oxygen. Second, at least some larval stages of most helminths are aerobic with an active tricarboxylic acid cycle and CN-sensitive respir-ation. Therefore, a marked aerobic-anaerobic transition in their energy metabolism occurs during development of most helminths. For example, schistosome miracidia and cercaria are free-living and aerobic. However, after penetration of the definitive host, during development from schistosomulum to adult, the schistosome assumes an increasing reliance on glycolysis for energy-generation (80, 82). Similarly, in *F. hepatica*, the ratio of tricarboxylic acid cycle activity to acetate and propionate formation appears to decrease during the development of the excysted juvenile fluke to the adult (44, 45, 83). In fact, Tielens *et al.* (83) have suggested a direct correlation between tricarboxylic acid cycle activity and the surface area of the fluke and have postulated that diffusion may limit oxygen utilization by tissues deeper in the parenchyma. This concept is supported by cytochemical detection of cytochrome oxidase in the nematodes, *N. braziliensis*, *H. contortus* and *Ascaridia galli* (84), all three of which exhibit significant staining in hypodermal mitochondria found immediately beneath the cuticle, but staining is markedly decreased in the internal tissues of the larger helminths. Whether these observations also apply to *A. suum* has not been determined. *Ascaris suum* contains substantial amounts of hemoglobin in its muscle and perienteric fluid. It may function to maintain a low but constant Po_2 internally. Oxygen is clearly used for biosynthetic reactions, such as the synthesis of hydroxyproline from proline for collagen synthesis. Whether it may be used by tissues other than muscle for energy generation remains to be determined.

Ascaris suum exhibits a number of alterations in carbohydrate metabolism during development from unembryonated egg to adult. In general, helminth eggs leave the definitive host in widely varying stages of development. Many nematode eggs, such as those of *A. suum*, are undifferentiated and require oxygen for embryonation. Many cestode and trematode eggs leave fully embryonated suggesting that embryonation occurs in the microaerobic habitat of the host. In *A. suum*, unembryonated 'eggs' leave the host and develop to contain infective second-stage larva (L2) after about 21 days. This developmental process is fuelled by glycogen and especially triacylglycerols stored in the 'egg.' All of the enzymes of the tricarboxylic acid cycle and β-oxidation are present during this developmental process and metabolism is aerobic (85). *A. suum* is one of the few metazoans capable of net glycogen synthesis from triacylglycerols, and the presence of a functioning glyoxylate cycle has been demonstrated in its embryonating eggs (86). At about day 10 of development, the activities of malate synthase and isocitrate lyase, two key enzymes in the cycle, increase dramatically, triacylglycerol stores decrease, and glycogen, consumed earlier in development, is resynthesized (85, 86). Once the development of the L2 is complete, its metabolic rate decreases significantly and the infective 'eggs' may remain dormant for long periods until ingestion by the definitive host. A number of interesting questions remain unanswered. What is the relationship between the enzymes catalyzing β-oxidation during early larval development and the reversal of β-oxidation in adult muscle? The triacylglycerols of unembryonated eggs contain substantial amounts (about 12% by weight) of 2-methylbutanoate and 2-methylpentanoate, major products of carbohydrate metabolism in adult muscle. Additionally, factors initiating and regulating processes such as fluctuations in glyoxylate cycle activity or the decrease in respiratory rate associated with dormancy in the L2 are poorly understood. In fact, little is known about regulation of gene expression during ascarid development.

After *A. suum* L2 are ingested, they excyst in the host gut, molt to the third-stage (L3) in the liver, then migrate to the lung. Metabolism of both the L2 and L3 is aerobic and cyanide-sensitive (85, 87). However, after the L3 migrate from the lungs back to the small intestine, they molt to the fourth-stage (L4) and respiration becomes cyanide-insensitive. In addition, they begin to form the branched-chain fatty acids characteristic of the adult (88, 89). During this transition from L2 to adult, the activities of enzymes associated with aerobic metabolism decrease dramatically, whereas those associated with anaerobic pathways increase in an equally dramatic fashion (87–89). Many of the key enzymes involved in anaerobic metabolism appear to be expressed as stage-specific isozymes (68, 90). The activities of these anaerobic enzymes are already elevated in L3 isolated directly from rabbit lungs, even though their metabolism is aerobic and CN-sensitive (88, 89). Apparently, they are synthesized earlier in the development of the L3. The location of these enzymes is not clear nor is there an understanding of the factors regulating the switch to fermentative pathways. Interestingly, whereas nematodes exhibit cell constancy during growth, the molt from L3 to L4 in *A. suum* is followed by a dramatic increase in muscle cell division that appears to parallel the development of anaerobic pathways (88).

5. PERSPECTIVE

Of all the metabolic pathways operating in parasitic helminths, the reactions associated with carbohydrate metabolism and energy generation are, by far, the most thoroughly understood. However, in spite of years of intensive study, many key areas remain only partially defined. These include the study of internal and external factors regulating glycogen metabolism, glycolytic flux and end-product formation, the role of oxygen as a terminal electron acceptor, ionic fluxes across the inner mitochondrial membrane and the identification of tissue and stage specific differences in energy generation. Clearly, energy generation in parasitic helminths differs substantially from that of the host and still remains a potentially important site for chemotherapeutic intervention. Future work in this area will continue to identify and characterize differences between host and parasite at the molecular level and with the application of modern techniques of protein characterization, crystallography, molecular cloning and site-directed mutagenesis, it is, perhaps, only a matter of time before these differences can be successfully exploited in rational drug design.

6. REFERENCES

1. Bennet, E. M. and Bryant, C. (1984) Energy metabolism of adult *Haemonchus contortus in vitro*: a comparison of benzimidazole-sensitive and resistant strains. *Mol. Biochem. Parasitol.* **10**: 335–346.
2. Powell, J. W., Stables, J. N. and Watt, R. A. (1986) An NMR study of the effect of glucose availability on carbohydrate metabolism in *Dipetalonema viteae* and *Brugia pahangi*. *Mol. Biochem. Parasitol.* **19**: 265–271.
3. Yacoub, N. J., Allen, B. L., Payne, D. M., Masaracchia, R. A. and Harris, B. G. (1983) Purification and characterization of phosphorylase *B* from *Ascaris suum*. *Mol. Biochem. Parasitol.* **9**: 297–307.
4. Donahue, M. J., Yacoub, N. J., Kaeini, M. R. and Harris, B. G. (1981) Activity of enzymes regulating glycogen metabolism in perfused muscle-cuticle sections of *Ascaris suum* (Nematoda). *J. Parasitol.* **67**: 362–367.
5. Hannigan, L. L., Donahue, M. J. and Masaracchia, R. A. (1985) Comparative purification and characterization of invertebrate muscle glycogen synthase from the porcine parasite *Ascaris suum*. *J. Biol. Chem.* **260**: 16099–16105.
6. Ghosh, P., Heath, A. C., Donahue, M. J. and Masaracchia, R. A. (1989) Glycogen synthesis in the obliquely striated muscle of *Ascaris suum*. *Eur. J. Biochem.* **183**: 679–685.
7. Donahue, M. J., Yacoub, N. J., Kaeini, M. R., Masaracchia, R. A. and Harris, B. G. (1981) Glycogen metabolizing enzymes during starvation and feeding of *Ascaris suum* maintained in a perfusion chamber. *J. Parasitol.* **67**: 505–510.
8. Donahue, M. J., Yacoub, N. J., Michnoff, C. A., Masaracchia, R. A. and Harris, B. G. (1981) Serotonin (5-hydroxytryptamine): a possible regulator of glycogenolysis in perfused muscle segments of *Ascaris suum*. *Biochem. Biophys. Res. Commun.* **101**: 112–117.
9. Donahue, M. J., Yacoub, N. J. and Harris, B. G. (1982) Correlation of muscle activity with glycogen metabolism in muscle of *Ascaris suum*. *Am. J. Physiol.* **242**: R514–R521.
10. Donahue, M. J., Masaracchia, R. A. and Harris, B. G. (1983) The role of cyclic AMP-mediated regulation of glycogen metabolism in levamisole-perfused *Ascaris suum* muscle. *Mol. Pharmacol.* **23**: 378–383.
11. Tielens, A. G. M., and van den Bergh, S. G. (1987) Glycogen metabolism in *Schistosoma mansoni* worms after their isolation from the host. *Mol. Biochem. Parasitol.* **24**: 247–254.
12. Tielens, A. G. M., van den Heuvel, J. M. and van den Bergh, S. G. (1990) Substrate cycling between glucose 6-phosphate and glycogen occurs in *Schistosoma mansoni*. *Mol. Biochem. Parasitol.* **39**: 109–116.
13. Tielens, A. G. M., van den Meer, P., van den Heuvel, J. M. and van den Bergh, S. G. (1991) The enigmatic presence of all gluconeogenic enzymes in *Schistosoma mansoni* adults. *Parasitology.* **102**: 267–276.
14. Gentleman, S., Abrahams, S. L. and Mansour, T. E. (1976) Adenosine cyclic 3′, 5′-monophosphate in the liver fluke, *Fasciola hepatica*. *Mol Pharmacol.* **12**: 59–68.
15. Estey, S. J. and Mansour, T. E. (1987) Nature of serotonin-activated adenylate cyclase during development of *Schistosoma mansoni*. *Mol. Biochem. Parasitol.* **26**: 47–60.
16. Ribeiro, P. and Webb, R. A. (1986) Demonstration of specific high-affinity binding sites for [^{3}H] 5-hydroxytryptamine in the cestode *Hymenolepis diminuta*. *Comp. Biochem. Physiol.* **84C**: 353–358.
17. Chaudhuri, J. and Donahue, M. J. (1989) Serotonin receptors in the tissues of adult *Ascaris suum*. *Mol. Biochem. Parasitol.* **35**: 191–198.

18. Supowit, S. C. and Harris, B. G. (1976) *Ascaris suum* hexokinase: purification and possible function in compartmentation of glucose 6-phosphate in muscle. *Biochim. Biophys. Acta* **422**: 48–59.

19. Hofer, H. W., Allen, B. L., Kaeini, M. R., Pette, D. and Harris, B. G. (1982) Phosphofructokinase from *Ascaris suum*. Regulatory kinetic studies and activity near physiological conditions. *J. Biol. Chem.* **275**: 3801–3806.

20. Hofer, H. W., Allen, B. L., Kaeini, M. R. and Harris, B. G. (1982) Phosphofructokinase from *Ascaris suum*. The effect of phosphorylation on activity near-physiological conditions. *J. Biol. Chem.* **257**: 3807–3810.

21. Kamemoto, E. S. and Mansour, T. E. (1986) Phosphofructokinase in the liver fluke *Fasciola hepatica*. Purification and kinetic changes by phosphorylation. *J. Biol. Chem.* **261**: 4346–4351.

22. Kamemoto, E. S., Lan, L. and Mansour, T. E. (1989) *In vivo* regulation of phosphorylation level and activity of phosphofructokinase by serotonin in *Fasciola hepatica*. *Arch. Biochem. Biophys.* **271**: 553–559.

23. Srinivasan, N. G., Wariso, B. A., Kulkarni, G., Rao, G. S. J. and Harris, B. G. (1988) Phosphofructokinase from *Dirofilaria immitis*. Stimulation of activity by phosphorylation with cyclic AMP-dependent protein kinase. *J. Biol. Chem.* **263**: 3482–3485.

24. Cook, P. F., Rao, G. S. J., Hofer, H. W. and Harris, B. G. (1987) Correlation between hysteresis and allosteric properties for phosphofructokinase from *Ascaris suum*. *J. Biol. Chem.* **262**: 14063–14067.

25. Thalhofer, H. P., Daum, G., Harris, B. G. and Hofer, H. W. (1988) Identification of two different phosphofructokinase-phosphorylating protein kinases from *Ascaris suum* muscle. *J. Biol. Chem.* **263**: 952–957.

26. Thalhofer, H. P., Starz, W., Daum, G., Wurster, B., Harris, B. G. and Hofer, H. W. (1989) Purification and properties of the cyclic 3′, 5′-AMP-binding protein from the muscle of the nematode *Ascaris suum*. *Arch. Biochem. Biophys.* **271**: 471–478.

27. Iltsch, M. H. and Mansour, T. E. (1987) Partial purification and characterization of cAMP-dependent protein kinase from *Fasciola hepatica*. *Mol. Biochem. Parasitol.* **23**: 265–274.

28. Kulkarni, G., Rao, G. S. J., Srinivasan, N. G., Hofer, H. W., Yuan, P. M. and Harris, B. G. (1987) *Ascaris suum* phosphofructokinase. Phosphorylation by protein kinase and sequence of the phosphopeptide. *J. Biol. Chem.* **262**: 32–34.

29. Klein, R. D., Olson, E. R., Favreau, M. A. *et al.* (1991) Cloning of a cDNA encoding phosphofructokinase from *Haemonchus contortus*. *Mol. Biochem. Parasitol.* **48**: 17–26.

30. Dedman, J. R., Gracy, R. W. and Harris, B. G. (1974) A method for estimating sequence homology from amino acid comparisons. The evolution of *Ascaris* employing adolase and glyceraldehyde-3-phosphate dehydrogenase. *Comp. Biochem. Physiol.* **49B**: 715–731.

31. Wilkes, J., Cornish, R. A. and Mettrick, D. F. (1981) Purification and properties of phosphoenolpyruvate carboxykinase from *Hymenolepis diminuta*. *J. Parasitol.* **67**: 832–840.

32. Rohrer, S. P., Saz, H. J. and Nowak, T. (1986) Purification and characterization of phosphoenolpyruvate carboxykinase from the parasitic helminth *Ascaris suum*. *J. Biol. Chem.* **261**: 13049–13055.

33. Barrett, J., Mendis, A. H. W. and Butterworth, P. E. (1986) Carbohydrate metabolism in *Brugia pahangi*. *Int. J. Parasitol.* **16**: 465–469.

34. Bueding, E. and Fisher, J. (1982) Metabolic requirements of schistosomes. *J. Parasitol.* **68**: 208–212.

35. Mendis, A. H. W. and Townson, S. (1985) Evidence for the occurrence of respiratory electron-transport in adult *Brugia pahangi* and *Dipetalonema viteae*. *Mol. Biochem. Parasitol.* **14**: 337–354.

36. Paget, T. A., Fry, M., and Lloyd, D. (1989) Respiration in the filarial nematode, *Brugia pahangi*. *Int. J. Parasitol.* **19**: 337–343.

37. Lane, C. A., Pax, R. A. and Bennett, J. L. (1987) L-Glutamine: an amino acid required for maintenance of the tegumental membrane potential of *Schistosoma mansoni*. *Parasitology* **94**: 233–242.

38. MacKenzie, N. E., van de Waa, E. A. Gooley, P. R. *et al.* (1989) Comparison of glycolysis and glutaminolysis in *Onchocherca volvulus* and *Brugia pahangi* by ^{13}C nuclear magnetic resonance spectroscopy. *Parasitology* **99**: 427–435.

39. Van Oordt, B. E. P., van den Heuvel, J. M., Tielens, A. G. M. and van den Bergh, S. G. (1985) The energy production of the adult *Schistosoma mansoni* is for a large part aerobic. *Mol. Biochem. Parasitol.* **16**: 117–126.

40. Saz, H. J. (1971) Anaerobic phosphorylation in *Ascaris* mitochondria and the effect of anthelmintics. *Comp. Biochem. Physiol.* **39B**: 627–637.

41. Davies, K. P. and Kohler. P. (1990) The role of amino acids in the energy generating pathways of *Litomosoides carinii*. *Mol. Biochem. Parasitol.* **41**: 115–124.

42. Paget, T. A., Fry, M. and Lloyd, D. (1987) Effects of inhibitors on the oxygen kinetics of *Nippostrongylus braziliensis*. *Mol. Biochem. Parasitol.* **22**: 125–133.

43. Behm, C. A., Bryant, C. and Jones, A. J. (1987) Studies of glucose metabolism in *Hymenolepis diminuta* using ^{13}C nuclear magnetic resonance. *Int. J. Parasitol.* **17**: 1333–1341.

44. Lloyd, G. M. (1986) Energy metabolism and its regulation in the adult liver fluke *Fasciola hepatica*. *Parasitology* **93**: 217–248.

45. Tielens, A. G. M., van den Huevel, J. M. and van den Bergh, S. G. (1984) The energy metabolism of *Fasciola hepatica* during its development in the final host. *Mol. Biochem. Parasitol.* **13**: 301–307.

46. Rioux, A. and Komuniecki, R. (1984) 2-methylvalerate formation in mitochondria of *Ascaris suum* and its relationship to anaerobic energy-generation. *J. Comp. Physiol.* **154**: 349–354.

47. Saz, H. J. and Weil, A. (1960) The mechanism of formation of α-methylbutyrate from carbohydrate by *Ascaris lumbricoides* muscle. *J. Biol. Chem.* **235**: 914–918.

48. Sangster, N. C. and Prichard, R. K. (1985) The contributution of a partial tricarboxylic acid cycle to volatile end products in thiabendazole resistant and susceptible *Trichostrongylus colubriformis*. *Mol. Biochem. Parasitol.* **14**: 261–274.

49. Ward, P. F. V. and Huskisson, N. S. (1978) The energy metabolism of adult *Haemonchus contortus in vitro*. *Parasitology* **77**: 255–271.

50. Barrett, J. (1981) *Biochemistry of Parasitic Helminths*. Macmillan, London.

51. von Brand, T. (1973) *Biochemistry of Parasites*. Academic Press, New York.

52. Kohler, P. (1991) The pathways of energy-generation in filarial parasites. *Parasitol. Today* **7**: 21–25.

53. Fairbairn, D. (1970) Biochemical adaptation and loss of genetic capacity in helminth parasites. *Bio. Rev.* **45**: 29–72.

54. Takamiya, S., Furushima, R. and Oya, H. (1984) Electron transfer complexes of *Ascaris suum* muscle mitochondria. I. Characterization of NADH–cytochrome *c* reductase (complex I–III), with special reference to cytochrome localization. *Mol. Biochem. Parasitol.* **13**: 121–134.

55. Kita, K., Takamiya, S., Furushima, R., Ma, Y. and Oya, H. (1988) Complex II is a major component of the respiratory chain in the muscle mitochondria of *Ascaris suum* with high fumarate reductase activity. *Comp. Biochem. Physiol.* **89B**: 31–4.

56. Fioravanti, C. F. (1981) Coupling of mitochondrial NADPH:NAD transhydrogenase with electron-transport in adult *Hymenolepis diminuta*. *J. Parasitol.* **67**: 823–831.

57. Kita, K., Takamiya, S., Furishima, R. *et al.* (1988) Electron transfer complexes of *Ascaris suum* muscle mitochondria III. Composition and fumarate reductase activity of complex II. *Biochim. Biophys. Acta* **935**: 130–40.

58. Takamiya, S., Furushima, R. and Oya, H. (1986) Electron-transfer complexes of *Ascaris suum* muscle mitochondria. II. Succinate coenzyme Q reductase (complex II) associated with substrate-reducable cytochrome B_{558}. *Biochim. Biophys. Acta* **848**: 99–107.

59. Hata-Tanaka, A., Kita, K., Furushima, R., Oya, H. and Itoh, S. (1988) ESR studies on iron–sulfur clusters of complex II in *Ascaris suum* mitochondria which exhibits strong fumarate reductase activity. *FEBS Lett.* **242**: 183–186.

60. Kita, K., Takamiya, S., Furishima, R. *et al.* (1988) Indispensability of rhodoquinone in the NADH–fumarate reductase system of *Ascaris* muscle mitochondria. In *Biochemistry, Bioenenergetics and Clinical Applications of Ubiquinone* (ed. L. Giorgio), Taylor and Francis, London.

61. Barrett, J. and Beis, I. (1973) The redox state of the free nicotinamide adenine dinucleotide couple in the cytoplasm and mitochondria of muscle tissue from *Ascaris lumbricoides* (Nematoda). *Comp. Biochem. Physiol.* **44A**: 331–340.

62. Komuniecki, R., Campbell, T. and Rubin, N. (1987) Anaerobic metabolism in *Ascaris suum*: acyl CoA intermediates in isolated mitochondria synthesizing 2-methyl branched-chain fatty acids. *Mol. Biochem. Parasitol.* **24**: 147–154.

63. Komuniecki, R., Komuniecki, P. R. and Saz, H. J. (1979) Purification and properties of the *Ascaris* pyruvate dehydrogenase complex. *Biochim. Biophys. Acta* **571**: 1–11.

64. Thissen, J., Desai, S., McCartney, P. and Komuniecki, R. (1986). Improved purification of the pyruvate dehydrogenase complex from *Ascaris suum* body wall muscle and characterization of PDH kinase activity. *Mol. Biochem. Parasitol.* **21**: 129–138.

65. Komuniecki, R., Rhee, R., Bhat, D., Duran, E., Sidawy. E. and Song, H. (1992) The pyruvate dehydrogenase complex from the parasitic nematode, *Ascaris suum*: novel subunit composition and domain structure of the dihydrolipoyl transacetylase component. *Arch. Biochem. Biophys.* **296**: 115–121.

66. Komuniecki, R. and Thissen, J. (1989) The pyruvate dehydrogenase complex from the parasitic nematode, *Ascaris suum*: stoichiometry of phosphorylation and inactivation. *Ann. N. Y. Acad. Sci.* **573**: 175–182.

67. Song, H. B., Thissen, J. and Komuniecki, R. (1991) Novel pyruvate dehydrogenase phosphatase activity from mitochondria of the parasitic nematode, *Ascaris suum*. *Mol. Biochem. Parasitol.* **48**: 101–104.

68. Johnson, K. R., Komuniecki, R., Sun, Y. and Wheelock, M. J. (1992) Characterization of cDNA clones for the alpha subunit of pyruvate dehydrogenase from *Ascaris suum*. *Mol. Biochem. Parasitol.* **51**: 37–47.

69. Thissen, J. and Komuniecki, R. (1988) Phosphorylation and inactivation of the pyruvate dehydrogenase from the anaerobic parasitic nematode, *Ascaris suum*: stoichiometry, and amino acid sequence around the phosphorylation sites. *J. Biol. Chem.* **263**: 19092–19097.

70. Komuniecki, R., Fekete, S. and Thissen, J. (1985) Purification and characterization of the 2-methyl-branched chain acyl CoA dehydrogenase, an enzyme involved in the enoyl CoA reduction in anaerobic mitochondria of the nematode: *Ascaris suum*. *J. Biol. Chem.* **260**: 4770–4777.

71. Komuniecki, R., McCrury, J., Thissen, J. and Rubin, N. (1989) Electron-transfer flavoprotein from anaerobic *Ascaris suum* mitochondria and its role in NADH-dependent 2-methyl branched-chain enoyl CoA reduction. *Biochim. Biophys. Acta.* **975**: 127–131.

72. Ma, Y., Funk. M., Dunham, W. and Komuniecki, R. (1993) Purification and characterization of electron-transfer flavoprotein: rhodoquinone oxidoreductase from anaerobic mitochondria of the adult parasitic nematode, *Ascaris suum*. *J. Biol. Chem.* **268**: 20360–20365.

73. Kohler, P. and Bachmann, R. (1980) Mechanisms of respiration and phosphorylation in *Ascaris* muscle mitochondria. *Mol. Biochem. Parasitol.* **1**: 75–92.

74. Saz, H. J. and Pietrzak, S. M. (1980) Phosphorylation associated with succinate decarboxylation to propionate in *Ascaris* mitochondria. *Arch. Biochem. Biophys.* **202**: 399–404.

75. Chojnicki, K., Dudzinska, M., Wolanska, P. and Michejda, J. (1987) Membrane potential in mitochondria of *Ascaris*. *Acta Parasit. Pol.* **32**: 67–78.

76. Wolstenholme, D. R., MacFarlane, J., Okimoto, R., Clary, D. O. and Waleithner, J. A. (1986) Bizarre tRNAs inferred from DNA sequences of mitochondrial genomes of nematode worms. *Proc. Nat. Acad. Sci. USA* **84**: 1324–1328.

77. Saz, H. J. and deBruyn, B. S. (1987) Separation and function of two acyl CoA transferases from *Ascaris lumbricoides* mitochondria. *J. Exp. Zool.* **242**: 241–245.

78. Suarez de Mata, Z., Arevalo, J. and Saz, H. J. (1991) Propionyl-CoA condensing enzyme from *Ascaris* muscle mitochondria II. Coenzyme A modulation. *Arch. Biochem. Biophys.* **258**: 166–171.

79. Paget, T. A., Fry, M. and Lloyd, D. (1988) The O_2-dependence of respiration and H_2O_2 production in the parasitic nematode *Ascaridia galli*. *Biochem. J.* **256**: 633–639.

80. Thompson, D. P., Morrison, D. D., Pax, R. A. and Bennett, J. L. (1984) Changes in glucose metabolism and cyanide-sensitivity in *Schistosoma mansoni* during development. *Mol. Biochem. Parasitol.* **13**: 39–51.

81. Fioravanti, C. F. and Reisig, J. M. (1990) Mitochondrial hydrogen peroxide formation and fumarate reductase in *Hymenolepis diminuta*. *J. Parasitol.* **76**: 457–463.
82. Horemans, A. M. C., Tielens. A. G. M. and van den Bergh, S. G. (1991) The transition from an aerobic to an anaerobic metabolism in transforming *Schistosoma mansoni* cercariae occurs exclusively in the head. *Parasitology* **102**: 259–265.
83. Tielens, A. G. M., van den Huevel, J. M. and van den Bergh, S. G. (1987) Differences in intermediary metabolism between juvenile and adult *Fasciola hepatica*. *Mol. Biochem. Parasitol.* **24**: 273–281.
84. Fry, M. and Beesley, J. E. (1985) Cytochemical localization of cytochrome oxidase in tissues of parasitic nematodes. *Parasitology* **90**: 145–156.
85. Barrett, J. (1976) Intermediary metabolism in *Ascaris* eggs. In: *Biochemistry of Parasites and Host-Parasite Relationships* (ed. Van den Bossche, H.), Elsevier, Amsterdam, pp. 117–123.
86. Barrett, J., Ward, C. W. and Fairbairn, D. (1970) The glyoxylate cycle and the conversion of triglycerides to carbohydrates in developing eggs of *Ascaris lumbricoides*. *Comp. Biochem. Physiol.* **35**: 577–586.
87. Saz, H. J., Lescure, O. L. and Bueding, E. (1968) Biochemical observations of *Ascaris suum* lung-stage larvae. *J. Parasitol.* **54**: 457–461.
88. Komuniecki, P. R. and Vanover, L. (1987) Biochemical changes during the aerobic–anaerobic transition in *Ascaris suum* larvae. *Mol. Biochem. Parasitol.* **22**: 241–248.
89. Vanover-Dettling, L. and Komuniecki, P. R. (1989) Effect of gas phase on carbohydrate metabolism of *Ascaris suum* larvae. *Mol. Biochem. Parasitol.* **36**: 29–40.
90. Duran, E., Komuniecki, R., Komuniecki, P. *et al.* (1993) Characterization of cDNA clones for the 2-methyl branched chain enoyl CoA reductase: an enzyme involved in branched-chain fatty acid synthesis in anaerobic mitochondria of the parasitic nematode, *Ascaris suum*. *J. Biol. Chem.* **268**: 22391–22396.

5 Amino Acid and Protein Metabolism

MICHAEL J. NORTH[1] and BARBARA C. LOCKWOOD[2†]

[1]*Department of Biological and Molecular Sciences, University of Stirling, Stirling, UK and* [2]*Laboratory for Biochemical Parasitology, Department of Zoology, University of Glasgow, Glasgow, UK*

SUMMARY

A variety of pathways of amino acid metabolism are encountered among parasites. Many parasites can utilize amino acids as energy sources during at least part of their life cycle and most amino acids can be used by at least one group of parasites. In some cases, as with proline catabolism in trypanosomatids, the pathway is identical to that found in higher animals although its relative importance may be greater. Other pathways, such as the dihydrolase pathway for arginine catabolism in anaerobic protozoa, are distinct from those of the host. Further ways in which parasite and host amino acid metabolism differ include the extent of alanine formation in many protozoan parasites and aspects of sulfur amino acid metabolism in anaerobic protozoa and helminths. Protein degradation can provide a source of amino acids for many parasites. Many of the proteolytic enzymes likely to be involved are now well characterized and the roles played by some of these in specific proteolytic processes are being identified. For example, hemoglobin degradation by *Plasmodium falciparum* trophozoites is an ordered process involving aspartic and cysteine proteinases, whereas digestion of blood proteins by various helminths is dependent on cysteine proteinases. Although many of the features of parasite amino acid and protein metabolism resemble those of the hosts, there are marked differences with respect to the properties of the enzymes involved, the relative importance of common pathways and the presence of parasite-specific pathways.

†Deceased

Biochemstry and Molecular Biology of Parasites
ISBN 0-12-473345-X

1. INTRODUCTION

The study of amino acid metabolism in parasites has not been conducted in a systematic manner and the overall picture is somewhat diffuse. This chapter focuses on recent findings and pathways that appear to be especially active in parasites or have unique features which distinguish the parasite's metabolism. Many of the features of helminth amino acid metabolism, for example, appear to be very similar to those of mammals and are not discussed here. An excellent review of amino acid metabolism in helminths has recently appeared (1) and further details on protozoan amino acid metabolism can be found in earlier texts (2, 3).

Like all animals, protozoa and helminths have a restricted capacity to synthesize *de novo* their own amino acids, and essential amino acids must be obtained exogenously from the diet as free amino acids or from proteins by proteolysis. Keto acids, by transamination, can also function as amino acid sources. Parasites, especially protozoa, have a requirement for the majority of the twenty amino acids which are needed for protein synthesis.

The amino acids available in the diet do not always match those required for biosynthesis. Some will be present in excess and can be catabolized, allowing amino acids to fulfil their second function as sources of energy. Individual amino acids also have additional roles, for example in osmoregulation or as neurotransmitters. As the major nitrogenous compounds of the cell amino acids are also used as precursors for the synthesis of a range of other nitrogen-containing compounds including polyamines (see Chapter 7) and, especially in helminths, biogenic amines (4) (see Chapter 14) which may themselves be subject to further metabolism (5).

2. AMINO ACID POOLS AND UPTAKE

2.1. Amino Acid Pools

In protozoa, alanine is generally the amino acid present at the highest concentration. High levels of glycine, serine and threonine have been reported. Amino acids other than those required for protein synthesis have also been detected. These include 2-aminobutyrate, 2-aminoadipate, hydroxylysine, 3-methylhistidine, norleucine, norvaline, ornithine and taurine. Amino acid levels fluctuate according to culture conditions.

A wide range in the free amino acid contents of helminths has been reported (1). Concentrations are usually higher than those of vertebrates but lower than in many other invertebrates. In contrast to vertebrates, helminths have one or two dominant free non-essential amino acids, usually alanine, proline, glutamate, serine and/or glycine. Non-protein amino acids detected include 3-alanine, 2-aminoisobutyrate, 3-aminoisobutyrate, 3-aminobutyrate, 4-aminoisobutyrate, 4-aminobutyrate, allo-lysine, allo-leucine, citrulline, ornithine, taurine, creatinine, homoserine, hydroxy-tryptophan, hydroxykynurine, norleucine, norvaline and sarcosine.

2.2. Amino Acid Transport

The ability to utilize exogeneous amino acids implies that parasites possess uptake systems. The earlier literature on amino acid uptake has been reviewed by Aomine (6) and Barrett (7) for protozoa and helminths, respectively. Further details of amino acid transport in helminths can be found in Chapter 12 whereas specific features of the transport of proline, glutamine, alanine and cysteine in protozoa are considered below.

3. AMINO ACID METABOLISM

3.1. Glutamate Dehydrogenase and Aminotransferases

Glutamate dehydrogenase (GDH) and aminotransferases have a central role in the catabolism and biosynthesis of amino acids. Aminotransferases redistribute nitrogen from amino acids to keto acids. When amino acids are catabolized the first step is frequently catalyzed by an aminotransferase which transfers the amino group to α-ketoglutarate to form glutamate which can be deaminated by GDH to give α-ketoglutarate and ammonia. For biosynthesis of amino acids these steps are reversed.

GDH activity has been reported in a wide variety of protozoa. Unlike mammalian GDH the protozoal enzymes are specific for either NAD or NADP, although not all species have been shown to contain both. The most detailed studies have been conducted on the enzymes of *Trypanosoma cruzi* (8, 9). Immunological comparisons on *Plasmodium falciparum* GDH (10) and sequence information for NADP-GDH from both *Giardia lamblia* (11) and *T. cruzi* (9) suggest a closer relationship of the protozoal enzymes to eubacterial and fungal GDHs than to mammalian GDH.

In many protozoa oxidative deamination of glutamate involving a cytosolic NADP-linked GDH is probably the most important route for the conversion of amino nitrogen into ammonia. The enzyme is also responsible for the catabolism of glutamate derived exogenously or from other amino acids such as proline or glutamine (see below) and may provide a major source of reduced NADP for biosynthetic reactions. The properties of a number of the enzymes, including those of *Trichomonas vaginalis* (12) and *P. falciparum* (10), are consistent with their primarily catabolic role. In *Trypanosoma cruzi* a biosynthetic role has been proposed for NADP-GDH with NAD-GDH fulfilling a catabolic function (9, 13). Only a few examples of NAD-dependent GDH have been described in parasitic protozoa. In addition to *T. cruzi* its presence has been reported in *Leishmania mexicana* (14). Observations on the developmental regulation of *L. mexicana* GDH levels are consistent with a role for the mitochondrial NAD-dependent enzyme in oxidative deamination and in the high rate of excretion of ammonia by promastigotes. A cytosolic NADP-dependent enzyme is detectable only in amastigotes. Malaria parasites may have a mitochondrial NAD-GDH to account for the ability of intact mitochondria to oxidize proline, although no such activity has yet been detected (15). In *T. cruzi* both the NAD-GDH and NADP-GDH are likely to play a role in coupling alanine production to the maintenance of the redox balance during glycolysis (9) (see below).

Many aminotransferase activities have been identified in parasitic protozoa; the most frequently reported are alanine and asparate aminotransferases. In parasites such as *Trichomonas vaginalis* (16) and *Trypanosoma cruzi* (9) in which detailed surveys have been undertaken a number of different aminotransferases have been identified. Although the names used for aminotransferases, e.g. alanine aminotransferase, often imply a high level of specificity, many use a broader range of amino acids as amino donors than their names suggest, e.g. asparate aminotransferase of *Trichomonas vaginalis* will also catalyse the transamination of aromatic amino acids (16), and *Trypanosoma cruzi* tyrosine aminotransferase also acts as an alanine aminotransferase (9). High levels of alanine aminotransferase activity are often associated with the formation of alanine as a major end-product of glucose metabolism (see below) but the precise role of other aminotransferases is not known. For example, gostatin, an irreversible inhibitor of aspartate aminotransferase does not affect the growth of *Trichomonas vaginalis* although it elevates aspartate concentrations.

GDH activity has been demonstrated in many helminths; both cytosolic and mitochondrial enzymes have been described (1). However, information on the details of their regulation is incomplete and difficult to interpret. The role of GDH in helminths is not yet established.

Many examples of aminotransferase activity have been described in helminths, although the enzymes have rarely been characterized (1, 17). Most are specific for α-ketoglutarate as the amino acceptor but have broad specificity with respect to the amino donor. There is considerable variation in the potential amino acid donors from one species to another. Those aminotransferases that have been characterized show some similarities to rat liver enzymes. As in protozoa, alanine aminotransferase is ubiquitous and, together with asparate aminotransferase, is usually the most active member of this group. Isoenzymes of alanine aminotransferase have been found in both cytoplasm and mitochondria. The results of a survey of nematode alanine aminotransferases suggest that there is a greater overall capacity for the synthesis of alanine than for its catabolism, the opposite of the situation with the rat liver enzymes (17).

3.2. Amino Acids as Energy Sources

Amino acids can be used as a source of energy by many parasites. In protozoa, in which amino acids may be the major and perhaps only energy source at certain stages of the life cycle, the picture of energy production derived from earlier studies which focused entirely on carbohydrate metabolism was often too simple. In parasitic helminths. amino acids probably play a lesser role in energy metabolism, and the general lack of catabolism may relate to the heavy commitment to synthetic processes.

The initial step of amino acid catabolism frequently involves the loss of the amino nitrogen to yield a carbon skeleton in the form of a keto acid, either as a result of transamination or by deamination. In general, energy is derived from amino acids by the oxidation of the carbon skeletons after entry into the intermediary metabolic pathways.

3.2.1. Proline metabolism

Proline is a particularly important energy source for certain protozoa (2). The food of blood-feeding insects is largely proteinaceous and during the insect stages parasite develop in the digestion products derived from proteins. Proline is a major energy source of insects and large quantities of this amino acid are present in the haemolymph A number of parasites, especially haemoflagellates, have evolved the capacity to exploit the availability of proline and utilize it as a source of energy in insect and culture stages

Proline catabolism follows the pathway shown in Fig. 5.1. When trypanosomatid utilize proline, products other than CO_2 are formed indicating that the TCA cycle i

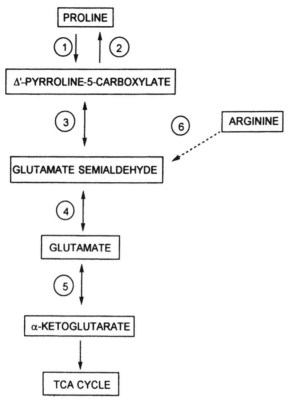

FIG. 5.1 Proline metabolism. 1, Proline oxidase; 2, Δ'-pyrroline-5-carboxylate reductase; 3, spontaneous reaction; 4, Δ'-pyrroline-5-carboxylate dehydrogenase; 5, glutamate dehydrogenase; 6, see Fig. 5.2.

incomplete. These other products can include succinate, alanine and aspartate, the latter two being formed by transamination of pyruvate and oxaloacetate, respectively. The high alanine aminotransferase activity of *Trypanosoma brucei* appears to be critical for growth on proline. The ability to oxidize proline is stage-specific and correlates with the availability of proline at different stages of the life cycle. In contrast to culture forms, bloodstream forms of trypanosomes have little or no proline oxidase activity.

The uptake of proline has been studied in detail in *Leishmania* species. Until recently all the evidence supported the view that it is an active process driven by the proton motive force across the membrane (18, 19). Both promastigotes and amastigotes take up proline, but promastigotes accumulate more, which is consistent with the preferential utilization of proline as an energy source at this stage. Different systems of proline transport in the two stages have been proposed: uptake is optimal at pH 7–7.5 in promastigotes and pH 5.5 in amastigotes reflecting the different environments of the parasites in the sandfly gut (slightly alkaline) and the macrophage lysosomes (acidic). However, an analysis of proline uptake in chemostat cultures of *L. donovani* has provided no evidence that proline is actively transported under these conditions (20, 21).

Proline oxidase has not been demonstrated in *Plasmodium* but the mitochondria of *P. yoelii* and *P. falciparum* are capable of oxidizing proline. This too may represent an adaptation for the insect phase of the life cycle (15).

Proline metabolism has not been studied in detail in helminths. The nematodes *Heligmosomoides polygyrus* and *Panagrellus redivivus* have activities similar to those of proline metabolism in rat liver (22). However, *Fasciola hepatica* has neither Δ'-pyrroline-5-carboxylate dehydrogenase nor proline oxidase activity and so cannot metabolize proline. Because of the lack of the former enzyme Δ'-pyrroline-5-carboxylate formed during arginine catabolism cannot be metabolized to glutamate and is diverted to proline which then accumulates (see below). *F. hepatica* does have high levels of Δ'-pyrroline-5-carboxylate reductase, the enzyme responsible for proline synthesis, and the step may be important for the reoxidation of NADH. The high levels of proline produced are probably responsible for bile duct hyperplasia and may also induce anaemia (7). A similar situation occurs with *Schistosoma mansoni* (1).

3.2.2. Arginine metabolism

Arginine catabolism in higher animals and most parasites follows the pathway shown in Fig. 5.2; steps 1–3. Ornithine can be diverted for polyamine biosynthesis via the ornithine decarboxylase reaction to form putrescine although there is some evidence for an alternative route from arginine to putrescine via agmatine in *T. cruzi* and *Ascaris lumbricoides* (see Chapter 7). Arginase is present in some protozoa, but two other pathways of arginine catabolism have been described. The γ- guanidinobutyramide pathway (Fig. 5.2; steps 4–8) is present in *L. donovani* (23), whereas the anaerobic protozoa *G. lamblia* (*intestinalis*), *Trichomonas vaginalis* and *Tritrichomonas foetus* use the dihydrolase pathway (Fig. 5.2; steps 9–11) which is unusual in eukaryotes. The pathway includes a substrate-level phosphorylation catalysed by carbamate kinase, and the production of ATP means that arginine can act as an energy source under anaerobic conditions. Certain aspects of the dihydrolase pathway which deal specifically with polyamine synthesis are expanded in Chapter 7. Arginine is present in both vaginal fluid and duodenal juice and it has been speculated that even though arginine is metabolized more slowly than glucose *in vitro*, it may be a major energy source *in vivo* (24, 25).

Arginase has been widely reported among helminths and there is evidence that arginine can be catabolized to CO_2 (1). In some trematodes, such as *F. hepatica* and *S. mansoni*, proline is the end product of arginine catabolism.

3.2.3. Glutamine metabolism

An adequate supply of glutamine, the most abundant amino acid in plasma, is necessary for optimal intraerythrocytic growth of *Plasmodium falciparum in vitro*. Trophozoites utilize exogenous L-glutamine in increasing amounts during maturation from the ring-form (26). Erythrocytes are normally relatively impermeable to glutamine and also have low permeability to glutamate. The elevated utilization of glutamine is made possible by a marked and selective parasite-induced increase in the permeability of the host membrane to glutamine (26). An enhanced influx of glycine, proline, alanine and asparagine has also been reported (27). It is possible that the increase in transport may also facilitate the release of osmotically active products of hemoglobin digestion.

Certain helminths survive for longer in basic medium if provided with glutamine as the sole exogenous carbon source (28, 29). Glutamine also promotes better survival if

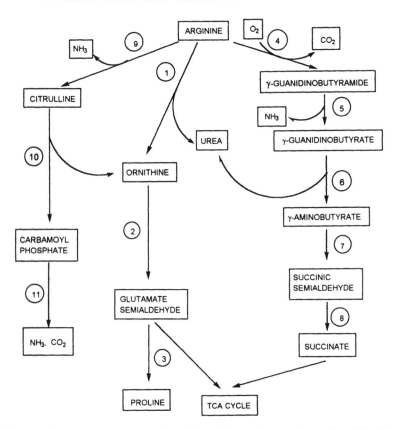

FIG. 5.2 Arginine metabolism. Steps 1–3: 1, arginase; 2, ornithine aminotransferase; 3, see Fig. 5.1. Steps 4–8, γ-guanidinobutyramide pathway: 4, arginine decarboxyoxidase; 5, γ-guanidinobutyramide aminohydrolase; 6, γ-guanidinobutyrate amidinohydrolase; 7, γ-aminobutyrate aminotransferase; 8, succinic semialdehyde dehydrogenase. Steps 9–11, Arginine dihydrolase pathway: 9, arginine deiminase; 10, catabolic ornithine transcarbomoylase; 11, carbamate kinase.

added to glucose/balanced salt solutions and enhances motility. Filariids, such as *Litomosoides carinii, Onchocerca gibsoni, Onchocerca volvulus* and *Brugia pahangi*, and schistosomes metabolize glutamine by aerobic terminal oxidation involving mitochondrial electron transport with CO_2 produced in the TCA cycle (see Chapter 4). The provision of TCA cycle intermediates provides a probable explanation for the stimulation of glucose utilization by glutamine. Some organic acids, lactate, acetate, succinate, glyoxylate and glutamate are produced as minor end-products.

The precise route of glutaminolysis has still to be established, but ^{13}C-NMR studies of *O. volvulus* and *B. pahangi* suggest it may involve glutamate, γ-aminobutyrate and succinate as intermediates and thus resembles the γ-aminobutyrate bypass (GABA shunt), a pathway found predominantly in mammalian brain (30) (Fig. 5.3; see Chapter 14). Glutaminases, which would convert glutamine directly into glutamate, are widely distributed in helminths (although not in *L. carinii* which may use the glutamine synthase reaction in reverse). Although glutamate decarboxylate and γ-aminobutyrate aminotransferase have been detected in helminths, no succinic

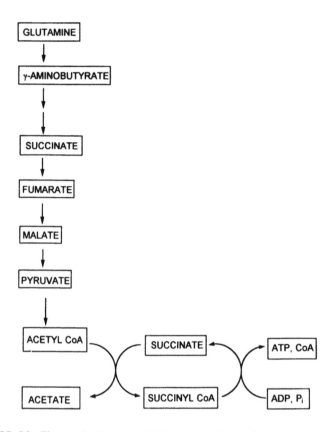

FIG. 5.3   The γ-aminobutyrate (GABA) shunt pathway of glutaminolysis.

semialdehyde dehydrogenase has been reported and the evidence for a functional γ-aminobutyrate bypass is incomplete (1).

3.2.4. Methionine metabolism

The pathways of sulphur amino acid metabolism of some parasites differ significantly from those of the host (see below). In *Trichomonas vaginalis* methionine is rapidly catabolized to methanethiol, α-ketobutyrate and ammonia by the activity of methionine γ-lyase (31, 32) which achieves in one step a reaction for which mammals and aerobic microorganisms require two (transamination and dethiomethylation). The enzyme is also responsible for the breakdown of homocysteine to hydrogen sulfide, α-ketobutyrate and ammonia. The formation of α-ketobutyrate from methionine may generate an additional source of energy; this could explain the unexpectedly high rate of methionine breakdown in *T. vaginalis*. It is probable that pyruvate:ferredoxin oxidoreductase catalyses the conversion of α-ketobutyrate to propionyl CoA which could be used for substrate level phosphorylation (Chapter 3). Methionine γ-lyase has only been found, in anaerobes, namely *Entamoeba histolytica*, *T. vaginalis* and some rumen ciliates. It is lacking, however, in *E. invadens*, *G. lamblia* (*intestinalis*), other trichomonads and all helminths examined to date.

3.2.5. Threonine and serine metabolism

Threonine dehydratase, which converts threonine into α-ketobutyrate and ammonia, has been detected in a number of anaerobic protozoa including trichomonads, *Entamoeba* and *Giardia*. It is involved in threonine catabolism rather than isoleucine biosynthesis (31). In other protozoa the metabolism of threonine not only provides energy but may be important for lipid synthesis. In culture forms of *Trypanosoma brucei*, although not in *Leishmania*, metabolism of exogenous threonine yields glycine and acetate, probably via the threonine dehydrogenase reaction (in which 2-amino-3-ketobutyrate is formed) and aminoacetone synthase reactions. Threonine dehydratase is not present. The acetate formed is not oxidized to CO_2 but is incorporated into fatty acids (33). The process is strongly inhibited by tetraethyliuram disulphide, an inhibitor of aldehyde dehydrogenase and a potent inhibitor of trypanosome growth.

Threonine and serine dehydratases have been detected in nematodes (1).

3.2.6. Branched-chain amino acid metabolism

Little is known about the metabolism of branched-chain amino acids in parasites. Addition of leucine to *Trichomonas vaginalis* results in the production and release of α-hydroxyisocaproate synthesized via the corresponding keto acid, α-ketoisocaproate (31). It is not yet known whether the latter can be metabolized to obtain energy. In the nematodes *H. polygyrus* and *Panagrellus redivivus* branched-chain amino acid metabolism follows the same pathways as in mammalian liver (22).

3.2.7. Other catabolic pathways

A variety of other enzymes involved in amino acid catabolism have been detected in both protozoa and helminths. These include deaminases such as histidase, decarboxylases, some of which are involved in biosynthesis of amines and related compounds, and hydroxylases of proline, tryptophan and tyrosine. These additional enzymes have mostly been reported in helminths (1). L-Amino acid oxidases and D-amino acid oxidases are also present and the availability of the latter would allow D-amino acids to be metabolized in the absence of amino acid racemases.

3.3. Amino Acids and Biosynthesis

The amino acids synthesized by parasites are generally restricted to those at the ends of short pathways originating at precursors in intermediary metabolic pathways (e.g. alanine from pyruvate, glutamate from α-ketoglutarate, asparate from oxaloacetate and serine and glycine from phosphoglycerate), or at other essential amino acids (e.g. cysteine from methionine, tyrosine from phenylalanine). In protozoa, very few amino acids are non-essential. Helminths may have a capacity to synthesize a greater number of amino acids, including some normally essential amino acids, but the involvement of contaminating microorganisms in these studies cannot be ruled out (1). In those cases in which parasites synthesize their own amino acids the pathways are almost certainly identical to those of mammalian cells (1).

3.4. Other Aspects of Amino Acid Metabolism

3.4.1. Alanine metabolism

Alanine is an end-product of glucose catabolism in *G. lamblia*, *T. vaginalis* and trypanosomatids. Its formation is linked to redox balance, alanine aminotransferase and GDH participating in the cyclic dissipation of reducing equivalents formed during glycolysis (Fig. 5.3). Alanine dehydrogenase, which could catalyse the same overall reaction, has never been detected in protozoa at high activity. In trypanosomatids like *Trypanosoma cruzi* oxidation of glycosomal NADH could be coupled to cytosolic alanine formation through reactions catalyzed by malate dehydrogenase, malic enzyme and NADH-GDH (9). A role for aminotransferases in alanine formation is supported by the fact that the addition of aminotransferase inhibitors such as L-cycloserine and carboxymethylamine to *Giardia* cultures blocks both parasite growth and alanine production without affecting the other glucose fermentation products, ethanol and acetate (34). In *T. cruzi* there is evidence for two different pools of alanine, an excreted pool derived from glycolytic phosphoenolpyruvate via oxaloacetate, malate and pyruvate (possibly a cytosolic pool) and one derived more directly from glycolysis,

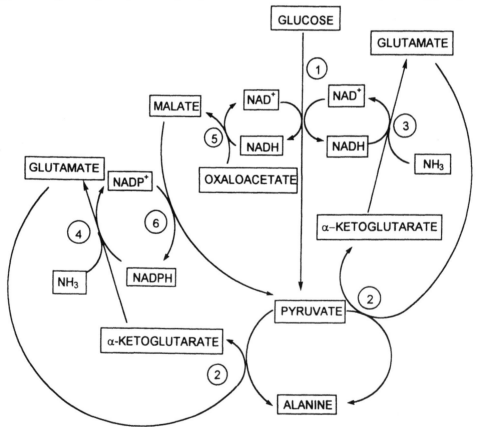

FIG. 5.4   Alanine synthesis and the reoxidation of glycolytic NADH. 1, Glycolysis; 2, alanine aminotransferase; 3, NAD-dependent glutamate dehydrogenase; 4, NADP-dependent glutamate dehydrogenase; 5, malate dehydrogenase; 6, malic enzyme (cytosolic).

perhaps via a bound (mitochondrial) form of pyruvate (35). There are two forms of alanine aminotransferase which could be responsible for generating the separate pools.

When alanine is a major end-product of glucose catabolism, substantial amounts of nitrogenous compounds must be available to contribute the amino group. In a number of species alanine formation is accompanied by proteolysis which would release a pool of amino acids from which the nitrogen could be derived. In *G. lamblia* and *Trichomonas vaginalis* arginine catabolism is a likely provider of nitrogen for alanine synthesis. The efflux of alanine from *G. lamblia* is due to an alanine antiport which is also responsible for alanine uptake and acts to maintain a balance between intracellular and extracellular alanine concentrations (36).

Alanine metabolism can be regulated by a number of factors. In *Leishmania* species alanine is excreted under some conditions, but in other circumstances is consumed. Promastigotes are capable of oxidizing alanine at rates comparable with those for glucose (37). Whether alanine is metabolized or synthesized is dependent on the level of oxygen; consumption occurs at higher oxygen levels and production when the level of oxygen is reduced. For *L. major* the transition occurs at a pO_2 of 6%. In *G. lamblia* alanine formation only occurs in the absence of oxygen (38). In *L. major* there is also a relationship between alanine metabolism and osmoregulation, with hypo-osmotic stress accompanied by alanine release (39). Under iso-osmotic conditions the internal pool of alanine is largely consumed but in hypertonic buffer the pool does not change and increases if glucose is present.

In helminths alanine is an end-product of anaerobic glycolysis and may be formed from glutamine during aerobic metabolism (28). Alanine catabolism has also been demonstrated. In addition to CO_2, products such as lactate, acetate and ethanol are formed indicating only partial oxidation and a relatively low energy efficiency (28).

3.4.2. Sulfur amino acid metabolism

Methionine is required by growing cells. As a precursor of S-adenosylmethionine (SAM) (40), it is involved in a number of synthetic processes such as transmethylations and polyamine biosynthesis (Chapter 7). Malaria parasites, for example, require exogenous methionine in amounts over and above those obtainable from the proteolysis of hemoglobin, the major nitrogen source of the intraerythrocytic stage. The ability to recycle methionine apparently evolved to avoid expensive synthesis and to deal with fluctuations in the dietary supply. Methionine is salvaged from methylthioadenosine (MTA) formed from decarboxylated SAM (Chapter 7). In mammals the conversion of MTA into 5-methylthioribose-1-phosphate is achieved in a single step catalyzed by MTA phosphorylase. This enzyme has been reported in *Trypanosoma brucei*. Other protozoa, including *G. lamblia*, *E. histolytica* and *Plasmodium falciparum* have no MTA phosphorylase. In its place are two enzymes, MTA nucleosidase and 5-methylthioribose (MTR) kinase, which catalyse the conversion of MTA into 5-methylthioribose-1-phosphate in two steps via MTR (Fig. 5.5). MTR kinase has been found only in bacteria and protozoa and its unique activity can be exploited for the design of potential drugs (41,42).

Nothing is known about methionine recycling in helminths, although cell-free extracts of *A. lumbricoides* can convert 2-hydroxy-4-methylthiobutyrate and 2-keto-4-

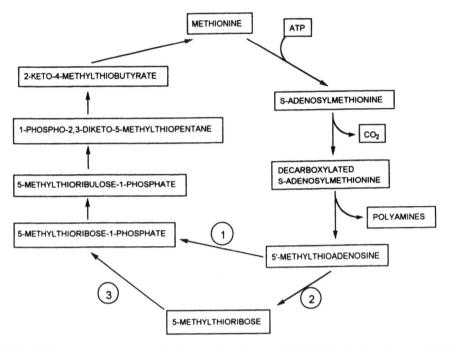

FIG. 5.5 Methionine salvage pathway via 5'-methylthioadenosine (MTA). 1, MTA phosphorylase; 2, MTA nucleosidase; 3, 5-methylthioribose kinase.

methylthiobutyrate to methionine. This may involve a 2-hydroxyacid dehydrogenase and a transaminase (1), the latter being part of the recycling pathway.

Cystathionine β-synthase from several organisms catalyses the synthesis of cystathionine from homocysteine and serine, the penultimate step in the cysteine biosynthetic pathway from methionine, and the reversible replacement of the β-SH group of cysteine with water to form serine and hydrogen sulfide. The enzyme of *Trichomonas vaginalis* and a similar enzyme found in nematodes display 'activated serine sulphydrase' activity in which the production of hydrogen sulfide from cysteine is stimulated by a second thiol substrate, a property which distinguishes them from their mammalian counterpart. Indeed all anaerobic protozoa and almost all nematodes contain this 'activated serine sulphydrase' activity, although it is absent from cestodes and trematodes (43). The protozoan and nematode enzymes are inhibited by a number of agents such as bithionol, dichlorophene and hexachlorophene which have antitrichomonal effects (44) and are anthelmintics (45). The physiological significance of the reaction is not known.

Giardia lamblia has a unique growth requirement for high concentrations of cysteine which cannot be made *de novo* (46). The parasite has a cysteine uptake system which, in a unique host–parasite interaction, is stimulated by human insulin-like growth factors, especially IGF-II (47).

A number of enzymes involved in cysteine metabolism have been described in the tapeworm *Hymenolepis diminuta* (48). In addition to a high cystathionine β-synthase activity it has low γ-cystathionase activity, and contains cysteine aminotransferase, cysteine dioxygenase and cysteine sulphinate aminotransferase; the latter two could be involved in the oxidation of cysteine.

3.5. Nitrogen Excretion

The catabolism of amino acids results in the release of nitrogen. Many parasites excrete the excess nitrogen in the form of ammonia. In protozoa there is no evidence for a complete urea cycle which would allow the conversion of ammonia into urea, and urea released by some species (49, 50) is likely to be formed solely as a product of arginine catabolism. Urea makes up between 2 and 10% of the total nitrogenous end-product of helminths, but evidence for a functional urea cycle in parasitic species has been equivocal (1). Arginase and ornithine carbamoyltransferase do occur widely but the other urea cycle enzymes do not. Small quantities of uric acid are excreted by some cestodes and trematodes.

Ammonia, urea and uric acid are not the only nitrogenous materials to be excreted. Parasites may release amino acids, such as proline and alanine (see above), polyamines (see Chapter 7) and other amines. Excretion/secretion of peptides and proteins also accounts for a significant proportion of released nitrogen, especially in helminths, although this clearly has other functional significance.

4. PROTEIN METABOLISM

4.1. Introduction

All parasites should have the ability to obtain amino acids from the digestion of protein. In many cases this may be restricted to processes involving endogenous protein which provide a mechanism for the removal of unwanted proteins and for the continued supply of amino acids for energy or the synthesis of new proteins during periods of starvation or transformation from one developmental form of the parasite to another. However, many parasites also utilize exogenous protein as an essential or supplementary source of amino acids. The proteolytic enzymes involved in digestion may be extracellular (in helminths they could function within the digestive tract or extracorporeally), yielding amino acids and peptides which then have to be absorbed, or intracellular within lysosomes to which proteins or protein-containing material are delivered by pinocytosis or phagocytosis. Most of the well-characterized proteolytic enzymes of helminths are extracellular, whereas the majority of the protozoan enzymes are intracellular although not all of these are lysosomal.

The function of parasite proteolytic enzymes is by no means restricted to the provision of amino acids. Many other key processes are dependent on proteolysis including stage-specific protein processing, the invasion of host tissues and penetration of cells (for details see Chapter 16), countering host defense mechanisms and developmental processes including egg hatching and moulting in helminths and excystment in protozoa. Some of these involve highly specific enzymes, whereas the proteolytic enzymes responsible for the breakdown of dietary protein are likely to have a broad specificity and degrade proteins to small peptides and amino acids.

Recent studies on proteolysis in parasites have tended to focus more on the enzymes than on the processes in which they participate (51). A great deal of information is available on the *in vitro* properties of individual enzymes, but the complexity of the proteolytic systems of some parasites, for example cysteine proteinases are often present in multiple forms, may make it difficult to establish precisely which enzyme(s) is responsible for which parasite function.

4.2. Proteolytic Enzymes

Proteolytic enzymes can exhibit exopeptidase (peptidase) activity, in which only peptide bonds at or adjacent to the amino or carboxyl terminus of a protein or peptide are cleaved, or endopeptidase (proteinase) activity in which hydrolysis occurs at internal peptide bonds. Some enzymes have both types of activity. Most studies of parasite enzymes have concerned endopeptidases (referred to here as proteinases). The majority of proteinases can be placed in one of four classes, aspartic, cysteine, metallo, or serine, according to their catalytic mechanism, which may distinguished from one another by the use of inhibitors (52). The assay methods used have tended to favour the detection of high activity proteinases which may have broad specificity, that is enzymes that are more likely to be involved in protein breakdown than in processing. Peptidases have been detected in a wide range of protozoa and helminths and many are likely to play a role in the digestion of peptide fragments generated by proteinases.

4.2.1. Proteolytic enzymes of protozoa

Proteinases have been analysed in most of the major species of parasitic protozoa (51, 53). Cysteine proteinases are the predominant type detected and have been found in most species, although they are not necessarily present at all stages of the life cycle (52). In trypanosomatids higher levels are usually associated with the stages in the human host (51), for example in *L. mexicana* most cysteine proteinases are only expressed in stationary-phase promastigotes and, more predominantly, in amastigotes (54). Stage-specific expression also occurs in *Plasmodium falciparum*, in which the cysteine proteinase gene is expressed only at the ring stage (55). Molecular cloning and sequencing have shown that the protozoal enzymes are homologous to cysteine proteinases of higher plants and animals (56). With two exceptions, the group D enzymes of *L. mexicana* (57), those cysteine proteinases for which sequence information is available are all more closely related structurally to mammalian cathepsins L and H than to cathepsin B.

In most protozoa the cysteine proteinases are intracellular and located in lysosomes. In *L. mexicana* highest expression correlates with the appearance of megasomes, large lysosome-like vacuoles found only in the amastigotes of this species: cysteine proteinases have proved more difficult to detect in other species of *Leishmania* which lack megasomes (54). A few species do release cysteine proteinases. In cultures of *T. vaginalis* and *Tritrichomonas foetus* over 50% of the total activity of axenic cultures is extracellular (58). There is evidence for the association of the major cysteine proteinase of *Trypanosoma cruzi*, cruzipain, with the surface of epimastigotes and infective amastigote-like forms and for its secretion through the flagellar pocket region (59). The cysteine proteinase of *E. histolytica* is considered by some workers to be secreted (60), although a more localized form of release upon parasite–host cell contact has been suggested by others (61). During infection of humans cysteine proteinases of *E. histolytica*, *Trichomonas vaginalis* and *Trypanosoma cruzi* can elicit antibody responses (56).

Less information is available on other classes of protozoal proteinase. Aspartic proteinases have been described only among sporozoa. Two genes encoding distinct aspartic proteinases have been cloned from *P. falciparum* (62, 63) and one from *Eimeria acervulina* (64). Metalloproteinases have been detected in a number of species but the only enzyme which has been well characterized is the promastigote surface proteinase (PSP or gp63) found in all species of *Leishmania* (65) (see Chapter 11). Serine proteinases have been detected in trypanosomatids, in which the enzymes appear to be cytosolic (66), *Plasmodium* species (67–69) and *E. tenella* (70).

4.2.2. Proteolytic enzymes of helminths

Proteinases of all four major classes have been identified in helminths. Although there are few examples of aspartic proteinases, a number of cysteine, metallo- and serine proteinases have been well characterized. Many of those studied are excretory/secretory products and have attracted attention not only because of their likely biological function but also because they are major antigens which induce an easily detectable immune response.

Molecular cloning and sequencing of genes encoding cysteine proteinases of the digenean *S. mansoni* (71) and the nematodes *Haemonchus contortus* (72) and *Ostertagia ostertagi* (73) have shown that they belong to the cathepsin B branch of the papain superfamily. However, cathepsin L-related cysteine proteinases have been cloned from the digeneans *Paragonimus westermani* (74) and *F. hepatica* (75). Indeed, the liver fluke has genes for both cathepsin L-like and cathepsin B-like cysteine proteinases (75). In these and other species the cysteine proteinases are mainly associated with the adult stage and are expressed in the gut. A gene encoding an antigenic protein homologous to the large subunit of calpains, calcium-dependent cysteine proteinases, has also been cloned from *S. mansoni* (76). Serine and metalloproteinases are frequently found in the excretory/secretory material of nematodes and digeneans. These enzymes are generally able to digest a range of proteins including some connective tissue proteins. The gene for the serine proteinase of *S. mansoni* (cercarial elastase) has been cloned (77), and the sequence of the enzyme reveals that it is related to the trypsin family, especially to rat pancreatic elastase I and II.

The stage-specific elaboration of helminth proteolytic enzymes is best exemplified in *S. mansoni* (78–80). The pre- and postacetabular glands of the cercariae produce an elastase (serine proteinase), other broad-specificity serine proteinases may have a role during the transition of the cercariae to schistosomula and the schistosomula have new metalloproteinases and peptidases. Schistosomula and adult worms produce a hemo-globinase, a cysteine proteinase, the expression of which in developing schistosomula has been correlated with digestive tract development. The egg also has associated proteolytic enzymes, including cysteine proteinases, and contains the highest levels of a membrane-bound leucine aminopeptidase which is present throughout the life cycle. Other proteinases may be present in the miracidia.

4.2.3. Proteinase inhibitors

In mammals polypeptides which inhibit proteinases probably have an important role in modulating proteolytic activity, and may play both a regulatory and protective role.

Similar proteinase inhibitors are produced by several nematodes and cestodes and these may interfere with processes dependent on host proteinases. For example, *Ascaris* spp. produce inhibitors of pepsin, trypsin, chymotrypsin and carboxypeptidase (7) which may provide protection against the digestive enzymes of the host. Recent studies suggest an involvement in host species specificity (81). A proteinase inhibitor taenia-statin is among peptides released by *Taenia* spp. which are capable of interfering with the host response including recruitment of effector cells and complement activation (82). *Echinococcus granulosus* releases a similar protein with some resemblance to α_1-anti-trypsin which inhibits elastase activity and neutrophil chemotaxis (83). A gene for a cysteine proteinase inhibitor, onchocystatin, has now been cloned from *Onchocerca volvulus* (84) as has a serpin (serine proteinase inhibitor) gene from *S. haematobium* (85).

Very little is known about proteinase inhibitors in protozoa, although some cystatin-like cysteine proteinase inhibitors have been detected (86). Whether any of these inhibitors modulate parasite-induced proteolysis is not yet known.

Some parasites can acquire proteinase inhibitors from their host (78, 87) which may provide protection from exogenous or endogenous proteinases but could also be important in masking antigenic extracellular parasite proteinases from the host immune system.

4.3. Proteolytic Processes and Nutrition

Because of the difficulties of growing most parasites in simple defined media which lack protein components such as serum, the extent to which parasites are dependent on exogenous protein for their amino acids is not well established. It is clear, however, that many parasites can utilize protein as a source of amino acids. Some protozoa, for example *Trypanosoma cruzi*, are able to grow on medium containing protein rather than amino acids as the only nitrogenous material (8). This parasite can store exogenous protein in special organelles called reservosomes for mobilization under conditions of nutritional stress (13). Uptake by pinocytosis and digestion of exogenous protein has also been demonstrated in other species (88, 89). The best example of protein utilization by a protozoan parasite is seen in *Plasmodium* trophozoites which avidly ingest and digest hemoglobin within infected erythrocytes. Between 25% and 75% of the hemoglobin in the infected erythrocyte is degraded within a period of a few hours. The effect of proteinase inhibitors on hemoglobin digestion in culture (68, 90) and in purified acidic digestive vacuoles (90, 91) indicates an ordered process within the vacuoles in which two aspartic proteinases and one cysteine proteinase are involved. The process is initiated by the cleavage of one specific peptide bond (α33Phe-34Leu) in the hemoglobin by the aspartic proteinase hemoglobinase I (90). The surface metallo-proteinase of *Leishmania* may function in the proteolytic digestion of hemoglobin in the midgut of the hematophagous sandfly vector (65).

Since cysteine proteinases play a major role in intracellular protein degradation in mammalian cells, it might be expected that proteolysis is important in protozoa with high intracellular levels of cysteine proteinases. In most cases, however, this is not yet known, nor have physiological substrate proteins been identified. Trichomonads have multiple cysteine proteinases of high activity (58), and the high capacity to hydrolyze proteins may be linked to their ability to use a range of amino acids for energy metabolism. Ammonia and amines derived from amino acids released by proteolysis in

Trichomonas vaginalis may help to neutralize the acidic pH within the vagina. The expression of cysteine proteinases in *L. mexicana* amastigotes has been linked to a role in antagonizing the microbicidal activity of the host cell, perhaps through amine production (54). Endogenous protein degradation has been investigated in *L. tropica* promastigotes (92), in which the rate of proteolysis increases during periods of nutrient deprivation. Vacuolar cysteine proteinases seem unlikely to be involved in this, however. Indeed, the role of cytosolic proteinases in protein utilization by protozoa has yet to be explored.

As in protozoa the best examples of protein utilization in helminths concern those parasites which feed on blood components. Many blood-feeding and blood-sucking helminths produce hemoglobin-digesting enzymes. The hemoglobinase of *S. mansoni*, which is responsible for the initial degradation of hemoglobin released from lysed erythrocytes, has been identified as a cysteine proteinase (71). In many other species the adult worms probably use cysteine proteinases to digest proteins in the digestive tract. It is noteworthy that cysteine proteinases are not found in the digestive tracts of higher animals, but this is not an adaptation to parasitism as the free-living *Caenorhabditis elegans* also has a gut cysteine proteinase (93). By degrading fibrinogen, proteinases can also act as anticoagulants and assist in the feeding process. The expression of a fibrinogen digesting cysteine proteinase of *H. contortus* correlates with blood feeding (72).

In cases where proteinases have been implicated in other functions in which extensive protein degradation is required, for example in tissue penetration by helminths in which metallo- and serine proteinases are involved (Chapter 16), it is not yet known whether the same enzymes might also play a role in feeding, although the possibility can not be excluded.

5. CONCLUDING REMARKS

In many respects the basic features of amino acid and protein metabolism of parasitic protozoa and helminths resemble those of their mammalian hosts. Proteins can be broken down extracellularly or within lysosomes, and amino acids taken up and used for biosynthesis or energy metabolism. The pathways of amino acid metabolism are mostly the same as those used by animal cells. There are, nevertheless, some significant differences. Although no unique class of proteinase has been found in either protozoa or helminths, differences in specificity between parasite and host enzymes mean that inhibitors can be designed which selectively inactivate parasite enzymes. Because proteolysis is so important to many parasites at various stages of their life cycle, proteinase inhibitors are being studied and tested as potential antiparasitic agents (79, 94, 95). With respect to amino acid metabolism some of the differences between host and parasite are somewhat subtle. Some parasite enzymes, for example, have properties which are clearly distinct from those of their mammalian counterparts, such as the cofactor dependence and regulation of glutamate dehydrogenases. The utilization of specific amino acids such as proline or the accumulation of others such as alanine reflects a difference in the relative importance of the pathways between parasite and host. Some differences are particularly striking, especially among anaerobic protozoa in which features of arginine and sulphur amino acid metabolism, for example, are more similar to those of certain bacteria. Thus in its details parasite amino acid and protein metabolism is sufficiently different from that of the host and would offer a number of potential targets for chemotherapy.

6. REFERENCES

1. Barrett, J. (1991) Amino acid metabolism in helminths. *Adv. Parasitol.* **30**: 39–105.
2. Gutteridge, W. E. and Coombs, G. H. (1977) *Biochemistry of Parasitic Protozoa.* Macmillan, London.
3. Von Brand, T. (1979) *Biology and Physiology of Endoparasites.* Elsevier/North Holland, Amsterdam.
4. Smart, D. (1989) What are the functions of the catecholamines and 5-hydroxytryptamine in parasitic nematodes? In: *Comparative Biochemistry of Parasitic Helminths* (eds Bennet, E. M., Behm, C. and Bryant, C.) Chapman and Hall, London, pp. 25–34.
5. Isaac, R. E., Eaves, L., Muimo, R. and Lamango, N. (1991) *N*-acetylation of biogenic amines in *Ascaridia galli. Parasitology* **102**: 445–450.
6. Aomine, M. (1981) The amino acid absorption and transport in protozoa. *Comp. Biochem. Physiol.* **68**A: 531–540.
7. Barrett. J. (1981) *Biochemistry of Parasitic Helminths.* Macmillan, London.
8. Cazzulo, J. J. (1984) Protein and amino acid catabolism in *Trypanosoma cruzi. Comp. Biochem. Biophys.* **79B**: 309–320.
9. Cazzulo, J. J. (1994) Intermediate metabolism in *Trypanosoma cruzi. J. Bioenerg. Biomembr.* **26**: 157–165.
10. Ling, I. T., Cooksley, S., Bates, P. A., Hempelmann, E. and Wilson, R. J. M. (1986) Antibodies to the glutamate dehydrogenase of *Plasmodium falciparum. Parasitology* **92**: 313–324.
11. Yee, J. and Dennis, P. P. (1992) Isolation and characterization of a NADP-dependent glutamate dehydrogenase gene from the primitive eucaryote *Giardia lamblia. J. Biol. Chem.* **267**: 7539–7544.
12. Turner, A. C. and Lushbaugh, W. B. (1988) *Trichomonas vaginalis:* characterization of its glutamate dehydrogenase. *Exp. Parasitol.* **67**: 47–53.
13. Urbina, J. A. (1994) Intermediary metabolism of *Trypanosoma cruzi. Parasitol. Today* **10**: 107–110.
14. Mottram, J. C. and Coombs G. H. (1985) *Leishmania mexicana:* enzyme activities of amastigotes and promastigotes and their inhibition by antimonials and arsenicals. *Exp. Parasitol.* **59**: 151–160.
15. Fry, M. (1991) Mitochondria of *Plasmodium.* In: *Biochemical Protozoology* (eds Coombs, G. H. and North, M. J.), Taylor and Francis, London, pp. 154–167.
16. Lowe, P. N. and Rowe, A. F. (1986) Aminotransferase activities in *Trichomonas vaginalis. Mol. Biochem. Parasitol.* **21**: 65–74.
17. Walker, J. and Barrett, J. (1991) Studies on alanine aminotransferase in nematodes. *Int. J. Parasitol.* **21**: 377–380.
18. Glaser, T. A. and Mukkada, A. J. (1992) Proline transport in *Leishmania donovani* amastigotes: dependence on pH gradients and membrane potential. *Mol. Biochem. Parasitol.* **51**: 1–8.
19. Zilberstein, D. (1991) Adaptation of *Leishmania* species to an acidic environment. In: *Biochemical Protozoology* (eds Coombs, G. H. and North, M. J.), Taylor and Francis, London, pp. 349–358.
20. ter Kuile, B. H. and Opperdoes, F. R. (1992) A chemostat study on proline uptake and metabolism of *Leishmania donovani. J. Protozool.* **39**: 555–558.
21. ter Kuile, B. H. (1993) Glucose and proline transport in kinetoplastids. *Parasitol. Today* **9**: 206–210.

22. Grantham, B. D. and Barrett J. (1986) Amino acid catabolism in the nematodes *Helig-mosomoides polygyrus* and *Panagrellus redivivus*. 2. Metabolism of the carbon skeleton. *Parasitology* **93**: 495–504.

23. Bera, T. (1987) The γ-guanidinobutyramide pathway of L-arginine catabolism in *Leishmania donovani* promastigotes. *Mol. Biochem. Parasitol.* **23**: 183–192.

24. Schofield, P. J., Edwards, M. R., Matthews, J. and Wilson, J. R. (1992) The pathway of arginine catabolism in *Giardia intestinalis*. *Mol. Biochem. Parasitol.* **51**: 29–36.

25. Edwards, M. R., Schofield, P. J., O'Sullivan, W. J. and Costello, M. (1992) Arginine metabolism during culture of *Giardia intestinalis*. *Mol. Biochem. Parasitol.* **53**: 97–104.

26. Elford, B. C. (1986) L-Glutamine influx in malaria-infected erythrocytes: a target for antimalarials. *Parasitol. Today* **2**: 309–312.

27. Elford, B. C., Pinches, R. A., Ellory, C. and Newbold, C. I. (1991) Substrate specificities for malaria-inducible red cell transport in relation to parasite metabolism. In: *Biochemical Protozoology* (eds Coombs, G. H. and North M. J.), Taylor and Francis, London, pp. 377–386.

28. Davies, K. P. and Köhler, P. (1990) The role of amino acids in the energy generating pathways of *Litomosoides carinii*. *Mol. Biochem. Parasitol.* **41**: 115–124.

29. Köhler, P. (1991) The pathways of energy generation in filarial parasites. *Parasitol. Today* **7**: 21–25.

30. MacKenzie, N. E., Van De Waa, E. A., Gooley, P. R. *et al.* (1989) Comparison of glycolysis and glutaminolysis in *Onchocerca volvulus* and *Brugia pahangi* by ^{13}C nuclear magnetic resonance spectroscopy. *Parasitology* **99**: 427–435.

31. Lockwood, B. C. and Coombs, G. H. (1991) Amino acid catabolism in anaerobic protists. In: *Biochemical Protozoology* (eds Coombs, G. H. and North, M. J.), Taylor and Francis, London, pp. 113–122.

32. Lockwood, B. C. and Coombs G. H. (1991) Purification and characterization of methionine γ-lyase from *Trichomonas vaginalis*. *Biochem. J.* **279**: 675–682.

33. Cross, G. A. M., Klein, R. A. and Linstead, D. J. (1975) Utilization of amino acids by *Trypanosoma brucei* in culture: L-threonine as a precursor for acetate. *Parasitology* **71**: 311–326.

34. Edwards, M. R., Gilroy, F. V., Jimenez, B. M. and O'Sullivan, W. J. (1989) Alanine is a major end product of metabolism by *Giardia lamblia:* a proton nuclear magnetic resonance study. *Mol. Biochem. Parasitol.* **37**: 19–26.

35. Frydman, B., de los Santos C., Cannata, J. J. B. and Cazzulo, J. J. (1990) Carbon-13 nuclear magnetic resonance analysis of [1-^{13}C]glucose metabolism in *Trypanosoma cruzi*. Evidence of the presence of two alanine pools and of two CO_2 fixation reactions. *Eur. J. Biochem.* **192**: 363–368.

36. Nygaard, T., Bennett, C. C., Grossman, G., Edwards, M. R. and Schofield, P. J. (1994) Efflux of alanine by *Giardia intestinalis*. *Mol. Biochem. Parasitol.* **64**: 145–152.

37. Darling, T. N., Balber, A. E. and Blum, J. J. (1989) Metabolic interactions between glucose, glycerol, alanine, and acetate in *Leishmania braziliensis panamensis*. *J. Protozool.* **36**: 217–225.

38. Paget, T. A., Raynor, M. H., Shipp, D. W. E. and Lloyd, D. (1990) *Giardia lamblia* produces alanine anaerobically but not in the presence of oxygen. *Mol. Biochem. Parasitol.* **42**: 63–68.

39. Blum, J. J. (1991) Intermediary metabolism of *Leishmania*. In: *Biochemical Protozoology* (eds Coombs, G. H. and North, M. J.), Taylor and Francis, London, pp. 123–133.

40. Lawrence, F. and Robert-Gero, M. (1991) S-Adenosylmethionine metabolism in parasitic protozoa. In: *Biochemical Protozoology* (eds Coombs, G. H. and North, M. J.), Taylor and Francis, London, pp. 436–449.

41. Riscoe, M. K., Ferro, A. J. and Fitchen, J. H. (1989) Methionine recycling as a target for antiprotozoal drug develoment. *Parasitol. Today* **5**: 330–333.

42. Riscoe, M. K., Tower, P. A., Peyton, D. H., Ferro, A. J. and Fitchen, J. H. (1991) Methionine recycling as a target for antiprotozoal drug development. In: *Biochemical Protozoology* (eds Coombs, G. H. and North, M. J.), Taylor and Francis, London, pp. 451–457.

43. Walker, J. and Barrett, J. (1991) Cystathionine β-synthase and γ-cystathionase in helminths. *Parasitol. Res.* **77**: 709–713.

44. Thong, K. W. and Coombs, G. H. (1987) *Trichomonas* species: homocysteine desulphurase and serine sulphydrase activities. *Exp. Parasitol.* **63**: 143–151.
45. Walker, J. and Barrett, J. (1992) Biochemical characterisation of the enzyme responsible for 'activated L-serine sulphydrase' activity in nematodes. *Exp. Parasitol.* **74**: 205–215.
46. Lujan, H. D. and Nash, T. E. (1994) The uptake and metabolism of cysteine by *Giardia lamblia* trophozoites. *J. Euk. Microbiol.* **41**: 169–175.
47. Lujan, H. D., Mowatt, M. R., Helman, L. J. and Nash, T. E. (1994) Insulin-like growth factors stimulate growth and L-cysteine uptake by the intestinal parasite *Giardia lamblia*. *J. Biol. Chem.* **269**: 13069–13072.
48. Gomez-Bautista, M. and Barrett, J. (1988) Cysteine metabolism in the cestode *Hymenolepsis diminuta*. *Parasitology* **97**: 149–159.
49. Coombs, G. H. and Sanderson, B. E. (1985) Amine production by *Leishmania mexicana*. *Ann. Trop. Med. Parasitol.* **79**: 409–415.
50. Yoshida, N. and Camargo, E. P. (1978) Ureotelism and ammoniotelism in trypanosomatids. *J. Bacteriol.* **136**: 1184–1186.
51. Coombs, G. H. and North, M. J. (1991) *Biochemical Protozoology*. Taylor and Francis, London.
52. North, M. J., Mottram, J. C. and Coombs, G. H. (1990) Cysteine proteinases of parasitic protozoa. *Parasitol. Today* **6**: 270–275.
53. McKerrow, J. H., Sun, E., Rosenthal, P. J. and Bouvier, J. (1993) *Annu. Rev. Microbiol.* **47**: 821–853.
54. Coombs, G. H., Robertson, C. D. and Mottram, J. C. (1991) Cysteine proteinases of leishmanias. In: *Biochemical Protozoology* (eds Coombs, G. H. and North, M. J.) Taylor and Francis, London, pp. 208–220.
55. Rosenthal, P. and Nelson, R. G. (1992) Isolation and characterization of a cysteine proteinase gene of *Plasmodium falciparum*. *Mol. Biochem. Parasitol.* **51**: 143–152.
56. North, M. J. (1992) Characteristics of cysteine proteinases of parasitic protozoa. *Biol. Chem. Hoppe-Seyler* **373**: 401–406.
57. Robertson, C. D. and Coombs, G. H. (1993) Cathepsin B-like cysteine proteases of *Leishmania mexicana*. *Mol. Biochem. Parasitol.* **62**: 271–280.
58. North, M. J. (1991) Proteinases of trichomonads and *Giardia*. In: *Biochemical Protozoology* (eds Coombs, G. H. and North, M. J.) Taylor and Francis, London, pp. 234–244.
59. Souto-Padrón, T., Campetella, O. E., Cazzulo, J. J. and de Souza, W. (1990) Cysteine proteinase in *Trypanosoma cruzi*: immunocytochemical localization and involvement in parasite–host cell interaction. *J. Cell Sci.* **96**: 485–490.
60. Keene, W. E., Hidalgo, M. E., Orozco, E. and McKerrow, J. H. (1990) *Entamoeba histolytica*: correlation of the cytopathic effect of virulent trophozoites with secretion of a cysteine proteinase. *Exp. Parasitol.* **71**: 199–206.
61. Schulte, W. and Scholze, H. (1989) Action of the major protease from *Entamoeba histolytica* on proteins of the extracellular matrix. *J. Protozool.* **36**: 538–543.
62. Francis, S. E., Gluzman, I. Y., Oksman, A. *et al.* (1994) Molecular characterization and inhibition of a *Plasmodium falciparum* aspartic hemoglobinase. *EMBO J.* **13**: 306–317.
63. Dame, J. B., Reddy, G. R., Yowell, C. A., Dunn, B. M., Kay, J. and Berry, C. (1994) Sequence, expression and modeled structure of an aspartic proteinase from the human malaria parasite *Plasmodium falciparum*. *Mol. Biochem. Parasitol.* **64**: 177–190.
64. Laurent, F., Bourdieu, C., Kaga, M. *et al.* (1993) Cloning and characterization of an *Eimeria acervulina* sporozoite gene homologous to aspartyl proteinases. *Mol. Biochem. Parasitol.* **62**: 303–312.
65. Etges, R. (1991) The promastigote surface proteinase of *Leishmania*. In: *Biochemical Protozoology* (eds Coombs, G. H. and North M. J.), Taylor and Francis, London, pp. 221–233.
66. Kornblatt, M. J., Mpimbaza, G. W. N. and Lonsdale-Eccles, J. D. (1992) Characterization of an endopeptidase of *Trypanosoma brucei brucei*. *Arch. Biochem. Biophys.* **293**: 25–31.
67. Barale, J.-C., Langsley, G., Mangel, W. F. and Braun-Breton, C. (1992) Malarial proteases — assignment of function to activity. *Res. Immunol.* **142**: 672–681.
68. Rosenthal, P. (1991) Proteinases of malaria parasites. In: *Biochemical Protozoology* (eds Coombs, G. H. and North, M. J.) Taylor and Francis, London, pp. 257–269.

69. Schrével, J., Deguercy, A., Mayer, R. and Monsigny, M. (1990) Proteases in malaria-infected red blood cells. *Blood Cells* **16**: 563–584.
70. Michalski, W. P., Crooks, J. K. and Prowse, S. J. (1994) Purification and characterisation of a serine-type protease from *Eimeria tenella* oocysts. *Int. J. Parasitol.* **24**: 189–195.
71. Klinkert, M.-Q., Felleisen, R., Link, G., Ruppel, A. and Beck, E. (1989) Primary structures of Sm31/32 diagnostic proteins of *Schistosoma mansoni* and their identification as proteases. *Mol. Biochem. Parasitol.* **33**: 113–122.
72. Pratt, D., Armes, L. G., Hageman, R., Reynolds, V., Boisvenue, R. J. and Cox, G. N. (1992) Cloning and sequence comparisons of four distinct cysteine proteases expressed by *Haemonchus contortus* adult worms. *Mol. Biochem. Parasitol.* **51**: 209–218.
73. Pratt, D., Boisvenue, R. J. and Cox, G. N. (1992) Isolation of putative cysteine protease genes of *Ostertagia osteragi*. *Mol. Biochem. Parasitol.* **56**: 39–48.
74. Yamamoto, M., Yamakami, K. and Hamajima, F. (1994) Cloning of a cDNA encoding a neutral thiol protease from *Paragonimus westermani* metacercariae. *Mol. Biochem. Parasitol.* **64**: 345–348.
74. Heussler, V. T. and Dobbelaere, D. A. E. (1994) Cloning of a protease gene family of *Fasciola hepatica* by the polymerase chain reaction. *Mol. Biochem. Parasitol.* **64**: 11–23.
76. Andresen, K., Tom, T. D. and Strand, M. (1991) Characterization of cDNA clones encoding a novel calcium-activated neutral proteinase from *Schistosoma mansoni*. *J. Biol. Chem.* **266**: 15085–15090.
77. Newport, G. R., McKerrow, J. H., Hedstrom, R. *et al.* (1988) Cloning of the proteinase that facilitates infection by schistosome parasites. *J. Biol. Chem.* **263**: 13179–13184.
78. Doenhoff, M. J., Curtis, R. H. C., Ngaiza, J. and Mohda, J. (1990) Proteases in the schistosome life cycle: a paradigm for tumour metastasis. *Cancer Metastasis Rev.* **9**: 381–392.
79. McKerrow, J. H. (1989) Parasite proteases. *Exp. Parasitol.* **68**: 111–115.
80. McKerrow, J. H. and Doenhoff, M. J. (1988) Schistosome proteases. *Parasitol. Today* **4**: 334–339.
81. Hawley, J. H., Martzen, M. R. and Peanasky, R. J. (1994) Proteinase inhibitors in *Ascardia*. *Parasitol. Today* **10**: 308–313.
82. Leid, R. W., Suquet, C. M., Grant, R. F., Tanigoshi, L., Blanchard, D. B. and Yilma, T. (1987) Taeniastatin, a cestode proteinase inhibitor with broad host regulatory activity. In: *Molecular Strategies of Parasitic Invasion* (eds Agabian, N., Goodman, H. and Nogueira, N.) Alan R. Liss, New York, pp. 653–663.
83. Shepherd, J. C., Aitken, A. and McManus, D. P. (1991) A protein secreted *in vivo* by *Echinococcus granulosus* inhibits elastase activity and neutrophil chemotaxis. *Mol. Biochem. Parasitol.* **44**: 81–90.
84. Lustigman, S., Brotman, B., Huima, T., Prince, A. M. and McKerrow, J. H. (1992) Molecular cloning and characterization of onchocystatin, a cysteine proteinase inhibitor of *Onchocerca volvulus*. *J. Biol. Chem.* **267**: 17339–17346.
85. Blanton, R. E., Licate, L. S. and Aman, R. A. (1994) Characterization of a native and recombinant *Schistosoma haematobium* serine protease inhibitor gene product. *Mol. Biochem. Parasitol.* **63**: 1–11.
86. Irvine, J. W., Coombs, G. H. and North, M. J. (1992) Cystatin-like cysteine proteinase inhibitors of parasitic protozoa. *FEMS Microbiol. Lett.* **96**: 67–72.
87. Peterson, K. M. and Alderete, J. F. (1983) Acquisition of α_1- antitrypsin by a pathogenic strain of *Trichomonas vaginalis*. *Infect. Immun.* **40**: 640–646.
88. Langreth, S. G. and Balber, A. E. (1975) Protein uptake and digestion in bloodstream and culture forms of *Trypanosoma brucei*. *J. Protozool.* **22**: 40–53.
89. Peterson, K. M. and Alderete, J. F. (1984) *Trichomonas vaginalis* is dependent on uptake and degradation of human low density lipoproteins. *J. Exp. Med.* **160**: 1261–1272.
90. Gluzman, I. Y., Francis, S. E., Oksman, A., Smith, C. E., Duffin, K. L. and Goldberg, D. E. (1994) Order and specificity of the *Plasmodium falciparum* hemoglobin degradation pathway. *J. Clin. Invest.* **93**: 1602–1608.
91. Goldberg, D. E., Slater, A. F. G., Beavis, R., Chait, B., Cerami, A. and Henderson, G. B. (1991) Hemoglobin degradation in the human malaria parasite *Plasmodium falciparum*: a catabolic pathway initiated by a specific aspartic protease. *J. Exp. Med.* **173**: 961–969.

92. Simon, M. W. and Mukkada, A. J. (1983) Intracellular protein degradation in *Leishmania tropica* promastigotes. *Mol. Biochem. Parasitol.* **7**: 19–26.
93. Ray. C. and McKerrow, J. H. (1992) Gut-specific and developmental expression of a *Caenorhabditis elegans* cysteine protease gene. *Mol. Biochem. Parasitol.* **51**: 239–250.
94. Cohen, F. E., Gregoret, L. M., Amiri, P., Aldape, K., Railey, J. and McKerrow, J. H. (1991) Arresting tissue invasion of a parasite by protease inhibitors chosen with the aid of computer modelling. *Biochemistry* **30**: 11221–11229.
95. North, M. J. (1991) Proteinases of parasitic protozoa: an overview. In: *Biochemical Protozoology* (eds Coombs, G. H. and North, M. J.), Taylor and Francis, London, pp. 180–190.

6 Purine and Pyrimidine Metabolism

RANDOLPH L. BERENS[1], EDWARD C. KRUG[1]
and J. JOSEPH MARR[2]

[1]Division of Infectious Diseases, University of Colorado Health Sciences
Center, Denver, CO, USA and [2]Ribozyme Pharmaceuticals, Inc.,
Boulder, CO, USA

SUMMARY

Purine and pyrimidine nucleotides are fundamental to life as they are involved in
nearly all biochemical processes. Purine and pyrimidine nucleotides are the
monomeric units of both DNA and RNA, ATP serves as the universal cellular
energy source, adenine nucleotides are components of three key coenzymes
(NAD^+, FAD and CoA), they are used to form activated intermediates, such as
UDP-glucose, and they serve as metabolic regulators.

Over the last fifteen years there has been a vast increase in the understanding
of the metabolism of purines and pyrimidines in parasitic protozoa and helminths.
The results of studies by numerous investigators have shown that purine metabolism in these pathogens differs significantly from that of their mammalian hosts.
Whereas mammalian cells can synthesize the purine ring *de novo* from simple
precursors, all the parasitic protozoa and helminths that have been investigated
lack *de novo* purine synthesis and must utilize preformed purines obtained from
their host environment. These pathogens use various purine salvage pathways to
supply their purine nucleotide requirements. These pathways can be relatively
simple as in the case of *Giardia lamblia* or complex as found in the leishmania.
These differences in purine metabolism between host and parasite is an area of
potential chemotherapeutic intervention that is actively being explored.

Unlike purines, pyrimidine metabolism, with the exception of the amitochondrial protists *Giardia lamblia*, *Trichomonas vaginalis* and *Tritrichomonas foetus*, is

Biochemistry and Molecular Biology of Parasites
ISBN 0-12-473345-X

similar to mammalian cells. With the above exceptions, all pathogenic protozoa and helminths appear to be able to synthesize pyrimidines *de novo*. Like mammalian cells, their ability to salvage pyrimidines is limited when compared to purine salvage.

1. PURINE METABOLISM

In mammals, purine ribonucleotides are synthesized *de novo* from amino acids, ribose, carbon dioxide and formate as well as from preformed purine bases and nucleosides through salvage pathways. The general route for *de novo* biosynthesis is the same in those species of mammals, birds, yeasts and bacteria that have been studied (1). Parasitic protozoans and helminths cannot synthesize purines *de novo* and thus rely solely on salvage pathways (2–12).

The salvage pathways utilized by these pathogens are similar to those used in mammalian cells (Fig. 6.1). In many cases the pathways differ in certain aspects, perhaps in response to the purine composition of the parasite's host environment. These routes permit cells to utilize exogenous purines derived from the degradation of nucleic acids or nucleotides. In mammalian cells and many parasitic organisms, purine bases

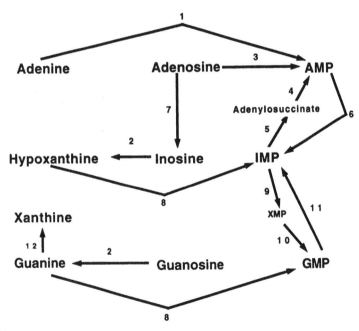

FIG. 6.1 Mammalian purine salvage and interconversion pathways. Enzymes listed in Figs. 6.1–6.11 are as follows: (1) adenine PRTase; (2) nucleoside phosphorylase; (3) purine (adenosine) nucleoside kinase; (4) adenylosuccinate lyase; (5) adenylosuccinate synthetase; (6) AMP deaminase; (7) adenosine deaminase; (8) hypoxanthine–guanine PRTase; (9) IMP dehydrogenase; (10) GMP synthetase; (11) GMP reductase; (12) guanine deaminase; (13) AMP kinase; (14) GMP kinase; (15) ribonucleotide reductase; (16) nucleoside diphosphokinase; (17) xanthine PRTase; (18) adenine deaminase; (19) purine nucleoside hydrolase; (20) HGXPRTase; (21) purine nucleoside phosphotransferase; (22) guanine PRTase; (23) purine deoxyribonucleoside kinase; (24) deoxyribonucleoside phosphotransferase.

are salvaged by reacting with phosphoribosyl-1-pyrophosphate (PRPP) to yield nuc-leoside-5'-monophosphates. These reversible reactions are catalyzed by individual phosphoribosyl transferases (PRTases). Since the pyrophosphate released in these reactions is hydrolyzed rapidly within the cell, the synthesis of purine mononucleotides proceeds irreversibly. Other salvage pathways exist that involve the conversion of purine bases into nucleosides and then to nucleotides. The former reactions are catalyzed by purine nucleoside phosphorylases, in which ribose-1-phosphate serves as the ribose donor and inorganic phosphate is released along with the purine nucleoside. Conversion of the nucleoside into the nucleotide occurs through phosphorylation at the 5' position on the ribose moiety by a purine nucleoside kinase or a nucleoside phosphotransferase. The former is usually relatively specific for a given purine nucleo-side whereas the latter is relatively non-specific with respect to both the donor and acceptor of the phosphate group. The only significant purine nucleoside kinase activity in mammalian cells is adenosine kinase. Purine nucleotides, which are released from RNA or DNA degradation, are catabolized to nucleosides and inorganic phosphate by non-specific phosphatases or nucleotidases. The nucleosides are acted upon by nucleo-side cleaving enzymes which can be either phosphorylases (require phosphate as a participant in the reaction to yield a purine and ribose-1-phosphate) or hydrolases (use water to produce the base and free ribose). The former are present in mammalian cells whereas the latter are not.

In addition to salvaging purines, most cells interconvert adenine and guanine nucleotides. Inosine monophosphate (IMP), is the common intermediate. IMP is converted into AMP by a two-step reaction catalyzed by adenylosuccinate synthetase and adenylosuccinate lyase. Guanine nucleotides are formed in a two-step reaction in which IMP is converted into xanthine monophosphate (XMP) and then aminated to GMP. Both GMP and AMP can be reconverted into IMP. Mammalian cells can also deaminate adenosine to inosine and guanine to xanthine (Fig. 6.1).

To maintain relatively constant internal purine nucleotide levels despite continual *de novo* synthesis and dietary intake, mammals catabolize and excrete excess purines as uric acid (man and higher primates) or allantoin (other mammals). Purine catabolism begins with conversion into either hypoxanthine or xanthine, both of which are then degraded to uric acid by xanthine oxidase. In most mammals, uric acid is further degraded to allantoin by urate oxidase. In parasitic protozoans and helminths there is no apparent catabolism of purines due to the lack of xanthine oxidase.

Purine deoxyribonucleotides are derived primarily from the respective ribonucleo-tide (Fig. 6.2). Intracellular concentrations of deoxyribonucleotides are very low compared to ribonucleotides; usually about 1% that of ribonucleotides. Synthesis of deoxyribonucleotides is by enzymatic reduction of ribonucleotide-diphosphates by ribonucleotide reductase. One enzyme catalyzes the conversion of both purine and pyrimidine ribonucleotides and is subject to a complex control mechanism in which an excess of one deoxyribonucleotide compound inhibits the reduction of other ribonuc-leotides. Whereas the levels of the other enzymes involved with purine and pyrimidine metabolism remain relatively constant through the cell cycle, ribonucleotide reductase level changes with the cell cycle. The concentration of ribonucleotide reductase is very low in the cell except during S-phase when DNA is synthesized. While enzymatic pathways, such as kinases, exist for the salvage of pre-existing deoxyribosyl compounds, nearly all cells depend on the reduction of ribonucleotides for their deoxyribonucleotide

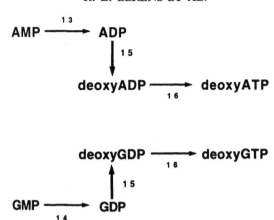

FIG. 6.2 Purine deoxynucleotide synthesis pathways. For identity of enzymes see legend to Fig. 6.1.

requirements. With the exception of *G. lamblia* and *T. vaginalis*, all the parasitic organisms discussed in this chapter possess ribonucleotide reductase and can synthesize purine deoxynucleotides based on one or more of the following criteria: organisms multiply in the absence of deoxynucleosides, reductase activity can be detected in cell extracts, bases or ribonucleosides are incorporated into DNA or organisms are sensitive to growth inhibition by hydroxyurea (a ribonucleotide reductase inhibitor).

1.1. Protozoa

1.1.1. Amitochondriates

1.1.1.1. Tritrichomonas foetus.   The salvage of purines by *T. foetus* (Fig. 6.3) is similar to that of leishmania (Fig. 6.7). Adenine, guanine, hypoxathine and xanthine are all incorporated into nucleotides. Adenine is the most efficiently incorporated followed by hypoxanthine, guanine, and xanthine. Of the nucleosides, inosine is incorporated at a rate similar to that of hypoxanthine; adenosine and guanosine are incorporated at a rate about half that of inosine. Hypoxanthine competes strongly with inosine for incorporation; guanine and adenine compete less effectively with their respective nucleosides. These results suggest that inosine salvage occurs primarily by conversion into hypoxanthine, whereas some adenosine and guanosine are salvaged by direct conversion into nucleotides (2). Salvage of hypoxanthine, inosine and adenine all give similar patterns of incorporation into nucleotides. There is interconversion between adenine and guanine nucleotides.

Phosphoribosyltransferase activity was found for hypoxanthine, guanine and xanthine but not for adenine (2). Adenine and guanine deaminase activities are present. Phosphorylase activities were found for adenosine, guanosine and inosine. Also present were adenosine kinase and a guanosine phosphotransferase; neither inosine kinase nor phosphotransferase activity was present. The IMP dehydrogenase differs from the mammalian enzyme in that it does not require K^+ for activity and it is more sensitive to inhibition by mycophenolic acid (13).

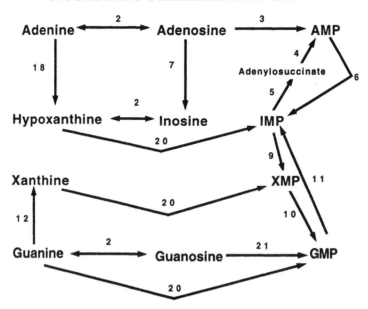

FIG. 6.3 *Tritrichomonas foetus* purine salvage and interconversion pathways. For identity of enzymes, see legend to Fig. 6.1.

The HGXPRTase activity appears to be the major route of purine salvage. Adenine is salvaged by deamination to hypoxanthine whereas adenosine can be either phosphorylated directly or deaminated. Guanosine is directly phosphorylated to GMP or cleaved to guanine.

1.1.1.2. Giardia, Trichomonas *and* Entamoeba. These parasitic protozoans differ from the other protozoans discussed in this chapter in that they are all incapable of interconversion between their guanine and adenine nucleotide pools. They are dependent on their host environment to supply them with both guanine and adenine. With the exception of *E. histolytica*, these parasites lack ribonucleotide reductase. This requires that the host also supply purine and pyrimidine deoxynucleosides.

G. lamblia cannot salvage hypoxanthine, inosine, or xanthine (3) (Fig. 6.4). Only adenine and adenosine are used for adenine nucleotide synthesis; only guanine and guanosine are used for guanine nucleotide synthesis. This implies that *G. lamblia* lacks XPRTase and HPRTase and that IMP is not a salvage intermediate. Radioisotope incorporation studies indicated that adenosine and guanosine are first converted into their respective bases and then phosphoribosylated. Substantial levels of PRTase activities were found for adenine and guanine but not for xanthine or hypoxanthine. These PRTases are highly specific (3, 14). Although no evidence has been found for phosphorylation of salvaged purine ribonucleosides, nucleoside analogs resistant to cleavage (7-deaza-adenosine, 9-deaza-adenosine and 9-deazainosine) are phosphorylated by both whole cells and crude cell-free extracts using phosphotransferase conditions (15). Guanine arabinoside is also phosphorylated by intact cells but not by cell lysates (16).

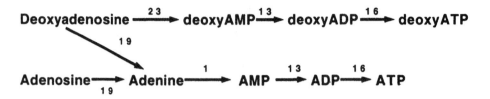

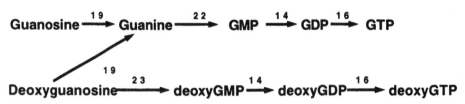

FIG. 6.4 *Giardia lamblia* purine salvage and interconversion pathways. For identity of enzymes see legend to Fig. 6.1.

There is no detectable ribonucleotide reductase in *G. lamblia*. Purine ribosides are incorporated into RNA but not DNA; purine deoxyribosides are readily incorporated into both RNA and DNA. A purine deoxynucleoside kinase activity was found in crude cell-free extracts (17).

T. vaginalis (4) and *E. histolytica* (5,18) lack the purine PRTase activities found in *G. lamblia* and salvage adenine and guanine by first converting them into their ribonucleosides which are in turn phosphorylated to their ribonucleotide mono-phosphates (Figs 6.5 and 6.6). Assays for the various salvage enzyme activities in *E. histolytica* and *T. vaginalis* have been performed and the results agree with the metabolic studies (4, 19–21).

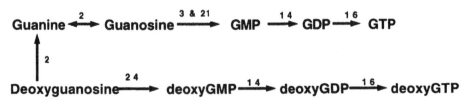

FIG. 6.5 *Trichomonas vaginalis* purine salvage and interconversion pathways. For identity of enzymes see legend to Fig. 6.1.

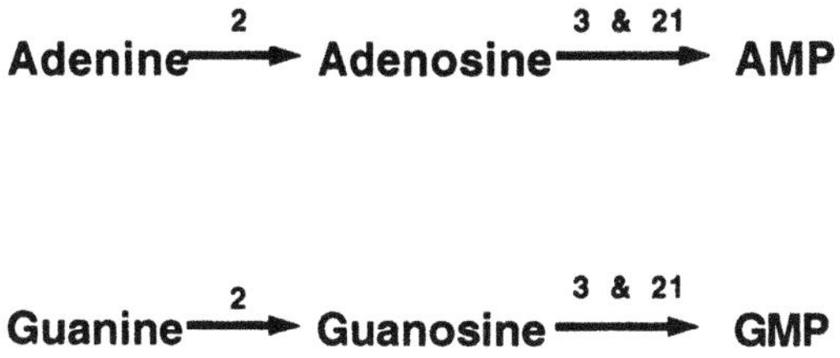

Adenine $\xrightarrow{\text{2}}$ Adenosine $\xrightarrow{\text{3 & 21}}$ AMP

Guanine $\xrightarrow{\text{2}}$ Guanosine $\xrightarrow{\text{3 & 21}}$ GMP

FIG. 6.6 *Entamoeba histolytica* salvage and interconversion pathways. For identity of enzymes see legend to Fig. 6.1.

Like *Giardia, T. vaginalis* lacks ribonucleotide reductase and is dependent on its host environment for deoxyribonucleotides (22). There is a deoxyriboside phosphotransferase which will phosphorylate the deoxyribosides of adenine and guanine but not their ribosides (Fig. 6.5). In this respect, *T. vaginalis* differs from *Giardia* where it was found that a nucleoside kinase phosphorylates these deoxyribosides (17). Whereas *T. vaginalis* lacks ribonucleotide reductase activity, *E. histolytica* can reduce ribonucleotides to 2′-deoxyribonucleotides based on growth inhibition by hydroxyurea (20).

1.1.2. Kinetoplastida

Differences in the metabolic sequences for purine metabolism in the various morphological forms of the pathogenic kinetoplastids appears to be only quantitative and not qualitative with the exception of *Leishmania*. *Leishmania* amastigotes and promastigotes differ with respect to particular enzymes which metabolize adenine and adenosine. Fig. 6.7 shows the major salvage and interconversion pathways of the Kinetoplastida.

1.1.2.1. Leishmania. *Leishmania* promastigotes convert adenine and inosine into a common intermediate, hypoxanthine, before being used for nucleotide synthesis (6). Adenine is converted into hypoxanthine by adenine deaminase (23). This enzyme is absent in *T. cruzi* (7), *T. b. gambiense* or *T. b. rhodesiense* (8) but has been found in other parasitic protozoans as described below.

The distribution of radiolabel is similar in *L. donovani* and *L. braziliensis* promastigotes irrespective of whether adenine, hypoxanthine, or inosine is used since hypoxanthine is the common precursor (6, 24). An efficient salvage pathway for guanine and xanthine exists, via the hypoxanthine–guanine PRTase (HGPRTase) and a xanthine XPRTase. Most of the guanine incorporated into the purine pool enters after deamination to xanthine by guanine deaminase. The ability to salvage xanthine is a feature not found in mammalian cells and represents a significant modification in purine salvage. Some adenine is salvaged directly by an adenine PRTase (APRTase) whereas the majority is deaminated to hypoxanthine. Recovery of radiolabel from both guanine and xanthine in adenine nucleotides indicates the presence of GMP reductase.

In vivo purine flow measured in *L. donovani* promastigotes confirms the active role of the adenine and guanine deaminases which lead to the production of hypoxanthine and xanthine, respectively (24). These two bases are metabolized to both adenine and guanine nucleotides. For this reason, although three different PRTases exist, the major route of adenine metabolism in promastigotes is through adenine deaminase to

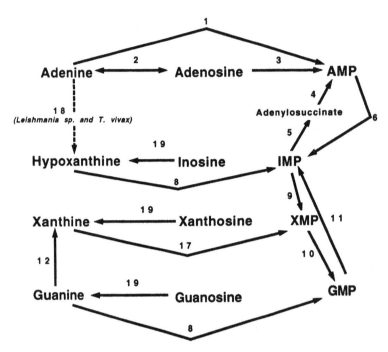

FIG. 6.7   Kinetoplastida purine salvage and interconversion pathways. For identity of enzymes see legend to Fig. 6.1.

hypoxanthine and for guanine through guanine deaminase to xanthine. Both hypoxanthine and xanthine are then activated by the respective PRTases.

Distinct 3'- and 5'-nucleotidases are present on the *L. donovani* surface membrane (25). These enzymes, combined with a very active surface-associated phosphatase (26), enable these parasites to use extracellular nucleotides and nucleic acids.

Less is known about purine metabolism in *Leishmania* amastigotes. Purine metabolism in the amastigotes is similar to that of promastigotes except for adenine and adenosine metabolism (27). Adenine is activated to AMP by an APRTase; there is no adenine deaminase. This differs significantly from the promastigote where virtually all the adenine is converted into hypoxanthine before incorporation into nucleotides. Adenosine is a precursor of hypoxanthine in both *L. donovani* promastigotes and amastigotes although the pathways differ. In amastigotes, adenosine is deaminated to inosine by adenosine deaminase. In promastigotes, adenosine is cleaved to adenine and then converted into hypoxanthine by adenine deaminase. Adenosine also is phosphorylated directly to AMP by adenosine kinase in both forms, but the specific activity is higher in amastigotes. The existence of two metabolic pathways for adenosine in amastigotes is supported by *in vivo* incorporation studies (27, 28). Purine salvage in *L. mexicana* differs from *L. donovani* (29). Adenine deaminase activity is present in both forms and no stage-specific differences in adenosine metabolism are found. Nucleoside kinase is found in both forms for guanosine, inosine, and xanthosine as well as adenosine. *L. donovani* lacks the ability to phosphorylate the former three nucleosides. The metabolism of guanine, xanthine, and their respective nucleosides is similar to that described for the promastigote of both *L. donovani* and *L. mexicana*.

There are three distinct PRTases in *L. donovani* (30). One has its major activity with hypoxanthine and guanine; a second with adenine; and a third with xanthine. Pyrazolo (3,4-d)pyrimidines, such as allopurinol, are efficient substrates for the HGPRtase. Promastigotes accumulate large quantities of allopurinol ribonucleoside-5′-phosphate when exposed to allopurinol (31). A separate PRTase for xanthine is unusual in a eukaryotic cell. XPRTase is also present in *L. mexicana* and *L. amazonensis* as well as in four non-pathogenic trypanosomatids (32, 33).

Leishmania adenylosuccinate synthetase has a narrow substrate specificity but accepts several IMP analogs which include allopurinol ribonucleotide (34). The GMP reductase from *L. donovani* is quite different from the human GMP reductase (35) and IMP analogs are more potent inhibitors for it. Other leishmanial enzymes that have been investigated include IMP dehydrogenase (36), nucleoside hydrolase and phosphorylase activities (37, 38), adenosine kinase (39), nucleotidases (40) and the adenylosuccinate lyase (34).

1.1.2.2. Trypanosoma cruzi.   T. cruzi epimastigotes incorporate both purine bases and ribonucleosides; adenine and guanine nucleotides are readily interconvertible (7, 41). With the exception of adenine, *T. cruzi* salvages and interconverts purines in a manner identical to *Leishmania* (Fig. 6.7). *T. cruzi* epimastigotes, unlike *Leishmania* promastigotes, do not possess an adenine deaminase but, like *Leishmania*, do possess high levels of guanine deaminase activity. The culture forms prefer purine bases to their respective ribonucleosides (7, 41). Purine ribonucleosides are remarkably stable in this parasite (7) when compared to *Leishmania* (24) and the African trypanosomes (8, 42) which have relatively high nucleoside cleaving activities.

T. cruzi amastigotes prefer purine bases to ribonucleosides; there is minimal salvage of ribonucleotides; and adenine and guanine nucleotide pools are readily interconvertible (41).

Individual enzymes of purine salvage are similar to those of *Leishmania*. PRTase activities were found for adenine, hypoxanthine, and guanine in the three forms (43). As in *Leishmania*, there is also a separate xanthine PRTase. Nucleoside kinase activities were found for adenosine, inosine, and guanosine (43), nucleoside hydrolase activities for inosine and guanosine and a nucleoside phosphorylase activity for adenosine. There are both nucleoside hydrolase and phosphorylase activities in epimastigotes (44, 45). The adenylosuccinate synthetase and adenylosuccinate lyase are essentially identical to those found in *L. donovani* (46).

It has been reported that *T. cruzi* epimastigotes do not possess either adenine deaminase or adenosine deaminase (7, 47). However, Gutteridge and Davies (43) reported that the latter was present in all three forms of this parasite. Whether these conflicting results are due to strain differences or an artifact resulting from carry-over from serum components of the growth media is unclear. *T. cruzi* epimastigotes do possess high levels of guanine deaminase which is sensitive to inhibition by the plant growth regulator 6-methylaminopurine (7, 47).

1.1.2.3. Trypanosoma brucei *complex.*   Unlike *Leishmania* and *T. cruzi*, the mammalian (bloodstream) forms of the *T. brucei* complex are much easier to obtain. Thus more is known on how purine metabolism in the insect (procyclic) forms compares to that of mammalian bloodstream forms.

Although procyclic forms of *T. brucei gambiense* are incapable of *de novo* synthesis, the last two enzymes of the pathway are present since the intermediate 5-amino-imidazolo-4-carboxamide ribosylphosphate (AICAR) will maintain viability for over 6 weeks (8). Purine base incorporation is qualitatively identical to that found for *T. cruzi*; both trypanosomes lack the adenine deaminase which is found in *Leishmania* (Fig. 6.7). Purine ribonucleosides were utilized slightly more efficiently than their corresponding purine bases.

The bloodstream forms of *T.b. gambiense* and *T.b. rhodesiense* rely on purine salvage (42, 48). *T.b. gambiense* and *T.b. rhodesiense* isolated from infected mice utilize purine bases in a manner similar to that described for the procyclic forms of *T.b. gambiense* (48). PRTase activities are present for all four bases. Ribonucleosides are salvaged by phosphorylation via a nucleoside kinase, phosphotransferase activity, or by ribonucleoside cleavage and subsequent activation by a PRTase.

Few detailed studies have been done on the purine salvage enzymes of procyclic African trypanosomes. *T.b. gambiense* has high levels of guanine deaminase and lacks adenine and adenosine deaminase activities (8). *T.b. brucei, T.b. gambiense* and *T.b. rhodesiense* convert allopurinol into aminopyrazolopyrimidine nucleotides and incorporates these into RNA (49). This indicates that HPRTase, succino-AMP synthetase, and succino-AMP lyase are present. At least three nucleoside cleavage activities are present (Berens, unpublished results): two are hydrolases, of which one is specific for purine ribonucleosides and the other is specific for purine deoxyribonucleosides. The third nucleoside cleavage activity is a methylthioadenosine/adenosine phosphorylase. The adenosine kinase is similar to that of *L. donovani* (Berens, unpublished results).

T.b. gambiense bloodstream forms have APRTase, HGPRTase, adenosine kinase and adenylosuccinate synthetase but lack adenosine deaminase. Two phosphorylase activities have been described for the bloodstream forms of *T.b. brucei* (42, 50). One catalyzes the reversible phosphorolysis of adenosine, inosine and guanosine and the other is specific for adenosine and methylthioadenosine. Guanine deaminase is present whereas both adenosine and adenine deaminase are absent (8). Similar results have been reported for *T. congolense* (51). *T. vivax* is unique among the other trypanosomes in that it has an adenine deaminase (51).

Thus with the exception of *T. vivax*, the African trypanosomes metabolize purines in a manner similar to *T. cruzi*. The major difference between the trypanosomes and *Leishmania* is the former's lack of adenine deaminase.

1.1.3. Apicomplexa

1.1.3.1. Plasmodium species.   Much of the information on purine metabolism in malaria parasites has been obtained by comparing the salvage abilities of uninfected erythrocytes to infected erythrocytes. Although current emphasis is on the human malarial parasite, *P. falciparum*, much of the information on purine metabolism was obtained using intraerythrocytic forms of rodent (*P. berghei*) (52) and avian (*P. lophurae*) (53) parasites. Since no major differences seem to exist between these species and *P. falciparum*, the latter will be used as the representative species.

Like its host cell, the intraerythrocytic forms of the malaria parasite are incapable of *de novo* purine synthesis (9) and are dependent on preformed purines (Fig. 6.8). Infection of the red blood cell (RBC) with *Plasmodium* sp. results in a greatly enhanced

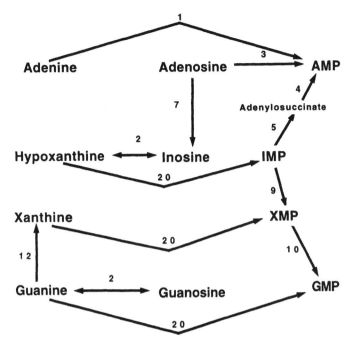

FIG. 6.8 Malaria purine salvage and interconversion pathways. For identity of enzymes, see legend to Fig. 6.1.

ability to salvage purines. When the specific activities of the salvage enzymes of free parasites were compared to those of uninfected human RBCs, it was found that *P. falciparum* has high activities of hypoxanthine, guanine and xanthine PRTase, inosine phosphorylase and adenosine deaminase (54). Relative to uninfected RBCs, the parasite's activities are 63-, 20-, 4800-, 4- and 800-fold higher, respectively. Adenine PRTase, adenosine kinase and guanosine phosphorylase are essentially the same as found for the host cell. *P. falciparum* appears to be capable of synthesizing both adenine and guanine nucleotides from hypoxanthine, inosine or adenosine. It is not clear if this parasite has the ability to interconvert between adenine and guanine nucleotides. However, hypoxanthine, adenine, guanine, inosine or adenosine can supply the purine requirements of *P. falciparum* growing in a purine-defined minimal medium suggesting the ability to interconvert between GMP and AMP (55). Hypoxanthine, as in *Leishmania*, is the preferred substrate for purine biosynthesis (54). It can be salvaged directly from the environment or by conversion of adenosine and inosine. The formation of hypoxanthine could occur either in the parasite or in the cytosol of the infected RBC as the corresponding enzymes are present in both. The finding that the growth of *P. falciparum* is inhibited when xanthine oxidase is used to deplete the supply of hypoxanthine in infected erythrocytes confirms the critical role of hypoxanthine (56). These findings are in agreement with previous studies with rodent (52) and avian malaria (53).

The adenosine deaminase, purine nucleoside phosphorylase, HGXPRTase and APRTase from *P. falciparum* have been characterized. The APRTase exhibits several properties that differentiate it from the human erythrocytic enzyme (57). The parasite

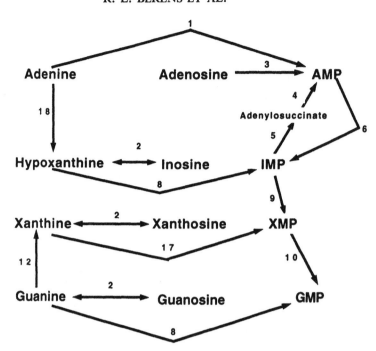

FIG. 6.9 *Toxoplasma gondii* purine salvage and interconversion pathways. For identity of enzymes see legend to Fig. 6.1.

enzyme is smaller (18 kDa vs 34 kDa), it is not inhibited by sulfhydryl reagents, it is inhibited by 6-mercaptopurine and 2,6-diaminopurine, and the K_m values for adenine and PRPP are significantly lower. The HGXPRTase differs from the RBC enzyme primarily in that its substrate K_m values are at least tenfold lower and that xanthine is an efficient substrate (58). However, the amino acid sequence of the *P. falciparum* enzyme shows extensive homology to the mammalian one (59). HGXPRTase is not found in the kinetoplastids. *Plasmodium* adenosine deaminase (ADA) also differs from that of the host RBC (60). Whereas the K_m for adenosine is about the same for both enzymes, the *Plasmodium* enzyme is not inhibited by erythro-9-(2-hydroxy-3-nonyl)adenine (EHNA), a potent inhibitor of human ADA.

1.1.3.2. Toxoplasma gondii. The salvage and interconversion capabilities of free *T. gondii* tachyzoites are summarized in Fig. 6.9. Of all the purines, AMP is used the most efficiently although both ADP and ATP are incorporated (10, 61, 62). These nucleotides are dephosphorylated to adenosine before being salvaged. In tachyzoites there are high levels of a nucleoside triphosphate hydrolase with a wide substrate specificity for both ribo- and deoxyribonucleotides (63). This is consistent with the hypothesis that host cell ATP is used as the dominant purine source by this parasite (10, 61). When ATP is dephosphorylated to adenosine and taken up, the predominant pathway for nucleotide synthesis appears to be direct phosphorylation by adenosine kinase. In addition, the organisms are capable of salvaging hypoxanthine, adenine, guanine, xanthine and their respective nucleosides. As in Kinetoplastida and *Plasmodium*, the ability to salvage xanthine is a major difference in purine salvage from that of mammalian host cells. Like

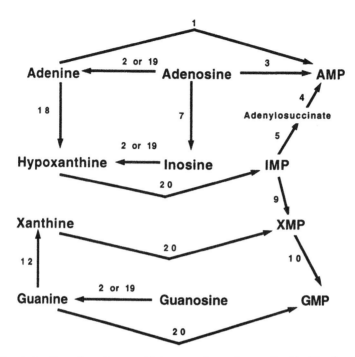

FIG. 6.10 *Eimeria tenella* purine salvage and interconversion pathways. For identity of enzymes see legend to Fig. 6.1.

Leishmania, Toxoplasma has adenine deaminase activity (62), supporting the suggestion that hypoxanthine salvage plays a central role in purine metabolism (61). Consistent with this is the finding that hypoxanthine incorporation is second only to that of adenosine. The majority of adenine, as in *Leishmania*, is salvaged by deamination to hypoxanthine.

Interconversion between adenine and guanine nucleotides occurs in *Toxoplasma* only in the direction of adenine to guanine (62). This suggests the absence of GMP reductase. That *T. gondii* can convert from adenine into guanine nucleotides but not the reverse places the parasite between trypanosomes and *Leishmania* which readily interconvert between their adenine and guanine nucleotides pools and organisms such as *Giardia* and *Trichomonas* which show no interconversion between the pools.

In the host cell adenine nucleotides appear to be the preferred purine source. These are dephosphorylated to cross the parasite membrane and then rephosphorylated by the parasite adenosine kinase.

1.1.3.3. Eimeria tenella.   The salvage of purines in *E. tenella* (Fig. 6.10) is similar to that found in *Toxoplasma* (11, 64). Both sporozoites and merozoites incorporate radioactive adenine, hypoxanthine and their respective ribonucleosides into adenine and guanine nucleotides (64). Guanine and xanthine are incorporated only into guanine nucleotides. Thus, *E. tenella*, like toxoplasma, lack GMP reductase activity and cannot satisfy purine requirements by salvage of guanine or xanthine alone. Adenine, adenosine, hypoxanthine, or inosine can meet the needs of this parasite. Purine bases are salvaged principally by phosphoribosylation (11, 64).

Purine nucleosides, with the exception of adenosine, are salvaged by converting them into the base followed by phosphoribosylation. Adenosine is phosphorylated directly by adenosine kinase or deaminated by adenosine deaminase to inosine. Adenine and adenosine deaminase are present in the sporozoite and merozoite forms (64); the former is not in extracts from unsporulated oocysts (11) but the latter has apparently not been looked for. The ability to deaminate both adenine and adenosine allows this parasite to synthesize guanine nucleotides in the absence of AMP deaminase. The ratio of labeled adenine nucleotides to guanine nucleotides is about 20% higher when both adenine and adenosine are the precursors compared to the ratio obtained when hypoxanthine or inosine was used (64). This indicates that although the major route of salvage for adenine and adenosine is by conversion into hypoxanthine, there is some direct conversion of these compounds into AMP.

Two salvage enzymes from *Eimeria* have been characterized. HGXPRTase is present in crude extracts of oocysts of *E. tinella* (11). Adenosine kinase is found in sporulated oocysts of three avian species of *Eimeria* (65).

1.2. Helminths

Purine metabolism of parasitic helminths has been studied in one trematode, two nematodes and two cestodes and in less detail in several non-parasitic helminths.

The trematode *Schistosoma mansoni* has been most thoroughly characterized of the helminths (Figure 6.11). Neither the adult worm nor schistosomule has *de novo* purine biosynthesis (12, 66). As with protozoan parasites, xanthine oxidase is absent.

Schistosomules of *S. mansoni* incorporate all purine bases and nucleosides with the exception of xanthine; xanthosine was not tested (12). Adenine was incorporated at the

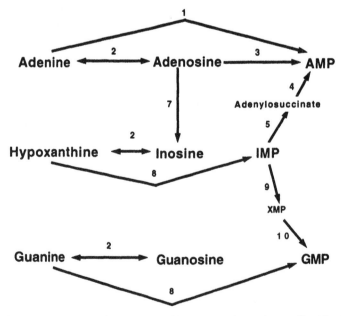

FIG. 6.11 *Schistosoma mansoni* purine salvage and interconversion pathways. For identity of enzymes see legend to Fig. 6.1.

highest rate, followed by hypoxanthine, then guanine. Nucleosides were salvaged in the same order. Interconversion between the adenine and guanine nucleotide pools is minimal. AMP deaminase and adenine deaminase were not detected. Thus, adenine nucleotides will not compensate for a deficiency in guanine nucleotides and vice versa. Based on these metabolic results it has been proposed that purine nucleoside kinase activities are functionally absent and all the purine nucleosides are rapidly converted into the corresponding bases before incorporation by the respective PRTases.

There is a relatively high activity of adenosine phosphorylase and low levels of adenosine kinase, deaminase and phosphotransferase (66). The adenosine kinase is responsible for incorporation of adenosine analogs such as tubercidin (67).

There are major quantitative differences in adenosine salvage in the adult form as compared to the schistosomules (66, 67). Adenosine kinase activity has been reported and is responsible for about one third of the adenosine salvaged. Adenosine can also be deribosylated to adenine by an adenosine phosphorylase. Unlike schistosomules, adenosine salvage in the adult worm involves more than one pathway. It can be salvaged through deamination first to inosine or via phosphorylase or a kinase to AMP. Guanine deaminase is present at a level which is 70% of that of the guanine PRTase activity; there is no evidence of xanthine PRTase activity. Interconversion between the adenine and guanine nucleotide pools has not been investigated in the adult worm.

The schistosomule HGPRTase has been purified and characterized (68). It is twice as active with guanine as with hypoxanthine and cannot use xanthine. The gene for this protein has been cloned and expressed at a high level in *Escherchia coli* (69).

In the tetrathyridia form of *Mesocestoides corti*, one of the two cestodes studied, *de novo* synthesis of purines was not found (70). *Hymenolepis diminuta* will salvage and incorporate [^{14}C]adenosine into a 70% ethanol-insoluble fraction (71).

Ascaris lumbricoides has adenosine and guanosine kinase activities, and is capable of breaking down AMP, adenine, xanthine, uric acid and allantoin to ammonia and urea, but is unable to catabolize guanine (72). The nematode *Ascaridia galli* has been reported to have AMP deaminase (72). The purine-catabolizing enzymes are largely unaddressed in parasitic nematodes but adenosine, adenine and guanine deaminases, xanthine oxidase and urate oxidase have been found in the free-living nematode *Panagrellus redivivus* (72).

Little work has been done on purine metabolism in other parasitic helminths. The dog heart worm, *Dirofilaria immitis* (73) and *Brugia pahangi*, a filarial worm of cats and primates, lack *de novo* purine biosynthesis (74). Thus the parasitic helminths is a group where purine metabolism is largely unexplored.

2. PYRIMIDINES

Most parasites, with the exceptions of *T. foetus* (75), *T. vaginalis* (76, 77), and *G. lamblia* (78, 79) are capable of synthesizing the pyrimidine ring *de novo*. The pyrimidine ring is built in five steps to give orotate which is then phosphoribosylated to give orotate monophosphate (OMP) (Fig. 6.12).

In mammalian cells *de novo* synthesis of UMP involves two multifunctional enzymes which are cytoplasmic and a single enzyme which is associated with mitochondria

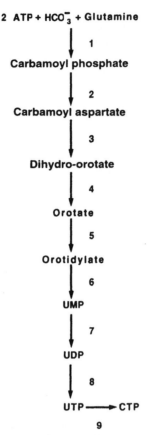

FIG. 6.12 Pyrimidine *de novo* synthesis pathway. Enzymes are as follows: (1) carbamoyl-phosphate synthetase II; (2) asparate carbamoyl-transferase; (3) dihydro-orotase; (4) dihydro-orotate oxidase; (5) orotate phosphoribosyltransferase; (6) orotidine-5'-phosphate decarboxylase; (7) nucleoside monophosphate kinase; (8) nucleotide diphospho kinase; (9) CTP synthetase.

(1, 80). The first three reactions are catalyzed by a trifunctional protein which contains carbamoyl-phosphate synthetase II, aspartate carbamoyltransferase and dihydro-orotase. This set of reactions begins with the synthesis of carbamoyl phosphate followed by its condensation with aspartic acid. The third step involves the closure of the ring through the removal of water by the action of dihydro-orotase to yield dihydro-orotate. The fourth enzyme, dihydro-orotate oxidase, oxidizes dihydro-orotate to orotate and is a mitochondrial flavoprotein enzyme located on the outer surface of the inner membrane and utilizes NAD^+ as the electron acceptor. The synthesis of UMP from orotate is catalyzed by a bifunctional protein which comprises orotate PRTase and orotidine 5'-phosphate (OMP) decarboxylase. The former phosphoribosylates orotate to give OMP; the latter decarboxylates OMP to UMP, the immediate precursor for the other pyrimidine nucleotides. It is interesting to note that whereas five molecules of ATP (including the ATP used in the synthesis of PRPP) are used in the *de novo* synthesis of IMP, no net ATP is used in the *de novo* synthesis of UMP. In *de novo* pyrimidine synthesis, two ATP molecules are used to synthesize carbamoyl phosphate and one ATP is needed to synthesize the PRPP used by orotate PRTase but 3 ATPs

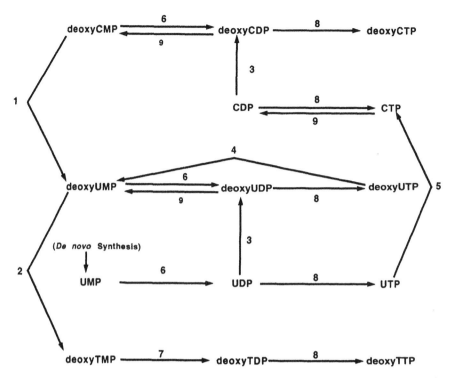

FIG. 6.13 Mammalian pyrimidine salvage and interconversion pathways. Enzymes listed in Figs 6.13–6.17 are as follows: (1) deoxyCMP deaminase; (2) thymidylate synthase; (3) ribonucleotide reductase; (4) deoxyuridine triphosphatase; (5) CTP synthetase; (6) nucleotide kinase; (7) deoxyTMP kinase; (8) nucleotide diphosphokinase; (9) non-specific phosphatase or nucleotidase; (10) cytidine kinase; (11) pyrimidine phosphorylase or hydrolase; (12) uracil PRTase; (13) cytidine deaminase; (14) thymidine kinase; (15) cytidine phosphotransferase; (16) uridine phosphotransferase; (17) thymidine phosphotransferase; (18) deoxyribonucleotide phosphotransferase; (19) cytosine PRTase.

can be synthesized from the NADH produced by dihydro-ororate oxidase reaction by oxidative phosphorylation. The need to invest energy for the *de novo* synthesis of purines may have resulted in more selective pressure to lose this pathway compared to the *de novo* pyrimidine pathway. In support of this contention is the finding that the amitochondriates *G. lamblia, T. vaginalis* and *T. foetus* do not synthesize pyrimidines *de novo.*

The synthesis of deoxyuridine, cytidine, deoxycytidine and thymidine nucleotides from UMP (Fig. 6.13) involves three reactions: CTP synthetase, ribonucleotide reductase, and thymidylate synthase (80). The first enzyme converts UTP into CTP and the second catalyzes the conversion of CDP, UDP, ADP and GDP into their respective deoxyribonucleotides. The last enzyme, thymidylate synthase, catalyzes the reductive methylation of deoxyUMP at the C-5 position giving deoxyTMP. The human enzyme has been extensively studied as it is a target enzyme in cancer chemotherapy. Besides these three enzymes, two other enzymes are involved in pyrimidine nucleotide synthesis and interconversion. DeoxyCMP deaminase converts deoxyCMP into deoxyUMP and deoxyUTP triphosphatase converts deoxyUTP into deoxyUMP. *Giardia lamblia,* and *Trichomonas vaginalis* lack both ribonucleotide reductase and thymidylate synthase and must meet their deoxypyrimidine requirements by salvage. *Tritrichomonas foetus* has a ribonucleotide reductase but lacks thymidylate synthase; thymidine nucleotides are synthesized from salvaged thymidine. The remaining parasites discussed in this chapter, like mammalian cells, are able to synthesize deoxypyrimidine nucleotides from UMP.

The ability of mammalian cells to salvage preformed pyrimidines is limited compared to their ability to salvage purines. In general, pyrimidine bases are poor substrates for nucleotide synthesis and under physiological conditions appear to be degraded rather than salvaged (1, 80). There is, however, some salvage of pyrimidine nucleosides by pyrimidine nucleoside kinase activities. Uridine and cytidine are phosphorylated by uridine kinase. This enzyme has a wide substrate range as it will phosphorylate a number of pyrimidine nucleoside analogs such as 5-fluorouridine and 5-fluorocytidine. Thymidine kinase can also phosphorylate deoxyuridine and its 5-halogenated analogs. Deoxycytidine kinase converts not only deoxycytidine but also deoxyguanosine and deoxyadenosine, into their respective deoxynucleotides. Pyrimidine catabolism involves cleavage of nucleosides to bases which are either excreted or broken down via ring cleavage to NH_4^+, CO_2 and β-amino acids (1).

2.1. Protozoa

2.1.1. Amitochondriates

2.1.1.1. Tritrichomonas foetus. *T. foetus* lacks *de novo* pyrimidine synthesis and must salvage exogenous pyrimidines (75) (Fig. 6.14). Uracil is incorporated at the highest rate. Uridine and cytidine are incorporated at a rate 10% that of uracil; thymidine is incorporated at 1% that of uracil; cytosine and thymine are not incorporated. Thymidine is incorporated only into the parasite's DNA; uracil and uridine are incorporated only into RNA.

Consistent with the metabolic data, there is no dihydrofolate reductase/thymidylate synthase activity (75). Thymidine is salvaged by a phosphotransferase. Uracil PRTase, uridine phosphorylase, cytidine deaminase and uridine and cytidine phosphotransferases were found. The major pyrimidine salvaged is uracil via its PRTase. The lack of incorporation of salvaged uracil into DNA and the lack of thymidylate synthase indicates that this parasite cannot synthesize TMP from UMP. It is dependent on salvage for its thymidine requirements. This parasite possesses a hydroxyurea-resistant ribonucleotide reductase and can synthesize deoxycytidine nucleotides from cytidine nucleotides.

2.1.1.2. Trichomonas vaginalis. *T. vaginalis* does not incorporate radiolabel from ^{14}C-labeled aspartate, orotate or bicarbonate into its pyrimidines, suggesting the absence of *de novo* synthesis (76, 77). Although there are conflicting reports as to the presence or absence of the specific enzymes of *de novo* synthesis (19, 77, 81), *T. vaginalis* appears to be incapable of *de novo* pyrimidine synthesis.

The salvage pathways of *T. vaginalis* differ from those of *T. foetus*. Both parasites can salvage uracil, uridine, cytidine, and thymidine but their preferences for these four pyrimidines differ significantly (Figs 6.14 and 6.15). Uracil is the most actively salvaged pyrimidine by *T. foetus*, whereas *T. vaginalis* preferentially salvages cytidine (77). Radiolabel from thymidine is not transferred into other pyrimidine nucleotides.

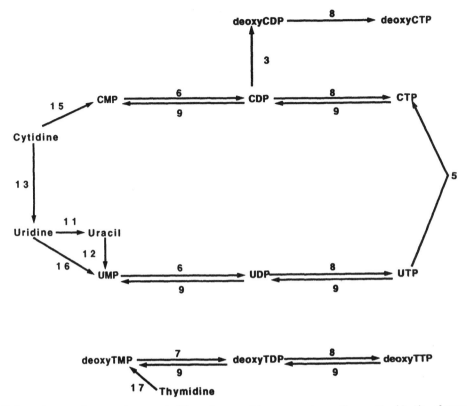

FIG. 6.14 *Tritrichomonas foetus* pyrimidine salvage and interconversion pathways. For identity of enzymes see legend to Fig. 6.13.

Radiolabeled cytidine is recovered in both cytidine and uridine nucleotides but not from DNA. Uracil and uridine are used predominantly for uridine nucleotide synthesis.

The salvage activities of *T. foetus* and *T. vaginalis* also differ (22, 77). Unlike *T. foetus*, the level of uracil PRTase activity is very low. Uracil is converted into uridine by a uridine phosphorylase; uridine is then phosphorylated by a uridine phosphotransferase to UMP (Fig. 6.15). Cytidine and thymidine also are converted into their nucleotide monophosphates by phosphotransferase activities. There is no detectable pyrimidine nucleoside kinase activity and the only significant interconversion among salvaged pyrimidines is catalyzed by cytidine deaminase to form uridine.

This parasite also differs from *T. foetus* in that it lacks nucleotide reductase activity. Pyrimidine deoxynucleotides are obtained by salvage of deoxynucleosides by the deoxyribonucleoside phosphotransferase which acts not only on pyrimidine but also purine deoxyribonucleosides (22).

2.1.1.3. Giardia lamblia. *Giardia lamblia* lacks *de novo* pyrimidine biosynthesis but has extensive interconversion among the salvage pathways (78, 79). As *Giardia* does not have a ribonucleotide reductase, it must salvage deoxycytidine as well as thymidine for DNA synthesis. The major route for thymidine salvage appears to be by a phospho-

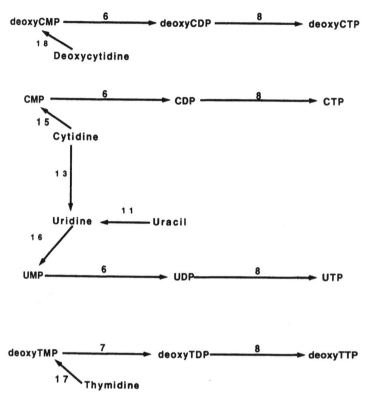

FIG. 6.15 *Trichomonas vaginalis* pyrimidine salvage and interconversion pathways. For identity of enzymes see legend to Fig. 6.13.

transferase (79), which contrasts with the deoxypurine salvage which is by kinase activity (17). The mode of deoxycytidine salvage has not been addressed (Fig. 6.16).

2.1.1.4. Entamoeba histolytica. There are several lines of evidence that *E. histolytica* is capable of pyrimidine *de novo* biosynthesis. It will incorporate orotate into nucleic acids (82), it has aspartate transcarbamoylase (83), and it will grow in medium that has been depleted of pyrimidines (84). It apparently can also meet its pyrimidine requirements by salvage (85). Its salvage pathways are similar to those of *T. vaginalis*, with the exception of the ability to salvage thymine (Fig. 6.15). Ribonucleotide reductase and thymidylate synthase activities are apparently present, since this parasite will grow in a pyrimidine deficient medium (84).

2.1.2. Kinetoplastida

Pyrimidine metabolism in the Kinetoplastida is similar to that of mammalian cells although there are some differences in the cellular location of enzyme activities.

2.1.2.1. Leishmania. The presence of *de novo* pyrimidine synthesis in *Leishmania* was first suggested by the finding that preformed pyrimidines were not needed for promas-

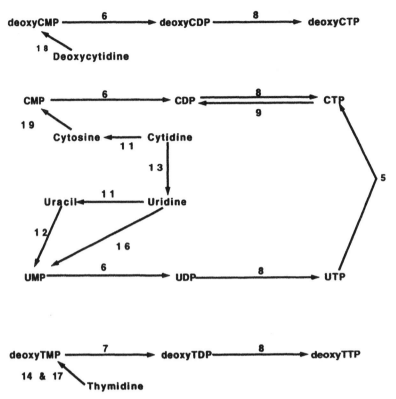

FIG. 6.16 *Giardia lamblia* pyrimidine salvage and interconversion pathways. For identity of enzymes see legend to Fig. 6.13.

tigote growth in defined growth media (86). All the enzymes needed for the synthesis of UMP have been found in *L. major* (81), *L. mexicana amazonensis* (87), *L.m. mexicana* (88) and *L. enriettii* (89). The last two enzymes of *de novo* synthesis, orotate PRTase and OMP decarboxylase, are not free in the cytoplasm but are located on the external surface of glycosomes (87,90). In addition, the parasite dihydroorotate oxidase is not associated with the mitochondria as in mammalian cells.

Unlike mammalian cells, the thymidylate synthase of *Leishmania* is a bifunctional protein which combines the synthase with dihydrofolate reductase (91).

Pyrimidine salvage and interconversion in *Leishmania* is less active than purine salvage (Fig. 6.17). Orotate is converted into OMP in *L. donovani* promastigotes and is used for the synthesis of CTP and UTP (24). Thymine salvage appears to be absent in *Leishmania* (81); thymidine is salvaged directly by a kinase. Uridine cleavage and uracil PRTase activities have been reported in *L.m. amazonensis* (87); uridine kinase was not found. Unlike mammalian cells, the *Leishmania* uracil PRTase is distinct from the orotate PRTase; the former is cytoplasmic and the latter is glycosomal (90). Cleavage activities for pyrimidine nucleosides, cytidine deaminase, and orotate PRTase were detected in four strains of *Leishmania* (41).

Little is known about pyrimidine metabolism in *Leishmania* amastigotes. Pyrimidine nucleoside cleavage and orotate PRTase activities have been found in *L.m. mexicana*

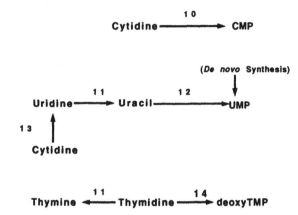

FIG. 6.17 Kinetoplastida pyrimidine salvage and interconversion pathways. For identity of enzymes see legend to Fig. 6.13.

amastigotes (41). The dihydro-orotate oxidase of this parasite is essentially identical to that of the promastigote (88).

2.1.2.2. T. cruzi. Pyrimidine metabolism in *T. cruzi* is very similar to that of *Leishmania* (Fig. 6.17). The presence of *de novo* pyrimidine synthesis enzymes in epimastigotes is well established (41, 81, 87, 92). The thymidylate synthase of *T. cruzi* is a bifunctional protein which combines the synthase with dihydrofolate reductase as in *Leishmania* (91). Pyrimidine interconversion and salvage by *T. cruzi* epimastigotes is similar to that in *Leishmania* (87, 90, 93).

Pyrimidine metabolism of trypomastigotes and amastigotes is identical to that of the epimastigote with the exception of uracil salvage (92). Although both trypomastigotes and epimastigotes have uracil PRTase activity, this activity may be absent in amastigotes (92).

2.1.2.3. Trypanosoma brucei *complex.* African trypanosomes of the *Trypanosoma brucei* complex metabolize pyrimidines in a manner similar to that of *Leishmania* and *T. cruzi* (Fig. 6.17). All six enzyme activities for the synthesis of UMP were detected in homogenates of blood trypomastigotes of *T.b. brucei* (87). In addition, uracil PRTase, cytidine deaminase, and pyrimidine cleavage activities have been detected in cell-free homogenates; no uridine kinase activity was detected (94).

In summary, pyrimidine metabolism in the kinetoplastids is functionally similar to that found in mammalian cells. In the *de novo* synthesis of UMP, these parasites differ in two respects: dihydro-orotate oxidase is cytoplasmic and not mitochondrial; and the last two enzymes of UMP synthesis are glycosomal instead of cytoplasmic. Pyrimidine salvage by these parasites is more diverse than that of mammalian cells but, unlike the purines, *de novo* synthesis plays the major role.

2.1.3. Apicomplexa

2.1.3.1. Plasmodium *species.* The mature erythrocyte does not have a requirement for pyrimidine nucleotides but it may contain low levels of *de novo* and salvage enzymes.

Plasmodium infection introduces the enzymes needed for *de novo* synthesis of UMP, CMP and TMP (54, 81, 95–98).

The enzymes are functionally similar to their mammalian counterparts. The dihydro-orotate oxidase, unlike that of the kinetoplastids, is associated with the parasite mitochondrion (99). However, the last two enzymes, the cytoplasmic orotate PRTase and OMP decarboxylase, are separate enzymes instead of being arranged in a bifunctional complex as is found in mammalian cells (99, 100). Metabolic studies indicate that CTP synthetase must be present.

The ability of *P. falciparum* and *Plasmodium* species to salvage preformed pyrimidines is minimal (54). The only salvage activity detectable is uracil PRTase but this may be the result of the broad substrate specificity of the parasite orotate PRTase (99).

Metabolic incorporation experiments indicate the presence of nucleoside reductase and thymidylate synthase activities. There is a bifunctional thymidylate synthase–dihydrofolate reductase similar to that in the kinetoplastids (91).

2.1.3.2. Toxoplasma gondii. *Toxoplasma gondii* synthesizes pyrimidines *de novo* (81, 101, 102). Like malaria, toxoplasma is limited in its ability to salvage preformed pyrimidines. Pfefferkorn and coworkers found that a uracil PRTase-deficient mutant is unable to access the pyrimidine pool of the host cell (102). The mutant grew normally, indicating that pyrimidine salvage was not essential and *de novo* synthesis must be present. It was also noted that wild-type cells can not salvage cytosine, cytidine or deoxycytidine. This parasite, like malaria, funnels all pyrimidine salvage through uracil (102).

2.1.3.3. Eimeria tenella.   It appears that the sporulated oocysts of *E. tenella* are capable of *de novo* pyrimidine synthesis since sporozoites incorporate ^{14}C-labeled aspartate and orotate into pyrimidine nucleotides (103). Besides *de novo* synthesis, *E. tenella* is also capable of pyrimidine salvage. *E. tenella* will incorporate uracil, cytidine and uridine but not thymidine into its nucleic acids (104). This parasite is dependent on UMP for the synthesis of TMP and thymidylate synthase activity is present in extracts of *E. tenella* (105). As with both *Plasmodium* and the kinetoplastids, the thymidylate synthase of *E. tenella* exists as a bifunctional protein (91).

2.2. Helminths

All helminths that have been studied can synthesize pyrimidines *de novo*.

All of the enzyme activities involved in *de novo* UMP biosynthesis have been identified in the trematodes *Schistosoma mansoni* (81, 106, 107) and *S. japonicum* (108). Carbamoyl phosphate synthetase II, aspartate transcarbamoylase and dihydro-orotase have been partially purified from the cytosol and appear to exist, as in mammalian cells, as a multienzyme protein (81, 106). Dihydro-orotate oxidase is membrane bound (81), but its electron acceptor has not been identified. Orotate PRTase and OMP decarboxylase are also cytosolic. In mammalian cells these last two enzymes exist as part of a multienzyme protein that efficiently 'channels' orotate to UMP since little free OMP is present in the cell. In *S. mansoni* they also exist as a multienzyme protein but apparently the 'channeling' is less efficient since levels of free OMP are significantly higher than those of UMP (107).

S. mansoni is capable of salvaging preformed pyrimidines, but interconversion is limited. Adults can salvage cytidine, uridine, thymidine and deoxycytidine, but of the pyrimidine bases only orotate and uracil are salvaged (66, 109).

The presence of various nucleoside phosphorylase, phosphotransferase and nucleoside kinase activities has been described (109, 110). Phosphoribosyltransferase activity was found only for uracil and orotate.

Ribonucleotide reductase is present in *S. mansoni* as indicated both by the incorporation of labeled uracil into deoxycytidine or thymidine (109) and the reduction of DNA synthesis by hydroxyurea treatment (111).

Evidence for pyrimidine *de novo* biosynthesis has also been found in other trematodes. Its enzymes are present in *Fasciola gigantica* (81), *Paragonimus ohirai* (112) and *Clonorchis sinensis* (112). *De novo* biosynthesis was verified by demonstrating the incorporation of [^{14}C]bicarbonate into the C-2 position of uracil in *P. ohirai* (112). The pyrimidine salvage enzymes, uridine kinase and thymidine kinase, are found in *P. ohirai* and *C. sinensis* extracts (112). Aspartate transcarbamoylase is found in *Fasciola hepatica* and *Paramphistomum cervi* extracts (113).

There is good evidence for *de novo* pyrimidine synthesis in cestodes. Five of the six enzymes needed for UMP synthesis are present in *Hymenolepis diminuta* and aspartate transcarbamoylase activity has been found in *Moniezia benedeni* (81). Salvage of preformed pyrimidines by a cestode was first reported in *Mesocestoides corti* (70). Thymidine kinase is the only cestode (*H. diminuta*) pyrimidine salvage enzyme that has been characterized (114).

In the nematodes the five soluble enzymes of pyrimidine *de novo* biosynthesis were found in crude extracts of *Nippostrongylus brasiliensis* and *Trichuris muris* (81). The first enzyme of *de novo* synthesis, glutamine-dependent carbamoyl-phosphate synthetase, was detected in *Ascaris* ovary extracts (115), and aspartate transcarbamoylase activity, the second enzyme of *de novo* synthesis, was found in *A. suum* (113).

Nematodes also appear to salvage preformed pyrimidines. The microfilariae of *D. immitis* incorporates uracil and uridine into RNA (73). Thymidylate synthase activity is also present (116, 117). Similar results were found with *Pangrellus silusiae* (118), *Caenorhabditis briggsae* (119) and *Brugia pahangi* (116).

In summary, all the helminths appear to synthesize pyrimidines *de novo* but the ability to salvage varies with the organism.

3. ACKNOWLEDGEMENTS

The authors thank Ms Betty Richardson for help with the preparation of the manuscript and Dr Lucy Ghoda for her suggestions and critique.

4. REFERENCES

1. Henderson, J. F. and Paterson A. R. P. (1973) *Nucleotide Metabolism, An Introduction*, Academic Press, New York.
2. Wang, C. C., Verham, R., Rice, A. and Tzeng, S. (1983) Purine salvage by *Tritrichomonas foetus. Mol. Biochem. Parasitol.* **8**: 325–337.

3. Wang, C. C. and Aldritt, S. (1983) Purine salvage networks in *Giardia lamblia*. *J. Exp. Med.* **158**: 1703–1712.
4. Heyworth, P. G., Gutteridge, W. E. and Ginger, C. D. (1982). Purine metabolism in *Trichomonas vaginalis*. *FEBS Lett.* **141**: 106–110.
5. Boonlayangoor, P., Albach, R. A. and Booden, T. (1980) Purine nucleotide synthesis in *Entamoeba histolytica*: a preliminary study. *Arch. Invest. Med.* **11** (Mex, 1 Suppl.): 83–88.
6. Marr, J. J., Berens, R. L. and Nelson, D. J. (1978) Purine metabolism in *Leishmania donovani* and *Leishmania braziliensis*. *Biochim. Biophys. Acta* **544**: 360–371.
7. Berens, R. L., Marr, J. J., LaFon, S. W. and Nelson, D. J. (1981) Purine metabolism in *Trypanosoma cruzi*. *Mol. Biochem. Parasitol.* **3**: 187–196.
8. Fish, W. R., Marr, J. J. and Berens, R. L. (1982) Purine metabolism in *Trypanosoma brucei gambiense*. *Biochim. Biophys. Acta* **714**: 422–428.
9. Gero, A. M. and O'Sullivan, W. J. (1990) Purines and pyrimidines in malarial parasites. *Blood Cells* **16**: 467–484; discussion 485–498.
10. Pfefferkorn, E. R. and Pfefferkorn, L. C. (1977) *Toxoplasma gondii:* specific labeling of nucleic acids of intracellular parasites in Lesch–Nyhan cells. *Exp. Parasitol.* **41**: 95–104.
11. Wang, C. C. and Simashkevich, P. M. (1981) Purine metabolism in the protozoan parasite *Eimeria tenella*. *Proc. Natl. Acad. Sci. USA* **78**: 6618–6622.
12. Dovey, H. F., McKerrow, J. H. and Wang, C. C. (1984) Purine salvage in *Schistosoma mansoni schistosomules*. *Mol. Biochem. Parasitol.* **11**: 157–167.
13. Verham, R., Meek, T. D., Hedstrom, L. and Wang, C. C. (1987) Purification, characterization, and kinetic analysis of inosine 5'-monophosphate dehydrogenase of *Tritrichomonas foetus*. *Mol. Biochem. Parasitol.* **24**: 1–12.
14. Aldritt, S. M. and Wang, C. C. (1986) Purification and characterization of guanine phosphoribosyltransferase from *Giardia lamblia*. *J. Biol. Chem.* **261**: 8528–8533.
15. Berens, R. L. and Marr, J. J. (1986) Adenosine analog metabolism in *Giardia lamblia*. Implications for chemotherapy. *Biochem. Pharmacol.* **35**: 4191–4197.
16. Miller, R. L., Nelson, D. J., LaFon, S. W., Miller, W. H. and Krenitsky, T. A. (1987) Antigiardial activity of guanine arabinoside. Mechanism studies. *Biochem. Pharmacol.* **36**: 2519–2525.
17. Baum, K. F., Berens, R. L., Marr, J. J., Harrington, J. A. and Spector, T. (1989) Purine deoxynucleoside salvage in *Giardia lamblia*. *J. Biol. Chem.* **254**: 21087–21090.
18. Lo, H. S. and Wang, C. C. (1985) Purine salvage in *Entamoeba histolytica*. *J. Parasitol.* **71**: 662–669.
19. Miller, R. L. and Lindstead, D. (1983) Purine and pyrimidine metabolizing activities in *Trichomonas vaginalis* extracts. *Mol. Biochem. Parasitol.* **7**: 41–51.
20. Austin, C. J. and Warren, L. G. (1983) Induced division synchrony in *Entamoeba histolytica*. Effects of hydroxyurea and serum deprivation. *Am. J. Trop. Med. Hyg.* **32**: 507–511.
21. Lobelle-Rich, P. A. and Reeves, R. E. (1983) The partial purification and characterization of adenosine kinase from *Entamoeba histolytica*. *Am. J. Trop. Med. Hyg.* **32**: 976–979.
22. Wang, C. C. and Cheng, H. W. (1984) The deoxyribonucleoside phosphotransferase of *Trichomonas vaginalis*. A potential target for anti-trichomonal chemotherapy. *J. Exp. Med.* **160**: 987–1000.
23. Kidder, G. W., Dewey, V. C. and Nolan, L. L. (1977) Adenine deaminase of a eukaryotic animal cell, *Crithidia fasciculata*. *Arch. Biochem. Biophys.* **183**: 7–12.
24. LaFon, S. W., Nelson, D. J., Berens, R. L. and Marr, J. J. (1982) Purine and pyrimidine salvage pathways in *Leishmania donovani*. *Biochem. Pharmacol.* **31**: 231–238.
25. Dwyer, D. M. and Gottlieb, M. (1984) Surface membrane localization of 3'- and 5'-nucleotidase activities in *Leishmania donovani* promastigotes. *Mol. Biochem. Parasitol.* **10**: 139–150.
26. Gottlieb, M. and Dwyer, D. M. (1981) *Leishmania donovani*: surface membrane acid phosphatase activity of promastigotes. *Exp. Parasitol.* **52**: 117–128.
27. Looker, D. L., Berens, R. L. and Marr, J. J. (1983) Purine metabolism in *Leishmania donovani* amastigotes and promastigotes. *Mol. Biochem. Parasitol.* **9**: 15–28.
28. Konigk, E. and Putfarken, B. (1980) Stage-specific differences of a perhaps signal-transferring system in *Leishmania donovani*. *Tropenmed. Parasitol.* **31**: 421–424.

29. Hassan, H. F. and Coombs, G. H. (1985) *Leishmania mexicana:* purine-metabolizing enzymes of amastigotes and promastigotes. *Exp. Parasitol.* **59**: 139–150.
30. Allen, T., Henschel, E. V., Coons, T., Cross, L., Conley, J. and Ullman, B. (1989) Purification and characterization of the adenine phosphoribosyltransferase and hypoxanthine–guanine phosphoribosyltransferase activities from *Leishmania donovani. Mol. Biochem. Parasitol.* **33**: 273–281.
31. Nelson, D. J., Bugge, C. J., Elion, G. B., Berens, R. L. and Marr, J. J. (1979) Metabolism of pyrazolo(3,4-d)pyrimidines in *Leishmania braziliensis* and *Leishmania donovani.* Allopurinol, oxipurinol, and 4-aminopyrazolo(3,4-d)pyrimidine. *J. Biol. Chem.* **254**: 3959–3964.
32. Hassan, H. F. and Coombs, G. H. (1985) Purine phosphoribosyltransferases of *Leishmania mexicana mexicana* and other flagellate protozoa. *Comp. Biochem. Physiol.* [B] **82**: 773–779.
33. Kidder, G. W. and Nolan, L. L. (1982) Xanthine phosphoribosyltransferase in *Leishmania:* divalent cation activation. *J. Protozool.* **29**: 405–409.
34. Spector, T., Jones, T. E. and Elion, G. B. (1979) Specificity of adenylosuccinate synthetase and adenylosuccinate lyase from *Leishmania donovani.* Selective amination of an antiprotozoal agent. *J. Biol. Chem.* **254**: 8422–8426.
35. Spector, T. and Jones, T. E. (1982) Guanosine 5'-monophosphate reductase from *Leishmania donovani.* A possible chemotherapeutic target. *Biochem. Pharmacol.* **31**: 3891–3897.
36. Wilson, K., Collart, F. R., Huberman, E., Stringer, J. R. and Ullman, B. (1991) Amplification and molecular cloning of the IMP dehydrogenase gene of *Leishmania donovani. J. Biol. Chem.* **266**: 1665–1671.
37. Koszalka, G. W. and Krenitsky, T. A. (1979) Nucleosidases from *Leishmania donovani.* Pyrimidine ribonucleosidase, purine ribonucleosidase, and a novel purine 2'-deoxyribonucleosidase. *J. Biol. Chem.* **254**: 8185–8193.
38. Konigk, E. (1978) Purine nucleotide metabolism in promastigotes of *Leishmania tropica:* inhibitory effect on allopurinol and analogues of purine nucleosides. *Tropenmed. Parasitol.* **29**: 435–438.
39. Datta, A. K., Bhaumik, D. and Chatterjee, R. (1987) Isolation and characterization of adenosine kinase from *Leishmania donovani. J. Biol. Chem.* **262**: 5515–5521.
40. Campbell, T. A., Zlotnick, G. W., Neubert, T. A., Sacci, Jr, J. B. and Gottlieb, M. (1991) Purification and characterization of the 3'-nucleotidase/nuclease from promastigotes of *Leishmania donovani. Mol. Biochem. Parasitol.* **47**: 109–117.
41. Gutteridge, W. E. and Gaborak, M. (1979) A re-examination of purine and pyrimidine synthesis in the three main forms of *Trypanosoma cruzi. Int. J. Biochem.* **10**: 415–422.
42. Schmidt, G., Walter, R. D. and Konigk, E. (1975) A purine nucleoside hydrolase from *Trypanosoma gambiense,* purification and properties. *Tropenmed. Parasitol.* **26**: 19–26.
43. Gutteridge, W. E. and Davies, M. J. (1981) Enzymes of purine salvage in *Trypanosoma cruzi. FEBS Lett.* **127**: 211–214.
44. Miller, R. L., Sabourin, C. L., Krenitsky, T. A., Berens, R. L. and Marr, J. J. (1984) Nucleoside hydrolases from *Trypanosoma cruzi. J. Biol. Chem.* **259**: 5073–5077.
45. Miller, R. L., Sabourin, C. L. and Krenitsky, T. A. (1987) *Trypanosoma cruzi* adenine nucleoside phosphorylase. Purification and substrate specificity. *Biochem. Pharmacol.* **36**: 553–560.
46. Spector, T., Berens, R. L. and Marr, J. J. (1982) Adenylosuccinate synthetase and adenylosuccinate lyase from *Trypanosoma cruzi,* specificity studies with potential chemotherapeutic agents. *Biochem. Pharmacol.* **31**: 225–229.
47. Nolan, L. L. and Kidder, G. W. (1980) Inhibition of growth and purine-metabolizing enzymes of trypanosomid flagellates by N_6-methyladenine. *Antimicrob. Agents Chemother.* **17**: 567–571.
48. Fish, W. R., Looker, D. L., Marr, J. J. and Berens, R. L. (1982) Purine metabolism in the bloodstream forms of *Trypanosoma gambiense* and *Trypanosoma rhodesiense. Biochim. Biophys. Acta* **719**: 223–231.
49. Fish, W. R., Marr, J. J. and Berens, R. L. *et al.* (1985) Inosine analogs as chemotherapeutic agents for African trypanosomes: metabolism in trypanosomes and efficacy in tissue culture. *Antimicrob. Agents Chemother.* **27**: 33–36.
50. Ghoda, L. Y., Savarese, T. M. and Northup, C. H. *et al.* (1988) Substrate specificities of 5'-deoxy-5'-methylthioadenosine phosphorylase from *Trypanosoma brucei brucei* and mammalian cells. *Mol. Biochem. Parasitol.* **27**: 109–118.

51. Ogbunude, P. O. and Ikediobi, C. O. (1983) Comparative aspects of purine metabolism in some African trypanosomes. *Mol. Biochem. Parasitol.* **9**: 279–287.

52. Manandhar, M. S. P. and Van Dyke, K. (1975) Detailed purine salvage metabolism in and outside the free malarial parasite. *Exp. Parasitol.* **37**: 138–146.

53. Yamada, K. A. and Sherman, I. W. (1981) Purine metabolism by the avian malarial parasite *Plasmodium lophurae. Mol. Biochem. Parasitol.* **3**: 253–264.

54. Reyes, P., Rathod, P. K., Sanchez, D. J., Mrema, J. E., Rieckmann, K. H. and Heidrich, H. G. (1982) Enzymes of purine and pyrimidine metabolism from the human malaria parasite, *Plasmodium falciparum. Mol. Biochem. Parasitol.* **5**: 275–290.

55. Geary, T. G., Divo, A. A., Bonanni, L. C. and Jensen, J. B. (1985) Nutritional requirements of *Plasmodium falciparum* in culture. III. Further observations on essential nutrients and antimetabolites. *J. Protozool.* **32**: 608–613.

56. Berman, P. A., Human, L. and Freese, J. A. (1991) Xanthine oxidase inhibits growth of *Plasmodium falciparum* in human erythrocytes *in vitro. J. Clin. Invest.* **88**: 1848–1855.

57. Queen, S. A., Vander Jagt, D. L. and Reyes, P. (1986) Characterization of adenine phosphoribosyltransferase from the human malaria parasite, *Plasmodium falciparum. Biochim. Biophys. Acta* **996**: 160–165.

58. Queen, S. A., Vander Jagt, D. and Reyes, P. (1988) Properties and substrate specificity of a purine phosphoribosyltransferase from the human malaria parasite, *Plasmodium falciparum. Mol. Biochem. Parasitol.* **30**: 123–133.

59. King, A. and Melton, D. W. (1987) Characterisation of cDNA clones for hypoxanthine–guanine phosphoribosyltransferase from the human malarial parasite, *Plasmodium falciparum*: comparisons to the mammalian gene and protein. *Nucleic Acids Res* **15**: 10469–10481.

60. Daddona, P. E., Wiesmann, W. P., Lambros, C., Kelley, W. N. and Webster, H. K. (1984) Human malaria parasite adenosine deaminase. Characterization in host enzyme–deficient erythrocyte culture. *J. Biol. Chem.* **259**: 1472–1475.

61. Schwartzman, J. D. and Pfefferkorn, E. R. (1982) *Toxoplasma gondii*: purine synthesis and salvage in mutant host cells and parasites. *Exp. Parasitol.* **53**: 77–86.

62. Krug, E. C., Marr, J. J. and Berens, R. L. (1989) Purine metabolism in *Toxoplasma gondii. J. Biol. Chem.* **264**: 10601–10607.

63. Asai, T., O'Sullivan, W. J. and Tatibana, M. (1983) A potent nucleoside triphosphate hydrolase from the parasitic protozoan *Toxoplasma gondii*. Purification, some properties, and activation by thiol compounds. *J. Biol. Chem.* **258**: 6816–6822.

64. LaFon, S. W. and Nelson, D. J. (1985) Purine metabolism in the intact sporozoites and merozoites of *Eimeria tenella. Mol. Biochem. Parasitol.* **14**: 11–22.

65. Miller, R. L., Adamczyk, D. L., Rideout, J. L. and Krenitsky, T. A. (1982) Purification, characterization, substrate and inhibitor specificity of adenosine kinase from several Eimeria species. *Mol. Biochem. Parasitol.* **6**: 209–223.

66. Senft, A. W. and Crabtree, G. W. (1983) Purine metabolism in the schistosomes: potential targets for chemotherapy. *Pharmacol. Ther.* **20**: 341–356.

67. Dovey, H. F., McKerrow, J. H. and Wang, C. C. (1985) Action of tubercidin and other adenosine analogs on *Schistosoma mansoni* schistosomules. *Mol. Biochem. Parasitol.* **16**: 185–198.

68. Dovey, H. F., McKerrow, J. H., Aldritt, S. M. and Wang, C. C. (1986) Purification and characterization of hypoxanthine–guanine phosphoribosyltransferase from *Schistosoma mansoni. J. Biol. Chem.* **261**: 944–948.

69. Craig, S. P. 3d, Yuan, L., Kuntz, D. A., McKerrow, J. H. and Wang, C. C. (1991) High level expression in *Escherichia coli* of soluble, enzymatically active schistosomal hypoxanthine/guanine phosphoribosyltransferase and trypanosomal ornithine decarboxylase. *Proc. Natl. Acad. Sci. USA* **88**: 2500–2504.

70. Heath, R. L. and Hart, J. L. (1970) Biosynthesis *de novo* of purines and pyrimidines in *Mesocestoides* (Cestoda). II. *J. Parasitol.* **56**: 340–345.

71. Page, C. R. III and MacInnis, A. J. (1975) Characterization of nucleoside transport in Hymenolepidid cestodes. *J. Parasitol.* **61**: 281–290.

72. Barrett, J. (1981) *Biochemistry of Parasitic Helminths*, University Park Press, Baltimore.

73. Jaffe, J. J. and Doremus, H. M. (1970) Metabolic patterns of *Dirofilaria immitis* microfilariae *in vitro. J. Parasitol.* **56**: 254–260.

74. Chen, S. N. and Howells, R. E. (1981) *Brugia pahangi*: uptake and incorporation of nucleic acid precursors by microfilariae and macrofilariae *in vitro*. *Exp. Parasitol.* **51**: 296–306.
75. Wang, C. C., Verham, R., Tzeng, S. F., Aldritt, S. and Cheng, H. W. (1983) Pyrimidine metabolism in *Tritrichomonas foetus*. *Proc. Natl. Acad. Sci. USA* **80**: 2564–2568.
76. Heyworth, P. G., Gutteridge, W. E. and Ginger, C. D. (1984) Pyrimidine metabolism in *Trichomonas vaginalis*. *FEBS Lett.* **176**: 55–60.
77. Wang, C. C. (1984) Purine and pyrimidine metabolism in Trichomonadidae and Giardia. In: *Molecular Parasitology* (ed. August, J. T.), Academic Press, Orlando, pp. 217–230.
78. Lindmark, D. G. and Jarroll, E. L. (1982) Pyrimidine metabolism in *Giardia lamblia* trophozoites. *Mol. Biochem. Parasitol.* **5**: 291–296.
79. Aldritt, S. M., Tien, P. and Wang, C. C. (1985) Pyrimidine salvage in *Giardia lamblia*. *J. Exp. Med.* **161**: 437–445.
80. Keppler, D. and Holstege, A. (1982) Pyrimidine nucleotide metabolism and its compartmentation. In: *Metabolic Compartmentation* (ed. Sies, H.), Academic Press, London, pp. 147–203.
81. Hill, B., Kilsby, J., Rogerson, G. W., McIntosh, R. T. and Ginger, C. D. (1981) The enzymes of pyrimidine biosynthesis in a range of parasitic protozoa and helminths. *Mol. Biochem. Parasitol.* **2**: 123–134.
82. Booden, T., Albach, R. A. and Boonlayangoor, P. (1978) Uptake of selected pyrimidines by axenically grown *Entamoeba histolytica*. *Arch. Invest. Med.* **9** Suppl. 1(Mex): 133–140.
83. Reeves, R. E. (1984) Metabolism of *Entamoeba histolytica* Schaudinn, 1903. *Adv. Parasitol.* **23**: 105–142.
84. Reeves, R. E. and West, B. (1980) *Entamoeba histolytica*: nucleic acid precursors affecting axenic growth. *Exp. Parasitol.* **49**: 78–82.
85. Booden, T., Boonlayangoor, P. and Albach, R. A. (1976) Incorporation of purines and pyrimidines in axenically grown *Entamoeba histolytica*. *J. Parasitol.* **62**: 641–643.
86. Steiger, R. F. and Black, C. D. V. (1980) Simplified defined media for cultivating *Leishmania donovani* promastigotes. *Acta Trop.* **37**: 195–198.
87. Hammond, D. J. and Gutteridge, W. E. (1982) UMP synthesis in the kinetoplastida. *Biochim. Biophys. Acta* **718**: 1–10.
88. Gero, A. M. and Coombs, G. H. (1982) *Leishmania mexicana*: conversion of dihydroorotate to orotate in amastigotes and promastigotes. *Exp. Parasitol.* **54**: 185–195.
89. Gutteridge, W. E., Dave, D. and Richards, W. H. (1979) Conversion of dihydroorotate to orotate in parasitic protozoa. *Biochim. Biophys. Acta* **582**: 390–401.
90. Hammond, D. J., Gutteridge, W. E. and Opperdoes, F. R. (1981) A novel location for two enzymes of *de novo* pyrimidine biosynthesis in trypanosomes and *Leishmania*. *FEBS Lett.* **1128**: 27–29.
91. Ivanetich, K. M. and Santi, D. V. (1990) Bifunctional thymidylate synthase–dihydrofolate reductase in protozoa. *FASEB J.* **4**: 1591–1597.
92. Hammond, D. J. and Gutteridge, W. E. (1980) Enzymes of pyrimidine biosynthesis in *Trypanosoma cruzi*. *FEBS Lett.* **118**: 259–262.
93. Kidder, G. W. (1984) Characteristics of cytidine aminohydrolase activity in *Trypanosoma cruzi* and *Crithidia fasciculata*. *J. Protozool.* **31**: 298–300.
94. Hassan, H. F. and Coombs, G. H. (1986) A comparative study of the purine- and pyrimidine-metabolising enzymes of a range of trypanosomatids. *Comp. Biochem. Physiol.* [B] **84**: 219–223.
95. Gutteridge, W. E. and Trigg, P. I. (1970) Incorporation of radioactive precursors into DNA and RNA of *Plasmodium knowlesi in vitro*. *J. Protozool.* **17**: 89–96.
96. Van Dyke, K., Tremblay, G. C., Lantz, C. H. and Szustkiewicz, C. (1970) The source of purines and pyrimidines in *Plasmodium berghei*. *Am. J. Trop. Med. Hyg.* **19**: 202–208.
97. Walsh, C. J. and Sherman, I. W. (1968) Purine and pyrimidine synthesis by the avian malaria parasite, *Plasmodium lophurae*. *J. Protozool.* **15**: 763–770.
98. Gero, A. M., Brown, G. V. and O'Sullivan, W. J. (1984) Pyrimidine *de novo* synthesis during the life cycle of the intraerythrocytic stage of *Plasmodium falciparum*. *J. Parasitol.* **70**: 536–541.

99. Rathod, P. K. and Reyes, P. (1983) Orotidylate-metabolizing enzymes of the human malarial parasite, *Plasmodium falciparum*, differ from host cell enzymes. *J. Biol. Chem.* **258**: 2852–2855.
100. Jones, M. E. (1980) Pyrimidine nucleotide biosynthesis in animals: genes, enzymes, and regulation of UMP biosynthesis. *Annu. Rev. Biochem.* **49**: 253–279.
101. Schwartzman, J. D. and Pfefferkorn, E. R. (1981) Pyrimidine synthesis by intracellular *Toxoplasma gondii*. *J. Parasitol.* **67**: 150–158.
102. Pfefferkorn, E. R. (1988) *Toxoplasma gondii* viewed from a virological perspective. In: *The Biology of Parasitism* (eds P. T. England and A. Sher), AR Liss, New York, pp. 479–501.
103. Krylov, Iu. M. (1982) Utilization by *Eimeria tenella* sporozoites (Coccidiida, Sporozoa) of [^{14}C] aspartic and [^{14}C] orotic acids for the biosynthesis of pyrimidine nucleotides. *Parazitologiia* **16**: 204–208.
104. Ouellette, C. A., Strout, R. G. and McDougald, L. R. (1973) Incorporation of radioactive pyrimidine nucleosides into DNA and RNA of *Eimeria tenella* (Coccidia) cultured *in vitro*. *J. Protozool.* **20**: 150–153.
105. Coles, A. M., Swoboda, B. E. and Ryley, J. F. (1980) Thymidylate synthetase as a chemotherapeutic target in the treatment of avian coccidiosis. *J. Protozool.* **27**: 502–506.
106. Aoki, T. and Oya, H. (1979) Glutamine-dependent carbamoyl-phosphate synthetase and control of pyrimidine biosynthesis in the parasitic helminth *Schistosoma mansoni*. *Comp. Biochem. Physiol.* **63**-B: 511–515.
107. Iltzsch, M. H., Niedzwicki, J. G., Senft, A. W., Cha, S. and el Houni, M. (1984) Enzymes of uridine 5′-monophosphate biosynthesis in *Schistosoma mansoni*. *Mol. Biochem. Parasitol.* **12**: 153–171.
108. Huang, Z.-Y., Chang, H.-L. and Chen, G.-Z. (1984) The metabolism of pyrimidine and purine in *Schistosoma japonicum*. *Chinese Med. J.* **97**: 698–706.
109. el Kouni, M. H. and Naguib, F. N. M. (1990) Pyrimidine salvage pathways in adult *Schistosoma mansoni*. *Int. J. Parasitol.* **20**: 37–44.
110. el Kouni, M. H., Naguib, F. N. M., Niedzwicki, J. G., Iltzsch, M. H. and Cha, S. (1988) Uridine phosphorylase from *Schistosoma mansoni*. *J. Biol. Chem.* **263**: 6081–6086.
111. den Hollender, J. E. and Erasmus, D. A. (1984) *Schistosoma mansoni*: DNA synthesis in males and females from mixed and single-sex infections. *Parasitology* **88**: 463–476.
112. Kobayashi, M., Yokogawa, M., Mori, M. and Tatibana, M. (1978) Pyrimidine nucleotide biosynthesis in *Clonorchis sinensis* and *Paragonimus ohirai*. *Int. J. Parasitol.* **8**: 471–477.
113. Kurelec, B. (1974) Aspartate transcarbamylase in some parasitic platyhelminths. *Comp. Biochem. Physiol.* **47**-B: 33–40.
114. Insler, G. D. and Halikias, F. J. (1991) Independent characterization of thymidine transport and subsequent metabolism in *Hymenolepis diminuta*-II. Purification and preliminary analysis of thymidine kinase. *Comp. Biochem. Physiol.* **98**-B: 181–186.
115. Aoki, T., Oya, H., Mori, M. and Tatibana, M. (1975) Glutamine-dependent carbamoyl-phosphatae synthetase in *Ascaris* ovary and its regulatory properties. *Proc. Japan Acad.* **51**: 733–736.
116. Jaffe, J. J., Comley, J. C. W. and Chrin, L. R. (1982) Thymidine kinase activity and thymidine salvage in adult *Brugia pahangi* and *Dirofilaria immitis*. *Mol. Biochem. Parastitol.* **5**: 361–370.
117. Jaffe, J. J. and Chrin, L. R. (1980) Folate metabolism in filariae: enzymes associated with 5,10-methylenetetrahydrofolate. *J. Parasitol.* **66**: 53–58.
118. Pasternak, J. and Samoiloff, M. R. (1970) The effect of growth inhibitors on postembryonic development in the free-living nematode, *Pangrellus silusiae*. *Comp. Biochem. Physiol.* **33**: 27–38.
119. Nonnenmacher-Godet, J. and Dougherty, E. C. (1964) Incorporation of tritiated thymidine in the cells of *Caenorhabditis briggsae* (Nematoda) reared in axenic culture. *J. Cell Biol.* **22**: 281–290.

7 Polyamine Metabolism

CYRUS J. BACCHI and NIGEL YARLETT

Haskins Laboratories and Department of Biology, Pace University, New York, NY, USA

SUMMARY

Polyamine metabolism by parasites differs in several significant ways from the mammalian host; these include, but are not limited to, enzyme half-life, turnover, substrate specificity, types and quantities of polyamines produced. The production of the novel *bis* glutathionyl spermidine adduct by trypanosomatids, its role as an antioxidant and the protein structure of trypanothione reductase is discussed with respect to the more conventional glutathione reductase system. The role of *S*-adenosylmethionine and decarboxylated *S*-adenosylmethionine as critical precursors in the biosynthesis of the higher polyamines is explored with respect to differences in the function and control of the pathway by various parasites. Polyamine biosynthesis in parasites is sufficiently different from that of the host to afford multiple opportunities for drug development, these may be aimed directly at circumventing polyamine biosynthesis or at inhibiting precursors necessary for polyamine synthesis.

1. INTRODUCTION

Polyamines are low-molecular-weight molecules which are found in both prokaryotes and eukaryotes, and which fulfil essential needs for growth, division and differentiation. The most common polyamines are: putrescine (1,4-diaminobutane; tetramethylene-diamine), spermidine (*N*-(3-aminopropyl)-1,4-butane diamine) and spermine (*N*,*N*'-bis(3-aminopropyl)-1,4-butane diamine). Cadaverine (1,5-diaminopentane) is also found in some micro-organisms (1, 2). Most prokaryotes do not contain spermine but all cells contain putrescine and spermidine, in most cases in millimolar concentrations.

Biochemistry and Molecular Biology of Parasites
ISBN 0-12-473345-X

The most compelling arguments for a polyamine requirement for cell growth stem from the cytostatic effect of inhibitors of polyamine synthesis on cultured cells (3, 4). Exploration of polyamine metabolism in parasites began only in the 1970s but these studies also indicated vital functions for these molecules as growth factors and in differentiation. Polyamines are positively charged and associate with negatively charged molecules such as phospholipids and nucleic acids through electrostatic interactions. Since polyamines, unlike inorganic ions, are synthesized by the cell, they can be expected to play more dynamic roles in cell metabolism (4).

Spermidine and spermine interact with DNA and also with tRNA. This binding is responsible in part for maintaining the conformation of these molecules (5). The interaction of polyamines with macromolecules, however, makes the estimation of intracellular bound and unbound pool sizes difficult since disruption of cells frequently destroys cell compartments, promoting reassociation of amines.

2. POLYAMINES IN PROTOZOA

Putrescine and spermidine have been found in all parasitic protozoa examined: *Giardia, Trichomonas, Leishmania, Trypanosoma, Naegleria, Entamoeba, Acanthamoeba, Plasmodium* and *Eimeria*. Some organisms also contain spermine, but this is not a constant feature of parasitic protozoa. The amine contents of selected parasites are listed in Table 7.1.

2.1. Metabolic Regulation

The common pathways of polyamine biosynthesis are depicted on Fig. 7.1. The amino acids arginine and methionine are the immediate precursors of polyamines. In anaerobic protozoa, e.g. *Trichomonas*, arginine is converted into ornithine via the arginine dihydrolase pathway (6). In aerobic organisms such as African trypanosomes, extracellular ornithine appears to be the starting point for conversion into putrescine. *Trypanosoma cruzi* utilizes arginine decarboxylase in synthesizing putrescine via agmatine through a pathway known from bacteria and plants (7–10). Production of spermidine and spermine is accomplished by addition of aminopropyl groups from decarboxylated S-adenosylmethionine (dc-AdoMet). AdoMet is itself synthesized from methionine and ATP. Specific synthases are responsible for aminopropyl group transfer to form spermidine and spermine. Some parasitic protozoa do not produce spermine, including African trypanosomes, *Trichomonas, Entamoeba* and *Giardia* (11, 12). *Leishmania* spp. and *Plasmodium* sp., however, appear to synthesize spermine, but only in small quantities (11).

Synthesis of polyamines in the majority of cells is initiated by ornithine decarboxylase (ODC: EC 4.1.1.17), which catalyses the conversion of ornithine to putrescine (Fig. 7.1) (4, 13). In mammalian cells ODC undergoes rapid induction and has a short half-life. In the parasitic protozoa so far examined (*T. b. brucei, P. falciparum, L. donovani* and *T. vaginalis*), ODC has a long half-life as judged from the rate of decay of activity in the presence of cycloheximide (14–17). In African trypanosomes the slow turnover of ODC appears to be attributable to the absence of a 36 amino acid region (PEST sequence) on the COOH-terminal end (18, 19). In mammalian cells, this is

TABLE 7.1 Polyamine content of some parasites[a]

Parasite[b]	Putrescine	Spermidine	Spermine	Putrescine/ spermidine
Trichomonas vaginalis Cl-NIH	38	3.5	1.3	10.9
965691	190	5.0	19.0	38.0
Clone 31	67	4.5	4.6	14.9
Clone G3	57	9.7	9.5	5.9
Tritrichomonas foetus F2	79	19.8	9.6	4.0
Trichomitus batrachorum B2	114	17.6	15.0	6.5
Tetrahymena thermophila[c]	10	3.0	0	3.1
Entamoeba histolytica	92	2.6	0.03	35.5
Giardia lamblia	10	9.6	0.8	1.0
Trypanosoma brucei	4	24.5	0	0.17
Trypanosoma rhodesiense	4	21.0	1.2	0.18
Leishmania donovani (promastigotes)	35	37.1	0	0.9
Leishmania donovani (amastigotes)	2	18.8	3.5	0.1
Trypanosoma cruzi	+	+	+	
Plasmodium falciparum[d]	9	33.0	8.0	0.3
Ancyclostoma ceylanicum	0.06	1.06	0.24	0.06
Hymenolepis diminuta	0.32	3.25	19.30	0.10
Nippostrongylus brasiliensis	ND	9.82	2.35	—
Brugia patei	0.19	40.0	61.50	0.005
Onchocerca volvulus	0.60	4.0	31.0	0.02
Romanomermis culicivorax[e]	2.74	6.90	0.12	0.40

[a]Values are expressed as: nmol/mg protein.
[b]Data are compiled from references 2, 12, 64 and 65.
[c]millimolar.
[d]pmol/10^6 parasitized erythrocytes.
[e]nmol/mg dry weight.

responsible for rapid intracellular degradation of proteins (20). The amino acid sequence of *L. donovani* ODC indicates that the COOH-terminus is eight amino acids longer than that of the trypanosome enzyme, but 28 residues shorter than the mouse enzyme (16). This lack of a COOH-terminal extension on the *Leishmania* enzyme is consistent with the long half-life observed (16). *Leishmania* ODC has a low EC_{50} for DL-α-difluoromethylornithine (DFMO) (30 μM) (21); hence the lack of *in vivo* activity of polyamine antagonists towards *Leishmania* spp. may be due to the intracellular location of the parasite and not to intrinsic differences in the organism. Both *P. falciparum* and *T. vaginalis* ODCs also have long half-lives (>1 h) in the presence of cycloheximide, but amino acid sequences have not yet been established.

AdoMet decarboxylase is the second key enzyme in polyamine biosynthesis. In mammalian cells it is activated by putrescine and is also characterized by the presence of a PEST sequence and a short half-life (13). Trypanosome AdoMet decarboxylase, like ODC, has a long (>6 h) half-life as shown by *in vitro* incubation studies with cycloheximide using procyclic and bloodstream trypomastigotes (Bacchi, unpublished results). The crude enzyme is only 50% stimulated by 1 mM putrescine (22) but in the partly purified state it is stimulated 8-fold by 2 mM putrescine (23). Trypanosome AdoMet decarboxylase, like other eukaryotic counterparts, contains a covalently

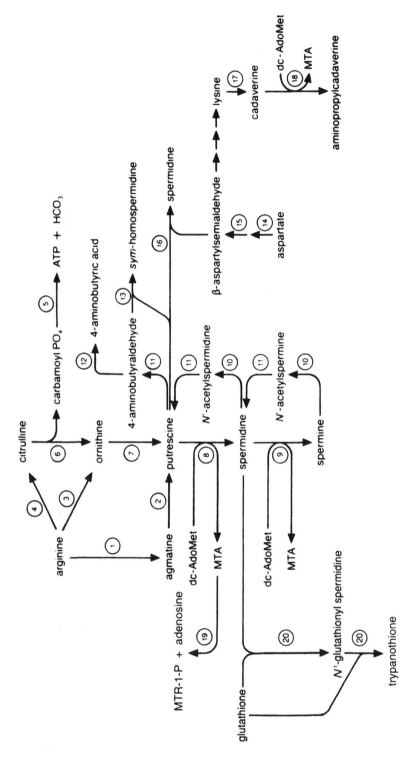

FIG. 7.1 Biosynthesis of polyamines and their analogs. 1, Arginine decarboxylase; 2, agmatine deiminase; 3, arginase; 4, arginine deiminase; 5, carbamoyl phosphate kinase; 6, catabolic ornithine carbamoyltransferase; 7, ornithine decarboxylase; 8, spermidine synthase; 9, spermine synthase; 10, polyamine acetylase; 11, diamine oxidase; 12, aldehyde dehydrogenase; 13, reactions occurring in some archaebacteria (76); 14, aspartate kinase; 15, aspartyl phosphate dehydrogenase; 16, aspartyl semialdehyde dehydrogenase; 17, lysine decarboxylase; 18, aminopropyl cadaverine synthase; 19, methylthioadenosine phosphorylase; 20, trypanothione synthetase. MTA, methylthioadenosine; MTR-1-P, methylthioribose-1-phosphate; dc-AdoMet, decarboxylated *S*-adenosyl-methionine.

122

bound pyruvate as cofactor, but its subunit composition may differ from that of other eukaryotes since antiserum against human AdoMet decarboxylase did not cross react with the partly purified trypanosome enzyme (23). Although the long half-lives of ODC and AdoMet decarboxylase in trypanosomes indicate lack of control of polyamine biosynthesis, there is evidence for some regulation. During the growth cycle of trypomastigotes in culture, a rapid rise in putrescine levels in early log phase is followed by an increase in spermidine levels and an abrupt fall in putrescine (24). These events parallel the rapid induction of ODC in log phase of procyclic forms (14) and are consistent with putrescine activation of AdoMet decarboxylase.

Although the half-life data of ODC and AdoMet decarboxylase are not known for many protozoa, there is evidence for some regulation of polyamine biosynthesis during the life cycle. *Plasmodium* spp. undergo dramatic changes in polyamine content during the life cycle and activities of the ODC and AdoMet decarboxylase peak during the early trophozoite stage (25). In addition, bis(benzyl)polyamine analogs inhibit growth of *P. falciparum in vitro* and *P. berghei in vivo* (26). These agents are debenzylated on entry into mammalian cells and repress ODC and AdoMet decarboxylase activities (27). Depletion of polyamines by the bis(benzyl)polyamines and ODC inhibitors prevents transformation of trophozoites to schizonts (25, 27–29). Similarly, inhibition of growth with bis(benzyl)polyamines has been observed for *Leishmania* (30). These observations strongly suggest that enzyme product interactions have a role in regulation of polyamine biosynthesis in *Plasmodium* spp. and *Leishmania* spp. However, *L. donovani* ODC, unlike the mammalian enzyme, is not negatively regulated by polyamines (16).

2.2. Trypanothione Metabolism

A unique feature of polyamine metabolism in trypanosomatids is the presence of a spermidine-containing peptide, N^1,N^8-bis(glutathionyl)spermidine (31). This unusual peptide is present only in kinetoplastid protozoa. Although *E. coli* contains N^1-glutathionylspermidine (32), it does not synthesize the bisglutathionyl conjugate. It is synthesized by addition of glutathionyl residues to the terminal amines of spermidine by a specific synthase (Fig. 7.1). Since trypanosomes lack catalase, glutathione peroxidase and glutathione reductase, trypanothione functions to detoxify free radicals (O_2^-, OH·) when the parasites are under oxidant stress (33) (see Chapter 9). Reduced trypanothione is used by the cell in the non-enzymatic reduction of oxidized glutathione. Trypanothione, in turn, is reduced by a specific FAD-containing trypanothione reductase which does not act on glutathione but exhibits 41% sequence identity with mammalian glutathione reductase (34, 35). The most notable differences are the replacement of three arginine residues of the glutathione reductase (residues 37, 28 and 347) by neutral residues in trypanothione reductase (34). In human glutathione reductase the arginine-37 and arginine-347 residues form an important ionic interaction with the α-carboxylate groups of glycine and glutamyl residues. Three acidic residues present in the trypanothione reductase active site (Glu-18, Asp-112 and Asp-116) have no counterpart in the glutathione reductase and may interact analogously with the positively charged spermidine moiety of trypanothione (34). This novel molecule and its dependence on the polyamine pathway make this a likely target for chemotherapy: its synthesis is depressed during exposure to the irreversible ODC inhibitor DL-α-difluoromethylornithine (DFMO) (36) and its thiol groups form stable bonds with the arsenic moieties of the trypanocide Melarsoprol, rendering it incapable of undergoing redox reactions and turning the adduct into an inhibitor of trypanothione reductase (37). In addition, trypanothione reductase from *T. cruzi* can reduce naphthoquinone and nitrofuran derivatives to auto-oxidizable products which in the presence of air, generate reactive oxygen species that kill the parasite (38) (see Chapter 9).

Arsenical-refractory clinical isolates of *T. b. rhodesiense* are resistant to lysis in the presence of pharmacologically achievable blood levels of arsenical drugs whereas clinically sensitive isolates are destroyed by $1-10\,\mu M$ concentrations of drug. Trypanothione levels in refractory isolates decrease minimally during drug exposure ($10\,\mu M$ for 3 h), whereas free trypanothione is almost undetectable in sensitive strains after exposure to a lower dose ($1\,\mu M$ for 2 h) (39). This study and another, which reveals the absence of an adenosine transporter in arsenical-resistant trypanosomes (40), suggest that the uptake of arsenical drugs is significantly reduced in refractory trypanosome isolates.

2.3. Arginine Dihydrolase Pathway in Anaerobic Protozoa

Polyamine biosynthesis in *Giardia intestinalis*, *Trichomonas vaginalis* and *Tritrichomonas foetus* is distinctly different from that occurring in aerobic protozoa. In these parasites arginine is the precursor of polyamine biosynthesis via the arginine dihydrolase pathway and ATP is also produced from this pathway (Fig. 7.1, and Chapter 5). The first enzyme of the pathway is arginine deiminase (EC 3.5.3.6) which is responsible for the liberation of ammonia from arginine with the resultant formation of citrulline. Citrulline undergoes phosphorolysis by catabolic ornithine carbamoyl transferase (carbamoyl phosphate: L-ornithine carbamoyl transferase, EC 2.1.3.3) to ornithine with the loss of the guanidine carbon atom as carbamoyl phosphate. Some anaerobic protozoa do not synthesize the pyrimidine ring hence the carbamoyl phosphate is not used for this purpose. Instead carbamoyl phosphate is broken down by a catabolic carbamate kinase (ATP: carbamate phosphotransferase, EC 2.7.2.2) to bicarbonate and ammonia with the concomitant phosphorylation of a molecule of ADP to ATP (Fig. 7.1). The ornithine formed is decarboxylated by ODC to form putrescine. The enzymes of this pathway, have been found in *T. vaginalis* (6), *T. foetus* (41) and *G. intestinalis* (42). The importance of this pathway is indicated by the rapid depletion of arginine from the growth medium and the large quantities of putrescine which result both *in vitro* and *in vivo* (12). This pathway is also assumed to be present in rumen ciliates. It has been reported that ciliates of the species *Entodinium*, *Diplodinium*, *Dasytricha* and *Isotricha* produce citrulline from arginine, ornithine from citrulline, putrescine and proline from ornithine and α-aminovalerate from proline (43). The importance of the pathway is consistent with putrescine as the major polyamine in these species also (Yarlett, Lloyd, Williams and Ellis, unpublished). The presence of the dihydrolase pathway has not been demonstrated in *Entamoeba*, although this organism also produces putrescine as the major polyamine.

The differences in putrescine biosynthesis between aerobic and anaerobic protozoa may be a consequence of the absence of mitochondria in the latter and may be linked to the need for alternative ATP-generating processes. This pathway has the advantage of producing a non-acidic end-product (putrescine), whereas the other major end-products (acetate, lactate and butyrate) are acidic. Support for the idea that high concentrations of putrescine may be beneficial to organisms growing under anaerobic conditions is highlighted by the finding that plant cells, when exposed to anoxic conditions also produce putrescine as the major polyamine. Spermine and spermidine biosynthesis are inhibited under these conditions (44). It appears though that *T. vaginalis* is unaffected by high concentrations of spermine (1) and is capable of growth in human semen (45) which contains in excess of 10 mM spermine (46).

2.4. AdoMet Metabolism in African Trypanosomes

AdoMet is a critical branch point metabolite. In its decarboxylated form, it serves as aminopropyl group donor in polyamine synthesis; as AdoMet it is the methyl group donor for most transmethylation reactions (Fig. 7.1). In eukaryotic cells AdoMet decarboxylase is activated by putrescine and is the rate-limiting step of polyamine biosynthesis. Since production of dc-AdoMet is irreversible (13) regulation of AdoMet decarboxylase by putrescine determines the overall rate of polyamine synthesis.

Intracellular AdoMet and dc-AdoMet increase in trypanosomes 70- and >1000-fold after DFMO treatment *in vivo* (47, 48). In comparison, mammalian cells treated with DFMO barely double their AdoMet content, and increase dc-AdoMet only 500-fold (49, 50). The large increase in trypanosome AdoMet and consequent >1000-fold rise in dc-AdoMet from almost undetectable levels, results in a combined intracellular pool of 5 mM — about 1400 times that of normal (47). The ratio of AdoMet to its demethylated product, *S*-adenosylhomocysteine (AdoHcy), indicates the relative potential of the cell to methylate molecules (methylation index) (51). In bloodform trypanosomes exposed to DFMO for 36 h the methylation index also becomes elevated from a normal level of 6.5 to 114 (47). The kinetics of the AdoMet synthetase appears to favor this increase since the trypanosome enzyme has a high K_i for AdoMet (240 μM) as compared to mammalian enzymes (2–40 μM) (52).

Disruption of AdoMet metabolism is a key element also in the action of 5'-{[(Z)-5-amino-2-butenyl]methylamino}5'-deoxyadenosine (MDL73811), an AdoMet decarboxylase inhibitor which is a highly effective trypanocide (53). Treatment of trypanosomes with MDL73811 caused a 20-fold increase in the trypanosome AdoMet level after only 1 h of exposure. In comparison, AdoMet levels in cultured mammalian cells rose only 1.5-fold after similar treatment with MDL73811 (48). This analog is rapidly concentrated by trypanosomes through the purine transport system (54).

In addition to elevated AdoMet levels, trypanosomes treated with DFMO *in vivo* experience a six-fold increase in the activity of protein methylase II but not I or III (55). Protein methylase II specifically methylates carboxyl groups of aspartate and glutamate residues. In mammalian cells, histone methylation may be involved in the condensation of euchromatin to heterochromatin prior to mitosis (56–58), and may therefore have a role in gene expression. Aspartate- and glutamate-rich histones have also been characterized from both *T. b. brucei* and *Crithidia fasciculata* making these likely substrates for protein methylase II activity (59, 60).

These observations indicate that synthesis of AdoMet in African trypanosomes is unregulated as compared to the mammalian host and that trypanosomes may therefore be more sensitive to imbalances in the AdoMet:AdoHcy ratio (methylation index). Morphological alterations (long slender to stumpy transformation) in African trypanosomes during DFMO treatment and accompanying changes in mitochondrial structure attributed to polyamine depletion (11,61) suggest that alterations in methylation patterns may have a significant role in the life cycle. Penketh *et al.* (62) found that methylating agents such as streptozotocin and procarbazine were specifically trypanocidal and caused synchronous differentiation of blood forms into short stumpy forms.

3. POLYAMINE METABOLISM IN HELMINTHS

Most data concerning polyamine synthesis and content of parasitic helminths concern adult worms. Helminths differ from the majority of eukaryotes in that they contain spermidine and spermine and lesser or undetectable amounts of putrescine (Table 7.1) (2, 63–66). Cadaverine has also been detected in some helminths (2), although an active lysine decarboxylase has not been conclusively demonstrated, and cadaverine has no known function in eukaryotic cells (67). The intracellular concentration of spermidine and spermine in helminths is similar to that of vertebrate cells. There are indications that helminths lack polyamine biosynthetic enzymes (ODC, ADC) and may be dependent on uptake from the host (64–66). However, helminths are capable of interconversion of polyamines taken up from the host, suggesting the involvement of a polyamine salvage pathway (64). Spermine is converted into N-acetylspermine and then spermidine via the combined actions of N-acetylspermine transferase and polyamine oxidase (Fig. 7.1). Spermidine is then converted into N^1- or N^8-acetylspermidine and putrescine via the combined actions of N^1- or N^8-acetylspermidine transferase and polyamine oxidase (Fig. 7.1) (64,65). In addition, helminths can convert putrescine into N-acetylputrescine, apparently a major excretory product (64,66). Most of the ornithine decarboxylating activity present in helminths is insensitive to DFMO but is sensitive to glutamate (65). It is likely that the release of CO_2 from ornithine is due to an ornithine aminotransferase which produces glutamic semialdehyde and then glutamate via glutamate semialdehyde dehydrogenase. Glutamate is subsequently decarboxylated to γ-aminobutyric acid (65,68). Two observations support this mechanism: (a) inhibition of the ornithine aminotransferase pathway with gabaculine (3-amino-2,3-dihydrobenzoic acid) a suicide inhibitor, abolishes CO_2 release from ornithine (64); (b) *Onchocerca volvulus*, a filariid nematode which lacks the ornithine aminotransferase pathway, does not liberate CO_2 from ornithine (64).

Although these studies indicate a lack of polyamine biosynthesis in helminths, it is important to note that larval stages have not been examined and that the potential for stage-specific activation should not be ignored.

4. CONSIDERATIONS IN THE INHIBITION OF POLYAMINE METABOLISM IN PARASITE CHEMOTHERAPY

The overall pathways of polyamine metabolism in pathogenic protozoa and helminths seem to be sufficiently different from the mammalian host to afford multiple opportunities for drug intervention. The important features are: the long half-life of ODC in kinetoplastids coupled with poor polyamine transport and the apparently unregulated synthesis of AdoMet. Some anaerobic protozoa such as *Trichomonas*, *Giardia* and *Entamoeba* spp. also excrete large quantities of putrescine, and may lack polyamine biosynthetic capabilities. These parasites rely in part on ATP production via the arginine dihydrolase pathway and may be particularly susceptible to arginine analogs or similar agents interfering with enzymes of the dihydrolase pathway. There is evidence that *T. cruzi* possesses arginine decarboxylase activity, which is sensitive to inhibition by difluoromethylarginine (9). In addition difluoromethylarginine is capable of reducing the infectivity of *T. cruzi* to mammalian cells (8, 10), whereas ODC inhibitors have no effect on infectivity (8). Specific inhibition of infectivity results from selective impairment of amastigote division without affecting the capacity of the invading trypomastigotes to transform into the replicative amastigote form (10). Thus, this pathway is a potential target for selective chemotherapy, both because mammalian cells lack arginine decarboxylase and because the enzyme appears to be critical for parasite invasion of host cells and amastigote multiplication (10).

As noted in Chapter 6, trypanosomes and the anaerobic protozoa are purine auxotrophs with significant capacity to transport (69–71), interconvert and salvage purines (72; see Chapter 6). Although chemotherapy based on purine salvage of these protozoa is now being developed (72), there is evidence that critical targets within AdoMet metabolism may also be vulnerable. AdoMet synthetase in kinetoplastids exhibits a high K_i for product inhibition, as well as a broad methionine requirement (52). The purine requirement for this enzyme is unknown and it is likely that an AdoMet analog based on purine or methionine could affect both polyamine and methylation reactions. In trypanosomatids, the salvage enzyme methylthioadenosine (MTA) phosphorylase has been found to have a broad substrate requirement (Fig. 7.1) (73). In African trypanosomes this enzyme also cleaves 5'-halo- and ethylthioribose derivatives of MTA to produce adenine plus the resulting ribose analog (74). The latter would normally be recycled to methionine (5). These agents are trypanocidal and their effects can be reversed with methionine or ketomethylthiobutyrate, an intermediate in the conversion of methylthioribose into methionine (74).

A critical detoxifying and salvage enzyme within the transmethylation pathway is AdoHcy hydrolase. This intermediate is highly toxic to cells and is rapidly hydrolyzed to adenine and homocysteine (75). African trypanosomes have high levels of AdoHcy hydrolase (47) and are susceptible to the AdoHcy analog Neplanocin A (75). Although this analog is toxic to mammalian cells, the range of substrate specificity for this enzyme is unknown and should be pursued for both trypanosomes and other protozoa.

5. REFERENCES

1. Yarlett, N. and Bacchi, C. J. (1988) Effects of DL-α-difluoromethylornithine on polyamine synthesis and interconversion in *Trichomonas vaginalis* grown in a semi-defined medium. *Mol. Biochem. Parasitol.* **31**: 1–10.
2. Gordon, R., Cornect, M., Walters, B. M., Hall, D. E. and Brosnan, M. E. (1989) Polyamine synthesis by the mermithid nematode *Romanomermis culicivorax*. *J. Nematol.* **21**: 81–86.
3. Pegg, A. E. (1986) Recent advances in the biochemistry of polyamines in eukaryotes. *Biochem. J.* **234**: 249–262.
4. Pegg, A. E. and McCann, P. P. (1988) Polyamine metabolism and function in mammalian cells and protozoans. *ISI Atlas of Science Biochemistry*. ICI Publishers, Cleveland, Ohio, pp. 11–18.

5. Porter, C. W. and Sufrin, J. R. (1986) Interference with polyamine biosynthesis and/or function by analogs of polyamines or methionine as a potential anticancer chemotherapeutic strategy. *Anticancer Res.* **6**: 525–542.

6. Linstead, D. and Cranshaw, M. A. (1983) The pathway of arginine catabolism in the parasitic flagellate *Trichomonas vaginalis*. *Mol. Biochem. Parasitol.* **8**: 241–252.

7. McCann, P. P., Bacchi, C. J., Bitonti, A. J., Kierszenbaum, F. and Sjoerdsma, A. (1988) In: *Progress in Polyamine Research* (eds Zappia, V. and Pegg, A. E.), Plenum Press, New York, pp. 727–736.

8. Kierszenbaum, F., Wirth, J. J., McCann, P. P. and Sjoerdsma, A. (1987) Arginine decarboxylase inhibitors reduce the capacity of *Trypanosoma cruzi* to infect and multiply in mammalian host cells. *Proc. Natl. Acad. Sci. USA* **84**: 4278–4282.

9. Majumder, S., Wirth, J. J., Bitonti, A. J., McCann, P. P. and Kierszenbaum, F. (1992) Biochemical evidence for the presence of arginine decarboxylase activity in *Trypanosoma cruzi*. *J. Parasitol.* **78**: 371–374.

10. Yakubu, M. A., Basso, B. and Kierszenbaum, F. (1992) DL-α-difluoromethylarginine inhibits intracellular *Trypanosoma cruzi* multiplication by affecting cell division but not trypomastigote–amastigote transformation. *J. Parasitol.* **78**: 414–419.

11. Bacchi, C. J. and McCann, P. P. (1987) In: *Inhibition of Polyamine Metabolism* (eds McCann, P. P., Pegg, A. E. and Sjoerdsma, A.) Academic Press, New York, pp. 317–344.

12. Yarlett, N. and Bacchi, C. J. (1991) In: *Biochemical Protozoology; Basis for Drug Design* (eds Coombs, G. H. and North, M. J.), Taylor & Francis, London, pp. 458–468.

13. Pegg, A. E. (1988) Polyamine metabolism and its importance in neoplastic growth and as a target for chemotherapy. *Cancer Res.* **48**: 759–774.

14. Bacchi, C. J., Garofalo, J., Santana, A., Hannan, J. C., Bitonti, A. J. and McCann, P. P. (1989) *Trypanosoma brucei brucei*: Regulation of ornithine decarboxylase in procyclic forms of trypomastigotes. *Exp. Parasitol.* **68**: 392–402.

15. Assaraf, Y. G., Kahana, C., Spira, D. T. and Bachrach, U. (1988) *Plasmodium falciparum*: Purification, properties, and immunochemical study of ornithine decarboxylase, the key enzyme in polyamine biosynthesis. *Exp. Parasitol.* **67**: 20–30.

16. Hanson, S., Adelman, J. and Ullman, B. (1992) Amplification and molecular cloning of the ornithine decarboxylase gene of *Leishmania donovani*. *J. Biol. Chem.* **267**: 2350–2359.

17. Yarlett, N., Goldberg, B., Moharrami, M. A. and Bacchi, C. J. (1992) Inhibition of *Trichomonas vaginalis* ornithine decarboxylase by amino acid analogs. *Biochem. Pharmacol.* **44**: 243–250.

18. Phillips, M. A., Coffino, P. and Wang, C. C. (1987) Cloning and sequencing of the ornithine decarboxylase gene from *Trypanosoma brucei*. *Mol. Biochem. Parasitol.* **22**: 9–17.

19. Ghoda, L., Phillips, M. A., Bass, K. E., Wang, C. C. and Coffino, P. (1990) Trypanosome ornithine decarboxylase is stable because it lacks sequences found in the carboxyl terminus of the mouse enzyme which target the latter for intracellular degradation. *J. Biol. Chem.* **254**: 11823–11826.

20. Rogers, S., Wells, R. and Rechsteiner, M. (1986) Amino acid sequences common to rapidly degraded proteins: the PEST hypothesis. *Science* **234**: 364–368.

21. Coons, T., Hanson, S., Bitonti, A. J., McCann, P. P. and Ullman, B. (1990) Alpha-difluoromethylornithine resistance in *Leishmania donovani* is associated with increased ornithine decarboxylase activity. *Mol. Biochem. Parasitol.* **39**: 77–90.

22. Bitonti, A. J., Bacchi, C. J., McCann, P. P. and Sjoerdsma, A. (1986) Uptake of α-difluoromethylornithine by *Trypanosoma brucei brucei*. *Biochem. Pharmacol.* **35**: 351–354.

23. Tekwani, B. L., Bacchi, C. J. and Pegg, A. E. (1992) Putrescine activated *S*-adenosylmethionine decarboxylase from *Trypanosoma brucei brucei*. *Mol. Cell. Biochem.* **117**: 53–61.

24. Bacchi, C. J., Lipschik, G. Y. and Nathan, H. C. (1977) Polyamines in trypanosomatids. *J. Bacteriol.* **131**: 657–661.

25. Assaraf, Y. G., Golenser, J., Spira, D. T. and Bachrach, U. (1984) Polyamine levels and the activity of their biosynthetic enzymes in human erythrocytes infected with the malarial parasite *Plasmodium falciparum*. *Biochem. J.* **222**: 815–819.

26. Bitonti, A. J., Dumont, J. A., Bush, T. L. *et al.* (1989) Bis(benzyl) polyamine analogs inhibit the growth of chloroquine resistant human malaria parasites (*Plasmodium falciparum*) *in vitro*

and in combination with α-difluoromethylornithine cure malaria. *Proc. Nat. Acad. Sci. USA.* **86**: 651–655.

27. Bitonti, A. J., McCann, P. P. and Sjoerdsma, A. (1991) In: *Biochemical Protozoology* (eds Coombs, G. and North, M.) Taylor & Francis, London, pp. 517–523.

28. Bitonti, A. J., McCann, P. P. and Sjoerdsma, A. (1987) *Plasmodium falciparum* and *Plasmodium berghei:* Effect of ornithine decarboxylase inhibitors on *Erythrocytic schizogony. Exp. Parasitol.* **64**, 237–243.

29. Wright, P. S., Byers, T. L., Cross-Doersen, D. E., McCann, P. P. and Bitonti, A. J. (1991) Irreversible inhibition of S-adenosylmethionine decarboxylase in *Plasmodium falciparum-*infected erythrocytes: Growth inhibition *in vitro. Biochem. Pharmacol.* **41**: 1713–1718.

30. Baumann, R. J., Hanson, W. L., McCann, P. P., Sjoerdsma, A. and Bitonti, A. J. (1990) Suppression of both antimony-susceptible and antimony-resistant *Leishmania donovani* by a bis(benzyl) polyamine analog. *Antimicrob. Agents Chemother.* **34**: 722–727.

31. Fairlamb, A. H., Blackburn, P., Ulrich, P., Chait, B. T. and Cerami, A. (1985) Trypanothione: a novel bis(glutathionyl) spermidine cofactor for glutathione reductase in trypanosomatids. *Science* **227**: 1485–1487.

32. Tabor, H. and Tabor, C. W. (1975) Isolation, characterization, and turnover of glutathionyl-spermidine from *Escherichia coli. J. Biol. Chem.* **250**: 2648–2654.

33. Fairlamb, A. H. and Cerami, A. (1992) Metabolism and functions of trypanothione in the kinetoplastida. *Annu. Rev. Microbiol.* **46**: 695–729.

34. Henderson, G. B., Murgolo, N. J., Kuriyan, J. *et al.* (1991) Engineering the substrate-specificity of glutathione reductase toward that of trypanothionine reduction. *Proc. Natl. Acad. Sci. USA.* **88**: 8769–8773.

35. Hunter, W. N., Smith, K., Derewenda, Z. *et al.* (1990) Initiating a crystallographic study of trypanothionine reductase. *J. Mol. Biol.* **216**: 235–237.

36. Fairlamb, A. H., Henderson, G. B., Bacchi, C. J. and Cerami, A. (1987) *In vivo* effects of difluoromethylornithine on trypanothionine and polyamine levels in bloodstream forms of *Trypanosoma brucei. Mol. Biochem. Parasitol.* **24**: 185–191.

37. Fairlamb, A. H., Henderson, G. B. and Cerami, A. (1989) Trypanothione is the primary target for arsenical drugs against African trypanosomes. *Proc. Natl. Acad. Sci. USA* **86**: 2607–2611.

38. Henderson, G. B., Ulrich, P., Fairlamb, A. H. *et al.* (1988) 'Subversive' substrates for the enzyme trypanothionine disulfide reductase: alternative approach to chemotherapy of Chagas' disease. *Proc. Natl. Acad. Sci. USA.* **85**: 5374–5378.

39. Yarlett, N., Goldberg, B., Nathan, H. C., Garofalo, J. and Bacchi, C. J. (1991) Differential sensitivity of *Trypanosoma brucei rhodesiense* isolates to *in vitro* lysis by arsenicals. *Exp. Parasitol.* **72**: 205–215.

40. Carter, N. and Fairlamb, A. H. (1993). Arsenical-resistant trypanosomes lack an unusual adenosine transporter. *Nature* 361: 173–176.

41. Yarlett, N., Lindmark, D. G., Goldberg, B., Moharrami, M. A. and Bacchi, C. J. (1994) Subcellular localization of the enzymes of the arginine dihydrolase pathway in *Trichomonas vaginalis* and *Trichomonas foetus. J. Euk. Microbiol.* **41**: 554–559.

42. Schofield, P. J. and Edwards, M. R. (1991) In: *Biochemical Protozoology* (eds Coombs, G and North, M.), Taylor and Francis, London, pp. 102–112.

43. Onodera, R., Yamaguchi, Y. and Morimoto, S. (1983) Metabolism of arginine, citrulline, ornithine and proline by starved rumen ciliate protozoa. *Agric. Biol. Chem.* **47**: 821–828.

44. Reggiani, R., Hochkoeppler, A. and Bertani, A. (1989) Polyamines and anaerobic elongation of rice coleoptile. *Plant Cell Physiol.* **30**: 893–898.

45. Daly, J. J., Sherman, J. K., Gren, L. and Hostetler, T. L. (1989) Survival of *Trichomonas vaginalis* in human semen. *Genitourin. Med.* **65**: 106–108.

46. Williams-Ashman, H. G. (1989) In: *The Physiology of Polyamines* Vol. 1 (eds. Bachrach, U. and Heimer, Y. M.), CRC Press, Boca Raton, pp. 3–20.

47. Yarlett, N. and Bacchi, C. J. (1988) Effect of DL-α-difluoromethylornithine on methionine cycle intermediates in *Trypanosoma brucei brucei. Mol. Biochem. Parasitol.* **27**: 1–10.

48. Byers, T. L., Bush, T. L., McCann, P. P. and Bitonti, A. J. (1991) Antitrypanosomal effects of polyamine biosynthesis inhibitors correlate with increases in *Trypanosoma brucei brucei* S-adenosyl-L-methionine. *Biochem. J.* **274**: 527–533.

49. Mamont, P. S., Danzin, C., Wagner, J., Siat, M., Joder-Ohlenbusch, A. M. and Claverie, N. (1982) Accumulation of decarboxylated S-adenosyl-L-methionine in mammalian cells as a consequence of the inhibition of putrescine biosynthesis. *Eur. J. Biochem.* **123**: 499–504.
50. Pegg, A. E. and McCann, P. P. (1982) Polyamine metabolism and function. *Am. J. Physiol.* **243** (Cell Physiol. 12): C212–C221.
51. Ueland, P. M. (1982) Pharmacological and biochemical aspects of S-adenosylhomocysteine and S-adenosylhomocysteine hydrolase. *Pharmacol. Rev.* **34**, 223–253.
52. Yarlett, N., Garofalo, J., Goldberg, B. *et al.* (1993) S-Adenosylmethionine synthetase in bloodstream *Trypanosoma brucei. Biochim. Biophys. Acta* **1181**: 68–76.
53. Bitonti, A. J., Byers, T. L., Bush, P. J. *et al.* (1990) Cure of *Trypanosoma brucei brucei* and *Trypanosoma brucei rhodesiense* infections in mice with an irreversible inhibitor of S-adenosylmethionione. *Antimicrob. Agents Chemother.* **34**, 1485–1490.
54. Byers, T. L., Casara, P. and Bitonti, A. J. (1992) Uptake of the antitrypanosomal drug 5′-([(Z)-4-amino-2-butenyl]methylamino)-5′-deoxyadenosine (MDL 73811) by the purine transport system of *Trypanosoma brucei brucei. Biochem. J.* **283**, 755–758.
55. Yarlett, N., Quamina, A. and Bacchi, C. J. (1991) Protein methylase in *Trypanosoma brucei brucei:* Activities and response to DL-α-difluoromethylornithine. *J. Gen. Microbiol.* **137**: 717–724.
56. Tidwell, T., Allfrey, V. G. and Mirsky, A. E. (1968) The methylation of histones during regeneration of liver. *J. Biol. Chem.* **243**: 707–715.
57. Shepherd, G. R., Hardin, J. M. and Noland, B. J. (1971) Methylation of lysine residues of histone fractions in synchronized mammalian cells. *Arch. Biochem. Biophys.* **143**: 1–5.
58. Paik, W. K., Lee, H. W. and Morris, H. P. (1972) Protein methylases in hepatomas. *Cancer Res.* **32**: 37–40.
59. Bender, K., Betschart, B., Schaller, J., Kaupfer, U. and Hecker, H. (1992) Structural differences between the chromatin of procyclic *Trypanosoma brucei brucei* and of higher eukaryotes as probed by immobilized typsin. *Acta Trop.* **50**: 169–184.
60. Duschak, V. G. and Cazzulo, J. J. (1990) The histones of the insect trypanosomatid, *Crithidia fasciculata. Biochim. Biophys. Acta.* **1040**: 159–166.
61. Giffin, B. F. and McCann, P. P. (1989) Physiological activation of the mitochondrion and the transformation capacity of DFMO induced intermediate and short stumpy bloodstream form trypanosomes. *Am. J. Trop. Med. Hyg.* **40**, 487–493.
62. Penketh, P. G., Krishnamurthy, S., Divo, A. A., Patton, C. L. and Sartorelli, A. C. (1990) Methylating agents as trypanocides. *J. Med. Chem.* **33**: 730–732.
63. Singh, R. P., Saxena, J. K., Ghatak, O. P., Wittich, R.-M. and Walter, R. D. (1989) Polyamine metabolism in *Setaria cervi.* Parasitol. Res. **75**: 311–315.
64. Walter, R. D. (1988) Polyamine metabolism of filaria and allied parasites. *Parasitol. Today* **4**: 18–20.
65. Sharma, V., Tekawani, B. L., Saxena, J. K. *et al.* (1991) Polyamine metabolism in some helminth parasites. *Exp. Parasitol.* **72**: 15–23.
66. Muller, S., Wittich, R.-M. and Walter, R. D. (1988) In: *Progress in Polyamine Research* (eds Zappia, V. and Pegg, A. E.), Plenum Press, New York, pp. 737–743.
67. Kallio, A. and Janne, J. (1983) Role of diamine oxidase during the treatment of tumour-bearing mice with combinations of polyamine antimetabolites. *Biochem. J.* **212**: 895–898.
68. Sharma, V., Katiyar, J. C., Ghatak, S. N. and Shukla, O. P. (1988) Ornithine aminotransferase of hookworm parasites. *Indian J. Parasitol.* **12**: 237–242.
69. Jarroll, E. L. and Lindmark, D. G. (1990) Giardia metabolism. In: *Giardiasis* (ed. Meyer, E. A.), Elsevier, Amsterdam, pp. 61–76.
70. Wang, C. C., Verham, R., Rice, A. and Tzeng, S. (1983) Purine salvage by *Trichomonas foetus. Mol. Biochem. Parasitol.* **8**: 325–337.
71. Gottlieb, M. and Dwyer, D. M. (1988) In: *The Biology of Parasitism* (eds Englund, P. T. and Sher, A.), Alan R. Liss, New York, pp. 449–465.
72. Marr, J. J. and Docampo, R. (1986) Chemotherapy for Chagas' disease: perspective of current therapy and considerations for future research. *Rev. Infect. Dis.* **8**: 884–903.

73. Ghoda, L. Y., Savarese, T. M., Northup, C. H. *et al.* (1988) Substrate specificities of 5'-deoxy-5'-methylthioadenosine phosphorylase from *Trypanosoma brucei brucei*. *Mol. Biochem. Parasitol.* **27**, 109–118.

74. Bacchi, C. J., Sufrin, J., Nathan, H. C. *et al.* (1991) 5'-Alkyl-substituted analogs of 5'-methylthioadenosine as trypanocides. *Antimicrob. Agents Chemother.* **35**: 1315–1320.

75. Chiang, P. K. and Miura, G. A. (1986) In: *Biological Methylation and Drug Design* (eds Borchardt, R. T., Creveling, C. R. and Ueland, P. M.), Humana Press, Clifton, pp. 239–252.

76. Kneifel, H., Stetter, K. O., Andreeson, J. R., Wiegel, J., Konig, H. and Schoberth, S. M. (1986) Distribution of polyamines in representative species of archaebacteria. *System. Appl. Microbiol.* **7**: 241–245.

8 Lipid and Membrane Metabolism of the Malaria Parasite and the African Trypanosome

WALLACE R. FISH

Pediatric Endocrinology, State University of New York— Health Science Center at Syracuse, Syracuse, N.Y., USA.

SUMMARY

Comparison of the lipids and lipid and membrane metabolism of the malaria parasite and the African typanosome with that of their environments emphasizes the unique ways in which parasites are able to access the nutrients available, and modify them to their advantage.

The differences between the salivarian trypanosomes and the *Plasmodium* species highlight the distinctions between situations where the host has limited access to the parasite (the erythrocyte-sequestered malaria parasite), and a totally extracellular existence (the bloodstream and insect form African trypanosome). The intraerythrocytic stages of the malaria parasite must not only accommodate the lipid requirements of both differentiation and massive multiplication, but also must prepare infective merozoites for re-invasion of erythrocytes. The African trypanosome, which produces only one daughter cell while maintaining a very rapid metabolic rate, must anticipate the stresses of temperature and nutrient changes upon entering the tsetse fly midgut. This is partially accomplished through significant alteration in the lipid composition of the bloodstream stumpy form. In both cases the continuation of the life cycle is the primary goal.

Both organisms acquire host lipid, albeit through different paths, possess significant pathways to modify these lipids, and have enzymic activities, for example phospholipases or transporters with unusual properties, which are able to facilitate the process. The malaria parasite, sequestered within the host

Biochemistry and Molecular Biology of Parasites
ISBN 0-12-473345-X

erythrocyte, may manipulate the host membrane lipid composition and shows greater biosynthetic power than the trypanosome.

Once host lipid or lipid constituents are acquired, both organisms respond in a similar fashion—modification of the gross lipid composition and, in some cases, the fatty acyl chain composition of complex lipids. In most cases it is unknown why these changes are needed. The malaria parasite may be attempting to alter the permeability of the parasitized cell to allow more facile entry of nutrients. A significant part of the African typanosomes' lipid metabolism is directed toward synthesis of not only membrane lipid, but supply of the glycolipid anchor of its unique surface coat.

1. INTRODUCTION

Protozoan parasites must interact with the host at a variety of levels: the acquisition of nutrients; evasion or confusion of the host's immune response; establishment and maintenance of the infected state. This interface and the interactions are, by necessity, membrane mediated. Most parasites possess more than one life cycle stage, including those in transmission vectors, and exist at various temperatures and in different milieus; this requires significant membrane alterations. The success of parasitic organisms shows that these processes are well established. They are reflected in parasite lipid metabolism which includes the classical definition as well as 'remodeling' of host lipids and membranes. This chapter will discuss the Apicomplexan intracellular parasites *Plasmodium* and *Toxoplasma* and compare them with the extracellular African trypanosomes. The lipid and membrane metabolism of the Apicomplexa, particularly *Plasmodium*, has been extensively reviewed (1–4), and only phenomena of current investigation are presented. The lipid metabolism of the African trypanosomes will necessitate a more complete discussion. The reader is encouraged to supplement this information by correlating it with the discussion of glycoproteins in Chapter 11. Their location and position are largely dependent on lipid anchoring but their functions are quite specific to their structure. Their function is a result of the complementarity of lipid and glycoprotein biochemistry.

2. THE APICOMPLEXAN PARASITES

In Apicomplexan parasites such as *Plasmodium*, *Toxoplasma* and *Babesia* lipid metabolism may be involved in invasion of host cells, concomitant parasitophorous vacuole formation, and differentiation and growth throughout the life cycle. During invasion of host cells and formation of the parasitophorous vacuole considerable membrane deformation, reformation and growth occur. The parasitophorous vacuole membrane is formed very rapidly upon invasion of the host cell through the activity of organellar (rhoptry) lipids (5). A variety of polypeptides also participate (4, 6, 7). The two processes of merozoite invasion and parasitophorous vacuole formation may be considered as continuous although distinct steps are apparent. Mikkelsen *et al.* (8) observed erythrocytes during, and after, invasion by *P. falciparum* merozoites metabolically labeled with fluorescent phospholipids (PhL). Upon invasion, lipids in the merozoite apical end distributed to the merozoite–erythrocyte membrane junction and into the erythrocyte membrane. After establishment of intracellular infection, the label was observed only in the parasitophorous vacuole membrane and parasite plasma membrane. It was concluded that no significant membrane metabolism had occurred; parasitophorous vacuole membrane formation involved the insertion of parasite lipid into the host erythrocyte membranes. Dluzewski et al. (9) followed invasion by P. falciparum, and parasitophorous vacuole membrane formation, of erythrocytes labeled with tagged phosphatidylethanolamine (PE). The absence of the label from both the merozoite attachment site and the region containing the intraerythrocytic parasite indicated that the parasitophorous vacuole membrane was predominantly, or wholly, of parasite origin. Only one partially characterized Plasmodium-specific lipid, a phosphoglycolipid, has been reported to be involved directly in erythrocyte invasion (10). Immunolocalization showed concentration at the merozoite apical end and the lipid was present where membrane deformation occured during erythrocyte invasion. The antibody also inhibited merozoite invasion in vitro.

Other studies suggest that merozoite lipids are not the source of the parasitophorous vacuole membrane. Haldar and Uyetake (11) observed that when erythrocytes were labeled with a fluorescent lipophilic probe it was internalized upon invasion of erythrocytes by P. falciparum merozoites, but not during subsequent growth. Ward et al. (12) concluded that labeled lipids in the erythrocyte membrane are indistinguishable from those of the forming parasitophorous vacuole membrane. The deposition of fluorescent lipophilic probes in the erythrocyte membrane was followed during merozoite invasion, and the probe did not change apparent concentration in any membrane during invasion and parasitophorous vacuole membrane formation, indicating that there was no addition of exogenous lipid.

Ultrastructural observations suggested that during invasion Plasmodium rhoptries extrude copious membranous material (see ref. 6 and references therein), which may provide a matrix for rhoptry proteins (13). Some rhoptry proteins bind to both erythrocyte inside-out vesicles and to vesicles prepared from phospholipids predominant in the erythrocyte membrane inner leaflet, particularly phosphatidylserine (PS) and phosphatidylinositol (PI) (14,15). One has been shown to intercalate into the erythrocyte membrane (16). These proteins, the RHOP-H complex and SERA, are proposed to integrate differentially into the erythrocyte membrane upon invasion, causing an inward membrane expansion, to produce the parasitophorous vacuole.

On balance, the evidence points to rhoptry lipids being involved, at the least, in sequestering proteins which, upon introduction into the erythrocyte membrane with or without rhoptry lipids, are directly responsible for the destabilization of the erythrocyte membrane, and subsequent parasitophorous vacuole membrane formation. Whether rhoptry lipids actually become an integral part of the parasitophorous vacuole membrane, wholly or in combination with erythrocyte lipids, is not clear. Rhoptry lipids probably combine directly with erythrocyte lipids in the parasitophorous vacuole membrane.

Both lipids and polypeptides are implicated in host cell invasion by the related parasite Toxoplasma gondii (for review, see refs. 17,18). The rhoptry lipid composition of T. gondii is unusual (19). The cholesterol (Cho) to PhL ratio is high (1.48), and phosphatidylcholine (PC) accounted for greater than 75% of the rhoptry total lipids (TL). The remaining identifiable species were PE, phosphatidic acid (PA) and lysolipids. Notable was the absence of PS, PI and sphingomyelin (SM), which are found in whole parasites. If this unusual rhoptry lipid composition were transferred to the parasitophorous vacuole membrane, it could participate in the binding and stabilization of vimentin-type intermediate filaments to the *Toxoplasma* parasitophorous vacuole membrane (20). These filaments may play a part in the aggregation of host cell organelles around the parasitophorous vacuole and, thus, could participate in supplying metabolites for biosynthesis (20).

A parasite phospholipase (most likely PLA$_2$) may participate in host cell invasion by *T. gondii* (21). The presumptive PLA$_2$ enhanced host cell invasion. Phospholipase has been shown to solubilize rhoptry proteins of *P. falciparum* presumably involved in erythrocyte invasion (22). Solubilization of internal, latent, membrane-bound rhoptry proteins by lipid degradation may be a general method in the Apicomplexa for initiating active host cell penetration and parasitophorous vacuole membrane formation after attachment.

2.1. Intracellular Growth

Once established within the erythrocyte, Apicomplexan parasites undergo multiple transformations and growth. *Plasmodium* species are unable to synthesize fatty acids (FA) *de novo*, and their ability to saturate, desaturate, elongate or shorten fatty acids is severely limited or non-existent (2, 23). The parasite must acquire much of its lipid either from the parasitized host cell or the external milieu. How does the malaria parasite provide for the tremendous increase in lipid necessary to assemble the increased membranous material?

For *P. falciparum*, and the related parasite *Babesia divergens*, high density lipoproteins (HDL) are a major source of preformed lipid, in particular PC, significant amounts of which were converted into PE (24, 25). A unidirectional flux of PC to the erythrocyte membrane from HDL and subsequently to the intracellular parasite was proposed for both parasites. Malaria-infected erythrocytes can acquire phospholipids (PC, PE, PS) directly from lipid vesicles; this phenomenon, which is enhanced by PhL transfer proteins, may involve exchange with the erythrocyte membrane (26).

Pouvelle *et al.* (27) describe an organelle of parasite origin (the 'parasitophorous duct') within *P. falciparum*-infected erythrocytes that directly connects the erythrocyte and parasitophorous vacuole membranes. High density lipoproteins could be transported through this organelle by passive diffusion to the interior; the intraerythrocytic parasite may acquire HDL by fluid phase endocytosis. Specific antibodies against the two major HDL apolipoproteins do not localize within intracellular parasites, suggesting that lipids must be acquired from HDL directly, without endocytosis (24). However, the existence of the parasitophorous duct has been questioned (11, 12, 28).

Thus, the infected erythrocyte has access to host plasma lipids, predominantly from HDL. Further, the HDL, or its lipids, must be transported from the erythrocyte membrane and through the parasitophorous vacuole membrane to the developing parasite (for review see ref. 29). The importance of this for *Plasmodium* is illustrated by the work of Grellier *et al.* (30) who showed that HDL alone supports the differentiation and multiplication of the parasite. Very low density lipoproteins (VLDL), low density lipoproteins (LDL) or apolipoproteins do not support complete schizogony and subsequent erythrocyte reinvasion. Host lipid, therefore, is essential for parasite growth, differentiation and continuation of the life cycle. These lipids are remodeled by the parasite.

The uptake of exogenous choline for PC synthesis is accepted (31, 32), although the nature of transport is contentious (33, 34). The disparate observations are perhaps due to variance in experimental technique: age of the erythrocytes employed; cation imbalance; reaction termination conditions; in vitro choline concentration; and innate characteristics of the two Plasmodium species examined (33, 34). It is also possible that interspecies differences in the enzymes involved in the ensuing metabolic fate of choline, for example choline kinase (35) or choline phosphotransferase (36), also influence the results, as well as interactions between the alternative pathways of PC synthesis from PE and PS. In a study of regulation of PC biosynthesis in P. knowlesi-infected erythrocytes (37), the parasite showed a comprehensive ability to modulate the pathway of PC synthesis from external choline depending on the availability of precursors, or alternative pathways.

Incorporation of other PhL components by the malaria parasite is not as well studied as PC. Exogenous serine, inositol and ethanolamine are incorporated into the corresponding PhL, PE being converted into PC; inositol is only found in PI. Incorporation of ethanolamine ≫ serine > choline ~ inositol, compared to uninfected erythrocytes (38). The much higher use of ethanolamine is probably due to its role as a known source of PC by transmethylation. Serine shows a high incorporation into PE through decarboxylation of PS, and then subsequent conversion into PC; conversion of PE into PS, performed by polar head group exchange in other organisms, probably does not occur in Plasmodium (2). Both P. falciparum and P. knowlesi synthesize PI from CDP-diacylglycerol and inositol via a PI synthase. The metabolism of PI also includes PI-bisphosphate and the secondary messengers diacylglycerol (DAG) and inositol-1,4,5-triphosphate, as part of an operational polyphosphoinositide cycle (2, 39). A sphingomyelin synthase is present in P. falciparum, and appears developmentally regulated (40, 41). Localization to the perinuclear region is unusual, since it is a cis-Golgi resident activity in other cell types. The parasite is able, therefore, to acquire host PhL, and their components, for synthesis of most major PhL types.

P. falciparum cultivated in vitro incorporates free fatty acids (FFA) into the major PhL species and other complex lipids (42, 43). Infected erythrocytes contain significantly higher acyl-CoA synthetase activity than normal erythrocytes (44). Although Plasmodium fatty acid transfer proteins have not been described, the surface of P. falciparum merozoites shows a 75 kDa heat shock protein cognate, which are known to bind fatty acids (2).

The malaria parasite, therefore, is replete with pathways to complete de novo synthesis of phospholipids, as long as a continuing supply of fatty acids, DAG and PhL components is available from host lipid.

In P. falciparum and P. knowlesi synthesis of glycolipids increases with intracellular development of the parasite. There is a stage-specificity to some glycolipid classes and differences among P. falciparum strains (45, 46).

Distinct differences and changes in the complex lipid composition and their acylation patterns are found among isolated parasites and in the plasma membranes of infected and uninfected erythrocytes. This has been determined for P. chabaudi in the mouse (47), in vitro propagated P. falciparum (48, 49) and P. knowlesi in monkeys (50, 51).

Isolated parasites possessed variably increased levels of PC and PI, and greatly decreased SM and PS when compared with uninfected erythrocyte membranes. However, whereas PE levels in isolated parasites were essentially unchanged in *P. falciparum* and *P. chabaudi*, *P. knowlesi* showed significantly higher PE levels than uninfected erythrocytes, confirming earlier findings (52). Total lipid fatty acid composition was dissimilar among *P. falciparum*, *P. chabaudi* and *P. knowlesi* (47, 48, 53), especially the major fatty acids ($C_{16:0}$, $C_{18:0}$, $C_{18:1}$, $C_{18:2}$ $C_{20:4}$). Although the unsaturation index of *P. knowlesi*- and *P. chabaudi*-infected erythrocytes appeared constant, *P. falciparum* showed a significant decrease. *P. falciparum* and *P. knowlesi* showed a significant decrease in $C_{20:4}$. The fatty acid composition of individual *P. falciparum* phospholipid classes was also examined, and those from parasite and infected erythrocyte membranes were very similar but different from normal erythrocytes. Most striking was a significant increase in $C_{16:0}$ and $C_{18:2}$ and a reduction in $C_{20:4}$, in the PE fraction, particularly of isolated parasites. PE showed the greatest decrease in unsaturation index of any PhL species.

Therefore, the FA composition of individual lipid classes, as well as the gross PhL composition of the malaria parasite is substantially different from that of the normal erythrocyte. In addition, the host cell membrane is changed in *P. falciparum*, which in some cases extends to the PhL molecular species composition. The fatty acid molecular species composition of some PhL from membranes of *P. chabaudi*- and *P. knowlesi*-infected erythrocytes has been examined (26, 54). Erythrocyte membranes from *P. chabaudi*-infected cells were similar in PC molecular species composition to both cohort unparasitized erythrocytes, and erythrocytes from uninfected animals. All changes in PC molecular species composition in infected erythrocytes could be attributed to the parasite. The parasite showed increases in most species containing $C_{20:4}$, and decreases in those showing only saturated FA. Therefore, at least in *P. chabaudi*, parasite PC is much less saturated than the host cell membrane. No changes in plasma PC molecular species composition were seen in infected animals. In contrast, *P. knowlesi* showed a major increase in one PC species, $C_{16:0/18:2}$, and the host cell membranes from infected erythrocytes displayed alterations in molecular species composition similar to those of the isolated parasite (26).

The erythrocyte-bound malaria parasite acquires host lipids by diverse methods and transports to the area of celluar growth. There they are remodeled significantly to meet the parasite's needs, including phospholipid head group alteration, modification of the fatty acid molecular species composition of certain phospholipids, and transformation of the erythrocyte membrane lipid composition in both infected and uninfected cells.

3. EXTRACELLULAR PARASITES

As the only unicellular parasite to live free in the host's bloodstream the African, or salivarian, trypanosomes are unique. Unfortunately, there is little information on African trypanosome lipids and lipid metabolism. The life cycle of the African trypanosome involves bloodstream forms in the mammalian host and various insect forms in the tsetse fly vector. Of the insect forms, only the *in vitro* propagated procyclic form, equivalent to those in the tsetse fly midgut, has been examined. Early studies used complex culture medium that contained significant exogenous lipid. Later studies used more defined media, which usually included fetal bovine serum. Differences in blood-stream and procyclic form lipid composition can be due to differences in individual isolates and, in the case of procyclic forms, the medium used for cultivation.

Dixon and Williamson (55) analysed the lipid composition of *T. brucei rhodesiense* bloodstream and procyclic forms. Although sterol esters (SE), FFA and sterols were about equal between the two stages, procyclic forms contained much more triacyl-glycerol (TAG) than bloodstream forms. Both life cycle stages contained similar amounts of PC, but procyclic forms contained more PE and far less SM than bloodstream forms. When the fatty acid (FA) content of bloodstream and procyclic forms was compared with that of host plasma and culture medium, respectively, significant differences were seen. Bloodstream forms showed less $C_{16:0}$ and $C_{18:1}$ than their environment, but greatly increased polyunsaturated fatty acids, namely $C_{18:2}$ and $C_{22:6}$. Procyclic forms showed much lower $C_{16:0}$ and higher $C_{18:0}$ levels than unused medium whereas $C_{18:1}$ was decreased and $C_{18:2}$ was much higher than that supplied. Curiously, these forms also possessed $C_{22:5}$, which was not present in preculture medium. Therefore, both bloodstream forms and *in vitro* propagated procyclic forms significantly increase their overall FA unsaturation over that available, indicating an ability to desaturate acyl chains. Significant chain elongation was also apparent.

The complex lipid composition of *T. b. brucei* bloodstream forms has been examined (56). Phospholipids showed a distribution of PC≫PE≫SM > PI ∼ PS, whereas in neutral lipid, Cho≫FFA > TAG ∼ SE. In general this PhL distribution, except for SM, has been confirmed (57, 58). The total lipid FA composition of the Trypanozoon subspecies *T. b. brucei*, *T. b. gambiense* and *T. b. rhodesiense* are similar and all species showed a high degree of FA unsaturation (59).

Dixon *et al.* (60) assessed the incorporation of radiolabeled glucose, glycerol and acetate into the complex lipids and FA of *T. b. rhodesiense* bloodstream and procyclic forms. Procyclics showed a higher incorporation of glucose into total lipid than bloodstream forms and were able to incorporate acetate into FA, with incorporation into saturates≫trienes≫monoenes/dienes. Incorporation into different chain length FA showed $C_{18} > C_{22} \sim C_{14} > C_{16} > C_{20}$. Bloodstream forms were able to accomplish significant FA chain elongation and desaturation. Both bloodstream and procyclic forms were able to oxidize FA, with procyclic forms showing higher activity. Procyclics generally showed a greater capacity than bloodstream forms at incorporating FA into complex lipids and, in both forms, incorporation into PhL was generally greater than into neutral lipid. Bloodstream forms consistently incorporated more total FA into phospholipids than procyclic forms, whereas the latter showed higher levels in neutral lipid. In general, procyclic forms incorporated FA into PC≫PI > PE, whereas blood-stream forms reversed the order of PI and PE. These studies pointed out at an early date that the FA and complex lipid metabolism of the two life-cycle stages were significantly different.

Venkatesan and Ormerod (61) examined lipid composition differences between two different bloodstream form stages of *T. b. rhodesiense*, the long slender form that is actively dividing, and the 'stumpy' form, a non-dividing form that differentiates from long slender forms. Although similar in the percentage of some lipid classes, the stumpy forms showed elevated total lipid, total neutral lipid, sterol ester, and PhL; PI was decreased, compared to slender forms. Stumpy forms also contained plasmalogens (alk-1-enyl ether linkage of FA) in the phospholipid fraction. The total polar lipid FA of both forms were similar. The polar lipid of both forms, when compared to host plasma, showed significant decreases in $C_{16:0}$ and 20-carbon polyenoic FA, whereas $C_{18:2}$ and 22-carbon polyenoic FA were greatly increased. Stumpy forms showed cholesterol ester with $C_{16:0}$, $C_{18:0}$, $C_{18:1}$, $C_{18:2}$, $C_{22:4}$ and $C_{22:5}$ increased over that of the host; 20-carbon polyenoics and $C_{22:6}$ were decreased. The proportion of most FA could be modulated by feeding the host (rat) a fat-deficient diet. This supported earlier studies which suggested that the bloodstream form trypanosome obtains much of its lipid from the host and that it is significantly remodeled.

These observations are interesting because the stumpy forms of pleomorphic salivarian trypanosomes are thought to be 'primed' for differentiation into procyclic forms after the tsetse fly takes a bloodmeal. When tsetse flies were fed on *T. congolense* procyclic forms suspended in a medium containing normal or delipidated fetal bovine serum, comparable midgut infections developed (62). However, flies fed on lipid-free medium showed significantly lower infection levels in the labrum and hypopharynx (epimastigotes and metacyclics). Therefore, lipid is necessary to achieve successful metacyclogenesis and continuation of the life cycle.

The fatty acids of the bloodstream forms of salivarian trypanosomes appear, on the most part, to be qualitatively similar to their host, although the quantitative FA composition of some lipid classes is much different. The African trypanosome seems to possess the ability to rearrange the FA composition of individual complex lipid classes and, through chain elongation and desaturation to modify host-acquired FA.

Procyclic forms do not synthesize cholesterol *de novo* but can incorporate this sterol if supplied exogenously; *de novo* synthesis of ergosterol does occur. Bloodstream forms are able to acquire and incorporate the major mammalian bloodstream sterol into their membranes although they possess no ability to synthesize sterol *de novo* (63). The method of acquiring host cholesterol is of major significance.

Both life cycle stages of *T. b. brucei* showed a significant and biphasic uptake of FA from the external medium, which was promoted by the presence of bovine serum albumin. The rate of FA uptake by culture forms was almost 16-fold greater than bloodstream forms of the same clone. The reversibility of the process was noted in the presence or absence of bovine serum albumin, suggesting a direct interaction of the parasite with bovine serum albumin and FA–albumin complexes. The finding that procyclic trypanosomes take up significantly more exogenous FFA than bloodstream forms is even more curious when one examines the literature on *de novo* FA biosynthesis by procyclic forms. *T. brucei* procyclic forms utilized large amounts of threonine from a synthetic culture medium; the products of this process were glycine and acetate, which were excreted in nearly equimolar amounts. It was postulated that threonine was the preferred substrate for acetate production and, although much acetate was excreted, it could be used for FA synthesis. The enzymes responsible were on NAD^+-dependent L-threonine dehydrogenase and aminoacetone synthase (acetyl-CoA:glycine C-acetyltransferase). Subsequently it was shown (64, 65) that threonine was the preferred substrate for acetate production even in the presence of excess acetate or glucose. Incorporation into lipid was over the *in vitro* growth cycle, and threonine contributed the majority of carbon atoms to palmitic and stearic acids. A known inhibitor of L-threonine dehydrogenase decreased the production of acetate and glycine from L-threonine, decreased incorporation of label from threonine into lipid and was trypanocidal. Additional investigation into threonine catabolism in the African trypanosomes showed that bloodstream forms also possessed this pathway (66). Therefore, bloodstream forms could obtain acetate units from threonine which explains the fact that bloodstream forms are able to elongate fatty acids chains (67). The two enzymes must be closely linked *in vivo* and are found in the mitochondrial matrix (68).

There are few further assessments of the biosynthetic capabilities of procyclic forms. Opperdoes (69) performed a detailed study on the presence and intracellular localization of alkoxyphospholipid biosynthesis enzymes in *T. brucei* procyclic forms. The complete pathway was reported to be localized in the microbody or 'glycosome' fraction, as were all other enzymes essential for *de novo* synthesis of this lipid class from glycerol and acyl-CoA (see Chapter 2). These pathways also exist in bloodstream forms.

As obligate blood parasites with limited *de novo* synthetic capabilities, both the bloodstream form of African trypanosomes and the intracellular malaria parasite must acquire and remodel host complex lipids. *Plasmodium* sp. possesses enzymic activities, particularly those involving PC and PE, that allow them to synthesize complex phospholipids. The African trypanosome must also obtain and synthesize complex lipids but has evolved a different strategy, which has been admirably reviewed by Mellors and Samad (70).

Early studies were concerned with the anemia associated with trypanosomiasis, perhaps caused by free fatty acids and lysophospholipids, as the products of phospholipases. Phospholipase A_1 is found in plasma and tissue fluids of infected animals and cleaves acyl chains from both exogenous lysophospholipids and PC, yielding both free fatty acids and 2-lysophosphatidylcholine. The latter lipid is then rearranged to a 1-acyl derivative, and is again acted upon by phospholipase A_1. Procyclic forms show much less activity than bloodstream forms, where it is associated with the plasma membrane.

Samad *et al.* (71) identified considerable acyl-CoA-dependent acyltransferase and acyl-CoA hydrolase activities in bloodstream forms of *T. brucei*. The former enzyme can acylate exogenous lysophospholipid with the specificity: lysoPC > lysoPI > lysoPE > lysoPA > lysoPS. The enzyme is membrane-bound and shows a preference for CoA-acyl thioesters different from that of mammalian acyl transferases which are frequently specific for arachidonate ($C_{20:4}$). The acyl-CoA hydrolase activity is most likely cytoplasmic in origin and does not appear to be involved in PhL synthesis. Besides acyltransferase and phospholipase A_1, an acyl-CoA ligase has been described; all three enzymes are localized in the plasma membrane. Procyclic forms possess little activity of any of the three enzymes which is concordant with the greater biosynthetic power of the insect form (72). Interconversion of diacyl phospholipids is suggested by the above (70, 71), as is PC generated by transmethylation of PE, and PE by decarboxylation of PS, although the amounts produced are comparable to that from transacylation.

From these studies it emerges that the lipid biosynthetic pathways of the African trypanosomes are complex, differing both from the host and the malaria parasite, and rely heavily on an external source of preformed lipid. Host complex lipids bound to serum proteins, most likely lipoproteins and particularly HDL, are a source of lipid for African trypanosomes as they appear to be for *Plasmodium*.

Bloodstream forms of *T. b. brucei* propagated *in vitro* under axenic conditions will not continuously divide without exogenously supplied LDL or HDL (73); apolipoproteins do not support growth and anti-apolipoprotein antibodies inhibited growth. LDL- and HDL-bound cholesterol, cholesterol ester, cholesterol oleoyl ether and PC were all accumulated, and some metabolized, by the parasite. In most cases uptake was more efficient from HDL. Cholesterol ester was cleaved to free cholesterol and FA by actively dividing, long slender trypomastigotes, although this life-cycle stage possessed only a limited ability to re-esterify cholesterol. This suggests that cholesterol ester serves as a source for bulk sterol, and for acyl chains for the PhL remodeling reactions described above. In contrast, short stumpy forms of the parasite showed efficient esterification of acyl chains to cholesterol, supporting the earlier observations of Venkatesan and Ormerod (61) that stumpy forms contain increased amounts of SE. Vandeweerd and Black (74) also noted that lipoprotein-bound PC was acquired and metabolized by bloodstream forms and the conversion products did not show lysophosphatidylcholine.

Therefore, there is firm evidence that the salivarian trypanosomes acquire, and subsequently modify, host-derived lipids, the majority of which are obtained from LDL or HDL. Stage-specific differences in *de novo* lipid synthesis are well supported. Procyclic forms are able to synthesize, elongate and desaturate FA, and to synthesize sterol (ergosterol). The lack of sterol synthesis by bloodstream forms is firmly established (see also chapter 11).

4. ACKNOWLEDGEMENTS

This chapter is dedicated to the late Dr George G. Holz, Jr—mentor, colleague and friend. The author thanks Dr H. J. Vial (Montpellier, France) for prepublication manuscripts, and Drs I. W. Sherman (Irvine, CA, USA) and A. Mellors (Guelph, Ontario, Canada) for comments on this chapter. Mr Castor Wamukoya Kweyu (ILRAD, Nairobi, Kenya) is thanked for help in gathering the materials needed to write this chapter.

5. REFERENCES

1. Simoes, A. P., Roelofsen, B. and Op den Kamp, J. A. F. (1992) Lipid compartmentalization in erythrocytes parasitized by *Plasmodium*-spp. *Parasitol. Today* **8**: 18–21.
2. Vial, H. J. and Ancelin, M. L. (1992) Malarial lipids. An overview. *Subcell. Biochem.* **18**: 259–306.
3. Smith, J. D. (1993) Phospholipid biosynthesis in protozoa. *Prog. Lipid Res.* **32**: 47–60.
4. Holder, A. A., Blackman, M. J., Borre, M. *et al.* (1994) Malaria parasites and erythrocyte invasion. *Biochem. Soc. Trans.* **22**: 291–295.
5. Joiner, K. A. (1991) Rhoptry lipids and parasitophorous vacuole formation—a slippery issue. *Parasitol. Today* **7**: 226–227.
6. Bannister, L. H. and Dluzewski, A. R. (1990) The ultrastructure of red cell invasion in malaria infections: a review. *Blood Cells* **16**: 257–292.
7. Gratzer, W. B. and Dluzewski, A. R. (1993) The red blood cell and malaria parasite invasion. *Semin. Hematol.* **30**: 232–247.
8. Mikkelsen, R. B., Kamber, M., Wadwa, K. S., Lin, P.-S. and Schmidt-Ullrich, R. (1988) The role of lipids in *Plasmodium falciparum* invasion of erythrocytes: a coordinated biochemical and microscopical analysis. *Proc. Natl. Acad. Sci. USA* **85**: 5956–5960.
9. Dluzewski, A. R., Mitchell, G. H., Fryer, P. R., Griffiths, S., Wilson, R. J. M. and Gratzer, W. B. (1992) Origins of the parasitophorous vacuole membrane of the malaria parasite, *Plasmodium falciparum*, in human red blood cells. *J. Cell Sci.* **102**: 527–532.

10. Sjoberg, K., Hosein, Z., Wahlin, B. *et al.* (1991) *Plasmodium falciparum*–an invasion inhibitory human monoclonal antibody is directed against a malarial glycolipid antigen. *Exp. Parasitol.* **73**: 317–325.

11. Haldar, K. and Uyetake, L. (1992) The movement of fluorescent endocytic tracers in *Plasmodium falciparum* infected erythrocytes. *Mol. Biochem. Parasitol.* **50**: 161–177.

12. Ward, G. E., Miller, L. H. and Dvorak, J. A. (1993) The origin of parasitophorous vacuole membrane lipids in malaria-infected erythrocytes. *J. Cell Sci.* **106**: 237–248.

13. Stewart, M. J., Schulman, S. and Vanderberg, J. P. (1986) Rhoptry secretion of membranous whorls by *Plasmodium falciparum* merozoites. *Am. J. Trop. Med. Hyg.* **35**: 37–44.

14. Sam-Yellowe, T. Y. (1992) Molecular factors responsible for host cell recognition and invasion in *Plasmodium falciparum*. *J. Protozool.* **39**: 181–189.

15. Perkins, M. E. and Ziefer, A. (1994) Preferential binding of *Plasmodium falciparum* SERA and rhoptry proteins to erythrocyte membrane inner leaflet phospholipids. *Infect. Immun.* **62**: 1207–1212.

16. Sam-Yellowe, T. Y., Shio, H. and Perkins, M. E. (1988) Secretion of *Plasmodium falciparum* rhoptry protein into the plasma membrane of host erythrocytes. *J. Cell Biol.* **106**: 1507–1513.

17. Bonhomme, A., Pingret, L. and Pinon, J. M. (1992) Review: *Toxoplasma gondii* cellular invasion. *Parassitologia* **34**: 31–43.

18. Joiner, K. A., Beckers, C. J. M., Bermudes, D., Ossorio, P. N., Schwab, J. C. and Dubremetz, J. F. (1994) Structure and function of the parasitophorous vacuole membrane surrounding *Toxoplasma gondii*. *Ann. N Y Acad. Sci.* **730**: 1–6.

19. Foussard, F., Lerichie, M. A. and Dubremetz, J. F. (1991) Characterization of the lipid content of *Toxoplasma gondii* rhoptries. *Parasitology* **102**: 367–370.

20. Halonen, S. K. and Weidner, E. (1994) Overcoating of *Toxoplasma* parasitophorous vacuoles with host cell vimentin type intermediate filaments. *J. Euk. Microbiol.* **41**: 65–71.

21. Saffer, L. D. and Schwartzman, J. D. (1991) A soluble phospholipase of *Toxoplasma gondii* associated with host cell penetration. *J. Protozool.* **38**: 454–460.

22. Etzion, Z., Murray, M. C. and Perkins, M. E. (1991) Isolation and characterization of rhoptries of *Plasmodium falciparum*. *Mol. Biochem. Parasitol.* **47**: 51–62.

23. Vial, H. J., Ancelin, M.-L., Philippot, J. R. and Thuet, M. J. (1990) Biosynthesis and dynamics of lipids in *Plasmodium*-infected mature mammalian erythrocytes. *Blood Cells* **16**: 531–555.

24. Grellier, P., Rigomier, D., Clavey, V., Fruchart, J.-C. and Schrevel, J. (1991) Lipid traffic between high density lipoproteins and *Plasmodium falciparum*-infected red blood cells. *J. Cell Biol.* **112**: 267–277.

25. Valentin, A., Rigomier, D., Precigout, E., Carcy, B., Gorenflot, A. and Schrevel, J. (1991) Lipid trafficking between high density lipoproteins and *Babesia divergens*-infected human erythrocytes. *Biol. Cell* **73**: 63–70.

26. Simoes, A. P., Moll, G. N., Beaumelle, B., Vial, H. J., Roelofsen, B. and Op den Kamp, J. A. F. (1990) *Plasmodium knowlesi* induces alterations in phosphatidylcholine and phosphatidylethanolamine molecular species composition of parasitized monkey erythrocytes. *Biochim. Biophys. Acta* **1022**: 135–145.

27. Pouvelle, B., Spiegel, R., Hsiao, L. *et al.* (1991) Direct access to serum macromolecules by intraerythrocytic malaria parasites. *Nature* **353**: 73–75.

28. Sherman, I. W. and Zidovetzki, R. (1992) A parasitophorous duct in *Plasmodium*-infected red blood cells. *Parasitol. Today.* **8**: 2–3.

29. Haldar, K. (1992) Lipid transport in *Plasmodium*. *Infect. Ag. Dis.* **1**: 254–262.

30. Grellier, P., Rigomier, D. and Schrevel, J. (1990) Induction in vitro de la schizogonie de *Plasmodium falciparum* par les lipoproteines humaines de haute densite (HDL). *C R Acad. Sci. Paris* **311 Serie III**: 361–367.

31. Ancelin, M. L., Parant, M., Thuet, M. J., Philippot, J. R. and Vial, H. J. (1991) Increased permeability to choline in simian erythrocytes after *Plasmodium knowlesi* infection. *Biochem. J.* **273**: 701–709.

32. Kirk, K., Wong, H. Y., Elford, B. C., Newbold, C. I. and Ellory, J. C. (1991) Enhanced choline and Rb^+ transport in human erythrocytes infected with the malaria parasite *Plasmodium falciparum*. *Biochem. J.* **278**: 521–525.

33. Kirk, K., Poli de Figueiredo, C. E., Elford, B. C. and Ellory, J. C. (1992) Effect of cell age on erythrocyte choline transport: implications for the increased choline permeability of malaria-infected erythrocytes. *Biochem. J.* **283**: 617–619.

34. Ancelin, M. L. and Vial, H. J. (1992) Saturable and non-saturable components of choline transport in *Plasmodium*-infected mammalian erythrocytes: possible role of experimental conditions. *Biochem. J.* **283**: 619–621.

35. Ancelin, M. L. and Vial, H. J. (1986) Several lines of evidence demonstrating that *Plasmodium falciparum*, a parasitic organism, has distinct enzymes for the phosphorylation of choline and ethanolamine. *FEBS Lett.* **202**:217–223.

36. Vial, H. J., Thuet, J. J. and Philippot, J. R. (1984) Cholinephosphotransferase and ethanolaminephosphotransferase activities in *Plasmodium knowlesi*-infected erythrocytes. Their use as parasite-specific markers. *Biochim. Biophys. Acta* **795**: 372–383.

37. Ancelin, M. L. and Vial, H. J. (1989) Regulation of phosphatidylcholine biosynthesis in *Plasmodium knowlesi*-infected erythrocytes. *Biochim. Biophys. Acta* **1001**: 82–89.

38. Elabbadi, N., Ancelin, M. L. and Vial, H. J. (1992) Use of radioactive ethanolamine incorporation into phospholipids to assess in vitro antimalarial activity by the semiautomated microdilution technique. *Antimicrob. Agents Chemother.* **36**: 50–55.

39. Elabbadi, N., Acelin, M. L. and Vial, H. J. (1994) Characterization of phosphatidylinositol synthase and evidence of a polyphosphoinositide cycle in *Plasmodium*-infected erythrocytes. *Mol. Biochem. Parasitol.* **63**: 179–192.

40. Elmendorf, H. G. and Haldar, K. (1993) Identification and localization of ERD2 in the malaria parasite *Plasmodium falciparum*: separation from sites of sphingomyelin synthesis and implications for organization of the Golgi. *EMBO J.* **12**: 4763–4773.

41. Elmendorf, H. G. and Haldar, K. (1994) *Plasmodium falciparum* exports the Golgi marker sphingomyelin synthase into a tubulovesicular network in the cytoplasm of mature erythrocytes. *J. Cell Biol.* **124**: 449–462.

42. Vial, H. J., Ancelin, M. L., Thuet, M. J. and Philippot, J. R. (1989) Phospholipid metabolism in *Plasmodium*-infected erythrocytes: guidelines for further studies using radioactive precursor incorporation. *Parasitology* **98**: 351–357.

43. Dieckmann-Schuppert, A., Bender, S., Holder, A. A., Haldar, K. and Schwarz, R. T. (1992) Labeling and initial characterization of polar lipids in cultures of *Plasmodium falciparum*. *Parasitol. Res.* **78**: 416–422.

44. Beaumelle, B. D. and Vial, H. J. (1988) Acyl-CoA synthetase activity in *Plasmodium knowlesi* infected erythrocytes displays peculiar substrate specificities. *Biochim. Biophys. Acta* **958**: 1–9.

45. Sherwood, J. A. Spitalnik, S. L., Aley, S. B., Quakyi, I. A. and Howard, R. J. (1986) *Plasmodium falciparum* and *P. knowlesi*: initial identification and characterization of malaria synthesized glycolipids. *Exp. Parasitol.* **62**: 127–141.

46. Sherwood, J. A., Spitalnik, S. L., Suarez, S. C., Marsh, K. and Howard, R. J. (1988) *Plasmodium falciparum* glycolipid synthesis: constant and variant molecules of isolates and of strains with differing knob and cytoadherence phenotype. *J. Protozool.* **35**: 169–172.

47. Wunderlich, F., Fiebig, S., Vial, H. and Kleinig, H. (1991) Distinct lipid composition and host cell plasma membranes from *Plasmodium chabaudi*-infected erythrocytes. *Mol. Biochem. Parasitol.* **44**: 271–278.

48. Hsiao, L. L., Howard, R. J., Aikawa, M. and Taraschi, T. F. (1991) Modification of host cell membrane lipid composition by the intra-erythrocytic human malaria parasite *Plasmodium falciparum*. *Biochem. J.* **274**: 121–132.

49. Maguire, P. A. and Sherman, I. W. (1990) Phospholipid composition, cholesterol content and cholesterol exchange in *Plasmodium falciparum*-infected red cells. *Mol. Biochem. Parasitol.* **38**: 105–112.

50. Joshi, P., Dutta, G. P. and Gupta, C. M. (1987) An intracellular simian malarial parasite (*Plasmodium knowlesi*) induces stage-dependent alterations in membrane phospholipid organization of its erythrocyte. *Biochem. J.* **246**: 103–108.

51. Van der Schaft, P. H., Beaumelle, B., Vial, H., Roelofsen, B., Op den Kamp, J. A. F. and Van Deenen, L. L. M. (1987) Phospholipid organization in monkey erythrocytes upon *Plasmodium knowlesi* infection. *Biochim. Biophys. Acta* **901**: 1–14.

52. McClean, S., Purdy, W. C., Kabat, A., Sampugna, J. and DeZeeuw, R. (1976) Analysis of the phospholipid composition of *Plasmodium knowlesi* and rhesus erythrocyte membranes. *Anal. Chim. Acta* **82**: 175–185.

53. Beaumelle, B. D. and Vial, H. J. (1986) Modification of the fatty acid composition of individual phospholipids and neutral lipids after infection of the simian erythrocyte by *Plasmodium knowlesi*. *Biochim. Biophys. Acta* **877**: 262–270.

54. Simoes, A. P., Fiebig, S., Wunderlich, F., Vial, H., Roelofsen, B. and Op den Kamp, J. A. F. (1993) *Plasmodium chabaudi*-parasitized erythrocytes: phosphatidylcholine species of parasites and host cell membranes. *Mol. Biochem. Parasitol.* **57**: 345–348.

55. Dixon, H. and Williamson, J. (1970) The lipid composition of blood and culture forms of *Trypanosoma lewisi* and *Trypanosoma rhodesiense* compared with that of their environment *Comp. Biochem. Physiol.* **33**: 111–128.

56. Carroll, M. and McCrorie, P. (1986) Lipid composition of bloodstream forms of *Trypanosoma brucei brucei*. *Comp. Biochem. Physiol.* **83B**: 647–651.

57. Doering, T. L., Pessin, M. S., Hoff, E. F., Hart, G. W., Raben, D. M. and Englund, P. T. (1993) Trypanosome metabolism of myristate, the fatty acid required for the variant surface glycoprotein membrane anchor. *J. Biol. Chem.* **268**: 9215–9222.

58. Patnaik, P. K., Field, M. C., Menon, A. K., Cross, G. A. M., Yee, M. C. and Butikofer, P. (1993) Molecular species analysis of phospholipids from *Trypanosoma brucei* bloodstream and procyclic forms. *Mol. Biochem. Parasitol.* **58**: 97–105.

59. Breton, J. C., Jauberteau, M. O., Rigaud, M. *et al.* (1988) Les acides gras de *Trypanosoma brucei brucei*. Etude comparative de *T. b. rhodesiense* et *T. b. gambiense*. *Bull. Soc. Pathol. Exot. Filiales* **81**: 632–636.

60. Dixon, H., Ginger, C. D. and Williamson, J. (1971) The lipid metabolism of blood and culture forms of *Trypanosoma lewisi* and *Trypanosoma rhodesiense*. *Comp. Biochem. Physiol.* **39B**: 247–266.

61. Venkatesan, S. and Ormerod, W. E. (1976) Lipid content of the slender and stumpy forms of *Trypanosoma brucei rhodesiense*: a comparative study. *Comp. Biochem. Physiol.* **53B**: 481–487.

62. Maudlin, I., Kabayo, J. P., Flood, M. E. T. and Evans, D. A. (1984) Serum factors and the maturation of *Trypanosoma congolense* infections in *Glossina morsitans*. *Z. Parasitenk.* **70**: 11–19.

63. Dixon, H., Ginger, C. D. and Williamson, J. (1972) Trypanosome sterols and their metabolic origins. *Comp. Biochem. Physiol.* **41B**: 1–18.

64. Klein, R. A. and Cross, G. A. M. (1976) Threonine as a two-carbon precursor for lipid synthesis in *Trypanosoma brucei*. *Trans. R. Soc. Trop. Med. Hyg.* **69**: 267.

65. Klein, R. A. and Linstead, D. J. (1976) Threonine as a preferred source of 2-carbon units for lipid synthesis in *Trypanosoma brucei*. *Biochem. Soc. Trans.* **4**: 48–50.

66. Linstead, D. J., Klein, R. A. and Cross, G. A. M. (1977) Threonine catabolism in *Trypanosoma brucei*. *J. Gen. Microbiol.* **101**: 243–251.

67. Klein, R. A. and Miller, P. G. G. (1981) Alternate pathways in protozoan energy metabolism. *Parasitology* **82**: 1–30.

68. Opperdoes, F. R., Markos, A. and Steiger, R. F. (1981) Localization of malate dehydrogenase, adenylate kinase and glycolytic enzymes in glycosomes and the threonine pathway in the mitochondrion of cultured procyclic trypomastigotes of *Trypanosoma brucei*. *Mol. Biochem. Parasitol.* **4**: 291–309.

69. Opperdoes, F. R. (1984) Localization of the initial steps in alkoxyphospholipid biosynthesis in glycosomes (microbodies) of *Trypanosoma brucei*. *FEBS Lett.* **169**: 35–39.

70. Mellors, A. and Samad, A. (1989) The acquisition of lipids by African trypanosomes. *Parasitol. Today* **5**: 239–244.

71. Samad, A., Licht, B., Stalmach, M. E. and Mellors, A. (1988) Metabolism of phospholipids and lysophospholipids by *Trypanosoma brucei*. *Mol. Biochem. Parasitol.* **29**: 159–169.

72. Bowes, A. E., Samad, A. H., Jiang, P., Weaver, B. and Mellors, A. (1993) The acquisition of lysophosphatidylcholine by African trypanosomes. *J. Biol. Chem.* **268**: 13885–13892.

73. Black, S. and Vandeweerd, V. (1989) Serum lipoproteins are required for multiplication of *Trypanosoma brucei brucei* under axenic culture conditions. *Mol. Biochem. Parasitol.* **37**: 65–72.

74. Vandeweerd, V. and Black, S. J. (1989) Serum lipoprotein and *Trypanosoma brucei brucei* interactions *in vitro*. *Mol. Biochem. Parasitol.* **37**: 201–212.

9 Antioxidant Mechanisms

ROBERTO DOCAMPO

Department of Veterinary Pathobiology, University of Illinois, Urbana, Ill, USA

SUMMARY

Because they live in oxygen-poor environments, parasites are particularly sensitive to oxidative stress. Catalase is absent in many. Glutathione peroxidase activities are very low or absent, and in some cases, such as in trypanosomatids, replaced by trypanothione-dependent peroxidase activities. In any case, peroxidase activities are quite low when compared with the glutathione peroxidase activities in mammalian organs. Superoxide dismutases, however, are present in most parasites, and these are features typical of facultative aerobes. Since superoxide dismutase catalyzes the dismutation of superoxide anion to form hydrogen peroxide and there is a deficiency in hydrogen peroxide-metabolizing systems, this explains why superoxide anion is so toxic to parasites. Parasites may tolerate a slow endogenous rate of hydrogen peroxide generation in their natural habitats but they are quite sensitive to an increased steady-state concentration of hydrogen peroxide as a result of drug metabolism or phagocytic cell attack.

1. INTRODUCTION

In aerobic organisms, molecular oxygen is a final acceptor of electrons in different electron transport systems, such as those present in mitochondria. Much experimental evidence now supports the thesis that the favorable aspects of aerobic life are also linked to potentially dangerous oxygen-linked processes. Several highly unstable and reactive oxygen-derived free radicals are produced during oxygen metabolism and oxidative processes in living cells; these include the superoxide anion (O_2^-), produced by the univalent reduction of molecular oxygen, and the highly toxic hydroxyl radical

Biochemistry and Molecular Biology of Parasites
ISBN 0-12-473345-X

(OH·), which can be produced from hydrogen peroxide (H_2O_2) and the superoxide anion in the presence of transition metal ions.

Oxidative damage caused by reactive oxygen species has been called 'oxidative stress'. Biological systems contain powerful enzymatic and non-enzymatic antioxidant mechanisms, and oxidative stress denotes a shift in the pro-oxidant/antioxidant balance in favor of the former (1). Reactive oxygen species are important mediators of several forms of cell damage in parasites and this discussion will focus on the defense mechanisms of parasites against these species.

2. TOXICITY OF OXYGEN IN PARASITES

A number of parasites live in oxygen-poor environments and are particularly sensitive to oxidative stress. Obligate anaerobes, which lack defenses against reactive oxygen species, are killed readily by oxygen at tensions higher than those in their natural habitats. Only two groups of protozoa, rumen ciliates and termite flagellates (2), belong to this category. It has been shown that rumen protozoa are able to respire under low O_2 tensions (1–2 kPa (kiloPascals) of O_2), where they exhibit a relatively high apparent K_m for O_2, whereas at O_2 tensions above 2 kPa the respiration decreases rapidly, indicating O_2 toxicity. Since respiration at low O_2 tensions is stable for several hours, it may be of importance to the organism at the low O_2 concentrations of rumen fluid (3). Other protozoa are less sensitive to oxygen damage and consume oxygen when available; they can be cultivated for indefinite periods under completely or nearly anaerobic conditions. Examples are *Entamoeba* spp. and *Trichomonas* spp. However, growth of *Trichomonas vaginalis* is stimulated by the presence of low oxygen tensions (1–3 μM) when compared to growth under anaerobiosis; this characteristic defines this organism as a microaerophile (4). Trypanosomatids are also able to survive under anaerobic conditions but, in contrast to the aforementioned forms, they cannot multiply in anaerobiosis (2).

All parasitic helminths studied thus far are facultative aerobes, showing various degrees of tolerance to a lack of oxygen. Because *Fasciola hepatica* and large intestinal helminths (*Ascaris*, *Monieza*, and others) live in oxygen-poor environments, their oxygen consumption is strictly dependent on the oxygen tensions and on their K_m for O_2. Most authors assume that they lead a predominantly anaerobic life *in vivo* (2) (Chapter 4).

For experimental purposes, parasites are frequently tested in media with greater oxygen tensions than their oxygen-poor natural environments. This sudden increase in the partial pressure of oxygen may result in metabolic changes analogous to those that occur when aerobic organisms are placed in hyperbaric oxygen. In fact, any gas mixture with Po_2 greater than that in which organisms are adapted may be considered hyperbaric oxygenation (5).

The high sensitivity of parasites to reagent H_2O_2, to different O_2^- (such as autoxidation of thiols, xanthine–xanthine oxidase, dihydro-orotate–dihydro-orotate oxidase) and H_2O_2-generating systems (such as glucose–glucose oxidase), and organic peroxides (such as *tert*-butyl hydroperoxide) is well established (5). In addition, their sensitivity to both drug-generated and phagocyte-derived oxygen reactive species is also well documented (5,6). It seems possible that O_2^- and H_2O_2 are damaging not by virtue of direct attack upon parasite components, but rather because they can give rise to the extremely reactive hydroxyl radical (OH·) in the presence of transition metal ions (Fenton reaction):

$$O_2^- + M^{n+1} \rightarrow M^n + O_2$$

$$M^n + H_2O_2 \rightarrow M^{n+1} + OH^- + OH·$$

$$O_2^- + H_2O_2 \rightarrow O_2 + OH^- + OH·$$

Recently, a novel mechanism for hydroxyl radical production, which is not dependent on the presence of transition metal ions, has been proposed (7). This involves the production of peroxynitrite (ONOO$^-$) arising from the reaction of nitric oxide (NO·) with superoxide (O$_2^-$), as shown in the following reactions:

$$NO· + O_2^- \rightarrow ONOO^-$$

$$ONOO^- + H^+ \rightarrow ONOOH$$

$$ONOOH \rightarrow OH· + NO_2·$$

The combination of NO· with superoxide competes with the dismutation of superoxide:

$$2H^+ + O_2^- + O_2^- \rightarrow H_2O_2 + O_2$$

Peroxynitrite has a pKa of 7.5 and will be therefore substantially protonated at physiological pH, to produce peroxynitrous acid (ONOOH). One route for the decomposition of this molecule is thought to be homolytic cleavage to produce hydroxyl radical (OH·) and nitrogen dioxide (NO$_2$·).

Currently, there has been much interest in the biologic role of NO· as a widespread transduction mechanism leading to a variety of functions in different cells. It is now generally accepted that NO· is the endothelium-derived relaxing factor and that it also participates in the regulation of nervous and immune systems (8,9). Activated macrophages form NO$_2^-$ and NO$_3^-$ from the terminal guanidino nitrogen atom(s) of L-arginine by a process now known to proceed via the formation of NO·. It has been postulated that the macrophage leishmanicidal activity induced by interferon-γ (IFN-γ) and lipopolysaccharide (LPS) (10) or by calcium ionophore A23187 in the presence of LPS (11) is mediated by the L-arginine:NO· pathway and it has been demonstrated that the leishmanial parasite can be killed by NO· in cell-free suspensions (10). The trypanocidal activity of IFN-γ-treated macrophages against T. cruzi also involves an L-arginine-dependent, nitric oxide-mediated mechanism (12). A nitric oxide-mediated cytostatic activity on T. brucei has also been reported (13) and peroxynitrite has been shown to kill T. cruzi epimastigotes in a dose-dependent manner (14). The precise mechanism of the microbicidal effect of NO· is at present unclear. However, in addition to its capacity to generate hydroxyl radical, it has been shown that exposure of Fe–S groups to NO· results in iron–nitrosyl complex formation, which causes inactivation of several enzymes (15).

3. REACTIVE OXYGEN SPECIES IN SPECIFIC PARASITES

Reactive oxygen species are known to play important roles in many biochemical reactions that maintain normal cell functions. These species are produced as normal intermediates in mitochondrial and microsomal electron-transport systems, and some enzymes that produce substantial amounts of O_2^- or H_2O_2 have been identified in several parasites (5).

Tritrichomonas foetus aerobically excretes H_2O_2 (16). In contrast, respiring *Trypanosoma cruzi* does not release either O_2^- or H_2O_2 to the suspending medium, although *T. cruzi* homogenates supplemented with reduced nicotinamide adenine dinucleotides are effective sources of O_2^- and H_2O_2 (5). The mitochondrial, microsomal, and supernatant fractions of *T. cruzi* also generate O_2^- and H_2O_2 in the presence of NADH or NADPH. Mitochondrial preparations of the *T. brucei* bloodstream form produce 0.2–0.7 nmol H_2O_2 per mg protein/min, which corresponds to 1–3% of the oxygen consumed. Although initial reports set the internal H_2O_2 concentration of *T. brucei* bloodstream forms to a high level (70 μM), further work reported values below the limit of detection. This implies an H_2O_2-metabolizing activity in these cells (17).

H_2O_2 formation by mitochondrial preparations of the protozoan *C. fasciculata* is qualitatively the same as that by mitochondria from vertebrate tissues (18). In contrast, it has been postulated that in the mitochondria of *T. brucei* procyclic trypomastigotes (19) and *T. cruzi* epimastigotes (20) an NADH-fumarate reductase could use O_2 as electron acceptor instead of fumarate by converting it into O_2^- which dismutates to generate H_2O_2. A succinate-dependent site for H_2O_2 production, which may be the semiquinone of coenzyme Q_9, as in mammalian mitochondria, has also been postulated to exist in the mitochondria of *T. cruzi* epimastigotes (20).

Although no direct demonstration of O_2^- or H_2O_2 formation by *Plasmodium* has been reported, a *Plasmodium*-induced erythrocyte oxidant sensitivity has been postulated (5). During its development inside the erythrocyte the parasite ingests and degrades hemoglobin to obtain essential amino acids. Hemoglobin breakdown in the food vacuole releases heme which is then incorporated into an insoluble material called hemozoin or malaria pigment. It has been postulated that release of low levels of heme or other iron-containing products within the infected erythrocytes could lead to the formation of reactive oxygen species (21, 22).

In contrast to the mammalian succinate oxidase, this mitochondrial complex in large helminths such as *F. hepatica*, *Ascaris lumbricoides* and *Monieza expansa*, forms H_2O_2 from oxygen instead of water and is insensitive to cyanide, azide and antimycin. In addition, there is evidence that the mitochondria of these large helminths contain a functional branched respiratory chain system with two terminal oxidases: cytochrome *o* and *a₃*. The major pathway with cytochrome *o* as its terminal oxidase is apparently linked with H_2O_2 formation (23).

The utilization of NAD(P)H and succinate by *Hymenolepis diminuta* mitochondrial membranes also results in H_2O_2 formation, although the presence of a branched respiratory chain in this cestode has not been established (24). Fumarate reductase has also been implicated as participating in the formation of H_2O_2 by *M. expansa* by a mechanism similar to that described in trypanosomatids (25).

Non-enzymatic sources of O_2^- and H_2O_2, such as the autoxidation of ferredoxins (present in anaerobic parasites such as *Trichomonas vaginalis*), tetrahydropteridines, thiols or other soluble parasite constituents, are of unknown importance under physiologic conditions.

4. ENZYMATIC DEFENSES OF PARASITES AGAINST REACTIVE OXYGEN SPECIES

4.1. Catalase

Catalase exerts a dual function: (1) decomposition of H_2O_2 to give H_2O and O_2 (catalatic activity) and (2) oxidation of electron donors, e.g., methanol, ethanol, formic acid, phenols, with the consumption of 1 mol of peroxide (peroxidatic activity):

$$2H_2O_2 \rightarrow 2H_2O + O_2 \quad \text{(catalatic activity)}$$

$$ROOH + AH_2 \rightarrow H_2O + ROH + A \quad \text{(peroxidatic activity)}$$

Catalase is present in the cytoplasm of *T. foetus* (26) but absent from other *Trichomonas* species, *Giardia lamblia* (27) and *E. histolytica* (5).

Although catalase is absent in most pathogenic trypanosomatids such as *T. brucei*, *T. cruzi*, and most *Leishmania* spp., it is present in other mammalian trypanosomes as well as avian, amphibian and insect trypanosomes (2). Although a high content of catalase has been reported in the intracellular (amastigote) forms of *L. donovani* (28), it is difficult to exclude the possibility that it may be a host contaminant because amastigotes were obtained from hamster spleen homogenates and it is known that catalase can be easily adsorbed to cell membranes.

Catalase is localized in most cells in subcellular organelles such as the peroxisomes (microbodies) of liver and kidney or in much smaller aggregates such as the micro-peroxisomes found in a variety of other cells. There are few reports of peroxisomes or catalase-containing microbodies in parasites. Catalase activity has been detected and localized by biochemical techniques in the microbodies (glycosomes) of some trypanosomatids such as *Crithidia luciliae* (29), and *Phytomonas* sp. (30). Although a 3,3'-diamionobenzidine reaction (a cytochemical technique widely used to demonstrate peroxidase activity in microbodies) has been reported in the denser area (core) of hydrogenosomes (31) and in *T. cruzi* microbodies (32), biochemical determinations were negative and the presence of other peroxidase activities could have been responsible for those positive reactions. The 3,3'-diaminobenzidine-positive microbodies of typical structure are present in lower trypanosomatids. For example, *Crithidia fasciculata* possesses a high activity of catalase and their catalase-containing microbodies occupy about 2.5% of the total cell volume (5, 33).

A low catalase activity has been found in sporulated oocysts and in sporozoites of *Eimeria tenella* (34) but its location was not determined. Only low levels of catalase were found in *P. berghei*, probably as a result of contamination of the preparations with host cell material (35). *Toxoplasma gondii* tachyzoites have an abundant catalase activity (36).

Catalase activity is low in trematodes, cestodes and nematodes (2), although it has been detected in significant amounts in the nematodes *Ancylostoma ceylanicum* and

Nippostrongylus brasiliensis (37). It is undetectable in newborn larvae, adult worms, and muscle larvae of *Trichinella spiralis* (38) as well as in *F. hepatica* and *H. diminuta* (39). In contrast *Heligomosomoides polygyrus* contains detectable levels of the enzyme (40). Surprisingly, a substantial catalase activity has been found in isolated *A. suum* mitochondria (41).

4.2. Glutathione Peroxidase and Other Peroxidase Activities

Glutathione peroxidase catalyzes the reaction of hydroperoxides (ROOH) with reduced glutathione (GSH) to form oxidized glutathione disulfide (GSSG) and the reduction product of the hydroperoxides (ROH):

$$ROOH + 2GSH \rightarrow ROH + GSSG + H_2O$$

Glutathione peroxidase is specific for its hydrogen donor, reduced glutathione, and non-specific for the hydroperoxide, which may be H_2O_2 or organic hydroperoxides (42). Since many of the radical or non-radical reactions in cells may lead to thiol oxidation to the disulfide, i.e., the oxidation of glutathione to form GSSG, the regenerative reaction of rereduction to GSH catalyzed by GSSG reductase can become pivotal in antioxidant defense (1):

$$GSSG + NADPH + H^+ \rightarrow 2GSH + NADP^+$$

No glutathione peroxidase activity has been detected in *E. histolytica* and *G. lamblia* (6). In this regard, it has been pointed out that *E. histolytica* (43), *G. duodenalis* and *T. foetus* (44) lack glutathione metabolism, and that cysteine is the main thiol component present in these parasites. *T. vaginalis* also lacks detectable GSSG reductase activity (45). Other thiol-dependent peroxidase and reductase activities have been suggested to be present in *G. intestinalis* but the nature of the thiol cofactors is not known (46).

Trypanosomatids do not have glutathione peroxidase. Some of them have a peroxidase activity dependent on a glutathione-spermidine cofactor termed trypanothione (T(SH)$_2$) (47) (see Chapter 7). Oxidation of trypanothione leads to the formation of trypanothione disulfide (T(S)$_2$). A trypanothione peroxidase activity using either H_2O_2 or organic hydroperoxides has been measured in crude extracts from *C. fasciculata* and *T. brucei* but the enzyme is apparently very labile and has not yet been purified (48). The details of the reaction mechanisms are not clear. However, a system analogous to the host glutathione reductase/glutathione peroxidase system utilizing trypanothione has been proposed (48). The following reaction scheme has been proposed for the metabolism of hydroperoxides:

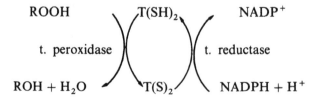

Recently, Carnieri *et al.* (49) reported the absence of a trypanothione peroxidase activity in *T. cruzi* and suggested that oxidation of endogenous reduced thiols accounted for the rates of H_2O_2 metabolism detected in intact cells. In contrast, trypanothione reductase has been purified to homogeneity from *C. fasciculata* (50), and *T. cruzi* (51), and X-ray structure studies have been performed with the enzymes from *C. fasciculata* (52–54) and *T. cruzi* (55–57). The enzyme, which has been found located only in the cytosol of *T. brucei* (58), catalyzes the NADPH-dependent reduction of trypanothione, but not glutathione. Ultimately, dihydrotrypanothione is capable of undergoing a rapid non-enzymatic disulfide exchange reaction with intracellular disulfides (RSSR), among them oxidized glutathione and cysteine (59). This reaction is summarized as:

$$T(SH)_2 + RSSR \rightarrow T(S)_2 + 2RSH$$

The cloned trypanothione reductase genes from *T. congolense* (60) and *T. cruzi* (61) have been expressed in *E. coli*, and the enzymes overproduced and purified. The trypanothione reductase genes from *Crithidia fasciculata* (62,63) and *T. brucei* (63) have also been cloned and sequenced. Site-directed mutagenesis of either the *E. coli* glutathione reductase (64) or the *T. congolense* trypanothione reductase expressed in *E. coli* (60) has been used to evaluate the role of different residues in the mutually exclusive specificities of trypanothione reductase and glutathione reductase.

In the first report (65) on the presence of the thiol-containing cofactor for GSH reduction in trypanosomes that was later identified as trypanothione (47), the authors reported that the addition of *L. mexicana* ultrafiltrates was able to restore the *T. brucei* glutathione (trypanothione) reductase activity, thus indicating the presence of trypanothione or a trypanothione-like compound in the leishmanias. An H_2O_2-dependent NADPH oxidation in the presence of trypanothione in ultrafiltrates of *L. mexicana amazonensis* promastigotes was reported later (66). More recently the gene for the *L. donovani* trypanothione reductase has been cloned and overexpressed in either *L. donovani* or *T. cruzi* (67). Overexpression of trypanothione reductase did not protect either *L. donovani* promastigotes or *T. cruzi* epimastigotes against oxidative stress caused by direct exposure of the cells to reagent H_2O_2 or to H_2O_2-generating drugs. It is noteworthy that cells overexpressing this enzyme did not show any difference in the amount of reduced thiols as compared to controls and in the metabolism of exogenously added H_2O_2 (67) thus confirming previous results (49) indicating that H_2O_2 decomposition in *T. cruzi* is due to non-enzymatic reactions of endogenous reduced thiols with peroxides, and that a true trypanothione peroxidase is absent in these cells.

It is difficult to evaluate other reports of the presence of glutathione peroxidase in *Leishmania* spp., such as one describing the presence of the enzyme in *L. tropica* promastigotes (67). A high content of glutathione peroxidase has also been reported in *L. donovani* intracellular forms (28), but it is difficult to rule out a contamination from the host for the same reasons indicated for the high catalase activity detected in these parasites.

Other protozoa do not contain trypanothione (65). A glutathione peroxidase activity has been reported in *T. gondii* tachyzoites (36). The glutathione metabolism of *P. falciparum*, *P. vinckei* and *P. berghei* has been studied. Glutathione peroxidase and glutathione reductase are present in the parasites as is a selenium-independent glutathione peroxidase (68). However, as occurs with the cyanide-sensitive superoxide dismutase present in malaria parasites, the possibility that the parasites store-up host glutathione reductase in the course of intraerythrocyte maturation could not be ruled out (69).

F. hepatica lacks glutathione peroxidase (39). Glutathione peroxidase activity in *Schistosoma mansoni* increases significantly as worms mature in their host and is positively correlated to the resistance to antioxidants (70). A glutathione peroxidase gene from *S. mansoni* has been cloned and sequenced (70). Newborn larvae of *T. spiralis* lack detectable glutathione peroxidase, whereas, adult worms and muscle larvae contain a high activity of this enzyme (38). In addition, *Ancylostoma ceylanicum* and *Nippostrongylus brasiliensis* contain significant glutathione peroxidase activities (37).

Other peroxidases less studied are an ascorbate peroxidase present in *T. cruzi*, cytochrome *c* peroxidases detected in mitochondrial preparations of *C. fasciculata* and in *A. lumbricoides* and *F. hepatica*, as well as peroxidases demonstrated in *F. hepatica*, *H. diminuta*, *M. expansa*, the hemocele fluid of *A. lumbricoides*, and the miracidium of *S. mansoni* (5). Their role in H_2O_2 detoxification is unknown.

4.3. Superoxide Dismutases

Superoxide dismutase (SOD) is a widely distributed enzyme that exists in a variety of forms. The copper–zinc enzyme (Cu,ZnSOD) is primarily located in the cytosol of eukaryotic cells. Mitochondria contain, in the matrix space, a distinctive cyanide-insensitive manganese-containing enzyme (MnSOD) similar to that found in pro-karyotes. In addition, a ferrienzyme (FeSOD) has been identified in bacteria that is also insensitive to cyanide. Amino acid sequence homologies indicate two families of superoxide dismutases. One of these is composed of the Cu,ZnSODs and the other of MnSODs and FeSODs. All these superoxide dismutases catalyze the same reaction $(2H^+ + O_2^- + O_2^- \rightarrow H_2O_2 + O_2)$ and with comparable efficiency.

Surprisingly, it has been reported that *G. intestinalis* lacks SOD activity (46). If these results are confirmed, it will be a notable exception among eukaryotic organisms.

In general, protozoa, as eukaryotic algae, lack the Cu,Zn enzyme. FeSODs have been found in cell lysates of *E. dispar* and *E. histolytica* and their genes have been cloned (71). A comparison of the two sequences revealed 5% nucleotide difference resulting in a single amino acid substitution. The primary structure of the enzymes showed the characteristics of FeSODs with approximately 55% similarity with other FeSODs sequences. The enzyme is encoded by single copy genes in both *E. dispar* and *E. histolytica*, and no correlation was found between pathogenic behavior of the amebae and the expression of FeSOD-related mRNA (71).

SOD activity is present in the facultative anaerobic flagellate *T. foetus* (72). About five-sixths of the activity of *T. foetus* is in the non-sedimentable portion of the cytoplasm. The rest is connected with the hydrogenosome. Dialyzed extracts of *T. foetus* have SOD at substantially higher levels than in other eukaryotic organisms (73). The activity is inhibited by H_2O_2 and azide but not by cyanide, suggesting that it is FeSOD(s). Three isozymes were seen by isoelectric focusing which appeared to be sensitive to inhibition by H_2O_2 (73).

A cyanide-sensitive SOD activity has been reported in *T. cruzi* (74). Cyanide-insensitive SODs are also present in trypanosomatids, such as *C. fasciculata* (75) and *Trypanosoma brucei* (76). The bimodal distribution of superoxide dismutase in *T. brucei* is similar to that described for other eukaryotes. Both SOD activities in *T. brucei* are insensitive to 1 mM cyanide (76). The SOD activities of *T. brucei*, *L. tropica* and *T. cruzi* are also cyanide-insensitive but peroxide- and azide-sensitive (77). SOD of *C. fas-ciculata* is located in the cytosol and exists in three forms, which may represent three distinct isozymes. Comparisons of the amino acid content of this SOD with those of SODs from other sources suggests that the crithidial enzyme is closely related to bacterial FeSOD and only distantly related to human MnSOD and Cu,ZnSOD and to the FeSOD of the alga *Euglena gracilis*. SOD activity has also been detected in *L. donovani*, although its nature has not been determined (5).

P. berghei isolated from mouse red blood cells also contains a cyanide-sensitive SOD activity. Plasmodial and mouse enzymes are indistinguishable electrophoretically. These results suggest that the malarial cyanide-sensitive SOD may be entirely of host origin and stored by the parasite. Accordingly, plasmodium isolated from mouse red blood cells contain mouse cyanide-sensitive SOD, whereas rat-derived parasites contain the rat enzyme (78). Another, peroxide-sensitive, apparently manganese-containing SOD, has been described in *P. falciparum* (79). Two *Babesia* species (*B. hylomysci* and *B. divergens*) have been reported to contain an endogenous superoxide dismutase that is cyanide-insensitive and inhibited by H_2O_2 indicating that iron is the cofactor metal (80). A similar situation has been found in *T. gondii* where the enzyme(s) are inhibited by azide and peroxide but not by cyanide (81).

Unsporulated oocysts of *E. tenella* have high SOD activity and contain several electrophoretically distinct forms of the enzyme, including two forms of Cu,ZnSOD, two forms of FeSOD and two forms of MnSOD. The SOD activity decreases during sporulation and oocysts sporulated for 48 h have low levels of SOD and contain only one form of the enzyme (MnSOD) (34).

S. mansoni possess a Cu,ZnSOD (82). This superoxide dismutase may be secreted since the product of *in vitro* translation of its mRNA is immunoprecipitable with sera pooled from patients with chronic schistosomiasis. Another Cu,ZnSOD, with homology to mammalian cytosolic forms of the enzyme has been recently cloned from *S. mansoni* (83). Extracts of *F. hepatica* show appreciable SOD activity (39).

SOD activities are also present in *H. diminuta* extracts and in the *A. lumbricoides* (5) but they have not been characterized. A cyanide-sensitive SOD has been purified from metacestodes of *Taenia taeniformis* (84).

SOD activity has also been detected in newborn larvae, adult worms, and muscle larvae of *T. spiralis*, the activity of newborn larvae being significantly lower than that of the other stages (38). A Cu,ZnSOD has been isolated and purified from muscle larvae of *T. spiralis* that is apparently secreted extracellularly (85). A Cu,ZnSOD from *Onchocerca volvulus* has been cloned (86). Other nematodes, such as *Trichostrongylus vitrinus*, *Trichostrongylus colubriformis*, *Nematodiurus battus*, *Teladorsagia circumcincta*, *Nippostrongylus brasiliensis* and *Haemonchus contortus* have higher superoxide dismutase activities in the third larval stages of each species and showed marked interspecies variations in isoenzyme profiles (87).

5. NON-ENZYMATIC DEFENSES OF PARASITES AGAINST REACTIVE OXYGEN SPECIES

Detoxication of reactive oxygen species is also possible through the action of several low-molecular-weight compounds. Some are soluble in water, such as ascorbic acid, thiols, pyruvate and urate, and some are lipid soluble, such as vitamin E and β-carotene (1).

GSH is absent in some parasites such as *E. histolytica* (43), *Tritrichomonas foetus* and *Giardia duodenalis* (44) where cysteine is the main thiol present. However, GSH is present in most parasites although its concentration varies widely. For example, it is low in trypanosomatids such as *T. cruzi* (88). In most trypanosomatids investigated thus far dihydrotrypanothione ($T(SH_2)$) is present in concentrations of about 0.15 mM (48). $T(SH_2)$ is the principal intracellular thiol during exponential growth of *C. fasciculata*, whereas monoglutathionylspermidine is the major thiol in stationary phase. The net result of this biochemical reshuffle of metabolites can be described as follows:

$$2GSH\text{-}SPD \rightarrow T(SH_2) + \text{spermidine}$$

which results in no net change in free GSH, but in the release of unconjugated spermidine. It has been suggested that the function of the redistribution of GSH-SPD into $T(SH_2)$ may be to provide higher levels of spermidine necessary for growth (89). $T(SH_2)$ has a standard redox potential of E': -0.242 ± 0.002 V (59) which is marginally more negative than glutathione (-0.230 V) but considerably less than other physiological dithiols such as lipoic acid and thioredoxin.

A role for pyruvate as H_2O_2 scavenger in trypanosomes has been postulated (90). However, results indicate that the H_2O_2-decomposing activity of pyruvate in *T. brucei* is negligible (17).

The presence of ascorbate peroxidase (32) in addition to a dehydroascorbate reductase in *T. cruzi* (91) has led to the suggestion of the occurrence of an enzymatic vitamin C redox cycle that would replenish ascorbic acid for reactive oxygen species removal (91).

Trypanosomatids, as well as *G. lamblia* and *Trichomonas* spp. lack xanthine oxidase and therefore uric acid, another free radical scavenger, is absent. In contrast, a xanthine oxidase activity has been reported in *Ancylostoma ceylanicum* and *Nippostrongylus brasiliensis* (37). There are no reports of the presence of vitamin E and β-carotene in parasites.

6. ACKNOWLEDGEMENTS

Work from the author received financial support from the UNDP/World Bank/World Health Organization Special Programme for Research and Training in Tropical Diseases.

7. REFERENCES

1. Sies, H. (1985) Oxidative stress: introductory remarks. In: *Oxidative Stress* (ed. Sies, H.), Academic Press, Orlando, pp. 1–8.
2. Von Brand, T. (1973) *Biochemistry of Parasites*, 2nd edn, Academic Press, New York, pp. 258–262.
3. Yarlett, N., Lloyd, D. and Williams, A. G. (1982) Respiration of the rumen ciliate *Dasytricha ruminantium* Schuberg. *Biochem. J.* **206**: 259–266.
4. Paget, T. A. and Lloyd, D. (1990) *Trichomonas vaginalis* requires traces of oxygen and high concentrations of carbon dioxide for optimal growth. *Mol. Biochem. Parasitol.* **41**: 65–72.
5. Docampo, R. and Moreno, S. N. J. (1984) Free-radical intermediates in the antiparasitic action of drugs and phagocytic cells. In: *Free Radicals in Biology* (ed. Pryor, W. A.), Academic Press, New York, pp. 243–288.

6. Docampo, R. (1990) Sensitivity of parasites to free radical damage by antiparasitic agents. *Chem.-Biol. Interact.* **73**: 1–27.

7. Beckman, J. S., Beckman, T. W., Chen, J., Marshall, P. A. and Freeman, B. (1990) Apparent hydroxyl radical production by peroxynitrite: implications for endothelial injury from nitric oxide and superoxide. *Proc. Natl. Acad. Sci. USA* **87**: 1620–1624.

8. Palmer, R. M. J., Ferrige, A. G. and Moncada, S. (1987) Nitric oxide release accounts for the biological activity of endothelium-derived relaxing factor. *Nature* **327**: 524–526.

9. Lowenstein, C. J. and Snyder, S. H. (1992) Nitric oxide: a novel biologic messenger. *Cell* **79**: 705–707.

10. Liew, F. T., Millott, S., Parkinson, C., Palmer, R. M. J. and Moncada, S. (1990) Macrophage killing of *Leishmania* parasite in vivo is mediated by nitric oxide from L-arginine. *J. Immunol.* **144**: 4794–4797.

11. Buchmüller-Rouiller, Y., Corradin, S. B. and Mauel, J. (1992) Macrophage activation for intracellular killing as induced by a Ca^{2+} ionophore. Dependence on L-arginine-derived nitrogen oxidation products. *Biochem. J.* **284**: 387–392.

12. Gazzinelli, R., Oswald, I. P., Hieny, S., James, S. L. and Sher, A. (1992) The microbicidal activity of interferon-γ-treated macrophages against *Trypanosoma cruzi* involves an L-arginine-dependent, nitrogen oxide-mediated mechanism inhibitable by interleukin-10 and transforming growth factor-β. *Eur. J. Immunol.* **22**: 2501–2506.

13. Vincendeau, P., Daulouede, S., Veyret, B., Darde, M. L., Bouteille, B. and Lemesre, J. L. (1992) Nitric oxide-mediated cytostatic activity on *Trypanosoma brucei gambiense* and *Trypanosoma brucei brucei*. *Exp. Parasitol.* **75**: 353–360.

14. Denicola, A., Rubbo, H., Rodriguez, D. and Radi, R. (1993) Peroxynitrite-mediated cytotoxicity to *Trypanosoma cruzi*. *Arch. Biochem. Biophys.* **304**: 279–286.

15. Hibbs, J. B., Jr, Taintor, R. R., Vavrin, Z. and Rachlin, E. M. (1988) Nitric oxide: a cytotoxic activated macrophage effector molecule. *Biochem. Biophys. Res. Commun.* **157**: 87–92.

16. Ninomiya, M. and Ziro, S. (1952) The metabolism of *Trichomonas vaginalis* with comparative aspects of trichomonads. *J. Biochem. (Tokyo)* **39**: 321–326.

17. Penketh, P. G. and Klein, R. A. (1986) Hydrogen peroxide metabolism in *Trypanosoma brucei*. *Mol. Biochem. Parasitol.* **20**: 111–121.

18. Kusel, J. P., Boveris, A. and Storey, B. T. (1973) H_2O_2 production and cytochrome *c* peroxidase activity in mitochondria isolated from the trypanosomatid hemoflagellate *Crithidia fasciculata*. *Arch. Biochem. Biophys.* **158**: 799–804.

19. Turrens, J. F. (1987) Possible role of NADH-fumarate reductase in superoxide anion and hydrogen peroxide production in *Trypanosoma brucei*. *Mol. Biochem. Parasitol.* **25**: 55–60.

20. Denicola-Seoane, A., Rubbo, H., Prodanov, E. and Turrens, J. F. (1992) Succinate-dependent metabolism in *Trypanosoma cruzi* epimastigotes. *Mol. Biochem. Parasitol.* **54**: 43–50.

21. Hebbel, R. P. and Eaton, J. W. (1989) Pathobiology of heme interaction with the erythrocyte membrane. *Semin. Hematol.* **26**: 136–149.

22. Atamna, H. and Ginsburg, H. (1993) Origin of reactive oxygen species in erythrocytes infected with *Plasmodium falciparum*. *Mol. Biochem. Parasitol.* **61**: 231–242.

23. Cheah, K. S. (1976) Electron transport system of *Ascaris* muscle mitochondria. In: *Biochemistry of Parasites and Host–Parasite Relationships* (ed. Van den Bossche, H.), Elsevier/North Holland Biomedical Press, Amsterdam, pp. 133–143.

24. Fioravanti, C. F. (1982) Mitochondrial NADH oxidase activity of adult *Hymenolepis diminuta* (Cestoda). *Comp. Biochem. Physiol.* **72B**: 591–594.

25. Cheah, K. S. (1987) The oxidase system of *Monieza expansa* (Cestoda). *Comp. Biochem. Physiol.* **23**: 277–302.

26. Müller, M. (1978) Biochemical cytology of trichomonad flagellates. I. Subcellular localization of hydrolases, dehydrogenases, and catalase in *Tritrichomonas foetus*. *J. Cell Biol.* **57**: 453–474.

27. Lindmark, D. G. (1980) Energy metabolism of the anaerobic protozoon *Giardia lamblia*. *Mol. Biochem. Parasitol.* **1**: 1–12.

28. Murray, H. W. (1982) Cell-mediated immune response in experimental visceral leishmaniasis. II. Oxygen-dependent killing of intracellular *Leishmania donovani* amastigotes. *J. Immunol.* **129**: 351–357.

29. Opperdoes, F. R., Borst, P. and Spits, H. (1977) Particle-bound enzymes in the bloodstream form of *Trypanosoma brucei*. *Eur. J. Biochem.* **76**: 21–28.

30. Sanchez-Moreno, M., Lasztity, D., Coppens, I. and Opperdoes, F. R. (1992) Characterization of carbohydrate metabolism and demonstration of glycosomes in a *Phytomonas* sp. isolated from *Euphorbia characias*. *Mol. Biochem. Parasitol.* **54**, 185–200.

31. Kulda, J., Kralova, J. and Vavra, J. (1977) The ultrastructure and 3,3′-diaminobenzidine cytochemistry of hydrogenosomes of trichomonads (Abstr). *J. Protozool.* **24**: 51A.

32. Docampo, R., Boiso, J. F. de, Boveris, A. and Stoppani, A. O. M. (1976) Localization of peroxidase activity in *Trypanosoma cruzi* microbodies. *Experientia* **32**: 972–975.

33. Bayne, R. A., Muse, K. E. and Roberts, J. F. (1969) Isolation of bodies containing the cyanide insensitive glycerophosphate oxidase of *Trypanosoma equiperdum*. *Comp. Biochem. Physiol.* **30**: 1049–1054.

34. Michalski, W. P. and Prowse, S. J. (1991) Superoxide dismutases in *Eimeria tenella*. *Mol. Biochem. Parasitol.* **47**: 189–196.

35. Fairfield, A. S., Eaton, J. W. and Meshnick, S. R. (1986) Superoxide dismutase and catalase in the murine malaria, *Plasmodium berghei*: content and subcellular distribution. *Arch. Biochem. Biophys.* **250**: 526–529.

36. Murray, H. W., Nathan, C. F. and Cohn, Z. A. (1980) Macrophage oxygen-dependent antimicrobial activity. IV. Role of endogenous scavengers of oxygen intermediates. *J. Exp. Med.* **152**: 1610–1624.

37. Batra, S., Singh, S. P., Katiyar, J. C. and Srivastava, V. M. L. (1990) Reactive oxygen intermediates metabolizing enzymes in *Ancyclostoma ceylanicum* and *Nipppostrongylus brasiliensis*. *Free Rad. Biol. Med.* **8**: 271–274.

38. Kazura, J. W. and Meshnick, S. R. (1984) Scavenger enzymes and resistance to oxygen mediated damage in *Trichinella spiralis*. *Mol. Biochem. Parasitol.* **10**: 1–10.

39. Barrett, J. (1980) Peroxide metabolism in the liver fluke *Fasciola hepatica*. *J. Parasitol.* **66**: 697.

40. Preston, C. M. and Barrett, J. (1987) *Heligomosomoides polygyrus*: peroxidase activity. *Exp. Parasitol.* **64**: 24–28.

41. Campbell, T., Rubin, N. and Komuniecki, R. (1989) Succinate-dependent energy generation in *Ascaris suum* mitochondria. *Mol. Biochem. Parasitol.* **33**: 1–12.

42. Chance, B., Sies, H. and Boveris, A. (1979) Hydroperoxide metabolism in mammalian organs. *Physiol. Rev.* **59**: 527–605.

43. Fahey, R. C., Newton, G. L., Arrick, B., Overdank-Bogart, T. and Aley, S. B. (1984) *Entamoeba histolytica*: a eukaryotic without glutathione metabolism. *Science* **224**: 70–72.

44. Brown, D. M., Upcroft, J. A. and Upcroft, P. (1993) Cysteine is the major low-molecular weight thiol of *Giardia duodenalis*. *Mol. Biochem. Parasitol.* **61**: 155–158.

45. Thong, K. W. and Coombs, G. H. (1987) Comparative study of ferredoxin-linked and oxygen-metabolizing enzymes of trichomonads. *Comp. Biochem. Physiol.* **87B**: 637–642.

46. Smith, N. C., Bryant, C. and Boreham, P. F. L. (1988) Possible roles for pyruvate:ferredoxin oxidoreductase and thiol-dependent peroxidase and reductase activities in resistance to nitroheterocyclic drugs in *Giardia intestinalis*. *Int. J. Parasitol.* **18**: 991–997.

47. Fairlamb, A. H., Blackburn, P., Ulrich, P., Chait, B. T. and Cerami, A. (1985) Trypanothione: a novel bis(glutathionyl)spermidine cofactor for glutathione reductase in trypanosomatids. *Science* **227**: 1485–1487.

48. Henderson, G., Fairlamb, A. H. and Cerami, A. (1987) Trypanothione dependent peroxide metabolism in *Crithidia fasciculata* and *Trypanosoma brucei*. *Mol. Biochem. Parasitol.* **24**: 39–45.

49. Carnieri, E. G. S., Moreno, S. N. J. and Docampo, R. (1993) Trypanothione-dependent peroxide metabolism in *Trypanosoma cruzi* different stages. *Mol. Biochem. Parasitol.* **61**: 79–86.

50. Shames, S. L., Fairlamb, A. H., Cerami, A. and Walsh, C. T. (1986) Purification and characterization of trypanothione reductase from *Crithidia fasciculata*, a newly discovered member of the family of disulfide-containing flavoprotein reductases. *Biochemistry* **25**: 3519–3526.

51. Krauth-Siegel, R. L., Enders, B., Henderson, G. B., Fairlamb, A. H. and Schirmer, R. H. (1987) Trypanothione reductase from *Trypanosoma cruzi*. Purification and characterization of the crystalline enzyme. *Eur. J. Biochem.* **164**: 123–128.

52. Hunter, W. N., Smith, K., Derewenda, Z. *et al.* (1990) Initiating a crystallographic study of trypanothione reductase. *J. Mol. Biol.* **216**: 235–237.

53. Kuriyan, J., Kong, X.-P., Krishna, T. S. R. *et al.* (1991) X-ray structure of trypanothione reductase from *Crithidia fasciculata* at 2.4A resolution. *Proc. Natl. Acad. Sci. USA* **88**: 8764–8768.

54. Bailey, S., Smith, K., Fairlamb, A. H. and Hunter, W. N. (1993) Substrate interactions between trypanothione reductase and N^1-glutathionylspermidine disulphide at 0.28-nm resolution. *Eur. J. Biochem.* **213**: 67–75.

55. Zhang, Y., Bailey, S., Naismith, J. H. *et al.* (1993) *Trypanosoma cruzi* trypanothione reductase. Crystallization, unit cell dimensions and structure resolution. *J. Mol. Biol.* **233**: 1217–1220.

56. Krauth-Siegel, R. L., Sticherling, C., Jost, I. *et al.* (1993) Crystallization and preliminary crystallographic analysis of trypanothione reductase from *Trypanosoma cruzi*, the causative agent of Chagas' disease. *FEBS Lett.* **317**: 105–108.

57. Lantwin, C. B., Schlichting, I., Kabsch, W., Pai, E. F. and Krauth-Siegel, R. L. (1994) The structure of *Trypanosoma cruzi* trypanothione reductase in the oxidized and NADPH reduced state. *Proteins* **18**: 161–173.

58. Smith, K., Opperdoes, F. R. and Fairlamb, A. H. (1991) Subcellular distribution of trypanothione reductase in bloodstream and procyclic forms of *Trypanosoma brucei*. *Mol. Biochem. Parasitol.* **48**: 109–112.

59. Fairlamb, A. H. and Henderson, G. (1987) Metabolism of trypanothione and glutathionyl-spermidine in trypanosomes. In: *Host–Parasite Molecular Recognition and Interaction in Protozoal Infections* (eds Chang, K. P. and Snary, D.), NATO ASI series, pp. 29–45.

60. Sullivan, F. X., Shames, S. L. and Walsh, C. T. (1989) Expression of *Trypanosoma congolense* trypanothione reductase in *Escherichia coli*: overproduction, purification and characteriz-ation. *Biochemistry* **28**: 4986–4992.

61. Sullivan, F. X. and Walsh, C. T. (1991) Cloning, sequencing, overproduction and purifica-tion of trypanothione reductase from *Trypanosoma cruzi*. *Mol. Biochem. Parasitol.* **44**: 145–148.

62. Field, H., Cerami, A. and Henderson, G. B. (1992) Cloning, sequencing, and demonstration of polymorphism in trypanothione reductase from *Crithidia fasciculata*. *Mol. Biochem. Parasitol.* **50**: 47–56.

63. Aboagye-Kwarteng, T., Smith, K. and Fairlamb, A. H. (1992) Molecular characterization of the trypanothione reductase gene from *Crithidia fasciculata* and *Trypanosoma brucei*: com-parison with other flavoprotein disulphide oxidoreductases with respect to substrate speci-ficity and catalytic mechanism. *Mol. Microbiol.* **6**: 3089–3099.

64. Henderson, G., Murgolo, N. J., Kuriyan, J. *et al.* (1991) Engineering the substrate specificity of glutathione reductase toward that of trypanothione reductase. *Proc. Natl. Acad. Sci. USA* **88**: 8769–8773.

65. Fairlamb, A. H. and Cerami, A. (1985) Identification of a novel, thiol-containing co-factor essential for glutathione reductase enzyme activity in trypanosomatids. *Mol. Biochem. Parasitol.* **14**: 187–198.

66. Penketh, P. G., Kennedy, W. P. K., Patton, C. L. and Sartorelli, A. C. (1987) Trypanosomatid hydrogen peroxide metabolism. *FEBS Lett.* **221**: 427–431.

67. Kelly, J. M., Taylor, M. C., Smith, K., Hunter, K. J. and Fairlamb, A. H. (1993) Phenotype of recombinant *Leishmania donovani* and *Trypanosoma cruzi* which over-express trypano-thione reductase. Sensitivity towards agents that are thought to induce oxidative stress. *Eur. J. Biochem.* **218**: 29–37.

67. Meshnick, S. R. and Eaton, J. W. (1981) Leishmanial superoxide dismutase: a possible target for chemotherapy. *Biochem. Biophys. Res. Commun.* **102**: 970–976.

68. Fritsch, B., Dieckmann, A., Menz, B. *et al.* (1987) Glutathione and peroxide metabolism in malaria-parasitized erythrocytes. *Parasitol. Res.* **73**: 515–517.

69. Hempelmann, E., Heiner-Schirmer, R., Fritsch, G., Hundt, E. and Groschel-Stewart, U. (1987) Studies on glutathione reductase and methemoglobin from human erythrocytes parasitized with *Plasmodium falciparum*. *Mol. Biochem. Parasitol.* **23**: 19–24.

70. Williams, D. L., Pierce, R. J., Cookson, E. and Capron, A. (1992) Molecular cloning and sequencing of glutathione peroxidase from *Schistosoma mansoni*. *Mol. Biochem. Parasitol.* **52**: 127–130.

71. Tannich, E., Bruchhaus, I., Walter, R. and Horstmann, R. D. (1991) Pathogenic and nonpathogenic *Entamoeba histolytica*: identification and molecular cloning of an iron-containing superoxide dismutase. *Mol. Biochem. Parasitol.* **49**: 61–72.

72. Lindmark, D. G. and Müller, M. (1974) Superoxide dismutase in the anaerobic flagellates *Tritrichomonas foetus* and *Monocercomonas* sp. *J. Biol. Chem.* **249**: 4634–4637.

73. Kitchener, K. R., Meshnick, S. R., Fairfield, A. S. and Wang, C. C. (1984) An iron-containing superoxide dismutase in *Tritrichomonas foetus*. *Mol. Biochem. Parasitol.* **12**: 95–99.

74. Boveris, A. and Stoppani, A. O. M. (1977) Hydrogen peroxide generation in *Trypanosoma cruzi. Experientia* **33**: 1306–1308.

75. Asada, K., Kanematsu, S., Okada, S. and Hayakawa, T. (1980) Phylogenetic distribution of three types of superoxide dismutase in organisms and cell organelles. In: *Chemical and Biochemical Aspects of Superoxide and Superoxide Dismutases* (eds Bannister, J. V. and Hill, H.), Elsevier, New York, pp. 136–153.

76. Opperdoes, F. R., Borst, P., Bakker, S. and Leene, W. (1977) Localization of glycerol-3-phosphate oxidase in the mitochondrion and particulate NAD^+-linked glycerol-3-phosphate dehydrogenase in the microbodies of the bloodstream form of *Trypanosoma brucei. Eur. J. Biochem.* **76**: 29–39.

77. Le Trang, N., Meshnick, S. R., Kitchener, K., Eaton, J. W. and Cerami, A. (1983) Iron-containing superoxide dismutase from *Crithidia fasciculata*. Purification, characterization, and similarity to leishmanial and trypanosomal enzymes. *J. Biol. Chem.* **258**: 125–130.

78. Fairfield, A. S., Meshnick, S. R. and Eaton, J. W. (1983) Malaria parasites adopt host cell superoxide dismutase. *Science* **221**: 764–766.

79. Ranz, A. and Meshnick, S. R. (1989) *Plasmodium falciparum*: inhibitor sensitivity of an endogenous superoxide dismutase. *Exp. Parasitol.* **69**: 125–128.

80. Becuwe, P., Slomianny, C., Valentin, A., Schrevel, J., Camus, D. and Dive, D. (1992) Endogenous superoxide dismutase activity in two *Babesia* species. *Parasitology* **105**: 177–182.

81. Sidbley, L. D., Lawson, R. and Weidner, E. (1986) Superoxide dismutase and catalase in *Toxoplasma gondii. Mol. Biochem. Parasitol.* **19**: 83–87.

82. Simurda, M. C., van Keulen, H., Rekosh, D. M. and LoVerde, P. T. (1988) *Schistosoma mansoni*: identification and analysis of an mRNA and a gene encoding superoxide dismutase. *Exp. Parasitol.* **67**: 73–84.

83. Cordeiro da Silva, A., LePresle, T., Capron, A. and Pierce, R. J. (1992) Molecular cloning of a 16-kilodalton Cu/Zn superoxide dismutase from *Schistosoma mansoni. Mol. Biochem. Parasitol.* **52**: 275–278.

84. Leid, W. R. and Suquet, C. M. (1986) A superoxide dismutase of metacestodes of *Taenia taeniformis. Mol. Biochem. Parasitol.* **18**: 301–311.

85. Rhoads, M. L. (1983) *Trichinella spiralis*: identification and purification of superoxide dismutase. *Exp. Parasitol.* **56**: 41–45.

86. Henkle, K. L., Liebau, E., Müller, S., Bergmann, B. and Walter, R. D. (1991) Characterization and molecular cloning of a Cu/Zn superoxide dismutase from the human parasite *Onchocerca volvulus. Infect. Immun.* **59**: 2063–2069.

87. Knox, D. P. and Jones, D. G. (1992) A comparison of superoxide dismutase (SOD, EC 1.15.1.1) distribution in gastro-intestinal nematodes. *Int. J. Parasitol.* **22**: 209–214.

88. Boveris, A., Sies, H., Martino, E. E., Docampo, R., Turrens, J. F. and Stoppani, A. O. M. (1980) Deficient metabolic utilization of hydrogen peroxide in *Trypanosoma cruzi. Biochem. J.* **188**: 643–648.

89. Shim, H. and Fairlamb, A. H. (1988) Levels of polyamines, glutathione, and glutathione-spermidine conjugates during growth of the insect trypanosomatid *Crithida fasciculata. J. Gen. Microbiol.* **134**: 807–817.

90. Fulton, J. and Spooner, J. (1956) Inhibition of the respiration of *Trypanosoma rhodesiense* by thiols. *Biochem. J.* **63**: 475–482.

91. Clarke, D., Albrecht, M. and Arévalo, J. (1994) Ascorbate variations and dehydroascorbate reductase activity in *Trypanosoma cruzi* epimastigotes and trypomastigotes. *Mol. Biochem. Parasitol.* **66**: 143–145.

10 Xenobiotic Metabolism

JAMES W. TRACY[1] and ELIZABETH A. VANDE WAA[2]

[1]Departments of Comparative Biosciences and Pharmacology and Environmental Toxicology Center and [2]Department of Comparative Biosciences, University of Wisconsin-Madison, Madison, WI, USA

SUMMARY

Over the past decade, work from several laboratories has demonstrated that parasites can metabolize xenobiotics (foreign chemicals), including some anti-parasitic drugs. In the case of the nitroheterocyclic agents metronidazole, nifurtimox, and niridazole, enzymatic nitroreduction within the target organism is responsible for the antiparasitic action of these drugs. Alternatively, xenobiotic metabolism can result in inactivation or detoxication of potentially toxic xenobiotics. For example, glutathione S-transferase-catalysed conjugation of electrophilic chemicals with glutathione has been found to play a role in cellular protection in several different parasites. Despite these few well-characterized examples, however, relatively little is known about the capacity of most parasites to metabolize xenobiotics. This chapter surveys the known pathways of xenobiotic metabolism in helminth and protozoan parasites and attempts to highlight areas where further investigation could prove useful. Elucidating the pathways by which parasites metabolize xenobiotics could provide important insights into parasite–drug interactions that may reveal novel strategies for rational drug design.

1. INTRODUCTION

Both free-living and parasitic organisms are exposed to xenobiotics (foreign chemicals) in their environment. Many of these chemicals, which by definition are not required for normal metabolic functions, are lipophilic and are readily absorbed into cells by passive

Biochemistry and Molecular Biology of Parasites
ISBN 0-12-473345-X

non-ionic diffusion. A few xenobiotics are eliminated unchanged and some undergo spontaneous chemical rearrangement or decomposition. The vast majority, however, undergo enzyme-catalyzed biotransformation into metabolites. These metabolites are generally more water soluble than the present compound and are more readily eliminated from the cell. Thus, xenobiotic metabolism is usually synonymous with detoxication. There are exceptions to this rule. Sometimes xenobiotics are metabolized to chemically reactive products that are toxic.

The pathways of xenobiotic metabolism can be divided into two groups of reactions. Phase I metabolism includes oxidation, reduction and hydrolysis. These are sometimes called 'functionalization' reactions, because a polar functional group (e.g., —OH, —COOH, —NH$_2$, etc.) is introduced into the molecule. Phase II metabolism includes various synthetic reactions between a polar functional group on a xenobiotic substrate and an endogenous cosubstrate. Because this results in formation of a new covalent bond, these reactions are collectively referred to as 'conjugation reactions' and the reaction products as 'conjugates'. Often the substrate of a phase II reaction is a phase I metabolite. The net result is usually an increase in the water solubility of the xenobiotic, which again favors its elimination from the cell. A detailed treatment of xenobiotic metabolism pathways can be found in monographs by Jakoby (1, 2) and Mulder (3). The chemistry of xenobiotic metabolism is covered in works by Jakoby *et al.* (4) and Anders (5).

In contrast to the information available on carbohydrate and energy metabolism in parasites, relatively little is known about the fate of xenobiotics, including clinically used antiparasitic drugs. This chapter deals primarily with xenobiotic metabolism in helminths, although where available, information on the fate of foreign compounds in protozoan parasites is also included.

2. PHASE I METABOLISM

2.1. Oxidation Reactions

2.1.1. Cytochromes P450

In most organisms, the prototypic phase I oxidation reaction is catalyzed by a family of microsomal hemoproteins, the cytochromes P450 (6). These enzymes are widely distributed in nature being found in vertebrates, insects, bacteria and plants. In parasites, however, the only cytochrome P450 detected thus far has been that in *Trypanosoma cruzi* epimastigotes (7). It was speculated that *T. cruzi* might utilize a cytochrome P450 system to metabolize foreign compounds, rendering this parasite insensitive to chemotherapeutic attack (7). However, that hypothesis has not been rigorously tested.

The apparent absence of cytochromes P450 from all other parasite species examined is therefore particularly enigmatic. A number of explanations have been proposed, but none is entirely convincing. These have been reviewed by Precious and Barrett (8). One caveat should be borne in mind. In all cases, the methods used to measure either cytochrome P450 content or catalytic activity have been relatively insensitive. Although mammalian liver is the richest source of cytochromes P450, there are now several examples of extrahepatic tissues that contain very low levels of cytochromes *P450* which are nevertheless important in xenobiotic metabolism (9). Thus, it remains to be seen whether the apparent absence of cytochromes *P450* in most parasites is merely due to the limitations of the methods used to detect them. The use of highly conserved cytochromes *P450* gene sequences (6) to search parasite genomes for this important group of xenobiotic metabolizing enzymes is another approach that has yet to be exploited.

2.1.2. Sulfoxidase

A few other oxidation reactions have been reported in helminths. The cytosolic fractions of *Moniezia expansa* and *Ascaris suum* catalyze the sulfoxidation of several xenobiotics, including the benzimidazole derivatives albendazole and fenbendazole, to the corresponding sulfoxides (10). The reaction requires NADPH and molecular oxygen, but unlike mammalian sulfoxidase, it is not associated with the microsomal fraction. Although parasite sulfoxidases have not been purified, their properties suggest they are soluble forms of flavin-linked monooxygenase.

2.1.3. Peroxidases

Glutathione (GSH) peroxidase (EC 1.11.1.9) is one of several selenoproteins that contain a unique selenocysteine residue at the active site (1). This enzyme catalyzes the GSH-dependent reduction of both hydrogen peroxide and organic hydroperoxides, including fatty acid hydroperoxides formed during lipid peroxidation. GSH peroxidase activity has been measured in both protozoan and helminth parasites (11, 12). The enzyme from *Schistosoma mansoni* has been cloned (13). The presence of the unique selenocysteine codon (UGA) confirms that it is a selenoprotein. Although GSH peroxidase activity appears to be absent from *Hymenolepsis diminuta* and *M. expansa* (11), these parasites do contain a peroxidase activity capable of oxidizing cytochrome *c* (cytochrome *c* peroxidase) (14). The latter enzyme also is present in *S. mansoni* (15), but its role in xenobiotic metabolism has yet to be established. Additional information regarding parasite peroxidases can be found in Chapter 9.

2.1.4. Xanthine oxidase

Xanthine oxidase (EC 1.2.3.2) catalyzes the formation of uric acid, an end-product of purine catabolism. The mammalian enzyme is a metalloflavoprotein composed of two subunits containing molybdenum, FAD and Fe/S clusters as prosthetic groups in a ratio of 1 : 1 : 4 per subunit (1). Besides its endogenous metabolic function, xanthine oxidase is also active toward a wide spectrum of oxidizable xenobiotic substrates. Although some cestodes and trematodes produce trace amounts of uric acid (16), the presence of xanthine oxidase activity in these organisms has not been demonstrated. Xanthine oxidase was found in the cytosolic fractions of the nematodes *Ancylostoma ceylanicum* and *Nippostrongylus brasiliensis* (17), but its activity toward xenobiotic substrates was not tested.

The role of xanthine oxidase in the generation of superoxide radical as a reaction product is often cited, and a xanthine–xanthine oxidase mixture is a convenient way to generate superoxide *in vitro* as a means of testing a parasite's antioxidant capacity (11). It should be noted that under normal conditions, xanthine oxidase is a dehydrogenase. Conversion of the dehydrogenase form into the oxidase often occurs inadvertently during enzyme purification. In mammalian tissues, prolonged ischemia causes the dehydrogenase to be converted into the oxidase form (18). Under that circumstance, superoxide radical is produced and extensive tissue damage may result. Whether an analogous process takes place in living parasites is unknown.

2.1.5. *Monoamine oxidase*

Monoamine oxidase (EC 1.4.3.4) activity is present in homogenates of adult *S. mansoni* (19). This flavoprotein, which catalyzes the oxidative deamination of amines, functions in the breakdown of endogenously produced neurotransmitter amines. Vertebrate monoamine oxidase also catalyzes the oxidation of xenobiotic amines, particularly in tissues such as the intestine (4). Whether the schistosome enzyme is also capable of metabolizing xenobiotics has not been investigated.

2.2. Reduction Reactions

Although oxidation reactions in parasites are relatively uncommon, there are several examples of parasite enzymes that catalyze the reductive metabolism of xenobiotics.

2.2.1. *Azoreductase and nitroreductase*

Helminth parasites have the capacity to reductively metabolize a variety of xenobiotics. Whereas all helminth species examined are able to reduce xenobiotics containing an azo functional group (R—N═N—R'), nitroreduction is not universally present. For example, *A. suum* and *M. expansa* reduce both azobenzene and 4-nitrobenzoic acid (20), whereas *H. diminuta* (21) and *Onchocerca gutturosa* (22) metabolize only the former. Although azoreductase and nitroreductase have not been purified from helminths, they appear to be confined to the soluble cell fraction and to use NADH, rather than NADPH, as the reduced pyridine cofactor (8). An exception is the nitroreductase from *S. mansoni* which displays greater activity with NADPH compared with NADH (23). A number of protozoan parasites including *Entamoeba histolytica*, *Giardia lamblia*, *Trichomonas vaginalis* and *T. cruzi* also reductively metabolize xenobiotics containing a nitro functional group (24).

2.2.2. *Role of nitroreduction in the metabolism and mode of action of nitroheterocyclic antiparasitic drugs*

A good illustration of how xenobiotic metabolism can result in toxic activation rather than detoxication is the role of parasite enzymes in the biotransformation of nitroheterocyclic drugs. Metronidazole, a 5-nitroimidazole derivative, is used for the treatment of *T. vaginalis* infections (25). The drug is reduced by trichomonads to form a nitroradical anion that can be demonstrated by electron spin resonance spectroscopy (24). The reaction is catalyzed by pyruvate:ferredoxin oxidoreductase localized in the hydrogenosome, an organelle unique to trichomonads (see Chapter 3). Under physiological conditions, metronidazole undergoes further metabolism in trichomonads to other reduced metabolites (nitroso and hydroxylamino derivatives) with eventual ring opening to yield acetamide and probably N-(2-hydroxyethyl)oxamic acid (26). The nitroradical anion does not react with cellular nucleophiles, but [^{14}C]metronidazole does bind covalently to proteins and nucleic acids in *T. foetus* (26). This is consistent with further reduction of the nitroradical anion. The interaction of these reactive intermediates with macromolecules is probably responsible for the ultimate cytotoxic action of metronidazole in trichomonads. Under aerobic conditions, redox cycling of metronidazole occurs leading to formation of superoxide radical (24). In *T. foetus*, reoxidation of the nitroradical with concomitant formation of superoxide has been suggested as a detoxication pathway to protect the organism from metabolites resulting from further reduction of the nitro radical anion. The superoxide radical and other reactive oxygen metabolites formed are then detoxicated by enzymes involved in protection against oxidative stress (see Chapter 9). Metronidazole is also effective against the anaerobic intestinal parasites *E. histolytica* and *G. lamblia*. Although the metabolic fate of the drug has not been studied in these species, its mechanism of action also is likely due to proximal nitroreduction.

Nifurtimox, a 5-nitrofuran derivative, is effective against *T. cruzi* infections. As with metronidazole, nifurtimox undergoes proximal nitroreduction within the target organism to form the nitroradical anion. Unlike the previous example, however, the toxicity of nifurtimox appears due to reactive oxygen metabolites formed as a result of nifurtimox redox cycling (24). Unlike trichomonads, *T. cruzi* is characterized by a deficiency in its enzymatic defenses against reactive oxygen species (see Chapter 9).

The antischistosomal drug niridazole, a 5-nitrothiazole derivative, undergoes enzyme-catalyzed nitroreduction within adult *S. mansoni* (23). In contrast to the situation with nifurtimox in *T. cruzi*, redox cycling with resulting formation of reactive oxygen metabolites appears to play a negligible role in the action of niridazole. In this case, the nitro radical anion undergoes further reduction by dismutation of the radical to form other reactive metabolites such as the nitroso (—N=O) and hydroxylamino (—NH—OH) derivatives that react with cellular nucleophiles (GSH, proteins, nucleic acids) to form covalent adducts. Ring opening of the hydroxylamino derivative takes place leading to formation of the immunosuppressive metabolite 1-thiocarbamoyl-2-imidazolidinone (27). This metabolite, however, has no antischistosomal activity. Drug disposition studies using [^{14}C]niridazole demonstrated that over 90% of the niridazole absorbed by *S. mansoni* is extensively metabolized (23). Nearly 30% of the radioactivity absorbed by the parasite was found bound to macromolecules and about 50% was present as unidentified water-soluble metabolites.

Metabolism of niridazole in schistosomes is quite distinct from that in the mammalian host, where the primary route of biotransformation involves cytochrome *P*450-catalyzed oxidation of the 2-imidazolidinone ring (28). Significantly, no oxidative metabolites are formed in schistosomes (23), consistent with the general view that these parasites are deficient in oxidative drug metabolism. That the metabolism of niridazole involved reduction to a nitro radical was supported by the observation that drug metabolism in cell-free extracts of *S. mansoni* was inhibited by molecular oxygen, but that consumption of reduced pyridine nucleotide was stimulated by niridazole. Although this is consistent with redox cycling of the nitro radical in the presence of oxygen, net reduction occurs in intact parasites even when they are incubated under aerobic conditions (23). That nitroreduction and subsequent covalent binding to parasite nucleophiles is involved in the activation of niridazole was demonstrated using the 4'-methyl analog. 4'-methyl[^{14}C]niridazole, which has no antischistosomal activity, was readily absorbed by schistosomes but was not metabolized and was not a substrate for schistosome nitroreductase (23).

The exact metabolic fate of a particular nitroheterocycle is dependent on both its redox potential and the nature of the nitroreductase found in a given parasite species. Furthermore, it is difficult to identify the sequence of secondary biochemical events that ultimately leads to parasite death. However, in each of the cases cited above, enzyme catalyzed proximal nitroreduction is an obligatory step in the metabolic activation and subsequent antiparasitic action of these drugs.

2.2.3. Other reduction reactions

Aldehyde reductases and ketone reductases are families of enzymes involved in reductive metabolism of both endogenously produced and xenobiotic aldehydes and ketones (1). Aldehyde reductase activity has been detected in cell-free preparations of *H. diminuta*, *Heligmosomoides polygyrus* and in the free-living nematode *Panagrellus redivivus* (8). Ketone reductase activity, on the other hand, appears absent from these organisms.

T. cruzi contains NADPH-cytochrome *c* reductase activity in both the microsomal and cytosolic-cell fractions (29). The cytosolic enzyme may act as an oxygenase, protecting the cell in times of hydrogen peroxide production, since trypanosomes are deficient in catalase (30). It may also metabolize drugs directly, possible conferring resistance to antitrypanosomal drugs (31). The particulate enzyme form probably functions in microsomal electron transport associated with trypanosomal cytochrome *P*450.

Despite the fact that quinones are ubiquitous in nature and many of them are cytotoxic (32), little is known of their metabolic fate in parasites. The cellular reduction of quinones proceeds via a one- or two-electron transfer catalyzed by a variety of flavoproteins. One-electron reduction to the semiquinone radical (typically catalyzed by NADPH-cytochrome *P*450 reductase or NADH–ubiquinone oxidoreductase) can initiate redox cycling and oxidative stress. In contrast, two-electron reduction (catalyzed by NAD(P)H:quinone oxidoreductase or DT-diaphorase) to form the corresponding hydroquinone is often viewed as a detoxication reaction (32). Quinone reductase activity has been reported in trypanosomes (12), and two distinct diaphorase activities have been detected in *E. histolytica* (33). Unlike the mammalian enzyme, however, the protozoan enzyme was insensitive to dicoumarol, a highly specific DT-diaphorase inhibitor. Bueding (34) observed that menadione (2-methyl-1,4-naph-thoquinone) was toxic to *S. mansoni*. It interfered with glucose metabolism and its effect was potentiated under aerobic conditions. Although this observation is consistent with the known redox cycling of menadione, the metabolism of that compound has not been investigated in schistosomes.

2.3. Hydrolysis Reactions

Parasites have the enzymatic capacity to hydrolyze a variety of xenobiotics. *M. expansa*, *A. suum* and *H. diminuta* contain *O*- and *N*-deacetylase activity, aryl sulfatase, and both acid and alkaline phosphatases (16, 21). The cestode *H. diminuta* can hydrolyze both α-

and β- glycosidic linkages. Esterase activity in these organisms can hydrolyze a number of model substrates. Significantly, however, helminths appear to lack the capacity to hydrolyze β-glucuronide substrates, indicating the absence of β-glucuronidase activity. *S. mansoni* contains esterase activity both toward the endogenous neurotransmitter acetylcholine and toward organophosphorus compounds such as metrifonate (35).

Epoxides are one class of electrophilic xenobiotics that are often toxic (4). Hydrolysis of the strained epoxide ring is catalyzed by epoxide hydrolase (EC 3.3.2.3) to give the corresponding *trans*-dihydrodiol that is not electrophilically active. Cell-free preparations of *O. gutturosa* contained low levels of epoxide hydrolase activity toward *trans*-stilbene oxide (22), but no activity was detected in *H. diminuta* (21). Since epoxide hydrolase in *O. gutturosa* was detected in a preparation containing both microsomes and cytosol, it is impossible to determine whether the enzyme is among the common microsomal epoxide hydrolases or whether it is cytosolic. Microsomal fractions prepared from *T. cruzi* epimastigotes showed epoxide hydrolase activity toward styrene oxide (36), which was increased three-fold when the organisms were exposed to the inducing agent phenobarbital. A more extensive survey would be useful to determine how common epoxide hydrolase is among protozoan and helminth parasites.

3. PHASE II METABOLISM

3.1. Conjugation with Glutathione

Conjugation of lipophilic compounds bearing an electrophilic center with the endogenous nucleophilic glutathione (GSH) is a major pathway of xenobiotic metabolism in most organisms (5, 37). Of all xenobiotic metabolism pathways characterized to date, conjugation with GSH is by far the most prevalent among both helminth and protozoan species. These reactions are catalyzed by a multifunctional family of mostly cytoplasmic enzymes, the GSH S-transferases (EC 2.5.1.18). These enzymes are highly specific for the endogenous cosubstrate GSH, but bind a broad spectrum of xenobiotics and endogenously produced electrophiles by virtue of a less-specific hydrophobic region at the active site (37).

GSH S-transferases are widely distributed in animals, being found in both vertebrates (37) and non-vertebrates (38), including helminth and protozoan parasites. Brophy and Barrett (39) have reviewed the distribution of GSH S-transferases in helminths and have observed that digean trematodes and intestinal cestodes generally have higher levels of total GSH S-transferase activity than do nematodes. Of the helminths examined to data, schistosomes show the highest level of activity, some 5–50-fold higher than others (40). GSH S-transferases have been identified and characterized from several protozoan species, including *T. cruzi* (41) and several *Acanthamoeba* species (42). No activity was detected in *E. histolytica*, *G. intestinalis* or *T. vaginalis* (42). The observation that *E. histolytica* lacks GSH S-transferase is consistent with the lack of GSH and its associated pathways in this organism (24). It is particularly curious that *T. cruzi* contains a typical GSH S-transferase, since trypanothione, rather than GSH, is the major intracellular thiol in these organisms (see Chapter 9).

Many organisms contain multiple forms of GSH S-transferase, but the isoenzymes of mammalian liver have been the most extensively studied (37). These abundant enzymes are characterized by distinct but overlapping substrate specificities that account for the liver's capacity to conjugate a structurally diverse group of chemicals. Based on primary structure, physicochemical properties and catalytic specificities, mammalian GSH S-transferases have been assigned to one of three classes (α, μ or π) that correspond three major gene families (43). The catalytically active enzymes exist as binary combinations of either identical (homodimer) or certain combinations of non-identical subunits (heterodimer); but heterodimers do not form between subunits of different classes (43).

The best characterized parasite GSH S-transferases are those of schistosomes. There are at least six biochemically distinct forms of the enzyme in adult S. mansoni. The three major isoenzymes (SmGST-1, SmGST-2 and SmGST-3), which comprise 2–4% of the total soluble protein (40), form an isoenzyme family composed of two homodimers (SmGST-1 and SmGST-3) and the corresponding heterodimer (SmGST-2) (44). Further evidence that SmGST-1 and SmGST-3 are closely related gene products comes from observations that an SmGST-3 cDNA hybridizes to messages encoding both subunits (45). Moreover, the ratio of two 28 kDa SmGST sequence variants in a cDNA library and the deduced isoelectric points of the encoded proteins (46) are consistent with the isoelectric points of SmGST-1 and SmGST-3 (40) and the ratio of the corresponding messages (45). From a comparison of its abundance, catalytic and physicochemical properties, SmGST-3 (subunit $M_r = 28\,500$) is equivalent to the isoenzyme character-ized as P28 and Sm28, which is being investigated as a candidate vaccine antigen (39, 47). The P28 GSH S-transferase was reported to be a surface antigen (47). However, a separate study failed to detect any GSH S-transferase within or on the tegument, including the dorsal spines of the male worm (48). Instead the enzymes diaplayed a punctate pattern of distribution that was restricted to the parenchyma (48).

Besides the SmGST-1/2/3 isoenzyme family, at least three other catalytically active SmGSTs are present in adult S. mansoni. SmGST-4 (subunit $M_r = 26\,000$), can be distinguished from the SmGST-1/2/3 family by its physicochemical properties, catalytic specificity, and its selective elution from GSH-agarose affinity columns with glutathione disulfide (49). SmGST-5 is a labile form of the enzyme that does not bind to GSH-agarose (40). It preferentially catalyzes the conjugation of epoxide substrates and the organophosphorus compound dichlorvos (50). Finally, SmGST-6 (subunit $M_r = 28\,000$) is an enzyme form that binds very weakly to GSH−agarose, with most of the activity being found in the unbound fraction. Unlike SmSGT-5, however, this isoenzyme is quite stable (Tracy and Vande Waa, unpublished observation).

cDNAs encoding several schistosome GSH S-transferases have been isolated (39, 45, 51, 52). Attempts have been made to include schistosome GSH S-transferases under the α, μ, π classification system (37). Although some short stretches of sequence similarity exist, it is clear that parasite and mammalian GSH S-transferases are at best distantly related (53). This difference is not too surprising in light of the ability of schistosome GSH S-transferases to stimulate host immunity.

Besides catalyzing the conjugation of numerous model xenobiotics, parasite GSH S-transferases are implicated in resistance to certain antiparasitic drugs. One form of S. mansoni GSH S-transferase (SmGST-5) catalyzes the GSH-dependent demethylation of dichlorvos, the active form of the antischistosomal drug metrifonate (50). The two metabolites that form, S- methylglutathione and des-methyldichlorvos are both phar-macologically inactive. The level of GSH S-transferase activity toward the model substrate 1-chloro-2,4-dinitrobenzene in the nematode H. contortus appears to correlate with resistance to cambendazole (39), although the conjugation of cambendazole was not demonstrated. GSH S-transferases can also play a role in protection against drug toxicity by acting as intracellular binding proteins (37). Several anthelmintics have been shown to bind to GSH S-transferases from M. expansa, H. diminuta and F. hepatica, although they do not conjugate these drugs (39). Those observations are consistent with the view that GSH S-transferases can participate in cellular protection by reducing the bioavailability of toxicants (43).

Although most studies have focused on the role of GSH S-transferases in xenobiotic metabolism, these enzymes also have multiple endogenous metabolic functions. For example, the biosynthesis of leukotriene C_4, the GSH conjugate of the epoxide leukotriene A_4, was shown to be catalyzed by the GSH S-transferase from Dirofilaria immitis (54). In addition to the selenium-dependent GSH peroxidase, eukaryotic cells have selenium-independent GSH peroxidase activity that is attributable to GSH S-transferase (37). As such, GSH S-transferases play a role in the reduction of fatty acid hydroperoxides formed during lipid peroxidation. In contrast to selenium-dependent GSH peroxidase, GSH S-transferases do not catalyze the reduction of hydrogen peroxide. Several helminth GSH S-transferases have been shown to catalyze the reduction of lipid and organic hydroperoxides (39, 49). GSH S-transferases catalyze the detoxication of 4-hydroxyalk-2-enals (37, 43). These α, β-unsaturated aldehydes are formed as byproducts of lipid peroxidation and are implicated as the ultimate cytotoxic products (55). The GSH S-transferases from S. mansoni catalyze the conjugation of a series of 4-hydroxyalk-2-enals, including 4-hydroxynon-2-enal, the isomer that is formed under physiological conditions (49). This observation supports the idea that these enzymes may play a role in protecting the adult parasites against host-mediated oxidative mechanisms.

The expression of numerous xenobiotic metabolizing enzymes can be induced by exposing the organism to a number of xenobiotics. For example, GSH S-transferase expression can be induced in mammalian tissues by treatment with various compounds, including the barbiturate phenobarbital, planar aromatic hydrocarbon carcinogens such as 3-methylcholanthrene, the environmental toxicant 2,3,7,8-tetrachlorodibenzo-p-dioxin (TCDD), phenolic antioxidants such as 2(3)-tert-butyl-4-hydroxyanisole (BHA), and Michael reaction acceptors that contain an electrophilic olefin such as α, β-unsaturated aldehydes and the antischistosomal drug oltipraz (37). Induction is characterized by an increase in enzyme content or specific activity and is paralleled by an increase in specific mRNA. Because not all GSH S-transferase subunits are induced by a given agent, only modest increases in total catalytic activity (< 2-fold) are often observed. This is because enzymatic activity measured in unfractionated cytosol with a single substrate such as 1-chloro-2,4-dinitrobenzene represents the average of all isoenzymes present, including those that are not induced.

Exposure to phenobarbital increases the specific activity of GSH S-transferase in Echinococcus granulosus and H. diminuta (39). In both cases, the level of induction was small, but statistically significant. Plumas-Marty et al. observed a twofold increase in GSH S-transferase activity in T. cruzi exposed to the same agent (56). Differential induction is more readily detected by measuring changes in the steady-state level of mRNA. Vande Waa *et al.* (45) were the first to demonstrate unequivocally that induction of a parasite xenobiotic metabolizing enzyme results in an increase in level of isoenzyme-specific message. Treatment of mice infected with *S. mansoni* with phenobarbital, 3-methylcholanthrene or BHA caused an increase of 5–10-fold in the steady-state level of mRNA hybridizing to pGT16.4, an *Sm*GST-3 cDNA. It has yet to be determined whether the increases in steady-state level of mRNA result from an increase in the rate of transcription or from a decrease in the rate of message degradation.

3.2. Mercapturic Acid Formation

In mammals, GSH *S*-conjugates, the metabolites of GSH *S*-transferase catalyzed reactions, are subject to further metabolism by a series of three enzyme-catalyzed reactions to form the corresponding *N*-acetylcysteine derivative or mercapturic acid, which is excreted in urine (3). The same group of three enzymes are also involved in recycling of GSH and in amino acid transport. The first step is catalyzed by γ-glutamyltransferase (EC 2.3.2.2), an enzyme localized on the outer surface of most cells. This enzyme has been detected in *A. suum* (57), *S. mansoni* (58) and *T. cruzi* epimastigotes (59). However, there is no evidence that parasites metabolize GSH *S*-conjugates to mercapturic acids. On the contrary, adult *S. mansoni* has been found to contain an energy-dependent transport system for excretion of the GSH *S*-conjugate, S-(2,4-dinitro[^{14}C]phenyl)glutathione (J. W. Tracy, unpublished observation). Inclusion of AT-125 (Acivicin), a highly selective inhibitor of γ-glutamyltransferase, did not affect recovery of the conjugate, suggesting that schistosomes do not metabolize GSH *S*-conjugates, but instead excrete them directly into the host environment.

3.3. Conjugation with Amino Acids

The metabolism of xenobiotic carboxylic acids to amino acid conjugates involves a two-step enzymatic pathway in which the xenobiotic is first activated in an ATP-dependent reaction to the corresponding coenzyme A thioester. This in turn serves as the substrate for an *N*-acyltransferase that catalyzes the formation of the peptide bond (3). Although glycine conjugates are the most commonly found metabolites, the specific amino acid acceptor depends on both the animal species and the chemical structure of the xenobiotic. Little is known about this conjugation pathway in parasites. *A. suum*, *M. expansa*, *H. diminuta* and *F. hepatica* failed to form hippurate, the glycine conjugate of benzoic acid (8), although small amounts of this product were detected in *M. benedeni* (60). Given that several other amino acids besides glycine can often serve as acyl acceptors particularly in invertebrate species (3), additional studies are needed before concluding that amino acid conjugation is not a pathway of parasite xenobiotic metabolism.

3.4. Glyoxylase

The glyoxylase system consists of two enzymes, glyoxylase I and glyoxylase II and GSH. Although these enzymes are widely distributed in many prokaryotic and eukaryotic species, their physiological function is poorly understood. It is generally believed that they are somehow involved in controlling cell proliferation (2), and play a role in the detoxication of 2-oxoaldehydes formed during lipid peroxidation. Glyoxylase I catalyzes the GSH-dependent conversion of 2-oxoaldehydes into the thiolester of GSH. This intermediate is then further metabolized by glyoxylase II to the corresponding 2-hydroxy acid. For example, methylglyoxal is biotransformed to lactic acid. There has been a single report of glyoxylase system in a helminth, the filarial worm *O. gutturosa* (22). Additionally, glyoxylase I and II activities have been reported in *Leishmania braziliensis* promastigotes (61). Because of the potential role of this xenobiotic metabolism pathway in protecting cells against potentially toxic products of lipid peroxidation, studies of the glycoxylase system may provide insights into how parasites resist the oxidative damage produced by effector cells of the host's immune system.

3.5. Other Conjugation Reactions

Although the number of parasites in which phase II reactions have been surveyed is quite limited, all investigations to date have revealed the apparent absence of several conjugation reactions commonly found in vertebrates. These include glucuronide formation (catalyzed by UDP-glucuronsyltransferases), glucoside formation (catalyzed by UDP-glucosyltransferases), formation of sulfuric acid esters (catalyzed by sulfotrans-ferases), *N*-acetylation (catalyzed by *N*-acetyltransferases) and methylation, which uses *S*-adenosylmethionine as a cosubstrate (2,3). One of the difficulties encountered in interpreting such data is that only a few parasite species have been examined using only a very limited number of potential xenobiotic substrates. Even with that caveat, however, it seems that the spectrum of phase II reactions in parasites is relatively limited. Since phase II reactions constitute an important enzymatic defense against toxic xenobiotics, the potential exists to exploit the absence of a particular pathway in the design of new drugs.

3.6. Xenobiotic Conjugation as a Pathway of Prodrug Activation

It is generally recognized that phase II metabolism represents a pathway of xenobiotic detoxication (3), although some notable exceptions to this rule are known (37). There is presently only one example in which conjugation of an antiparasitic drug ultimately leads to formation of a reactive metabolite that relates to its antiparasitic activity. That drug is hycanthone (1-2(2-(diethylamino)ethylamino)-4-hydroxymethylthioxanthen-9-one), a substituted aryl methanol derivative (62) that was once used extensively for treatment of human *S. mansoni* and *S. hematobium* infections. According to the current view of hycanthone's mechanism of action, the drug is first metabolized within the parasite to a phosphate ester conjugate (63), which is chemically unstable and decomposes to yield a reactive carbocation that in turn alkylates guanine residues in parasite DNA (64). Extensive efforts to demonstrate the presence of this ATP-dependent activating enzyme in a hycanthone-sensitive strain of *S. mansoni* maintained in this laboratory have been uniformly unsuccessful (Fialkowski and Tracy, unpublished observations). The reason for this discrepancy remains unclear.

4. CONCLUSION

Studies on antiparasitic drugs have historically focused on the effects of the drug on the parasite, but it is now apparent that parasites can metabolize xenobiotics and that such metabolism can relate to drug efficacy. In the case of nitroheterocyclic drugs metronidazole, nifurtimox and niridazole, enzymatic nitroreduction results in formation of chemically reactive metabolites that are essential for expression of antiparasitic drug efficacy. Alternatively, metabolism of a drug can result in its detoxication, leading to decreased efficacy or even drug resistance. Understanding the pathways by which parasites metabolize xenobiotics and comparing those pathways with those operating in the host organism should provide insights into parasite–drug interactions that may reveal novel strategies for rational drug design that take advantage of differences in the metabolic fate of a drug in host and parasite (28). The presence or absence of particular xenobiotic metabolism pathways could provide clues for design of new, more effective antiparasitic drugs. In spite of the obvious link between xenobiotic metabolism by parasites and the action of antiparasitic drugs, little attention has been paid to the capacity of parasites to metabolize xenobiotics. It would seem that this would continue to be a fruitful area of investigation, particularly among the parasites of human and veterinary importance, and that additional pathways of xenobiotic metabolism in parasites await discovery.

5. REFERENCES

1. Jakoby, W. B. (ed.) (1980) *Enzymatic Basis of Detoxication*, Vol. I. Academic Press, New York, 415 pp.
2. Jakoby, W. B. (ed.) (1980) *Enzymatic Basis of Detoxication*, Vol. II. Academic Press, New York, 369 pp.
3. Mulder, G. J. (ed.) (1990) *Conjugation Reactions in Drug Metabolism: An Integrated Approach.* Taylor and Francis, London, 413 pp.
4. Jaboby, W. B., Bend, J. R. and Caldwell, J. (eds) (1982) *Metabolic Basis of Detoxication.* Academic Press, New York, 375 pp.
5. Anders, M. W. (ed.) (1985) *Bioactivation of Foreign Compounds.* Academic Press, New York. 555 pp.
6. Nebert, D. W. and Gonzalez, F. J. (1987) P450 genes: structure, evolution and regulation. *Annu. Rev. Biochem.* **56**: 945–993.
7. Agosin, M., Naquira, C., Capdevila, J. and Paulin, J. (1976) Hemoproteins in *Trypanosoma cruzi* with emphasis on microsomal pigments. *Int. J. Biochem.* **7**: 585–593.
8. Precious, W. Y. and Barrett, J. (1989) Xenobiotic metabolism in helminths. *Parasitol. Today* **5**: 156–160.
9. Pottenger, L. and Jefcoate, C. R. (1990) Characterization of a novel cytochrome P450 from the transformable cell line C3H/10T1/2. *Carcinogensis* **11**: 321–327.
10. Douch, P. G. C. and Buchanan, L. L. (1979) Some properties of the sulphoxidases and sulfoxide reductases of the cestode *Moniezia expansa* and the nematode *Ascaris suum* and mouse liver. *Xenobiotica* **9**; 675–679.
11. Callahan, H. L., Crouch, R. K. and James, E. R. (1988) Helminth anti-oxidant enzymes: a protective mechanism against host oxidants? *Parasitol. Today* **4**: 218–225.
12. Docampo, R. and Moreno, S. N. J. (1984) Free-radical intermediates in the antiparasitic action of drugs and phagocytic cells. In: *Free Radicals in Biology* (ed. Pryor, W. A.), Vol. 6. Academic Press, New York, pp. 243–288.

13. Williams, D. L., Pierce, R. J., Cookson, E. and Capron, A. (1992) Molecular cloning and sequencing of glutathione peroxidase from *Schistosoma mansoni. Mol. Biochem. Parasitol.* **52**: 127–130.
14. Paul, J. M. and Barrett, J. (1980) Peroxide metabolism in the cestodes *Hymenolepis diminuta* and *Moniezia expansa. Int. J. Parasitol.* **10**: 121–124.
15. Mkoji, G. M., Smith, J. M. and Prichard, R. K. (1988) Antioxidant systems in *Schistosoma mansoni:* correlation between susceptibility to oxidant killing and the levels of scavengers of hydrogen peroxide and oxygen free radicals. *Int. J. Parasitol.* **18**: 661–666.
16. Barrett, J. (1981) *Biochemistry of Parasitic Helminths.* University Park Press, Baltimore, pp. 231–233.
17. Batra, S., Singh, S. P., Gupta, S., Katiyar, J. C. and Srivastava, V. M. L. (1990) Reactive oxygen intermediates metabolizing enzymes in *Ancylostoma ceylanicum* and *Nippostrongylus brasiliensis. Free Radical Biol. Med.* **8**: 271–274.
18. McCord, J. M. (1985) Oxygen-derived free radicals in postischemic tissue injury. *N. Engl. J. Med.* **312**: 159–163.
19. Nimmo-Smith, R. H. and Raison, C. G. (1968) Monoamine oxidase activity of *Schistosoma mansoni. Comp. Biochem. Physiol.* **24**: 403–416.
20. Douch, P. G. C. and Blair, S. S. B. (1975) The metabolism of foreign compounds in the cestode, *Moniezia expansa,* and the nematode, *Ascaris lumbricoides* var *suum. Xenobiotica* **5**: 279–292.
21. Munir, W. A. and Barrett, J. (1985) The metabolism of xenobiotic compounds by *Hymenolepsis diminuta. Parasitology* **91**: 145–156.
22. Pemberton, K. D. and Barrett, J. (1989) The detoxication of xenobiotic compounds by *Onchocerca gutturosa* (Nematoda: Filarioidea). *Int. J. Parasitol.* **19**: 875–878.
23. Tracy, J. W., Catto, B. A. and Webster, L. T., Jr (1983) Reductive metaboism of niridazole by *Schistosoma mansoni* correlation with covalent drug binding to parasite macromolecules. *Mol. Pharmacol.* **24**: 291–299.
24. Docampo, R. (1990) Sensitivity of parasites to free radical damage by antiparasitic drugs. *Chem. Biol. Interact.* **73**: 1–27.
25. Müller, M. (1986) Reductive activation of nitroimidazoles in anaerobic microorganisms. *Biochem. Pharmacol.* **35**: 37–41.
26. Beaulieu, Jr, B. B., McLafferty, M. A., Koch, R. L. and Goldman, P. (1981) Metronidazole metabolism in cultures of *Entamoeba histolytica* and *Trichomonas vaginalis. Antimicrob. Agents Chemotherap.* **20**: 410–414.
27. Catto, B. A., Tracy, J. W. and Webster, L. T., Jr (1984) 1-Thiocarbamoyl-2-imidazolidinone, a metabolite of niridazole in *Schistosoma mansoni. Mol. Biochem. Parasitol.* **10**: 111–120.
28. Webster, L. T., Jr, Tracy, J. W., Blumer, J. L., Catto, B. A. and Sissors, D. L. (1984) Relationships of niridazole metabolism to antiparasitic efficacy and host toxicity. In: *Proceedings of IUPHAR Ninth International Congress of Pharmacology.* Macmillan, London, pp. 363–367.
29. Kuwahara, T., White, R. A. and Agosin, M. (1984) NADPH-cytochrome *c* reductases of *Trypanosoma cruzi. Biochem. Biophys. Res. Commun.* **124**: 121–124.
30. Kuwahara, T., White, R. A. and Agosin, M. (1985). A cytosolic FAD-containing enzyme catalyzing cytochrome *c* reduction in *Trypanosoma cruzi.* I. Purification and some properties. *Arch. Biochem. Biophys.* **239**: 18–28.
31. Agosin, M. and Ankley, G. T. (1987) Conversion of *N,N*-dimethylaniline to *N,N*-dimethylaniline-*N*-oxide by a cytosolic flavin containing fraction from *Trypanosoma cruzi. Drug Metab. Dispos.* **15**: 200–203.
32. Monks, T. J., Hanzlik, R. P., Cohen, G. M., Ross, D. and Graham, D. R. (1992) Quinone chemistry and toxicity. *Toxicol. Appl. Pharmacol.* **112**: 2–16.
33. Weinbach, E. C., Harlow, D. R., Claggett, C. E. and Diamond, L. S. (1977) *Entamoeba histolytica:* diaphorase activities. *Exp. Parasitol.* **41**: 186–197.
34. Bueding, E. (1950) Carbohydrate metabolism of *Schistosoma mansoni. J. Gen. Physiol.* **33**: 475–495.
35. Reiner, E. (1981) Esterases in schistosomes: reaction with substrates and inhibitors. *Acta Pharmacol. Toxicol.* **49**, Suppl. 5: 72–78.

36. Yawetz, A. and Agosin, M. (1979) Epoxide hydrase in *Trypanosoma cruzi* epimastigotes. *Biochim. Biophys. Acta* **585**: 210–219.
37. Sies, H. and Ketterer, B. (eds) (1988) *Glutathione Conjugation: Mechanisms and Biological Significance*, Academic Press, London, 480 pp.
38. Clark, A. G. (1989) The comparative enzymology of the glutathione *S*-transferases from non-vertebrate organisms. *Comp. Biochem. Physiol.* **92B**: 419–446.
39. Brophy, P. M. and Barrett, J. (1990). Glutathione transferase in helminths. *Parasitology* **100**: 345–349.
40. O'Leary, K. A. and Tracy, J. W. (1988) Purification of three cytosolic glutathione *S*-transferases from adult *Schistosoma mansoni*. *Arch. Biochem. Biophys.* **264**: 1–12.
41. Yawetz, A. and Agosin, M. (1981) Purification of the glutathione *S*-transferase of *Trypanosoma cruzi*. *Comp. Biochem. Physiol.* **68B**: 237–243.
42. Dierickx, P. J., Almar, M. M. and De Jonckheere, J. F. (1990) Glutathione transferase activity in some flagellates and amoebae, and purification of the soluble glutathione transferases from *Acanthamoeba*. *Biochem. Int.* **22**: 593–600.
43. Mannervik, B. and Danielson, U. H. (1988) Glutathione transferases. Structure and catalytic activity. *CRC Crit. Rev. Biochem.* **23**: 283–337.
44. O'Leary, K. A. and Tracy, J. W. (1994) A *Schistosoma mansoni* glutathione *S*-transferase isoenzyme family composed of two homodimers and the corresponding heterodimer. *Comp. Biochem. Physiol.* (under review).
45. Vande Waa, E. A., Campbell, C. K., O'Leary, K. A. and Tracy, J. W. (1993) Induction of *Schistosoma mansoni* glutathione *S*-transferase by three classes of xenobiotics. *Arch. Biochem. Biophys.* **303**: 15–21.
46. Pierce, R. J., Khalide, J., Williams, D. L. *et al.* (1994) *Schistosoma mansoni*: characterization of sequence variants of the 28-kDa glutathione *S*-transferase. *Exp. Parasitol.* **79**: 81–84.
47. Taylor, J. B., Vidal, A., Torpier, G. *et al.* (1988) The glutathione transferase activity and tissue distribution of a cloned M_r28K protective antigen of *Schistosoma mansoni*. *EMBO J.* **7**: 465–472.
48. Holy, J. M., O'Leary, K. A., Oaks, J. A. and Tracy, J. W. (1989) Immunocytochemical localization of the major glutathione *S*-transferases in adult *Schistosoma mansoni*. *J. Parasitol.* **75**: 181–190.
49. O'Leary, K. A., Hathaway, K. M. and Tracy, J. W. (1992) *Schistosoma mansoni*: single-step purification and characterization of glutathione *S*-transferase isoenzyme 4. *Exp. Parasitol.* **75**: 47–55.
50. O'Leary, K. A. and Tracy, J. W. (1991) *Schistosoma mansoni*: glutathione *S*-transferase-catalyzed detoxication of dichlorvos. *Exp. Parasitol.* **72**: 355–361.
51. Henkle, K. J., Davern, K. M., Wright, M. D., Ramos, A. J. and Mitchell, G. F. (1990) Comparison of the cloned genes of the 26- and 28-kilodalton glutathione *S*-transferases of *Schistosoma japonicum* and *Schistosoma mansoni*. *Mol. Biochem. Parasitol.* **40**: 23–34.
52. Trottein, F., Kieny, M. P., Verwaerde, C. *et al.* (1990) Molecular cloning and tissues distribution of a 26-kilodalton *Schistosoma mansoni* glutathione *S*-transferase. *Mol. Biochem. Parasitol.* **41**: 35–44.
53. Brophy, P. M. and Pritchard, D. I. (1994) Parasitic helminth glutathione *S*-transferases: an update on their potential as targets for immuno- and chemotherapy. *Exp. Parasitol.* **79**: 89–96.
54. Weller, P. F., Longworth, D. L. and Jaffe, J. J. (1989) Leukotriene C_4 synthesis catalyzed by *Dirofilaria immitis* glutathione *S*-transferase. *Am. J. Trop. Med. Hyg.* **40**: 171–175.
55. Esterbauer, H., Zollner, H. and Schaur, R. J. (1988) Hydroxyalkenals: Cytotoxic products of lipid peroxidation. *ISI Atlas Biochem.* **1**: 311–317.
56. Plumas-Marty, B., Verwaerde, C., Loyens, M. *et al.* (1992) *Trypanosoma cruzi* glutathione-binding proteins: immunogenicity during human and experimental Chagas' disease. *Parasitology* **104**: 87–98.
57. Dass, P. D. and Donahue, M. J. (1986) γ-Glutatmyl transpeptidease activity in *Ascaris suum*. *Mol. Biochem. Parasitol.* **20**: 233–236.
58. Frappier, F., Azoulay, M. and Leroy, J.-P. (1988) Effect of oltipraz on the metabolism of glutathione in *Schistosoma mansoni*. *Biochem. Pharmacol.* **37**: 2864–2866.

59. Moncada, C., Repetto, Y., Aldunate, J., Letelier, M. E. and Morello, A. (1989) Role of glutathione in the susceptibility of *Trypanosoma cruzi* to drugs. *Comp. Biochem. Physiol.* **94B**: 87–91.
60. Kurelec, B. (1971) Metabolic path of benzoic acid in parasitic helminths *Fasciola hepatica* and *Moniezia benedeni*. *Int. J. Biochem.* **2**, 245–248.
61. Darling, T. N. and Blum, J. J. (1988) D-Lactate production by *Leishmania braziliensis* through the glyoxylase pathway. *Mol. Biochem. Parasitol.* **28**: 121–128.
62. Archer, S. (1985) The chemotherapy of schistosomiasis. *Ann. Rev. Pharmacol. Toxicol.* **25**: 485–508.
63. Pica-Mattoccia, L., Archer, S. and Cioli, D. (1992) Hycanthone resistance in schistosomes correlates with the lack of an enzymatic activity which produces the covalent binding of hycanthone to parasite macromolecules. *Mol. Biochem. Parasitol.* **55**: 167–176.
64. Archer, S., El-Hamouly, W., Seyed-Mozaffari, A., Butler, R. H., Pica-Mattoccia, L. and Cioli, D. (1990) Mode of action of the schistosomicide hycanthone: site of DNA alkylation. *Mol. Biochem. Parasitol.* **43**: 89–96.

11 Surface Constituents of Kinetoplastid Parasites

SALVATORE J. TURCO

Department of Biochemistry, University of Kentucky Medical Center, Lexington, KY, USA

SUMMARY

Cell surface components undoubtedly play key roles in the survival of protozoan parasites in hostile environments and in confrontation with host immune responses. In spite of their virtually certain relevance to disease processes that these parasites cause, the biochemistry, physiology and molecular biology of the surface constituents remain largely unexplored.

This chapter deals with membrane phenomena such as surface antigens, glycosylphosphatidylinositol anchors, surface glycoproteins, signal transduction, receptors, transporters and important extracellular glycoconjugates of Kinetoplastid parasites. A clearer understanding of the properties of these macromolecules should contribute to a more complete interpretation of the roles they play in pathogenesis. Extensive knowledge of the cell surface of the parasites may lead to rational approaches to parasite chemotherapy that exploit fundamental molecular differences between the mammalian host, the insect vector and the parasite.

1. IMPORTANT KINETOPLASTID MEMBRANE MACROMOLECULES

1.1. African Trypanosomes

1.1.1. Variant surface glycoprotein

In the bloodstream of the host, African trypanosomes, as best demonstrated with *T. brucei*, have the remarkable ability to conduct antigenic variation in which the parasite periodically switches its surface antigens and thereby evades its host's immune defenses.

Biochemistry and Molecular Biology of Parasites
ISBN 0-12-473345-X

The dominant surface molecule on trypanosomes is the variant surface glycoprotein (VSG). Ten million identical copies of this molecule crowd the surface of the bloodstream form of the parasite with a densely packed coat that covers the entire surface, masking other underlying molecules. The VSGs are highly immunogenic and, as a result, most of the parasites are recognized by antibodies of an immune system, and destroyed by complement-mediated lysis and by antibody-mediated phagocytosis. A few of the bloodstream parasites survive by expressing an antigenically new VSG coat and become progenitors of a new wave of parasites. It is believed that the organism has the genetic potential (reviewed in ref. 1) to express a thousand different VSGs, each with unique amino acid sequences. Conceivably, African trypanosomes could sustain an almost indefinite infection. VSG is a glycoprotein ($M_r = 60\,000$) of about 500 amino acids, comprising approximately 10% of the total cellular protein. VSGs form dimers and sometimes higher oligomers in solution, but it is uncertain whether these occur naturally.

VSG is synthesized in the endoplasmic reticulum (ER), undergoes several post-translational modifications during movement through the ER and Golgi network, and reaches the plasma membrane in approximately 15 minutes. As is typical of membrane proteins, an N-terminal signal peptide is removed during insertion of the VSG in the luminal side of the RER. All VSGs possess at least one and occasionally several N-glycosidically linked oligosaccharide chains. The structure of the asparagine-linked oligosaccharide from *T. brucei* has been reported to be a standard 'high mannose' type of carbohydrate chain (2).

Within one minute after completion of the primary translation product, a significant post-translational modification occurs that has profound implications for the biology of the parasite. The carboxyl end of the newly synthesized VSG contains a hydrophobic domain of about 15 amino acid residues. A preformed glycosylphosphatidylinositol (GPI) makes a nucleophilic attack on VSG, displacing the hydrophobic segment. The enzyme that catalyzes this transfer has not been characterized as yet, but the result is VSG polypeptide covalently linked to the GPI anchor in the membrane. The basic structure of the GPI anchor is conserved among different eukaryotes (3–5). The GPI anchor of the trypanosomal VSG was the first to be completely elucidated as ethanolamine-PO_4-Man(α1,2)Man(α1,6)Man(α1,6)GlcN-PI and it also may contain a side chain of galactosyl residues on the proximal mannose residue (6). A significant aspect of the trypanosomal GPI anchor of VSG is the exclusive presence of two ester-linked myristates (saturated C_{14} fatty acid) on the PI portion. No other eukaryotic GPI anchor possesses this unique PI structure.

The assembly of the preformed glycolipid precursor (7) first involves the addition of GlcNAc from UDP-GlcNAc to PI, forming GlcNAc-PI. The product is rapidly deacetylated to form GlcN-PI. Three mannose residues are donated sequentially from mannosylphosphoryldolichol (synthesized from GDP-Man) to form Man_3GlcN-PI. The phosphoethanolamine group is then donated from phosphatidylethanolamine to form the complete GPI anchor. It is highly likely that the assembly of this basic structure occurs similarly for GPI anchors in mammals as well since they possess the identical core GPI structure. However, the trypanosome goes further by remodeling the PI moiety. In a sequence of unusual exchange reactions, the PI portion is deacylated and reacylated with two myristate groups. The resultant product is believed to be the direct GPI precursor to VSG, but this has not yet been established by means of a cell-free system. After transfer of the GPI to the VSG polypeptide, galactosyl residues are then transferred, presumably from UDP-Gal, to the proximal mannose residue of the core region.

1.1.2. GPI-specific phospholipase C

A potent glycosyl phosphatidylinositol-specific phospholipase C (GPI-PLC) is present in bloodstream and metacyclic trypanosomes, which possess a VSG surface coat, but both are lacking in procyclic trypanosomes. As the name implies, the enzyme cleaves the VSG anchor between the diacylglycerol moiety and the phosphate group of the PI anchor (4, 5). The released protein product therefore contains a cyclic 1,2-phosphate on the inositol residue. Furthermore, GPI-PLC treatment exposes the glycan portion of the GPI anchor. Since this segment is conserved in all trypanosomal VSGs and is recognized by a given antibody, it is therefore referred to as the cross-reactive determinant (CRD).

GPI-PLC has been purified to homogeneity and has been found to be a single polypeptide of about 37 kDa (8–10). Approximately 30 000 molecules of the enzyme are present per parasite, representing about 0.04% of the cell's protein. The GPI-PLC differs from other known phospholipase Cs in that it does not require calcium and shows a high degree of specificity; it is specific for GPI structures and does not cleave phosphatidylinositol or other phospholipids. Interestingly, the GPI-PLC appears to be a membrane-bound enzyme, yet its deduced amino acid sequence lacks any obvious hydrophobic domain that might interact with membranes. The function of GPI-PLC is unknown. Since it has been established that the enzyme is associated with intracellular membranes, the enzyme is unlikely to be involved in releasing VSG from the surface of bloodstream trypanosomes. Rather the VSG molecules may have to be internalized to undergo turnover. This may explain the virtual lack of efficient release of surface VSG from bloodstream trypanosomes in culture. It is not clear how the VSG is made available to the compartmentalized GPI-PLC.

1.1.3. Procyclin

Procyclic forms of African trypanosomes lack the VSG coat. Instead, they express an abundant stage-specific surface glycoprotein that coats their entire surface. This glycoprotein is called procyclin and its apparent molecular weight is between 30 and 40 kDa (11). Procyclin has an unusual primary structure since approximately 40% of the protein sequence consists of a Glu-Pro repeat. Thus the protein is highly negatively charged which enables it to form an extended rod-like structure. Estimates of six million copies of procyclin per procyclic trypanosome are comparable to the number of VSGs per bloodstream trypanosome. Procyclins are believed to be anchored into the membrane via a GPI moiety, which is unlike the VSG derivative. Whether the distinct structures of the two GPI anchors have biological significance for the various stages of the parasite is unknown.

During the conversion of bloodstream forms of the trypanosome to the insect-stage procyclic forms, there is a significant rise in the levels of procyclin-specific mRNA. Procyclin glycoprotein is then present on the surface of the procyclic forms soon thereafter in spite of a considerable VSG coat. It has been determined that although the molecular mass of procyclin is less than a quarter of that of VSG, its extended structure enables is to protrude through the VSG coat. During the differentiation process, the VSG molecules are gradually replaced by procyclins. The net result is that the trypanosome is never uncoated, which otherwise would make it vulnerable to serum components ingested by the tsetse fly and the insect's digestive juices. Other than as a protective coat for the procyclic forms of the parasite, it is not known whether procyclin has additional functions.

1.2. American Trypanosomes

1.2.1. Lipopeptidophosphoglycan

The surface glycoconjugates of *Trypanosoma cruzi* are of interest by virtue of their association with host cell penetration and immunogenic properties. One of these molecules is lipopeptidophosphoglycan (LPPG), a major constituent of the plasma membrane of epimastigote forms (12). LPPG was found to be a ceramidephosphoinositol-anchored oligosaccharide. Phosphoceramide as an anchor has also been reported in a lipophosphonoglycan from *Acanthamoeba castellanii* and in a protein from *Dictyostelium discoideum*. LPPG contains the Man(α1,4)GlcN-*myo*-inositol-1-PO$_4$ motif, characteristic of glycolipid anchor proteins. Interestingly, the LPPG and the lipophosphonoglycan from *A. castellanii* possess 2-aminoethylphosphonic acid as a component. If the LPPG serves as a glycolipid anchor for proteins, this substituent may possibly be the site of protein attachment. The carbohydrate portion of LPPG consists of four mannose and two non-reducing galactofuranose residues.

The function of LPPG has yet to be determined. Analogous glycolipids have been described in plants, yeast and fungi and may be functionally similar to gangliosides in higher eukaryotes (13). Whether LPPG serves in any capacity in parasite–host cell interactions is unknown.

1.2.2. Neuraminidase–Sialyltransferase

Trypanosoma cruzi expresses a neuraminidase activity which is developmentally regulated. Enzyme activity is maximal in infective trypomastigotes, 10 to 560 times less in epimastigotes, and is absent in amastigotes. The neuraminidase, of an apparent molecular mass of 160–200 kDa, is located on the outer membrane and is released by a stage-specific phospholipase C. Of interest is the presence of sequences similar to low-density lipoprotein receptor and to the type III module of fibronectin (14).

Incubation of sialyl substrates, such as α-1 acid glycoprotein, fetuin and sialyl lactose, with either live parasites or with preparations of lysed trypanosomes results in the release of sialic acid at an optimum pH of 6.0 to 6.5. Moreover, incubation of live trypomastigotes with human erythrocytes causes a time-dependent release of sialic acid from the erythrocyte surface. Similarly, the level of desialylation of erythrocytes from *T. cruzi*-infected mice correlates with the degree of parasitemia. Further studies revealed that neuraminidase also removes sialic acid from the surface of myocardial and endothelial cells, indicating that this molecule may play a critical role in the pathology of Chagas' disease (15).

A striking observation was that inhibition of the neuraminidase activity with either antibodies or high- and low-density lipoproteins enhances infectivity of *T. cruzi* in fibroblast cells *in vitro*. This enhanced infectivity can be abrogated by the addition to the monolayers of *Vibrio cholerae* neuraminidase, indicating a role for sialic acid in the infection process.

Cell surface sialoglycoconjugate protects trypomastigotes from the lytic effect of complement. However, the parasite is unable to incorporate sialyl residues from cytidine monophosphate (CMP)–sialic acid or free sialic acid. Rather, these residues are acquired by direct transfer from extrinsic sialyl-containing molecules. Recently, the enzyme responsible has been characterized as a *trans*-sialidase which specifically transfers α(2-3)-linked sialic acid from mammalian cell surface glycoproteins to parasite surface molecules (16). Interestingly, the resulting sialylated molecules on the parasite are believed to be involved in target cell recognition, since they enhance invasion of host cells *in vitro* (17). Moreover, attachment of trypomastigotes to host cells can be inhibited by monoclonal antibodies that recognize sialic acid residues on a parasite surface molecule (16). Therefore, *T. cruzi* appears to express two distinct enzyme activities, a neuraminidase and a sialyl-transferase, which both modulate sialylation levels and infectivity in opposite manners. Indeed, whereas neuraminidase controls the infection by a negative mechanism, the *trans*-sialidase activity exerts a positive influence on the infectivity of *T. cruzi*. Thus notwithstanding the demonstrated importance of sialic acid in the interaction between *T. cruzi* and its host, several questions remain unanswered concerning the exact role of sialylation and its regulation by both the neuraminidase and the *trans*-sialidase.

1.3. *Leishmania*

The striking ability of *Leishmania* parasites to survive in hostile environments with impunity throughout their life cycle has implicated the importance of specialized glycoconjugates on their surface. These glycoconjugates are generally glycoproteins or complex carbohydrates anchored in the membrane by covalently linked glycosylphosphatidylinositol. The nature of the GPI anchor is distinct from the glycolipid anchoring the trypanosomal VSGs. The phosphatidylinositol portion of leishmanial GPI anchors characteristically contains a 1-*O*-alkyl hydrocarbon and may be esterified on the *sn*-2 position of the glycerol backbone with a fatty acid.

1.3.1. Lipophosphoglycan

Lipophosphoglycan (LPG) is the major cell surface glycoconjugate on the promastigote form of the parasite and is believed to be important for its uptake by and survival within macrophages (18). LPG consists of four domains: (i) a phosphatidylinositol lipid anchor; (ii) a phosphosaccharide core; (iii) a repeating phosphorylated saccharide region; and (iv) a small oligosaccharide cap structure. Structural analyses of LPG from several *Leishmania* species have indicated complete conservation of the lipid anchor, extensive conservation of the phosphosaccharide core, and variability of sugar composition and sequence in the repeating phosphorylated saccharide units and the cap structure.

LPG is anchored by the unusual phospholipid 1-*O*-alkyl-2-*lyso*-phosphatidyl (*myo*)inositol. In all species of *Leishmania* examined thus far, the aliphatic chain consists of either a C_{24} or C_{26} saturated, unbranched hydrocarbon. Similar to many GPI-anchored proteins, LPG can be hydrolyzed by bacterial phosphatidylinositol-specific phospholipase C producing 1-*O*-alkylglycerol and the entire polysaccharide chain. Attached to the inositol of the lipid anchor of LPG is the phosphosaccharide core region. The glycan core consists of an unacetylated glucosamine, two mannoses, a galactose-6-phosphate, a galactopyranose, and a galactofuranose. The presence of the latter is extremely unusual in eukaryotic glycoconjugates, especially since the furanose is internal in a carbohydrate chain. As with all other glycosylphosphatidylinositol-anchored proteins known, LPG possesses the Man(α1,4)GlcN(α1,6)*myo*-inositol-1-PO$_4$ motif. The LPG cores of *L. donovani* and *L. mexicana* possess a glucosyl-αl-phosphate attached in phosphodiester linkage to the C-6 hydroxyl of the proximal mannose residue. A substantial percentage of the *L. major* LPG also contains the identical glucosyl-αl-phosphate substitution, whereas the remainder does not. Another interesting sequence in the core region is the Gal(α1,3)Gal unit which is reported to be the epitope for circulating antibodies in patients with leishmaniasis (19, 20).

The salient feature of LPG is the repeating phosphorylated saccharide region. All LPG molecules reported thus far contain multiple units of a backbone structure of PO$_4$-6Gal(β1,4)Man(α1). One of the noteworthy features of the backbone is the 4-*O*-substituted mannose residue, which is not present in any other known eukaryotic glycoconjugate. The *L. donovani* LPG has no further substitutions of the backbone sequence, whereas in the *L. mexicana* LPG, approximately 25% of galactose residues are substituted at the C-3 hydroxyl with βGlc residues. The repeating units of the *L. major* LPG are the most complex in that approximately 87% of the galactose residues are further substituted with small saccharide side chains containing one to four residues of galactose, glucose, or the pentose arabinose. The presence of the common Gal-Man disaccharide backbone and the species-specific substitutions on the galactose residue could account for common and species-specific epitopes reported in serological observations. The number of repeating units per LPG molecule is directly dependent on the growth stage of the promastigote and is discussed below. LPG is terminated at the non-reducing end with one of several small, neutral oligosaccharides containing galactose or mannose.

The three-dimensional structure of the basic repeating PO$_4$-6Gal(β1,4)Man(α1) disaccharide units of LPG in solution has been recently determined (21). The repeating units, as expected, exist in solution with limited mobility about the Gal(β1,4)Man linkages. In contrast, a variety of stable rotamers exist about the Man(α1)-PO$_4$-6Gal linkages. An important feature of each of these low energy conformers is that the C-3 hydroxyl of each galactose residue is exposed and freely accessible. This is the particular position that is substituted with glucose in the LPG from *L. mexicana*, or galactose, glucose and arabinose in the LPG from *L. major*. Thus, these additional units could be accommodated without major conformational changes of the repeat backbone. Another intriguing finding from the molecular modeling studies is that each of the stable conformers of the Man(α1)-PO$_4$-6Gal linkages may exist in a different configuration within the same LPG molecule. These torsional oscillations confer an ability upon the LPG molecule to contract or expand in a manner reminiscent of a 'slinky spring', resulting in a molecule whose length is potentially 90 Å when fully contracted to 160 Å when fully extended, assuming an average of 16 repeat units.

A peculiar and significant observation is the modification in the structure of LPG that accompanies the process of metacyclogenesis (22). Analysis of the LPG derived from the infectious, metacyclic stage of the parasite revealed conservation of the *lyso*-1-*O*-alkylphosphatidylinositol lipid anchor and the phosphosaccharide core. The most striking difference, however, was an approximate doubling in size displayed by the metacyclic form of LPG, due to an approximate doubling in the number of repeating phosphorylated saccharide units. Such changes in LPG structure have profound implications on function and suggest important points of regulation of the glycosyl-transferases involved in LPG biosynthesis.

The uniqueness of the overall structure of LPG and its highly unusual domains would suggest that LPG has important functions for the *Leishmania* parasite in its life cycle. Indeed, evidence has been provided for a surprisingly large number of potential functions that enable the promastigote to survive in the hydrolytic environments it encounters (reviewed in ref. 18).

1.3.2. Glycosylphosphoinositides

The various parts of LPG have been shown to exist in *Leishmania* promastigotes as components of proteins or as distinct entities. Regarding the latter, a family of molecules termed glycosylphosphatidylinositol antigens (GPIs) (23, 24) or glyco-sylinositolphospholipids (GIPLs) (25) are present abundantly in *Leishmania*. Structural analyses of these glycolipids have indicated that they closely resemble the phosphosac-charide core-phosphatidylinositol region of LPG. In particular, the *L. major* GIPLs consist of a small mannose- and galactose-containing glycan which is glycosidically linked via an unacetylated glucosamine residue to either 1-*O*-alkyl-2-acyl-PI or *lyso*-1-*O*-alkyl-PI. The glycan parts of these molecules are completely identical to the analogous portions of LPG, including the salient Man(α1,4)GlcN(α1,6)*myo*-inositol and, in the larger GIPLs, the rare galactofuranose. Unlike *L. major*, promastigotes of *L. donovani* synthesize abundant GIPLs containing one to four mannose residues and are not galactosylated. Although *L. donovani* amastigotes do not synthesize LPG, they continue to synthesize GIPLs in quantities comparable to that reported in promas-tigotes. The *L. donovani* amastigote GIPLs containing one to three mannose residues, are structurally different from promastigote GIPLs and appear to be precursors to glycolipid anchors of proteins. In addition, ceramidephosphoinositides have been found in *L. donovani*. Whether these particular phosphosphingolipids can be further sub-stituted with carbohydrate residues as observed in *Trypanosoma cruzi*, yeast, fungi and plants has yet to be established.

1.3.3. Promastigote surface protease (GP63)

The major surface glycoprotein of *Leishmania* parasites is a glycoprotein with an apparent molecular weight of 63 kDa called gp63 (26). In promastigotes, gp63 covers the entire cell surface and the estimated 500 000 copies per cell account for 1% of the total cellular protein. In amastigotes, gp63 is expressed at a lower level, representing about 0.1% of the total cellular protein (27). Of interest, the bulk of gp63 in amastigotes is localized in the flagellar pocket region, although it is also present on the cell surface in a reduced number.

Gp63 is glycosylated with *N*-linked mannose oligosaccharide chains (28). Similar to the variant surface antigens of African trypanosomes, membrane anchoring of gp63 is mediated via a myristic-acid containing glycosylphosphatidylinositol (GPI) lipid an-chor (26). However, in amastigotes, the gp63 subpopulation present in the flagellar pocket lacks a membrane anchor (27).

Gp63 has been characterized as a zinc-metalloprotease (26, 28). The pH optimum of promastigote-derived gp63 appears to be neutral and cleavage is sequence-specific (28, 29). Although there is some confusion regarding the recognition sequence(s), the enzyme preferentially cleaves at the amino side of the amino acid residues (29). By virtue of its ability to cleave a peptide similar to its propeptide sequence, activation of gp63 may involve an autocatalytic mechanism (29).

The abundance and location of gp63 suggest that this enzyme may play a role in the infection process. Indeed, its expression is decreased in attenuated compared to virulent promastigotes, and is found in increased amounts in stationary phase *L. braziliensis*. By virtue of its activity, gp63 may interact with and proteolytically degrade host macromolecules. Indeed solubilized gp63 molecules covalently complex and cleave the human plasma protease inhibitor α_2-macroglobulin. The formation of α_2-macroglobulin–gp63 complexes results in a dose-dependent inhibition of the proteolytic activity of soluble gp63 on azocasein. However, α_2-macroglobulin may not interact with promastigotes in the bloodstream since gp63 at the surface of live promastigotes does not recognize α_2-macroglobulin. Gp63 may also participate in the attachment of promastigotes to macrophage surface receptors, in part because it binds the third complement component that can be cleaved to the CR3 ligand C3bi.

1.3.4. Acid phosphatase

Leishmanial parasites have two distinct types of acid phosphatases: a surface membrane-bound and an extracellular form of the enzyme. The membrane-bound acid phosphatase has been purified from *L. donovani* and could be resolved into one major and two minor isoenzymes (30, 31). The native enzyme ($M_r = 120\,kDa$) is believed to be composed of two identical 60 kDa subunits, has a pH optimum between 5.0 and 5.5, and possesses an extracellularly oriented active site. Some of the preferred natural phosphomonoester substrates for the enzyme include phosphotyrosine, fructose-1,6-bisphosphate and AMP.

Several biological roles of the membrane-bound acid phosphatase have been proposed. The enzyme may provide the parasite with a source of inorganic phosphate by hydrolyzing phosphomonoesters of metabolites. The acid phosphatase may also play an important role in suppressing toxic oxygen metabolite production (the oxidative burst) normally induced upon entry of the microbes into macrophages. The phosphatase might cause this inhibition by dephosphorylating critical substrates that enable the macrophage to become activated. Alternatively, it is well known that specific proteins, such as NADPH oxidase, are phosphorylated during macrophage activation. Phosphoproteins are not especially good substrates for the acid phosphatase. Therefore, it is unlikely that the enzyme acts as a phosphoprotein phosphatase, which would adversely affect the oxidative pathway involved in generating superoxide and hydrogen peroxide.

1.3.5. 3'-and 5'-Nucleotidases

Nucleotidases possessing 3'- and 5'-hydrolase activities are present on the surface of *L. donovani* and *L. mexicana* (32). Cell surface 5'-nucleotidases are commonly present on the plasma membrane of mammalian cells and are believed to be involved in purine uptake. Likewise, the biological role for the surface nucleotidases of *Leishmania* is believed to involve purine acquisition. The levels of 3′-nucleotidase can be increased up to 10-fold if the parasites are grown in medium lacking purines. In more extensive studies in *Crithidia*, the levels of 3′-nucleotidase are more severely affected by the concentration of purines in the medium. The enzyme levels can be induced as much as several orders of magnitude if the parasites are grown in purine-deficient medium. The elevated activity of the surface 3′-nucleotidase under purine depletion conditions presumably enables the parasites to derive purines from purine nucleotides and from nucleic acid sources. Regulation of the levels of the 3′-nucleotidase appears to be at the transcriptional level. The levels of the crithidial 5′-nucleotidase are not as affected by the amount of purines in the extracellular medium since there is only a 4 to 5-fold increase in the enzyme during conditions of purine depletion.

2. EXTRACELLULAR KINETOPLASTID MACROMOLECULES

2.1. Extracellular LPG-like Glycoconjugates of *Leishmania*

LPG-like substances, collectively termed excreted factor (EF), are present in conditioned medium from *Leishmania* parasites. The components of the EF can be organized into several categories. In one, LPG can form very tight complexes with albumin in the medium. Analysis of this form of LPG in the medium indicates that it is identical in all respects to the cell-associated LPG. One probable interpretation is that the lipid portion of LPG interacts with the hydrophobic binding pocket of albumin facilitating its release from the surface of the promastigote.

Another category of LPG-like substances is an extracellular phosphoglycan (exPG), which was purified from conditioned medium of *L. donovani* and characterized (33). Structural analysis indicated that the exPG consisted of the following structure: $(CAP) \rightarrow [PO_4\text{-}6Gal(\beta1,4)Man(\alpha1)]_{10-11}$. The 'cap' was found to be one of several small neutral oligosaccharides, the most abundant being the branched trisaccharide $Gal(\beta1,4)[Man(\alpha1,2)]Man(\alpha1)$. Thus, the exPG is identical to cellular-derived LPG with the important exceptions of a lack of a lipid anchor, the phosphosaccharide core and several repeating units. Further information indicates that the exPG originates from surface LPG. The mechanism of release of the exPG as well as its possible function, if any, are not known.

2.2. Secreted Acid Phosphatase of *Leishmania*

In addition to the membrane-bound acid phosphatase described above, *Leishmania* promastigotes, except those of *L. major*, secrete a soluble acid phosphatase. The repeating phosphorylated saccharide units of LPG have been shown to comprise a carbohydrate chain of the acid phosphatase secreted by *L. donovani*, *L. tropica* and *L. mexicana*. The number of repeating units per carbohydrate chain and the nature of the linkage to the polypeptide are unknown. The latter is not believed to involve a typical *N*-glycosidic linkage to an Asn residue since the repeating units were not removed from acid phosphatase by *N*-glycanase digestion. The enzymes secreted by *L. mexicana* and *L. donovani* have been purified and were shown to differ significantly. The secreted acid phosphatase of *L. mexicana* is a filamentous phosphoglycan polymer, whereas the enzyme from *L. donovani* is non-filamentous mono- or oligomer.

The major difference between the secreted and membrane-bound acid phosphatase is that the latter is approximately 40-fold more resistant to the inhibitory effects of sodium tartrate. Furthermore, non-specific antibodies to the extracellular enzyme do not cross-react with membrane-bound enzyme. The secreted enzyme also does not appear to inhibit neutrophil oxidative burst, unlike its membranous counterpart. The biological role of the secreted enzyme is unknown, but may be involved in modification of host activities.

2.3. Hemolysin of *Trypanosoma cruzi*

Following invasion of their host cell, several intracellular parasites remain inside the parasitophorous vacuole which may or may not fuse with lysosomes. Other pathogens, such as *Rickettsia* spp, *Listeria monocytogenes*, *Shigella flexneri*, *Theileria parva*, *Babesia bovis* and *Trypanosoma cruzi* must leave the phagocytic vacuole and establish themselves into the cytosolic compartment to survive successfully. Obviously, disruption of the phagocytic vacuole represents a critical step in this escape process (reviewed in ref. 34). The role of a hemolysin secreted by *T. cruzi* amastigotes is discussed in this section.

Shortly after phagocytosis, *T. cruzi* trypomastigotes are exposed to lysosomal enzymes and an acidified environment, and transform to the intracellular amastigote form. Then, within hours, most of the parasites are found in the cytoplasm of the host cell. Analysis of *T. cruzi* amastigote-conditioned medium revealed the presence of a hemolytic protein optimally active at pH 5.5 with undetectable activity at neutral pH, suggesting that it functions in acidic intracellular compartments (35). Sieving experiments showed that the hemolysin causes membrane lesions of about 10 nm diameter. Further characterization of this hemolysin revealed immunological cross reactivity with reduced and alkylated C9 (the ninth component of complement) and exhibits an apparent molecular weight of 75 kDa (36). The homology with C9 is of importance considering the pore-forming activity of this molecule.

In *T. cruzi*-infected macrophages, the hemolysin was localized inside the luminal space and closely apposed to the phagosome membrane (36), indicating that it is secreted into the acidic phagosome. Finally, the observation that the exit of *T. cruzi* from the phagosomes is inhibited by drugs that raise the pH of the intracellular compartment supports the notion that the acid-active hemolysin is involved in the destruction of the parasitophorous vacuoles, and hence allows the parasite to multiply in the appropriate compartment.

3. *N*-GLYCOSYLATION PATHWAY OF GLYCOPROTEINS

Similar to glycoproteins from other eukaryotic sources (37), glycoproteins from trypanosomatids possess *N*-glycosidically linked oligosaccharides. These asparagine-linked carbohydrate chains typically are the 'high mannose' type of oligosaccharides, containing up to nine mannose and two *N*-acetylglucosamine residues (38). Variations have been reported in the attachment of a non-reducing glucose residue in the leishmanial gp63, galactofuranose in *Crithidia* glycoproteins, and more exetensive processing of the high mannose chains to 'complex' type chains as found in glycoproteins in *T. cruzi*.

The biosynthesis of N-linked oligosaccharide chains containing GlcNAc and Man residues is believed to involve a cyclic sequence. The cycle consists of the preassembly of GlcNAc and Man residues bound in an activated state to the polyisoprenoid lipid dolichol pyrophosphate. Polyprenols have long been noted to serve as carriers of carbohydrates in glycan assembly. The polyprenols range in size from 11 isoprene units in bacteria (the C_{55} undecaprenol) to the larger 17–20 isoprene units in mammalian cells. In the latter, the polyprenol carrier is dolichol since the α-isoprene unit is saturated. The length of the trypanosomatid polyprenol carrier is approximately 11 isoprene units, but the α-isoprene unit is saturated. Thus, the lipid carrier involved in sugar transfer might be an evolutionary marker.

The cyclic sequence in N-linked oligosaccharide biosynthesis is initiated by the transfer of N-acetylglucosaminylphosphate (GlcNAc-P) from UDP-GlcNAc to dolichylphosphate (Dol-P) to form GlcNAc-P-P-Dol. The glycosyltransferase that catalyzes this step has been isolated and cloned from mammalian sources. The antibiotic tunicamycin inhibits the enzyme, which is also the case in trypanosomatids. Moreover, variants of *L. mexicana* have been isolated based on resistance to tunicamycin and have amplified expression of the enzyme. The variant cells are more virulent than the parental wild-type, suggesting that the expression of a gene(s) encoding this and possibly other related enzymes is the molecular basis of leishmanial virulence (28).

In the assembly of the dolichol-linked oligosaccharide intermediate, a second N-acetylglucosamine residue and then five mannose residues are added directly from the respective nucleotide-sugars. In other eukaryotic cells, four additional mannose residues are added to the growing lipid-linked oligosaccharide using mannosylphosphoryldolichol as the donor. The latter has been isolated from a variety of protozoa, but its participation in oligosaccharide-lipid assembly has only been implied. Not all trypanosomatids synthesize dolichol-linked oligosaccharides containing the full complement of nine mannose residues as occurs in higher eukaryotic cells. The dolichol-linked intermediates involved appear to be $Man_9GlcNAc_2$ in *T. cruzi*, $Man_7GlcNAc_2$ in *C. fasciculata*, and $Man_6GlcNAc_2$ in *L. mexicana*. Surprisingly, these lipid-bound oligosaccharides do not possess the three glucose residues that are characteristic in all other eukaryotic cells. $Glc_3Man_9GlcNAc_2$-P-P-dolichol has long been known to be the universal donor of asparagine-linked oligosaccharides in eukaryotic cells. The transfer of the completed oligosaccharide to nascent proteins generates dolichol pyrophosphate which is hydrolyzed to the monophosphate to reinitiate the cycle. In several protozoan parasites, however, the transferred mannosylated oligosaccharide on proteins undergoes transient glucosylation. The added glucose residue is believed to be donated by UDP-Glc instead of glucosylphosphoryldolichol and is removed shortly after transfer to the protein-linked oligosaccharide.

4. SIGNAL TRANSDUCTION

That any cell can grow, differentiate or execute a particular function, requires the ability to respond to extracellular signals encountered in the microenvironment. Such extracellular information has to be transmitted across the cellular membrane for subsequent conversion into intracellular signals. In mammalian cells, the elucidation of transmembrane signaling pathways and their implication in cell function regulation represents an important area of research (39, 40). Typically, activation of cell surface receptors triggers intracellular biochemical cascades involving either the generation of second messengers such as Ca^{2+}, phospholipid metabolites or cyclic nucleotides, or the activation of kinases and phosphorylation of regulatory proteins. Ultimately, these events lead to altered gene expression and phenotypic modifications.

Kinetoplastid parasites such as *Leishmania* and trypanosomes undergo important morphological and metabolical changes during their life cycle, which involve proliferation within insect and mammalian hosts. As discussed elsewhere, the ability of these parasites to transform rapidly from one developmental stage to the next may involve cell surface receptor-coupled signal transduction pathways. Although the biochemical details of the processes involved are still poorly understood, characterization of molecules homologous to components of mammalian cell signaling machinery have been recently reported in trypanosomes and *Leishmania*. Eventually, a better understanding of the mechanisms regulating adaptation and proliferation in mammalian hosts may provide specific targets for the design of better drug treatments. A brief list of some key membrane-associated molecules identified so far in these parasites is described below.

4.1. G Proteins

Most G proteins are heterotrimers consisting of α, β and γ subunits which function in close association with several transmembrane receptors. Their activity is regulated by the binding of GTP to the α subunit, followed by subsequent hydrolysis of GTP to GDP, and release of GDP (41). Functions regulated by G proteins include stimulation of phosphoinositide hydrolysis, modulation of adenylate cyclase activity and activity of ion channels (41). Therefore, G proteins play a crucial role in relaying information from cell surface receptors to the intracellular signal-transduction machinery.

L. donovani promastigotes express a 38 kDa membrane protein (p38) which strongly cross-reacts with an antiserum raised against the *C*-terminal sequence of the α subunit of mammalian G proteins (42). This p38 membrane protein specifically binds $[\alpha^{32}P]GTP$ in a manner similar to mammalian G proteins. Moreover, a fraction of the bound GTP is hydrolyzed, indicating that *L. donovani* p38 possesses GTPase activity. The interesting aspect of this membrane protein is that its expression appears to be developmentally regulated. Indeed, growth of promastigotes in conditions that favor the expression of amastigote stage-specific antigen (X) resulted in a decreased p38 expression.

4.2. Inositol Phosphate Metabolism

Receptor-mediated hydrolysis of membrane inositol phospholipids generates inositol-1,4,5-trisphosphate (InsP3) and diacylglycerol (DAG), which constitute a major common second messenger system in mammalian cells (39). InsP3 is released in the cytosol where its primary function is to mobilize Ca^{2+} from intracellular stores. On the other hand, DAG remains in the membrane where it activates Ca^{2+}-primed protein kinase C (39, 40).

In *T. cruzi* epimastigotes, [^{3}H]inositol and [^{32}P]P$_i$ are incorporated into several phosphatidylinositol lipids consisting of phosphatidylinositol (PI), phosphatidylinositol phosphate (PIP), and phosphatidylinositol bisphosphate (PIP2), as well as their derivatives inositol phosphate (InsP), inositol bisphosphate (InsP2), and InsP3 (43). Increasing intracellular Ca^{2+} concentrations in digitonin-permeabilized epimastigotes stimulate rapid accumulation of InsP2 and InsP3 and generation of DAG. In this particular system, Ca^{2+} is believed to activate a phospholipase C.

Similarly to *T. cruzi*, inositol phospholipids have been identified in *T. brucei brucei* (44). Moreover, two distinct PI synthase activities (CDP-diglyceride-dependent and -independent) are present in membrane fractions. In the absence of added lipid precursor, the incorporation of free inositol into PI is strongly stimulated by the addition of Mn^{2+} and less by Mg^{2+}. Addition of exogenous CDP-diglyceride to the system increases the uptake of inositol by a factor of 1.5 to 3.5. These inositol phosphates may mediate part of the *T. brucei brucei* response to external signals. Previous work have shown that transformation of bloodstream forms to procyclic cells can be induced by *cis*-aconitate or citrate. Analysis of inositol phosphates in [³H]inositol-labeled cells exposed to *cis*-aconitate revealed that levels of InsP3 increased rapidly (maximum levels at one hour) and subsequently returned to basal level. The levels of InsP and InsP2 were not or little affected by *cis*-aconitate.

4.3. Adenylate Cyclases

In eukaryotic cells, cyclic AMP is synthesized from ATP by the plasma membrane-bound adenylate cyclase, and modulates, through a cAMP-dependent protein kinase, metabolism and proliferation. A plasma membrane-associated adenylate cyclase from *T. brucei* has been purified and characterized (45). The enzyme from bloodstream forms has a neutral pH optimum, and kinetic analysis revealed a K_m for ATP of 1.75 mM and a K_m for Mg^{2+} of 4 mM. Inhibition studies indicated no effect with cAMP, but a profound inhibition by PP_i, which was competitive with respect to ATP. This observation suggests that the *T. brucei* adenylate cyclase interacts with the phosphate portion of the ATP molecule, in contrast to adenylate cyclase in rat liver plasma membranes. Other differences between the *T. brucei* and mammalian adenylate cyclases were observed. For instance, the parasite enzyme activity was not stimulated by glucagon or epinephrine. Furthermore, fluoride, a potent activator of mammalian adenylate cyclase, inhibited the activity of the *T. brucei* enzyme.

Similarly to *T. brucei*, an adenylate cyclase activity associated with the plasma membrane fraction was described in epimastigote forms of *T. cruzi* (46). Other investigators have observed that treatent of epimastigotes with the β-adrenergic agonist isoproterenol causes an increase in intracellular levels of cAMP, suggesting the stimulation of a receptor-mediated adenylate cyclase (47). Importantly, this increase in intracellular cAMP concentrations is accompanied by an inhibition of cell growth. This observation supports the finding that proliferating *T. lewisi* contain less intracellular cAMP than non-dividing adult forms. Furthermore, a comparison of intracellular cAMP levels in trypomastigotes and epimastigotes of *T. cruzi* revealed a fourfold increase in the latter developmental stage (48). Consistent with this observation, trypomastigote cAMP-binding protein showed a 2.5-fold increase in specific activity. *In toto*, these observations lend support to the notion that modulation of membrane-associated adenylate cyclase activity may mediate adaptation of these parasites to changes in environmental or growth conditions (i.e. hosts).

In *T. brucei* a family of genes has been identified with a high degree of homology to eukaryotic adenylate and guanylate cyclase genes. One of these genes, ESAG4, was present in the expression site also containing the active gene for the VSG, whereas the other genes were spread throughout the genome. The ESAG4 gene was able to complement adenylate cyclase minus yeast mutants. This gene was developmentally regulated and only expressed in bloodstream forms where it was mainly associated with the flagellum (49). The products of the other members of the gene family were expressed in the procyclic stage.

4.4. Kinases

One of the key components of mammalian signal transduction pathways is protein kinase C (PKC), which is activated by DAG and Ca^{2+} (40) and phosphorylates serine and threonine residues. PKC contains a catalytic domain containing an ATP-binding site, and a regulatory domain where the sites involved in Ca^{2+}, DAG and phospholipid binding reside. Whereas the inactive form of PKC is cytosolic, translocation to the plasma membrane is required for activity. This process is controlled by intracellular Ca^{2+} concentrations.

A PKC activity from *T. cruzi* epimastigotes has been characterized (50). Chromatography on DEAE-cellulose of a *T. cruzi* extract revealed one kinase activity peak which was activated by Ca^{2+} and phospholipids. This kinase has a maximum activity in the presence of Ca^{2+}, phosphatidylserine, and phorbol esters or DAG, which is a characteristic property of PKC. In addition, the activity could be inhibited by staurosporine or polymyxin B, two inhibitors of PKC. Most of the PKC activity (60–70%) is associated with particulate fractions. Since this form is considered as primed or activated, this finding implies that in *T. cruzi* epimastigotes, a large portion of PKC is already activated. Perhaps the most striking finding of this study is the difference in phosphorylation rates between pathogenic and a non-pathogenic strains of this parasite; PKC activity being higher in the pathogenic strain, suggesting that in *T. cruzi* epimastigotes, PKC modulates functions associated with virulence.

The presence of protein kinase C activity was also investigated in bloodstream and procyclic forms of *T. brucei* (51). Fractionation of cell homogenates over DEAE-cellulose revealed several kinase activities in each developmental form, which are more stable and more abundant in the bloodstream stage. Typical PKC activities (based on cofactors requirement) were detected only in the bloodstream stage. Accordingly, these PKC activities were inhibited by low concentrations (1 μM) of H7, a widely used inhibitor of PKC. Moreover, immunoblotting using an antibody specific for the hinge region of mammalian α and β PKC subspecies revealed a 98 kDa protein. This protein is not expressed in the procyclic stage, suggesting a differential regulation of protein kinase activities in the two stages of *T. brucei*. The involvement of PKC in physiological responses or processes remains to be established.

5. MEMBRANE TRANSPORT SYSTEMS

In addition to providing a physical barrier between the intracellular milieu of a cell and the surrounding environment, the plasma membrane carries out important physiologic functions, such as nutrient acquisition, secretion and excretion. In this section, three critical membrane transport systems described in *Leishmania* are discussed.

5.1. Plasma Membrane H⁺-ATPase

Studies on sugar and amino acid uptake by *L. donovani* promastigotes revealed that addition of D-glucose or L-proline caused a rapid influx of protons into these cells, indicating that both substrates are co-transported with protons (52). This active transport system involves a proton-motive force (pmf)-driven mechanism which requires the maintenance of a proton electrochemical gradient. Such a gradient is composed of the chemical gradient (ΔpH) and the membrane potential ($\Delta\psi$) (52).

The proton pumps that regulate the pmf may also function in the maintenance of pH homeostasis. This represents a particularly critical function since throughout their life cycle, developmental stages of *L. donovani* are exposed to environments with major differences in pH. In the gut of the sandfly, promastigotes proliferate at a neutral pH. On the other hand, intracellular amastigotes live in the acid pH of macrophage phagolysosomes, obviously requiring a mechanism for maintenance of cytoplasmic pH. The existence of a system regulating intracellular pH homeostasis was demonstrated by several observations. First, *L. donovani* promastigotes maintain intracellular pH values close to neutral between extracellular pH values of 5.0 and 7.4. Similarly, amastigote intracellular pH is maintained close to neutral at external pH values as low as 4.0 (53, 54). Second, maintenance of intracellular pH is an energy-requiring process, as incubation of promastigotes in the absence of glucose resulted in a reduced pH transmembrane gradient (53). Third, use of the ATPase inhibitor dicyclohexylcarbodiimide acidified the cytoplasm in both promastigotes and amastigotes, suggesting the participation of a proton-extruding ATPase in intracellular pH maintenance (53).

An Mg^{2+}-dependent, proton-translocating ATPase activity has been identified and characterized in surface membrane from *L. donovani* promastigotes (55). The enzyme has a pH optimum of 6.5, and is inhibited by orthovanadate and dicyclohexylcarbodiimide. Membrane vesicles, devoid of attached cytoskeletal elements, were shown to contain ATPase activity. These vesicles were incubated in the presence of acridine orange and ATP to demonstrate the proton-pumping activity of the surface membrane ATPase. This experiment revealed an ATP-driven acidification of the vesicles' internal space in those vesicles with an inside-out orientation.

By using an antiserum directed against the plasma membrane H^+-ATPase of *Saccharomyces cerevisiae*, a 66 kDa membrane protein of *L. donovani* promastigotes has been identified (56). The immunoprecipitated protein contained an ATP hydrolytic activity with an optimum pH of 6.5, and was sensitive to orthovanadate. The enzyme, which is also present on the flagellar surface and the flagellar pocket, appears to be a transmembrane protein with the active site localized on the inner face of the plasma membrane. Moreover, the H^+-ATPase is also associated with membranes of the Golgi apparatus and intracellular vacuoles.

By virtue of its proton-pumping activity, the plasma membrane H^+-ATPase regulates intracellular pH homeostasis, and thus is essential for parasite survival. This is supported by the observation that tricyclic drugs, which reduce the pmf in *L. donovani*, are toxic (55). Therefore, this enzyme represents an ideal target for drug therapy.

5.2. P-glycoprotein

In mammalian cells, multidrug resistance has been closely linked to the increased expression of a P-glycoprotein in the plasma membrane, which correlates with a decreased intracellular drug concentration (59). P-glycoprotein is a highly conserved transmembrane protein in eukaryotic cells, which consists of a tandem duplication of approximately 590 amino acids, linked via a 60-amino acid region. Each half of P-glycoprotein contains six transmembrane sequences and one ATP-binding site, consistent with the ATP requirement for activity (57). This protein is homologous to several bacterial transport proteins that regulate uptake and transport of molecules into bacteria. The fact that P-glycoprotein is also normally expressed in specific mammalian tissues and organs suggests a critical role for this protein family in membrane transport.

A P-glycoprotein homolog gene has been identified in amplified circular DNA from methotrexate-resistant *Leishmania tarentolae* and *L. major* promastigotes (58, 59). The amplified *L. tarentolae* P-glycoprotein gene encodes a protein which consists of two similar halves, each containing six putative transmembrane segments and a nucleotide binding site (58). In addition, the *L. tarentolae* genome appears to contain at least five P-glycoprotein genes. Of interest is the demonstration that both *L. major* and *L. tarentolae* P-glycoprotein genes (derived from H circles from methotrexate-resistant lines) confer resistance to heavy metals such as arsenite and trivalent antimonials, but not to pentavalent antimonials (59). In contrast to its mammalian counterpart, the *Leishmania* P-glycoprotein does not confer resistance to vinblastine and puromycin. Moreover, drugs that reverse multidrug resistance in mammalian cells do not reverse resistance to arsenite or trivalent antimonials.

The function of *Leishmania* P-glycoproteins during the life cycle of these parasites is not known. One possibility is that they protect the *Leishmania* against potentially toxic compounds encountered in both hosts.

5.3. Ca²⁺ Transport

Calcium plays a major regulatory role in the expression of cellular functions. Therefore, movement of Ca^{2+} ions across the plasma membrane and maintenance of intracellular levels represent crucial events for cell activity.

In *L. donovani* promastigotes, the concentration of free cytoplasmic Ca^{2+} is maintained at very low levels (100 nM), even in the presence of 1 mM Ca^{2+} in the medium. The maintenance of such a gradient is energy-dependent, since the concentration of cytosolic free calcium is increased by metabolic inhibitors such as KCN and H^+-ATPase inhibitors. Two transport systems regulate Ca^{2+} trafficking. One is respiration-dependent, indicating a mitochondrial localization. The second Ca^{2+} transport system is respiration-independent and requires ATP for its activity. This system exhibits a high affinity to Ca^{2+} ($K_m = 92$ nM), functions optimally at pH 7.1 and is sensitive to orthovanadate (60). Purification of the Ca^{2+}-ATPase from *L. donovani* promastigotes revealed that it consists of two subunits of 51 and 57 kDa, and exhibits an apparent molecular weight of 200 kDa. Moreover, the high affinity of the purified Ca^{2+}-ATPase for Ca^{2+} ($K_m = 35$ nM) is significantly enhanced by calmodulin from heterologous systems ($K_m = 12$ nM) (61). It has been determined that the Ca^{2+}-ATPase spans across the plasma membrane and has its catalytic site oriented towards the cytoplasmic face. Although there is presently no demonstrated evidence that Ca^{2+} regulates specific leishmanial functions, the plasma membrane Ca^{2+}-ATPase undoubtedly modulates critical processes in this parasite.

5.4. Glucose Transport

In all the trypanosomatids glucose plays a major role both as energy substrate and as carbon source (see Chapter 2). Glucose does not permeate freely through biological membranes and in several representatives of this family glucose transporters have been reported. There is a stereospecific glucose carrier in the plasma membrane of *T. brucei*. In the vertebrate stage D-glucose but not L-glucose enters the cell down a concentration gradient by facilitated diffusion (62–65). This carrier had a K_m of 2 mM for D-glucose. It differed from the mammalian counterpart in that it was also able to transport D-fructose. Fructose derivatives may, therefore, have potential as antitrypanosomal agents. In *T. brucei* the glucose carrier is the rate-limiting step in the glycolytic pathway as long as the external glucose concentration remained below the physiological concentration of glucose in blood of 5 mM (65). Under these conditions the intracellular glucose concentration remained very low. At an external concentration of 4–5 mM the capacity of the glucose carrier was just sufficient to saturate the subsequent glycolytic steps. Above 5 mM the transport of glucose into the cell was not rate-limiting and the cell accumulated glucose, because other steps, most likely the transport of glucose from the cytosol into the glycosome, or the phosphorylation of glucose by the glycosomal hexokinase reaction, became rate-limiting (65). In the case of procyclic insect stages, using 2-deoxyglucose as substrate, the presence of an active glucose transporter capable of accumulating glucose was suggested. The generation of a concentration gradient could not be confirmed, however, using D-glucose as substrate (66). Experiments suggesting the existence of an active glucose in trypanosomatids, when based solely on the accumulation inside the cell of labeled 2-deoxyglucose, should be considered with great caution, since such an accumulation could equally be explained by a combination of facilitated diffusion, followed by phosphorylation of 2-deoxyglucose to 2-deoxyglucose-6-phosphate by the trypansomatid hexokinase. Subsequent hydrolysis during the processing of the samples then leads to a dramatic overestimation of the intracellular free 2-deoxyglucose (66). In *Leishmania* promastigotes an active glucose carrier has been suggested. This was based on the observation of proton movement in association with glucose transport (52). However, the reported 80-fold accumulation of D-glucose against a concentration gradient in the case of *L. donovani* (52, 67) has been recently shown to be the result of a massive intracellular accumulation of label in the form of an as yet unidentified glucose metabolite (68). In this respect it is important to note that a massive conversion of glucose into a mannose-containing polysaccharide was reported for stationary-phase promastigotes of *L. donovani* (69).

Both in *Leishmania* and in *T. brucei* genes have been identified that have a significant degree of homology with the glucose transporter superfamily, a larger family of integral membrane proteins that exist in diverse organisms from bacteria to mammals and that transport various sugars. The transporter preferentially expressed in promastigotes of *Leishmania enriettii* is encoded by a family of eight tandemly clustered genes and was more closely related in sequence to glucose transporters than to the other known transporters and may thus be a *Leishmania* glucose transporter (70). In *L. donovani*, in addition to the gene described for *L. enriettii*, a second related, but structurally different, gene has been identified, which was not regulated during the parasite's life cycle. In *L. donovani*, these genes were not arranged in simple tandem repeats (71). In *T. brucei*, two highly related families of respectively two and five tandemly repeated genes were identified, which had about 46% identity with the *L. enriettii* gene and approximately 20% with the human glucose transporter (72). No definite proof exists that any of these gene products is actually involved in the transport of glucose and, if correct, whether they are involved in glucose transport over either the plasma or over the glycosomal membrane.

5.5. Pyruvate Transport

The major end-products of carbohydrate metabolism in most trypanosomatids studied to date, in addition to carbon dioxide, are lactate, alcohol, succinate and acetate. Bloodstream forms of *T. brucei* are unique among the trypanosomatidae in that they lack lactate dehydrogenase or other mechansims to further metabolize pyruvate and thus pyruvate is massively excreted (see Chapter 2). Pyruvic acid has a pK_a of 2.49 and at physiological pH is almost completely dissociated. Therefore, it cannot freely permeate through biological membranes. To enable the high rate of pyruvate excretion observed in *T. brucei*, this organism has incorporated in its plasma membrane a facilitated diffusion carrier, highly specific for pyruvate. Contrary to other monocarboxylic acid carriers in the mitochondrial or plasma membrane of other eukaryotic cells, the trypanosome's pyruvate carrier did not transport L-lactate (73). This carrier, which had a K_m for pyruvate influx of 2 mM and a V_{max} that was sufficient to allow for the rapid rate of efflux of pyruvate, also differed from all other known monocarboxylic acid carriers with respect to its sensitivity to inhibitors (73). Inhibition of this carrier led to a large accumulation of intracellular pyruvate, a significant drop of cytosolic pH and the subsequent swelling and lysis of the cells (Wiemer and Opperdoes, unpublished data). These observations together with the different specificity of the trypansome carrier towards both its substrate and inhibitors render this protein an interesting target for the development of new antitrypanosomal drugs.

5.6. Proline Transport

For many trypanosomatids it has been described that they are able to metabolize proline, which is abundantly present in the insect haemolymph, as the major nitrogen and carbon source. Proline uptake takes place via a carrier protein in the plasma membrane, but the nature of the transport is unclear. Active transport of proline into *Leishmania* spp. by means of a proton/proline symporter has been suggested using radioactive proline (52). The use of proline itself rather than a non-metabolizable analog, however, complicates the interpretation of the results since transport and metabolism cannot be separated and thus the intracellular proline concentration is easily overestimated. In chemostat-grown cells of *L. donovani* no evidence was found for any intracellular accumulation of this amino acid and it was suggested that proline was taken up by facilitated diffusion (74).

6. RECEPTORS FOR HOST-DERIVED MOLECULES

Proliferation in insect and mammalian hosts implies that some parasites come in contact with an incredible variety of host-derived molecules. These molecules may or may not influence the existence of these parasites within their hosts. In fact, most of the molecules are probably of no interest. On the other hand, some of them may be toxic and aimed, specifically or not, at the destruction of the intruders. This category of molecules includes, among others, oxygen radicals, nitric oxide, complement components and proteases. Alternatively, some host molecules may be used by parasites to promote their growth and survival, as well as cell and tissue invasion. In this section, the presence and implications of receptors for host-derived molecules on trypanosomes and *Leishmania* will be discussed.

6.1. Epidermal Growth Factor Receptor

Epidermal growth factor (EGF) is a hormone polypeptide involved in the growth regulation of several mammalian cell types. The action of EGF is mediated through specific binding to a cell surface receptor, triggering rapid phosphorylation events and modifications in gene expression patterns.

Recently, the expression of an EGF receptor homolog has been reported in both bloodstream and procyclic stages of *T. brucei* (75). Antibodies raised against the external domain of the human EGF receptor or the v-*erb*B protein recognized a surface protein of 135 kDa. The purified 135 kDa polypeptide possesses kinase activity, a feature of mammalian EGF receptors. Moreover, incubation of membrane-enriched fractions with EGF in the presence of $[^{32}P]ATP$ resulted in the phosphorylation of several polypeptides, including the 135 kDa EGF receptor homolog. Therefore, *T. brucei* membranes contain a protein kinase which can be activated by EGF, implying binding activity. This speculation was substantiated by the observation that biotinylated EGF bound to two membrane polypeptides, one of 135 kDa and one of 50 kDa. To determine whether EGF exerts any effect on the growth of trypanosomes, procyclic forms were grown in serum-free medium. Addition of 120 nM EGF stimulated growth. Thus, EGF not only binds to a trypanosome cell surface polypeptide, but also stimulates the growth of *T. brucei* cultures. Presently, there are no details concerning the signal transduction pathways stimulated by EGF in *T. brucei*. However, the identification in trypanosomes of molecules homologous to mammalian cell signaling machinery components (see Section 4) suggests that EGF triggers intracellular pathways resembling those described in mammalian cells.

6.2. Low-density Lipoprotein Receptor

Growth of bloodstream forms of *T. brucei* is dependent on the presence of several types of molecules they are unable to synthesize. For instance, despite the fact that cholesterol is the main membrane sterol, bloodstream forms of *T. brucei* are unable to synthesize their own cholesterol. This suggests that *T. brucei* incorporates intact cholesterol from the host. In mammals, serum cholesterol is associated mainly with low-density lipoproteins (LDL). Interestingly, uptake of LDL by bloodstream forms of *T. brucei* occurs via a receptor-mediated process, with a clearance of two to three orders of magnitude higher than that of serum albumin (76, 77). Degradation products of LDL incubated with *T. brucei* are rapidly detectable. Indeed, bloodstream forms of *T. brucei* incubated with $400\,\mu g$ LDL/ml degraded the protein at a rate of $20\,\mu g$ LDL per mg of cells per hour. This process parallels increased intracellular esterification of cholesterol (78). Therefore, trypanosomes are able to obtain cholesterol by uptake and degradation of host LDL. The importance of such an uptake mechanism for the growth of *T. brucei* stems from the observation that removal of LDL from culture medium or antibodies against purified LDL receptors inhibit cell growth (77, 79).

Binding of LDL to bloodstream forms is saturable, dependent on the presence of Ca^{2+}, and is specific, since labeled ligand could be displaced by the homologous protein. The receptors were localized on the membrane of the flagellar pocket only, where receptor-mediated endocytosis processes are believed to occur in trypanosomes. The search for a receptor revealed a glycoprotein made of a single subunit with an apparent molecular weight of 86 kDa, which was later identified as a degradation product of a 145 kDa receptor (77, 80). Each cell expresses 52 000 copies of low-affinity receptors (K_d of 250 nM) and 1800 copies of high affinity receptors (K_d of 5.7 nM). Despite the fact that these cells undergo antigenic variation with respect to their VSG, the receptors behaved as stable surface antigens and could be immunologically recognized in several variants (81). These receptors also cross-reacted with the mammalian LDL receptor.

More recently, a transmembrane protein localized in the flagellar pocket of *T. brucei* has been characterized (82). Based on the cDNA sequence, this protein has a predicted molecular weight of 130 kDa, and has a signal peptide, a transmembrane domain, and a 41-amino acid cytoplasmic extension The extracellular domain of the protein, termed cysteine-rich, acid integral membrane protein (CRAM), is acidic, containing a cysteine-rich 12-amino acid repeat. Sequence homologies comparison revealed structural similarities with eukaryotic cell surface receptors, with high homology to the human LDL receptor. However, based on their molecular weight, CRAM and the 85 kDa LDL receptor previously described are unlikely to be the same protein. Finally, CRAM was found to be expressed in both procyclic and bloodstream forms of *T. brucei* and to be highly conserved among trypanosomes.

6.3. Heparin Receptor

For intracellular parasites, invasion of host tissues and cells often represents the initial step towards successful establishment of infection. This process involves recognition molecules expressed by both the host and the invading organisms. Identification and characterization of these molecules is an area of keen interest and will eventually lead to a better understanding of parasitism.

A heparin-binding protein, named penetrin, has been characterized in *T. cruzi* (83). It is a 60 kDa protein located on the surface of trypomastigotes and adheres to heparin, heparan sulfate, and collagen, three components of the extracellular matrix (ECM). Coating of plastic surfaces with these molecules supported the attachment of *T. cruzi*. Therefore, penetrin may mediate interaction of *T. cruzi* with ECM and facilitate migration through basement membranes. In addition, penetrin binds to *T. cruzi* host cells, in a saturable and specific manner, suggesting that penetrin may also promote *T. cruzi* attachment to and invasion of host cells. Indeed, excess purified penetrin effectively prevents *T. cruzi* invasion. Similarly, infection could be prevented by the addition of excess heparin and heparan sulfate in a dose-dependent manner, whereas chondroitin sulfate and hyaluronic acid were without effect. Moreover, expression in *Escherichia coli* endowed the bacteria with the ability to adhere and penetrate non-phagocytic Vero cells, in a proteoglycan-inhibitable manner. This study suggests an important role for this *T. cruzi* heparin receptor in the initiation of the infection process.

Promastigotes of *L. donovani* possess cell surface receptors for heparin (84). Binding studies have indicated that there are an estimated 860 000 receptors per parasite, with a binding constant of 4.7×10^{-7} M. Bound [^{3}H]heparin was not displaced by hyaluronic acid or other glycosaminoglycans, indicating a relatively specific binding. The main consequence of heparin binding to promastigotes is the inhibition of a cell surface histone–protein kinase, suggesting that heparin, by regulating cell surface enzyme activity, may influence the host–parasite interaction. Further studies revealed that the cell surface heparin receptors are recruited from an internal pool after trypsinization (85). This internal pool contains at least 30% of the receptors normally present on the surface of promastigotes. *De novo* protein synthesis is not required for recruitment and cell surface expression of the internal receptor pool. A comparison of heparin binding affinities between the regenerated and the native receptors indicated similar binding constants. Finally, the heparin receptors are localized almost exclusively on the flagella of promastigotes.

The role of heparin receptors expressed on the surface of *L. donovani* promastigotes is not clear yet. Attachment of the parasites to host surfaces, such as the villus of the sandfly gut epithelium and the mammalian macrophages might involve these receptors. This hypothesis remains to be tested.

6.4. Transferrin Receptor

Transferrin is the major iron transport protein of mammalian cells and transferrin receptors are present on virtually all cells, their expression increasing with cell proliferation. The expression of receptors for human transferrin has been demonstrated in both *T. cruzi* and *L. infantum* (86, 87).

T. cruzi multiplies within mammalian cells as an amastigote, and iron is required for their growth. The possibility that transferrin mediates iron uptake was supported by the demonstration that binding of ^{125}I-labeled transferrin to amastigotes is concentration-dependent, saturable and specific. Mammalian cells express one type of receptor for transferrin with a K_d of 10^{-9} M and a molecular weight of 180 kDa. Similarly, amastigotes also express one class of receptors (8×10^4 receptors per cells) with a lower affinity (K_d of 2.8×10^{-6} M and a molecular weight of 200 kDa. Uptake of iron occurs via a receptor-mediated endocytosis process. Indeed, [^{59}Fe]transferrin bound to amastigotes at 4°C for 1 hour was dissociated by acid treatment. On the other hand, [^{59}Fe]transferrin was not dissociated by acid treatment when amastigotes were incubated at 37°C for 1 hour, indicating that transferrin has been internalized. Of interest, trypomastigotes, which do not multiply within the mammalian host, did not express detectable transferrin receptor.

A different approach has been taken to identify transferrin receptors on cultured *L. donovani* promastigotes. A monoclonal antibody, raised against promastigote membranes recognized a membrane-associated glycoprotein of 78 kDa. Subsequently, this monoclonal antibody was found to bind human transferrin, also of 78 kDa. The membrane-associated 78 kDa protein, which could be released by mild treatment with acetic acid, was characterized as an iron-containing protein and reacted with an antitransferrin antiserum. This molecule is likely to be human transferrin from the culture medium, which was supplemented with human hemolyzed blood.

Binding studies of ^{125}I-labeled human transferrin to membrane preparations revealed the presence of a high affinity saturable binding site ($K_d = 2.2 \times 10^{-8}$ M), similar to that of human transferrin receptor. Purified amastigotes of *L. infantum* and *L. mexicana amazonensis* also bind ^{125}I-labeled transferrin similarly to promastigote membrane preparations. It is thus likely that transferrin mediates iron transport in both life stages of *Leishmania* parasites.

Transferrin has been shown to be an essential growth factor for *T. brucei* bloodstream forms and deprivation of cells of transferrin led to growth arrest (88). A 42 kDa transferrin-binding protein of *T. brucei*, apparently unrelated to the transferrin receptor described above for *Leishmania*, has been purified and characterized and its corresponding gene was associated with the VSG expression site (89). This binding protein was attached to the plasma membrane via a glycolipid anchor and thus lacked a cytoplasmic domain. Thus far, it is unclear how such a protein could function as a receptor and interact with the intracellular clathrin-coated vesicles supposed to be involved in receptor-mediated uptake of transferrin by the trypanosomes (76).

7. REFERENCES

1. Donelson, J., and Rice-Ficht, A. (1985) Molecular biology of trypanosome antigenic variation. *Microbiol. Rev.* **49**: 107–125.
2. Zamze, S. E., Wooten, E. W., Ashford, D. A., Ferguson, M. A. J., Dwek, R. A. and Rademacher, T. W. (1990) Characterization of the asparagine-linked oligosaccharides from *Trypanosoma brucei* type-I variant surface glycoproteins. *Eur. J. Biochem.* **187**: 657–663.
3. Low, M. G. (1987) Biochemistry of the glycosyl-phosphatidylinositol membrane protein anchors. *Biochem. J.* **244**: 1–13.
4. Ferguson, M. A. J. and Williams, A. F. (1988) Cell-surface anchoring of proteins via glycosyl-phosphatidylinositol structures. *Annu. Rev. Biochem.* **57**: 285–320.
5. Cross, G. A. M. (1990) Glycolipid anchoring of plasma membrane proteins. *Annu. Rev. Cell. Biol.* **6**: 1–39.
6. Ferguson, M. A. J., Homans, S. W., Dwek, R. A. and Rademacher, T. W. (1988) Glycosyl-phosphatidylinositol moiety that anchors *Trypanosoma brucei* variant surface glycoprotein to the membrane. *Science* **239**: 753–759.
7. Doering, T. L., Masterson, W. J., Hart, G. W. and Englund, P. T. (1990) Biosynthesis of glycosyl phosphatidylinositol membrane anchors. *J. Biol. Chem.* **265**: 611–614.
8. Bulow, R. and Overath, P. (1986) Purification and characterization of the membrane-form variant surface glycoprotein hydrolase of *Trypanosoma brucei*. *J. Biol. Chem.* **261**: 11918–11923.
9. Herald, D., Krakow, J. L., Bangs, J. D., Hart, G. W. and Englund, P. T. (1986) A phospholipase C from *Trypanosoma brucei* which selectively cleaves the glycolipid on the variant surface glycoprotein. *J. Biol. Chem.* **261**: 13813–13819.
10. Fox, J. A., Duszenko, M., Ferguson, M. A. J., Low, M. G. and Cross, G. A. M. (1986) Purification and characterization of a novel glycan-phosphatidylinositol-specific phospholipase C from *Trypanosoma brucei*. *J. Biol. Chem.* **261**: 15767–15771.
11. Roditi, I. and Pearson, T. W. (1990) The procyclin coat of African trypanosomes (or the not-so-naked trypanosomes). *Parasitol. Today* **6**: 79–82.

12. Lederkremer, R. M., Tanaka, C. T., Alves, M. J. M. and Colli, W. (1977). Lipopeptidephosphoglycan from *Trypanosoma cruzi*. Eur. J. Biochem. **74**: 265–267.

13. Laine, R. A. and Hsieh, T. C. Y. (1987) Inositol-containing sphingolipids. *Methods Enzymol.* 138: 186–195.

14. Pereira, M. E. A., Mejia, J. S., Ortega-Barria, E., Matzilevich, D. and Prioli, R. P. (1991) The *Trypanosoma cruzi* neuraminidase contains sequences similar to bacterial neuraminidases, YWTD repeats of the low density lipoprotein receptor, and type III modules of fibronectin. *J. Exp. Med.* **174**: 179–191.

15. Libby, P., Alroy, J. and Pereira, M. E. A. (1986) A neuraminidase from *Trypanosoma cruzi* removes sialic acid from the surface of myocardial and endothelial cells. *J. Clin. Invest.* **77**: 127–135.

16. Schenkman, S., Jiang, M.-S., Hart, G. W. and Nussenzweig, V. (1991) A novel cell surface trans-sialidase of *Trypanosoma cruzi* generates a stage-specific epitope required for invasion of mammalian cells. *Cell* **65**: 1117–1125.

17. Piras, M. M., Henriques, D. and Piras, R. (1987) The effect of fetuin and other sialoglycoproteins on the *in vitro* penetration of *Trypanosoma cruzi* trypomastigotes into fibroblastic cells. *Mol. Biochem. Parasitol.* **22**: 135–143.

18. Turco, S. J. and Descoteaux, A. (1992) The lipophosphoglycan of *Leishmania* parasites. *Annu. Rev. Microbiol.* **46**: 65–94.

19. Avila, J. L., Rojas, M. and Garcia, L. (1988) Persistence of elevated levels of galactosyl-α(1,3)galactose antibodies in sera from patients cured of visceral leishmaniasis. *J. Clin. Microbiol.* **26**: 1842–1847.

20. Towbin, H., Rosenfelder, G., Wieslander, J. *et al.* (1987) Circulating antibodies to mouse laminin in Chagas disease, American cutaneous leishmaniasis, and normal individuals recognize terminal galactosyl(α1,3)galactose epitopes. *J. Exp. Med.* **166**: 419–432.

21. Homans, S. W., Melhert, A. and Turco, S. J. (1991) Solution structure of the lipophosphoglycan of *Leishmania donovani*. *Biochemistry* **31**: 654–661.

22. Sacks, D. L. (1989) Metacyclogenesis in *Leishmania* promastigotes. *Exp. Parasitol.* **69**: 100–103.

23. Rosen, G., Londner, M. V., Sevlever, D. and Greenblatt, C. L. (1988) *Leishmania major*: Glycolipid antigens recognized by immune human sera. *Mol. Biochem. Parasitol.* **27**: 93–100.

24. Sevlever, D., Pahlsson, P., Rosen, G., Nilsson, B. and Londner, M. V. (1991) Structural analysis of a glycosylphosphatidylinositol glycolipid of *Leishmania donovani*. *Glycoconj. J.* **8**: 321–329.

25. McConville, M. J. and Bacic, A. (1989) A family of glycoinositol phospholipids from *Leishmania major*. *J. Biol. Chem.* **264**: 757–766.

26. Bordier, C. (1987) The promastigote surface protease of *Leishmania*. *Parasitol. Today* **3**: 151–153.

27. Medina-Acosta, E., Karess, R. E., Schwartz, H. and Russell, D. G. (1989) The promastigote surface protease (gp63) of *Leishmania* is expressed but differentially processed and localized in the amastigote stage. *Mol. Biochem. Parasitol.* **37**: 263–274.

28. Chang, K.-P., Chadhuri, G. and Fong, D. (1990) Molecular determinants of *Leishmania* virulence. *Annu. Rev. Microbiol.* **44**: 449–454.

29. Bouvier, J., Schneider, P., Etges, R. and Bordier, C. (1990) Peptide substrate specificity of the membrane-bound metalloprotease of *Leishmania*. *Biochemistry* **29**: 10113–10119.

30. Gottlieb, M. and Dwyer, D. M. (1981) Protozoan parasite of humans: surface membrane with externally disposed acid phosphatase. *Science* **212**: 939–941.

31. Dwyer, D. M. (1987) The roles of surface membrane enzymes and transporters in the survival of *Leishmania*. In: *Host–Parasite Cellular and Molecular Interactions in Protozoal Infections* (eds Chang, K.-P. and Snary, D), NATO ASI Series. Springer-Verlag, Heidelberg, pp. 175–182.

32. Gottlieb, M. (1985) Enzyme regulation in a trypanosomatid: effect of purine starvation on levels of 3′-nucleotidase activity. *Science* **227**: 72–74.

33. Greis, K. D., Turco, S. J., Thomas, J. R., McConville, M. J., Homans, S. W. and Ferguson, M. A. J. (1992) Purification and characterization of an extracellular phosphoglycan from *Leishmania donovani*. *J. Biol. Chem.* **267**: 5876–5881.

34. Andrews, N. W. and Webster, P. (1991) Phagolysosomal escape by intracellular pathogens. *Parasitol. Today* **7**: 335–340.

35. Andrews, N. W. and Whitlow, M. B. (1989) Secretion by *Trypanosoma cruzi* of a hemolysin active at low pH. *Mol. Biochem. Parasitol.* **33**: 249–256.
36. Andrews, N. W., Abrams, C. K., Slatin, S. L. and Griffiths, G. (1990) A *T. cruzi*-secreted protein immunologically related to the complement component C9: evidence for memebrane pore-forming activity at low pH. *Cell* **61**: 1277–1287.
37. Kornfeld, R. and Kornfeld, S. (1984) Assembly of asparagine-linked oligosaccharides. *Annu. Rev. Biochem.* **54**: 631–664.
38. Mendelzon, D. H., Previakto, J. O. and Parodi, A. J. (1986) Characterization of protein-linked oligosaccharides in trypanosomatid flagellates. *Mol. Biochem. Parasitol.* **18**: 355–367.
39. Berridge, M. J. (1987) Inositol triphosphate and diacylglycerol: two interacting second messengers. *Annu. Rev. Biochem.* **56**: 159–193.
40. Nishizuka, Y. (1988) The molecular heterogeneity of protein kinase C and its implications for cellular regulation. *Nature* **334**: 661–665.
41. Gilman, A. G. (1987). G proteins: transducers of receptor-generated signals. *Annu. Rev. Biochem.* **56**: 615–649.
42. Cassel, D., Shoubi, S., Glusman, G., Cukierman, E., Rotman, M. and Zilberstein, D. (1991) *Leishmania donovani*: characterization of a 38-kDa membrane protein that cross-reacts with the mammalian G-protein transducin. *Exp. Parasitol.* **72**: 411–417.
43. Docampo, R., and Pignataro, O. P. (1991) The inositol phosphate/diacylglycerol signalling pathway in *Trypanosoma cruzi*. *Biochem. J.* **275**: 407–411.
44. Deshusses, J. M. P., Belet, M. and Champeaux, L. (1990) Multiple roles of inositol in trypanosomes. *Biochem. Soc. Trans.* **18**: 716–717.
45. Martin, B. R., Voorheis, H. P. and Kennedy, E. L. (1978) Adenylate cyclase in bloodstream forms of *Trypanosoma (Trypanozoon) brucei* sp. *Biochem. J.* **175**: 207–212.
46. Zingales, B., Carniol, C., Abrahamson, P. A. and Colli, W. (1979) Purification of an adenylate cyclase-containing plasma membrane fraction from *Trypanosoma cruzi*. *Biochim. Biophys. Acta* **550**: 233–244.
47. Oliveira, M. M., Antunes, A. and de Mello, F. G. (1984) Growth of *Trypanosoma cruzi* epimastigotes controlled by shifts in cyclic AMP mediated by adrenergic ligands. *Mol. Biochem. Parasitol.* **11**: 283–292.
48. Rangel-Aldao, R., Allende, O., Triana, F., Piras, R., Henriquez, D. and Piras, M. (1987) Possible role of cAMP in the differentiation of *Trypanosoma cruzi*. *Mol. Biochem. Parasitol.* **22**: 39–43.
49. Paindavoine, P., Rolin, S. van Assel, Geuskens, M. *et al.* (1992) A gene from the variant surface glycoprotein expression site encodes one of several transmembrane adenylate cyclases located on the flagellum of *Trypanosoma brucei*. *Mol. Cell. Biol.* **12**: 1218–1225.
50. Gomez, M. L., Erijman, L., Arauzo, S., Torres, H. N. and Tellez-Inon, (1989) Protein kinase C in *Trypanosoma cruzi* epimastigotes forms: partial purification and characterization. *Mol. Biochem. Parasitol.* **36**: 101–108.
51. Keith, K., Hide, G. and Tait, A. (1990) Characterization of protein kinase C like activities in *Trypanosoma brucei*. *Mol. Biochem. Parasitol.* **43**: 107–116.
52. Zilberstein, D. and Dwyer, D. M. (1985) Protonmotive force-driven active transport of D-glucose and L-proline in the protozoan parasite *Leishmania donovani*. *Proc. Natl. Acad. Sci. USA* **82**: 1716–1720.
53. Glaser, T. A., Baatz, J. E., Kreishman, G. P. and Mukkada, A. J. (1988) pH homeostasis in *Leishmania donovani* amastigotes and promastigotes. *Proc. Natl. Acad. Sci. USA* **85**: 7602–7606.
54. Zilberstein, D., Philosoph, H. and Gepstein, A. (1989) Maintenance of cytoplasmic pH and proton motive force in promastigotes of *Leishmania donovani*. *Mol. Biochem. Parasitol.* **36**: 109–118.
55. Zilberstein, D. and Dwyer, D. M. (1988) Identification of a surface membrane proton-translocating ATPase in promastigotes of the parasitic protozoan *Leishmania donovani*. *Biochem. J.* **256**: 13–21.
56. Liveanu, V., Webster, P. and Zilberstein, D. (1991). Localization of the plasma membrane and mitochondrial H^+-ATPases in *Leishmania donovani* promastigotes. *Eur. J. Cell Biol.* **54**: 95–101.

57. Endicott, J. A. and Ling, V. (1989) The biochemistry of P-glycoprotein-mediated multidrug resistance. *Annu. Rev. Biochem.* **58**: 137–171.
58. Ouellette, M., Fase-Fowler, F. and Borst, P. (1990) The amplified H circle of methotrexate-resistant *Leishmania tatrentolae* contains a novel P-glycoprotein gene. *EMBO J.* **9**: 1027–1033.
59. Callahan, H. L. and Beverley, S. M. (1991) Heavy metal resistance: a new role for P-glycoproteins in *Leishmania*. *J. Biol. Chem.* **266**: 18427–18430.
60. Philosoph, H. and Zilberstein, D. (1989) Regulation of intracellular calcium in promastigotes of the human protozoan parasite *Leishmania donovani*. *J. Biol. Chem.* **264**: 10420–10424.
61. Ghosh, J., Ray, M., Sarkar, S. and Bhaduri, A. (1990) A high affinity Ca^{2+}-ATPase on the surface membrane of *Leishmania donovani* promastigotes. *J. Biol. Chem.* **265**: 11345–11351.
62. Gruenberg, J., Sharma, P. R. and Deshusses, J. (1978) D-glucose transport in *Trypanosoma brucei*. D-glucose transport is the rate-limiting step of its metabolism. *Eur. J. Biochem.* **89**: 461–469.
63. Game, S., Holman, G. D. and Eisenthal, R. S. (1989) Sugar transport in *Trypanosoma brucei*: a suitable kinetic probe. *FEBS Lett.* **194**: 126–130.
64. Eisenthal, R. S., Game, S. and Holman, G. D. (1989) Specificity and kinetics of hexose transport in *Trypanosoma brucei*. *Biochim. Biophys. Acta* **985**: 81–89.
65. Ter Kuile, B. H. and Opperdoes, F. R. (1991) Glucose uptake by *Trypanosoma brucei*. *J. Biol. Chem.* **266**: 857–862.
66. Ter Kuile, B. H. and Opperdoes, F. R. (1992) Mutual adjustment of glucose uptake and metabolism in *Trypanosoma brucei* grown in a chemostat. *J. Bacteriol.* **174**: 1273–1279.
67. Ter Kuile, B. H. and Opperdoes, F. R. (1992) Comparative physiology of two protozoan parasites *L. donovani* and *Trypanosoma brucei* grown in chemostats. *J. Bacteriol.* **174**: 2929–2934.
68. Ter Kuile, B. H. and Opperdoes, F. R. (1993) Uptake and turnover of glucose by *Leishmania donovani*. *Mol. Biochem. Parasitol.* **60**: 313–321.
69. Keegan, F. P. and Blum, J. J. (1992) Utilization of a carbohydrate reserve comprised primarily of mannose by *Leishmania donovani*. *Mol. Biochem. Parasitol.* **53**: 193–200.
70. Stein, D. A., Cairns, B. R. and Landfear, S. M. (1990) Developmentally regulated transporter in *Leishmania* is encoded by a family of clustered genes. *Nucleic Acids Res.* **18**: 1549–1557.
71. Langford, C. K., Ewbank, S. A., Hanson, S. S., Ullman, B. and Landfear, S. M. (1992) Molecular characterization of two genes encoding members of the glucose transporter superfamily in the parasitic protozoan *Leishmania donovani*. *Mol. Biochem. Parasitol.* **55**: 51–64.
72. Bringaud, F. and Baltz, T. (1992) A potential hexose transporter gene expressed predominantly in the bloodstream form of *Trypanosoma brucei*. *Mol. Biochem. Parasitol.* **52**: 111–122.
73. Wiemer, E. A. C., Ter Kuile, B. H., Michels, P. A. M. and Opperdoes, F. R. (1992) Pyruvate transport across the plasma membrane of the bloodstream form of *Trypanosoma brucei* is mediated by a facilitated diffusion carrier. *Biochem. Biophys. Res. Commun.* **184**: 1028–1034.
74. Ter Kuile, B. H. and Opperdoes, F. R. (1992) A chemostat study on prolin uptake and metabolism in *L. donovani*. *J. Protozool.* **39**: 555–558.
75. Hide, G., Gray, A., Harrison, C. M. and Tait, A. (1989) Identification of an epidermal growth factor receptor homologue in trypanosomes. *Mol. Biochem. Parasitol.* **36**: 51–60.
76. Coppens, I., Opperdoes, F. R., Courtoy, P. J. and Bauhuin, P. (1987) Receptor-mediated endocytosis in the bloodstream form of *Trypanosoma brucei*. *J. Protozool.* **34**: 465–473.
77. Coppens, I., Baudhuin, P., Opperdoes, F. R. and Courtoy, P. J. (1988) Receptors for the host low density lipoproteins on the hemoflagellate *Trypanosoma brucei*: purification and involvement in the growth of the parasite. *Proc. Natl. Acad. Sci. USA* **85**: 6753–6757.
78. Gillett, M. P. T. and Owen J. S. (1987) *Trypanosoma brucei brucei* obtains cholesterol from plasma lipoproteins. *Biochem. Soc. Trans.* **15**: 258–259.
79. Black, S. and Vanderweerd, V. (1989) Serum lipoproteins are required for multiplication of *Trypanosoma brucei brucei* under axenic culture conditions. *Mol. Biochem. Parasitol.* **37**: 65–72.
80. Coppens, I., Ph. Bastin, P. J., Courtoy, P. J., Baudlin, P. and Opperdoes, F. R. (1991) A rapid method purifies a glycoprotein of Mr 145,000 as the LDL receptor of *Trypanosoma brucei*. *Biochem. Biophys. Res. Commun.* **178**: 185–191.

81. Coppens, I., Ph. Bastin, P. J., Opperdoes, F. R., Baudlin, P. and Courtoy, P. J. (1992) Antigenic stability of the LDL receptor and cross-reactivity with the host receptor. *Exp. Parasitol.* **74**: 77–86.
82. Lee, M. G.-S., Bihain, B. E., Russell, D. G., Deckelbaum, R. J. and Van der Ploeg, L. H. T. (1990) Characterization of a cDNA encoding a cystein-rich cell surface protein located in the flagellar pocket of the protozoan *Trypanosoma brucei*. *Mol. Cell. Biol.* **10**: 4506–4517.
83. Ortega-Barria, E. and Pereira, M. E. A. (1991) A novel *T. cruzi* heparin-binding protein promotes fibroblast adhesion and penetration of engineered bacteria into mammalian cells. *Cell* **67**: 411–421.
84. Mukhopadhyay, N. K., Shome, K., Saha, A. K., Hassell, J. R. and Glew, R. H. (1989) Heparin binds to *Leishmania donovani* promastigotes and inhibits protein phosphorylation. *Biochem. J.* **264**: 517–525.
85. Butcher, B. A., Shome, K., Estes, L. W. *et al. Leishmania donovani*: cell-surface heparin receptors of promastigotes are recruited from an internal pool after trypsinization. *Exp. Parasitol.* **71**: 49–59.
86. Lima, M. F. and Villalta, F. (1990) *Trypanosoma cruzi* receptors for human transferrin and their role. *Mol. Biochem. Parasitol.* **38**: 245–252.
87. Voyiatzaki, C. S. and Soteriadou, K. P. (1990) Evidence of transferrin binding sites on the surface of *Leishmania* promastigotes. *J. Biol. Chem.* **265**: 22380–22385.
88. Schell, D., Borowy, N. K. and Overath, P. (1991) Transferrin is a growth factor for the bloodstream form of *Trypanosoma brucei*. *Parasitol. Res.* **77**: 558–560.
89. Schell, D., Evers, D., Preis, D. *et al.* (1991) A transferrin binding protein of *Trypanosoma brucei* is encoded by one of the genes of the variant surface glycoprotein expression site. *EMBO J.* **10**: 1061–1066.

12 The Structure and Function of Helminth Surfaces

DAVID P. THOMPSON and TIMOTHY G. GEARY

Animal Health Discovery Research, Upjohn Laboratories, Kalamazoo, MI, USA

SUMMARY

The external surfaces of parasitic helminths, termed the tegument in cestodes and trematodes, and the cuticle in nematodes, are adapted to serve a wide range of biological functions. Though most research has focused on structural aspects or their roles in immune evasion (Chapter 16) and nutrient absorption, the external surfaces of helminths also serve important roles in locomotion, excretion and regulation of electrochemical and osmotic gradients.

The teguments of cestodes and trematodes are membrane-bounded syncytia that contain enzymes found in other transporting epithelia. Nutrients are absorbed across the tegument by mechanisms that are similar to those in vertebrate intestine. Glucose and amino acids are absorbed by Na^+-coupled, active processes that are inhibited by the same agents that inhibit nutrient transport in vertebrates. Cestode and trematode teguments are selectively permeable to K^+, and contain a ouabain-sensitive Na^+/K^+-ATPase that maintains a $-60\,mV$ electrical potential across the outward-facing membrane. Both cestodes and nematodes are weak osmoregulators. In cestodes, succinate and acetate excretion may be linked to osmotic regulation. In schistosomes, an Na^+-H^+ exchanger in the tegument is involved in maintaining the pH of interstitial fluids within the organism.

The nematode cuticle consists largely of cross-linked collagen-like proteins that are secreted by the hypodermis, an anatomical syncytium that is also specialized for transport. The quasi-elastic structure of the cuticle and the high turgor

Biochemistry and Molecular Biology of Parasites
ISBN 0-12-473345-X

pressure maintained across the body wall are essential aspects of locomotion in nematodes. The cuticle is traversed by negatively charged pores which, in *Ascaris suum*, are 15 Å in radius. These pores are large enough to accommodate passage of nutrient and waste molecules as well as inorganic ions and water. Extensive transport of these substances occurs across the cuticle/hypodermis complex, though little is known of proteins that mediate these processes. Cl^- and organic anions contribute to a 30–40 mV (inside negative) electrical potential that is maintained across the nematode body wall.

The absence of a gut in cestodes has simplified interpretation of functional properties of the external surface. The gastrodermis in trematodes and the intestine in nematodes are composed of cells with microvilli which greatly amplify the surface area for absorption. However, too little is known about transport across the internal surfaces of trematodes or nematodes to draw definitive conclusions about their importance to the parasites, or the extent to which they resemble analogous processes in other organisms.

1. INTRODUCTION

The external surfaces of parasitic helminths, termed the cuticle in nematodes and the tegument in trematodes and cestodes, serve many biological roles. The most obvious is that of a barrier which shields the organism from external conditions. Other roles include bidirectional transport of inorganic and organic molecules and structural support. The external structures of these organisms exhibit remarkable developmental changes and exhibit a complex biochemistry that is adapted for the parasitic life style. In nematodes and trematodes, the gut surface also plays an important role in digestion, nutrient absorption and transport. This chapter illustrates the structural and functional biology, biochemistry and, where available, molecular biology of these surfaces. Unfortunately, worm surfaces have received scant attention by classical biochemistry. Most of what is discussed is properly classified as physiology, especially transport physiology. Little work has been done on the proteins which mediate these physiological processes. On the other hand, advances in recombinant DNA technology have frequently bypassed the stage of protein biochemistry entirely. Genes that encode surface proteins have been cloned wholesale in a (so far) fruitless search for protective antigens. However, since these gene products have been selected on the basis of their immunogenicity rather than function, this considerable effort has shed only incidental light on the functional biochemistry of this important tissue.

Within each class of helminths, the best-studied organisms are described (i.e. *Hymenolepis diminuta*, *Schistosoma mansoni*, *Ascaris suum*) to establish general principles. Important exceptions are noted where appropriate. Comments are primarily restricted to adult stages, which generally cause the most important pathology (with some notable exceptions, e.g., *Echinococcus* spp. and *Onchocerca volvulus*) and represent the principal targets for chemotherapy.

2. CESTODES

Although cestodes have received less attention than the trematodes or nematodes, discussion begins with these parasites because of their anatomical simplicity. The absence of a gut simplifies the interpretation of functional properties of the external surfaces (reviewed in refs 2 and 4).

2.1. External Surface

2.1.1. Structural considerations

2.1.1.1. Gross and microscopic anatomy. All interactions between the cestode and its environment occur across the external surface, or tegument. In addition to the functional properties associated with this tissue in other helminths, the cestode surface must also be structurally adapted to perform all functions normally associated with intestinal tissue. Indeed, the cestode body plan has often been conceptualized as an 'inside-out intestine'. Consequently, research on the cestode surface has been somewhat biased toward features that account for its role in nutrition.

The surface of adult cestodes forms a coherent boundary between the fluid compartment of the fibrous interstitium and the environment. The tegument contains several structurally distinct components (Fig. 12.1).

FIG. 12.1 Schematic of the tegument and underlying tissues in *Hymenolepis diminuta*. Abbreviations: gc, glycocalyx; mi, microtriche; bl, basal lamellum; cc, cytoplasmic channel; cm, circular muscle; lm, longitudinal muscle; n, nucleus; cy, cyton. After ref. 3.

The outermost layer of the adult tegument is the glycocalyx. This mucopolysac-charide–glycoprotein coat adheres to the tegument plasma membrane, which is the limiting boundary of a syncytial tissue termed the distal cytoplasm. The external surface is evaginated into structures termed microtriches, reminiscent of the apical microvilli found on vertebrate intestinal epithelial cells. The microtriches amplify the functional surface area of the parasite 2–12-fold, depending on the species and stage of development.

The inner surface of the tegumental syncytium is bounded by a fibrillar basal lamellum. This syncytium, like that of trematodes, is elaborated by cell bodies (subtegumental cells, cytons) that lie beneath the basal lamellum. To the fribrillar components of the basal lamellum are anchored the somatic muscle cells, so that contraction is coupled to movement of the body. Our analysis of the cestode surface focuses on the material bounded by the basal lamellum and the glycocalyx.

The tegument of cestodes is not homogeneous; regions of specialized structures are found over most of the adult animal. These include adaptations which characterize the scolex; this region is involved in attachment to the host intestinal tract. Putative excretory and secretory openings interrupt the continuum of the distal cytoplasm and plasma membrane. Sensory nerve processes, which terminate in external cilia, also interrupt it, as does the genital atrium. Little is known about the biochemistry of these specialized structures; the discussion will focus on tegument typical of the proglottids.

2.1.1.2 Biochemistry and molecular biology of structural components. The cestode tegument contains numerous structural proteins and enzymes, some of which are clearly absorbed from host fluids. *In vitro* studies measuring uptake and incorporation of [^{14}C]leucine by *H. diminuta* revealed at least 17 proteins synthesized by the parasite within the microtriches of the tegument (4). Together, the glycocalyx and microtriches contain at least 30 distinct proteins and glycoproteins, with molecular masses of 12–237 kDa (5). These proteins impart a negative charge to the surface of the parasite. They are rapidly shed, exhibiting half-lives of only about 6 h, and are replaced by new proteins synthesized in the subtegumental cells. Among structural proteins, actin, tubulin, collagen and keratin are present in high levels in cestodes, though information on their relative abundance within the tegument is lacking.

The cestode tegument contains a variety of glycolipids and phospholipids. In *H. diminuta*, lipids comprise ~40% of the outer tegumental membrane, including the microtriches, with the major components being cholesterol, cardiolipids and phosphatidylethanolamine (6, 7). It is believed that lipid droplets are formed in the subtegumental cells from fatty acids and sterols which collect there following absorption from the host across the tegument (2). The predominant fatty acids in cestodes contain 16–18 carbons and are 50–60% unsaturated (8). Like other helminths, cestodes are unable to synthesize cholesterol or other long-chain fatty acids *de novo* from acetyl-CoA.

Little is known of the physiological or biochemical characteristics of the tegument of the developing cestode that distinguish them from adults. Endocytosis may play a more important role in the absorption of macromolecules in immature stages, though supporting data come only from studies on *Taenia crassiceps* (9). Marked structural changes in the tegument typically follow exsheathment. Little is known about the synthesis and assembly of the tegument in cestodes, but subtegumental cells are important (2). Some of the changes in the tegument following exsheathment may be regulated by neuropeptides released beneath the basal lamina. In *Diphyllobothrium dendritium*, these neuropeptides appear to be released in response to the body heat of the definitive avian host (10). The possibility that the tegument actively responds to such signals is strengthened by the presence of a typical G protein in the tegumental brush border membrane of *H. diminuta* (11).

2.1.2. Functional biology

2.1.2.1. Nutrient absorption. The kinetics and physiology of transport systems have been the focus of most cestode tegument research. Many of the systems that underlie nutrient absorption are far better characterized than analogous systems in trematodes and nematodes. One reason for this is that highly standardized methodologies introduced by Read *et al.* (12) for studying nutrient transport in *Hymenolepis diminuta* were used by subsequent investigators. This has facilitated mechanistic and quantitative comparisons between studies and among species. Also, the absence of any type of digestive tract allows unambiguous interpretation of data from transport studies.

Cestodes derive most of their energy from glucose, which is actively absorbed across the tegument by a single transport system (13). This system is Na^+-coupled and is blocked by phlorizin, a specific inhibitor of Na^+-coupled glucose transport in many organisms. Glucose is absorbed against a steep concentration gradient by cestodes, and the process is inhibited by agents that reduce ATP synthesis, including *p*-chloromercuribenzoate, iodoacetate and 2,4-dinitrophenol (14). These features are hallmarks of active transport. The process is also inhibited competitively by galactose and several other monosaccharides, indicating the presence of a common carrier system for hexoses (13). Though the mechanisms of carbohydrate absorption in cestodes and mammals are generally similar, it is more restricted in cestodes; only glucose and galactose are accumulated (13).

Glycerol absorption in *H. diminuta* occurs by passive diffusion at high concentrations (> 0.5 mM) and by a carrier-mediated process at lower concentrations. At lower concentrations, absorption of this lipid precursor is non-linear, dependent on temperature and pH, and competitively inhibited by glycerol and α-glycerophosphate. The existence of two distinct carriers for this molecule is suggested by studies which show that only about half the saturable component in *H. diminuta* is Na^+-sensitive and inhibitable by 1,2-propandiol (2).

In *H. diminuta*, two distinct carrier-mediated transport systems for fatty acid absorption have been partially characterized, one specific for short-chain fatty acids and the other for longer chain molecules. Acetate absorption in *H. diminuta* is mediated by the short-chain carrier at low concentrations, but passive diffusion occurs at high concentrations (15).

Amino acid absorption in adult cestodes occurs through multiple carriers that are saturable and sensitive to temperature and to a variety of inhibitors (2, 13). Amino acid accumulation typically occurs against steep concentration gradients (16), and the amino acid carrier systems in cestodes seem to be similar to those in mammals. There are separate carriers for acidic and basic amino acids and multiple carriers for the neutral amino acids, which overlap in their specificities. Unlike mammals, however, which show higher affinities for L-amino acids, cestode transporters are not stereoselective. Also, amino acid absorption in some cestodes is not dependent on co-transport of inorganic ions; replacement of Na^+ with choline or K^+ does not affect amino acid transport kinetics (16).

The available evidence suggests that cestodes can synthesize pyrimidines *de novo*, whereas purines must be absorbed from host fluids across the tegument (2). Both types of nitrogenous base are absorbed *in vitro*, mediated in part in *H. diminuta* by at least three distinct carriers, two of which contain multiple substrate binding sites (17). A passive diffusional component contributes to both purine and pyrimidine absorption (Chapter 6).

Endocytosis does not appear to play a major role in the absorption of proteins, though the capacity of cestodes to absorb macromolecules has not been clearly delineated. Horseradish peroxidase and ruthenium red appear to be endocytosed by the pseudophyllidean cestodes *Lingula intestinalis* and *Schistocephalus solidus* (18) and by *Hymenolepis nana* (19). The apparent lack of endocytosis detected in most studies that have addressed this topic may be attributable to the fact that only mature regions of the proglottids were examined (2).

2.1.2.2. Digestion. As in the epithelial lining of the vertebrate small intestine, the cestode tegument contains many enzymes usually associated with digestion, including alkaline and acid phosphatases, esterases and disacharidases (20, 21). It is not known, however, if digestive enzymes associated with the outer tegumental membrane are derived from the parasite or the host. The concept that these enzymes may participate in extracorporeal digestion of nutrient macromolecules, i.e., 'contact digestion' (21), is a matter of controversy (22).

Proteins within the glycocalyx and the membrane of the underlying microtriches are not digested by proteolytic enzymes in the lumen of the host intestine, and this may be due to a physical property of the glycocalyx. One possibility is that the glycocalyx helps maintain an unstirred water layer between the parasite and the contents of the host intestine. The organic acid end-products of carbohydrate metabolism excreted by cestodes could form a microenvironment in the immediate vicinity of the tegument that is too acidic for host proteases to function (23). Alternatively, the glycocalyx may contain factors, as yet undefined, that specifically inhibit host hydrolytic enzymes. For instance, *H. diminuta* inactivates trypsin and chymotrypsin by releasing a protease inhibitor (24).

2.1.2.3. Transport of inorganic ions. Homeostatic regulation of inorganic ion concentrations is a key function of all transport epithelia. Proteins that serve this function in vertebrates are also present in cestodes. The tegument contains a Ca^{2+}-ATPase activity which, by analogy with better characterized systems, is essential for maintaining intracellular Ca^{2+} at low concentrations. Two forms of the enzyme occur in the outer tegumental membrane of *H. diminuta*, one of which is calmodulin-dependent (25).

Na^+/K^+-ATPase, a key enzyme in the regulation of intracellular levels of these cations in most organisms, has been demonstrated in the tegument of *E. granulosus*, but has not yet been detected in other cestodes (3). The effect of ouabain, a specific inhibitor of this enzyme, on amino acid absorption by *H. diminuta* suggests that Na^+/K^+-ATPase is present and functional in other species.

2.1.2.4. Water balance and excretion. In addition to regulating essential electrical gradients across neuronal and muscle membranes, ion transport in cestodes, like other metazoans, may also be linked to the control of hydrostatic pressure gradients across cell membranes and across the body wall. Osmoconforming organisms behave as simple osmometers, and shrink or swell in direct relation to the osmolarity of their environment. Some cestodes are osmoconformers, adjusting internal osmotic pressure by varying fluid volume (2). A few species, including *H. diminuta*, regulate volume (i.e., maintain their weight at a constant level) over a fairly wide range of osmotic pressures; however, the worms rapidly swell or shrink in media which are below or above that range, respectively (23). How *H. diminuta* regulates its volume is not known, but the ability to excrete organic acid end-products of glucose metabolism (succinate and acetate) appears to be important. Osmoregulation in this cestode is glucose-dependent, and both glucose absorption and metabolism increase linearly with osmotic pressure of 210–335 mosmol/l, whereas the rates of uptake of leucine, Na^+ and Cl^- are unaffected by osmotic pressure. The high levels of succinate and acetate excreted as a consequence of osmotically induced glucose metabolism may acidify the medium surrounding the parasite (23, 26).

3. TREMATODES

Unlike cestodes, trematodes have an internal surface, the gastrovascular cavity (gastrodermis), which has a single opening that serves as both mouth and anus. The gastrodermis is likely to be an important organ for nutrient digestion and absorption, though definitive experimental support is lacking. The belief that tegument-associated antigens are the best candidates for vaccine development has motivated considerable research on tegument proteins and the effects of immune substances on the external surface, primarily in *Schistosoma* spp. Few of the surface proteins have been characterized biochemically. Focus on the external surface of trematodes has also been spurred by the belief that the tegument is the principal site of action of the important antischistosomal drug praziquantel.

3.1. External Surfaces

3.1.1. Structural considerations

3.1.1.1. Gross and microscopic anatomy. Like cestodes, trematodes are bounded by a layer of cytoplasm, undivided by internal membranes, termed a syncytial tegument (Fig. 12.2). In schistosomes the tegument is about 4 μm thick. It is limited externally by a heptalaminate membrane that consists of two closely apposed lipid bilayers. Based on the distribution of intramembranous particles observed in freeze-fracture studies, the inner bilayer is the true plasma membrane of the parasite. The membrane extends inward 1–2 μm along numerous pits that branch and interconnect to form a lattice which increases the surface area of the parasite at least 10-fold (28). The tegument of *F. hepatica* is much thicker and less pitted than that of schistosomes. It is bounded externally by a standard lipid bilayer from which extends a thin coat of glycoprotein

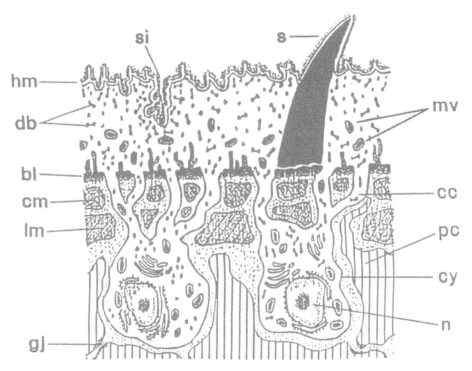

FIG. 12.2 Schematic of the tegument and underlying tissues in *Schistosoma mansoni*. Abbreviations: hm, heptalaminate membrane; db, discoid body; mv, multilaminate vesicle; s, spine; si, surface invagination; bl, basal lamellum; cm, circular muscle; lm, longitudinal muscle; gj, gap junction; cc,cytoplasmic channel; pc, parenchymal cell; cy, cyton; n, nucleus. After ref. 27.

with projecting side chains of gangliosides and oligosaccharides (29). The function of this thin glycocalyx is unknown.

The syncytial tegumental cytoplasm in schistosomes contains mitochondria and two major inclusions: discoid bodies and multilaminate vesicles. Discoid bodies are precursors of the dense ground substance of the cytoplasm as well as the surface bilayer (30). They may also break down to form crystalline spines which are most prominent on the dorsal surface of male schistosomes. Multilaminate vesicles consist of concentric whorls of membrane which migrate to the surface and form the outer bilayer. Mitochondria in the syncytial tegument are sparse, small and contain few cristae.

Tegumental nuclei, most mitochondria and ribosomes are contained in subtegumental cells (cytons) located beneath the layers of muscle that underlie the syncytium. The subtegumental cells are connected to the syncytium through cytoplasmic channels that contain numerous microtubules. Inclusion bodies synthesized in the subtegumental cells arise from Golgi bodies and are transported to the syncytium through these channels. This process is interrupted by agents which interfere with protein synthesis or microtubule assembly (31). The tegumental surface membrane is constantly replaced by multilaminate vesicles that migrate to the apical region of the syncytium and fuse with the plasma membrane (32). The half-life of the outer membrane is about 3 hours (33).

Much less is known about the synthesis and assembly of structural components in the tegument of *F. hepatica*.

3.1.1.2. Biochemistry and molecular biology of structural components. The schistosome tegument contains several phospholipids, the most abundant being phosphatidylcholine (57%), phosphatidylethanolamine (17%), sphingomyelin, lysophosphatidylcholine, phosphatidylinositol and cerebroside glycolipids (7, 34). Palmitic acid and oleic acid are the most abundant fatty acids (35). These molecules are derived from the host, as schistosomes are unable to synthesize cholesterol or long-chain fatty acids *de novo* (36). However, schistosomes can interconvert fatty acids and cleave the polar head from some phospholipids (37).

The protein composition of the surface membrane of the tegument of schistosomes or *F. hepatica* has not been fully characterized. In schistosomes, this membrane possesses a range of enzymatic activities, including alkaline phosphatase, Ca^{2+}-ATPase, and glycosyl transferase (38–40). In addition, ouabain binding sites on the surface of the tegument indicate the presence of Na^+/K^+-ATPase (41). A large number of surface-associated proteins in schistosomes have been detected immunologically. Some are embedded in the tegumental membrane, whereas others are anchored by glycosyl-phosphatidylinositol. At least one is dependent on palmitoylation for membrane attachment (42). Unfortunately, functional information on these surface-associated proteins is quite limited. In some cases, nucleotide sequence analysis of genes that encode immunoreactive proteins has evolved biochemistry from antigenicity; para-myosin was found to be a partially protective tegumental antigen (43). These studies also showed that the surface spines are composed of paracrystalline arrays of actin filaments. Glutathione *S*-transferase was identified in similar experiments. Members of this enzyme family are present in the tegumental cytoplasm of schistosomes (44) and *F. hepatica* (45), presumably acting to protect the animals from oxidants. Interestingly, the enzyme is also found in the intestinal epithelium of *F. hepatica* (45).

Carbohydrates exposed on the surface of schistosomes include mannose, glucose, galactose, *N*-acetylglucosamine, *N*-acetylgalactosamine and sialic acid. These sugars are linked to glycolipids and glycoproteins in the tegumental membrane, which also contains the enzymes required for their biosynthesis (46). Surface glycoproteins are anchored to the outer lipid bilayer by glycosylphosphatidylinositol through a process mediated by an endogenous phosphatidylinositol-specific lipase (47).

3.1.1.3. Developmental biology, synthesis and assembly. Schistosome cercariae undergo profound structural and biochemical transformation following penetration into the definitive host (48). During penetration, they lose their tails and begin to lose their fibrillar surface coat, or glycocalyx. Detachment of the glycocalyx follows secretion of proteases from the surface; these proteases may also play a role in tissue penetration (49, 50). By one hour after penetration, multilaminate vesicles begin to migrate to the outer surface of the tegument. During this period, numerous microvilli form from the outer membrane of the tegument, then are rapidly shed in a process that leads to the formation of the heptalaminate outer membrane (32). This protein shed-ding process may play an important role in the development of concomitant immunity, since it may prime the host to mount immune responses against subsequent challenge by cercariae(51). Formation of the heptalaminate membrane is complete within three hours after penetration. At this point, the tegument of the schistosomulum and adult stages are morphologically similar, as are their sterol and phospholipid compositions (52). Much higher levels of several surface proteins are expressed by immature stages than by adult schistosomes, and many surface proteins are differentially expressed on the surface of adult males and females (53). A Ca^{2+}-binding protein is found in the tegument of developing larvae and may play a role in surface remodeling (54).

3.1.2. Functional biology

3.1.2.1. Biological functions of the structure. The trematode tegument is structurally adapted for transport, immune evasion and communication with the neuromuscular system (via gap junctions). Numerous pits at the surface of the tegument markedly increase the surface area of the parasite, which is consistent with a transport function (see below). The fact that all of the pits in schistosomes examined *in situ* contain erythrocytes (55) suggests that these structures are open to the external environment. The tegument is less pitted in *F. hepatica*, but numerous invaginations of the surface effectively increase the surface area.

The double bilayer at the surface of the tegument in schistosomes is common only to blood-dwelling trematodes; species that inhabit other environments, including *F. hepatica*, are limited by a standard lipid bilayer (56). The double membrane in schistosomes may satisfy a structural requirement that allows rapid replacement of old or immunologically damaged membrane by new membrane derived from the multi-laminate vesicles. It may also permit host antigens, largely blood group glycolipids and glycoproteins, to be incorporated into the surface membrane of the parasite, thereby disguising it from immune attack (57). Some lipid components of this membrane, such as lysophosphatidylcholine, may protect the organism by lysing leukocytes (58).

Several structural components of the tegument suggest its involvement with the neuromuscular system. The tegument is electrically coupled by gap junctions to underlying muscle bundles in schistosomes (59), allowing changes in electrochemical gradients at the surface of the parasite to be transmitted directly to the muscle fibers below, presumably through gap junctions (60). The dorsal surface spines may serve as holdfasts to maintain the parasites within blood vessels or the bile duct. The tegument also contains acetylcholinesterase covalently anchored to the membrane through phosphatidylinositol (61). The surface location of this enzyme may destroy host acetylcholine in the circulation, protecting the parasite from its neuromuscular effects.

3.1.2.2. Nutrient absorption. A role for the tegument in nutrient absorption can be inferred on the basis of structural (above), biochemical and ecological considerations (62). Its numerous pits and channels provide an enormous surface area for absorption, and the tegument is constantly exposed to a nutrient-rich environment. Many enzymes that function in amino acid absorption are located in the tegument, and some of these, such as leucine aminopeptidase, are not present in the gut. Trematodes possess an incomplete digestive tract, and both *S. mansoni* and *F. hepatica* can survive extended *in vitro* incubations in the absence of detectable nutrient absorption across the intestine (63, 64). Glucose absorption in trematodes can be detected during immature stages of the life cycle, which lack an intestine (65).

Adult schistosomes depend solely on host plasma glucose for energy. Physiological experiments suggested that glucose flux across the tegument occurs by an active Na^+-dependent, carrier-mediated process (66,67). The inward-directed Na^+ gradient that is maintained, in part, by Na^+/K^+-ATPase in the tegument, was thought to provide the driving force for this process. However, cloning and functional expression of schistosome glucose transporters in *Xenopus laevis* oocytes revealed that transport is Na^+-independent and ouabain-insensitive (68). The discrepancy between physiological and molecular biological data may arise because manipulations of Na^+ in intact worms may lead to reductions in viability that are reflected in decreased glucose uptake. Alternatively, it is possible that Na^+-dependent glucose transporters predominate in adult animals, but were not identified in the cloning experiments. Nevertheless, this example illustrates the power of recombinant DNA technology to spur progress in understanding the basic biology of parasites.

Most glucose absorption occurs across the dorsal tegument in male worms; there is no evidence for regional differences in females (69). An intriguing aspect of glucose uptake by schistosomes is that, for worm pairs *in copula* (the normal condition *in vivo*), all or most glucose obtained by the female is supplied by the male partner (70). In addition, mated pairs absorb more glucose than unmated schistosomes, even though the total apparent surface area exposed to the external medium is much reduced. Increased glucose uptake in pairs may stem from increased rate or number of transport sites on the dorsal surface of the male (70).

Amino acid uptake by trematodes occurs primarily across the tegument, as mechanical ligation of the pharynx or drug treatments that induce regurgitation do not significantly affect uptake kinetics. Most evidence suggests that amino acid absorption by *F. hepatica* occurs by passive diffusion, as uptake is linear over a wide range of concentrations and there is no evidence for competitive inhibition among related amino acids (71). In schistosomes, it appears that some amino acids, such as methionine, glutamate, arginine and alanine, are absorbed across the tegument by a combination of passive diffusion and a carrier-mediated system (72). The number of distinct amino acid uptake systems in schistosomes is unknown, though there is evidence for at least two and possibly three. These systems have a high affinity for amino acids only, as methionine uptake is not inhibited by other organic compounds, and moving the amino group to carbons adjacent to the α-carbon reduces the affinity of the transport system (72).

Trematodes do not synthesize cholesterol *de novo*, but acquire it from the host. Uptake occurs via the tegument, primarily at the dorsal surface of males (73), and the cholesterol is then redistributed throughout the body (74). Cholesterol, like glucose, is transferred between partners *in copula* (75). Other lipids are also transported across the tegument, and regionally specified areas of this surface may be involved (74).

3.1.2.3. Water balance. Like cestodes, trematodes have a limited capacity to regulate the osmolarity of their body fluids. During the process of transformation from the cercarial stage to the schistosomulum stage following penetration of the host organism, schistosomes become much more permeable to water. When placed in hypo-osmotic medium, adult schistosomes (and other trematodes) gain weight rapidly, then exhibit a slow recovery to control weight levels. This response indicates a capacity for osmoregulation, and is interpreted as an initial osmotic uptake of water across the tegument followed by a slower efflux of ions from the tissue, with water following (76). In hyperosmotic medium, schistosomes slowly lose weight due to osmotic efflux of water (77).

How trematodes re-establish osmotic equilibrium is not known. They have a well-developed protonephridial system that may function in osmoregulation. However, there is no evidence that the protonephridial system responds to ionic or osmotic stress (78). Osmoregulation could also be mediated by active transport of ions across the tegument. In *S. mansoni*, the Na^+ concentration in extracellular fluid is approximately equal to that in the bathing medium over a wide range. The Na^+ concentration in the tegumental syncytium, however, is maintained at a constant level that is much lower than that in the extracellular fluid or the external medium (76).

3.1.2.4. Transport of inorganic ions. In both *S. mansoni* and *F. hepatica*, the low Na^+ concentration in the tegument is maintained by an Na^+/K^+-ATPase (27, 41). This enzyme contributes extensively to the maintenance of an electrochemical gradient across the outer tegumental membrane which, in *S. mansoni*, results in a membrane potential of $-60\,mV$ (79).

As indicated by the effect of changing external ion concentrations on tegumental membrane potential, the tegument is more permeable to K^+ than to Na^+ or Cl^-. This is a common feature of transporting epithelia in a wide range of vertebrates and invertebrates. In addition, the schistosome tegument contains at least one type of voltage-dependent ion channel (80). This channel has been studied using inside-out patches of outer tegumental membrane from adult female *S. mansoni*. It is cation-selective (i.e., conducts Na^+ and K^+, but not Cl^-) and exhibits a large unitary conductance.

3.1.2.5. Excretion. Trematodes excrete a number of metabolic end-products, including lactate, amino acids, NH_4^+ and H^+. Protonephridial organs and the tegument probably play a role in excretion. Levels of Na^+ within the tegumental syncytium become elevated after acidification of the external medium. Amiloride and low Na^+ medium interfere with recovery from an acid load, suggesting the existence of an Na^+-H^+ exchanger in the tegument (81).

3.2. Internal Surfaces

3.2.1. Structural considerations

3.2.1.1. Gross and microscopic anatomy. The digestive tract of trematodes consists of a mouth, pharynx and esophagus that branches to form a pair of intestinal caeca, which extend posteriorly and end blindly. The caeca consist of a single layer of epithelial cells supported by a thin layer of longitudinal and circular muscle fibers. The epithelial layer is syncytial in schistosomes, but is cellular in *F. hepatica* (27). The surface area of the caeca is amplified up to 100-fold by numerous digitiform microvilli or pleomorphic lamellae. The gut epithelial cells contain numerous mitochondria and ribosomes, and most of the cytoplasm is occupied by endoplasmic reticulum, Golgi bodies and secretory bodies.

3.2.1.2. Developmental biology. In schistosomes, the gut becomes functional within a few hours after penetration of the definitive host. *F. hepatica* begin to ingest host tissue immediately upon excystment of the metacercaria, with breakdown products of host tissue apparent in the caeca within a few hours. In newly excysted *F. hepatica*, the intestinal caeca appear to serve as reservoirs for enzymes used to penetrate the wall of the host gut during migration to the liver. Upon reaching the liver, the immature flukes begin to feed on hepatic cells, and the caeca are transformed into digestive/absorptive organs (46). Adult *F. hepatica* ingest mainly erythrocytes, but also other host tissues, including bile duct epithelium and mucus (27). Digestive enzyme activities associated with the gastrodermis of adult *F. hepatica* include those already listed for schistosomes plus amino peptidase and lipase (82) and a cathepsin L-like protease (83). Little is known about the biochemistry associated with development of the trematode gastrodermis. Erasmus (84) gives a detailed review of the functional morphology of the gastrodermis in larval and adult trematodes.

3.2.2. Functional biology

3.2.2.1. Nutrient digestion and absorption. The importance of the gut surface in trematodes as a site for nutrient digestion and absorption is suggested on the basis of several observations. Schistosomes begin to feed on host blood within 4–6 hours after penetration. *In vitro*, immature stages can survive several days without erythrocytes, but their growth and development is severely retarded (85). Females ingest many more erythrocytes than males, presumably in response to the demand for amino acids and proteins used in egg formation (27). Lysis and digestion of erythrocytes begins in the esophagus of schistosomes, and appears to be initiated by the release of dense secretory bodies from gland cells in the posterior esophagus (86).

A wide range of enzymes involved in digestion and processing of nutrients are found in gut epithelial cells of trematodes, including proteases, esterases, acid phosphatases and glucose-6-phosphatase (27). The most thoroughly studied is an abundant hemoglobinase present in gut epithelial cells of *S. mansoni* (Chapter 5). It degrades hemoglobin released from ingested erythrocytes and is a 31 kDa acid thiol protease similar to cathepsin B or L. The gene that encodes this enzyme has been cloned (87). Many other peptidases are also present in the schistosome gut (49), but their specific roles in digestion have not been fully elucidated. A cathepsin L-like protease has also been found in vesicles in the gut epithelia of *F. hepatica* (83).

The trematode gastrodermis epithelium is probably the principal site of absorption for large-molecular-weight nutrients. These nutrients seem to be critical for growth and long-term survival and for egg production by females (88a). The mechanisms of absorption are uncharacterized. Endocytosis occurs to a very limited extent in *F. hepatica*, but there is no evidence for this process in schistosomes. There is no information available on the role of the trematode gastrodermis in the transport of inorganic ions or water.

3.2.2.2. Excretion. There is little information on the role of the trematode gastrodermis in excretion of metabolic end-products. Most excretion is believed to occur by way of the highly developed protonephridial system, which is distinct from the tegument or the gastrodermis. In schistosomes and *F. hepatica*, the byproducts of erythrocyte digestion collect in the gastrodermis and are periodically regurgitated. In schistosomes, cellular organelles and macromolecules derived from the gastrodermis are constantly discharged across the internal surface into the lumen (86). This process has not been studied in *F. hepatica*.

4. NEMATODES

The rigid cuticle that is the external surface of nematodes is quite distinct from the tegument of cestodes and trematodes. Nematodes also possess a functional gut that ends in a muscularly controlled rectum. Although the various niches occupied by nematodes have led to evolutionary surface adaptations particularly suited for specific environments, the general features of their surface biology are well conserved.

More is known about the biochemistry of nematode surfaces than about the surfaces of other helminths, for several reasons. Parasitic nematodes pose a considerable challenge to veterinary and human medicine, and attention has been focused on them for economic reasons. The size of *Ascaris suum* has greatly facilitated physiological and biochemical studies. Finally, a great deal of work has been performed on the free-living nematode *Caenorhabditis elegans*. The power of genetics has illuminated a large number of basic biological and biochemical phenomena in this organism. Fortunately, *A. suum* and *C. elegans* provide data that can generally be extrapolated to other nematodes. Many reviews on the surfaces of nematodes are available (13, 88a–101).

4.1. External surfaces

4.1.1. Structural considerations

4.1.1.1. Gross and microscopic anatomy. Nematodes are bounded externally by a complex, multilayered cuticle that extends into and lines the pharynx, rectum, cloaca and other orifices. The cuticle represents an extracellular secretion of the hypodermis, an anatomical syncytium that forms a continuous cellular layer immediately beneath the cuticle and is specialized for transport and secretion. Though the cuticle and hypodermis are morphologically distinct, they represent a functional unit (Fig. 12.3).

Nematode cuticles are tremendously diverse in structure. Among species that parasitize vertebrates, the cuticle of the adult stage typically consists of four distinct layers: the epicuticle, cortex, medial layer and composite basal layer. The outermost layer, or epicuticle, is 6–30 nm thick and appears trilaminate in electron micrographs. Considerable controversy exists over whether this layer should be considered the limiting 'membrane' of nematodes (29, 93, 98). The epicuticle is not dissociated by treatments that typically dissolve membranes (99) and lipophilic markers do not exhibit the lateral mobility characteristic of membranes when inserted into the surface of nematodes (100a). Furthermore, the structure is devoid of intramembranous particles that are normally associated with cell membranes (100b). At least some stages and species of nematodes also exhibit a loosely associated surface coat. This coat is secreted through the excretory pore and from the esophagous (89).

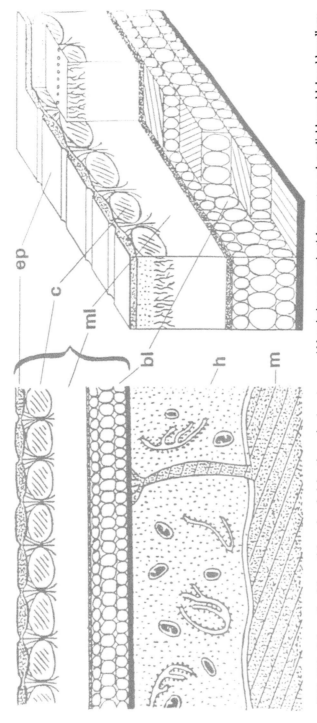

FIG. 12.3 Schematic of the cuticle and underlying tissues in *Ascaris suum*. Abbreviations: ep, epicuticle; c, cortex; ml, medial layer; bl, basal lamellum; h, hypodermis; m, muscle. After refs. 96 and 97.

217

The cortex is an amorphous and electron-dense layer composed of at least two sublayers which are abundant in keratin-like proteins and a highly insoluble protein termed cuticulin. The medial layer is a fluid-filled compartment that also contains fine collagenous fibers. In some species, the fluid contains hemoglobin. The composite basal layer contains crosslinked collagen fibers in two to three distinct sublayers which spiral around the nematode at an angle 75° to the longitudinal axis.

The cuticle may contain a range of gross structural elaborations including annulae, lateral and transverse ridges (also called alae) and spines. These structures are usually restricted to the cortex, though the medial layer is also sometimes involved (101). In some filariae, the cuticle is traversed by anatomical pores that extend to the hypodermis (91). There is no histological evidence for such pores in intestinal parasites, though the existence of functional pores is suggested by biophysical studies on the permeability characteristics of the cuticle (102).

The hypodermis lies immediately beneath the cuticle. It is multicellular or a syncytium, depending on the species. In most species, an outer hypodermal membrane is apparent, and it is likely that its absence in micrographs of other species is an artefact. The outward-facing membrane of the hypodermis probably represents the true limiting membrane of nematodes. Although typical cytoplasmic constituents are uniformly distributed in the hypodermis, nuclei are located only within lateral ridges that project into the pseudocoelom at the mid-ventral, mid-dorsal and lateral lines. The ventral and dorsal ridges also contain the major nerve cords, whereas the lateral ridges, in some species, contain small canals that may serve an excretory or secretory function. The hypodermis also contains fibers that anchor the somatic muscles to the cuticle.

4.1.1.2. Biochemistry and molecular biology of structural components.   The surface coat is composed primarily of carbohydrate, including negatively charged sugar residues (probably sulfated). Hydrophilic proteins known to be exposed to the environment have also been localized to this structure (89).

The precise chemical composition of the hydrophobic epicuticle is unknown. In some species, this layer is digested by elastase, suggesting that an elastin-like protein may be important (99). Most of the tissue in the underlying layers of the cuticle consists of collagen or collagen-like proteins. In *C. elegans*, two types of collagen have been identified that correspond to two types of extracellular matrix, the cuticle and basement membranes, including a layer that underlies the hypodermis (103). Cuticular collagens contain higher levels of proline and hydroxyproline than vertebrate collagens (104). About 75% of *C. elegans*, *A. suum* and *Haemonchus contortus* cuticular protein is soluble in 2-mercaptoethanol; the remainder consists of highly insoluble proteins such as a keratin-like protein (97, 103) and cuticulin, a protein crosslinked through tyrosine residues (as in cuticle collagen) (104–106). Tyrosine crosslinkages are achieved by peroxidation, leading to dityrosine residues (105), but also include an unusual iso-trityrosine component whose synthesis remains unresolved (105, 106). These insoluble proteins are located primarily in the cortex (106, 107).

There are approximately 100 collagen genes in *C. elegans* (103, 108), of which about 20 have been cloned and sequenced. Those associated with the cuticle typically encode short proteins (~ 300 amino acids) that fall into four families. Cuticle collagen genes in *A. suum* and *H. contortus* also form evolutionary conserved families (90, 108–110). Much information about cuticle collagen has been derived from the analysis of *C.*

elegans mutants (92, 103, 108). Mutations in collagen genes cause specific morphological abnormalities, which reflect the localization or function of the gene product (103). In addition, two cuticulin genes have been cloned from *C. elegans* (94). Antibodies have been used to localize this protein in the external cortical layer (94).

In addition to collagen and collagen-like proteins, nematode cuticle contains at least one glycoprotein and small amounts of hyaluronic acid and lipid. Lipid components, which are localized in the cortex of *A. suum* (111), have not been chemically characterized. Also present in the cuticle are chondroitin sulfate and sulfated mucopolysaccharides, which convey a negative charge to the cuticle (99).

4.1.1.3. Developmental biology, synthesis and assembly. The nematode cuticle is a dynamic structure that is partially or completely replaced during each of four molts. The biochemical events that stimulate and mediate molting are not well understood, but the hypodermis appears to be the site at which cuticle proteins are formed (95, 112). Selective laser ablation of *C. elegans* hypodermal cells before formation of the adult cuticle leads to abnormalities of the cuticle in regions directly above the lesions (113). In addition, protocollagen proline hydroxylase, an enzyme essential for cuticle protein synthesis, is present in the hypodermis of adult *A. suum* (114). Proteins recovered from the hypodermis of *A. suum* differ from those in the cuticle (115); considerable processing of proteins probably occurs in the cuticle layers prior to their incorporation.

Most species of nematode grow extensively between molts, and some continue to grow even after the final molt. To accommodate that growth, the cuticle must expand concurrently. In *A. suum*, cuticle thickness increases in direct proportion to increasing length of the parasite. Most of this growth in thickness occurs in the medial layer, which may become six to eight times thicker as the adult parasite grows from a length of 5 cm to 30 cm (101). How cuticular proteins permeate the scaffolding of the growing cuticle to reach their destinations and become appropriately integrated remains a major question in nematode biology.

4.1.2. Functional biology

4.1.2.1. Biological functions of the structure. Nematodes lack a rigid skeleton or circular muscles in the body wall that, in other metazoa, provide antagonistic systems against which longitudinal muscles can act to produce movement. They utilize instead a hydrostatic skeleton that is dependent on a high internal pressure within the pseudocoelom and limitations on expansion imposed by the cuticle. Internal pressures recorded from the pseudocoel of *A. suum* average 70 mmHg above ambient, and values as high as 275 mmHg are recorded during locomotion (116). These values are much higher than the transmural pressures recorded in other invertebrates. The biochemical mechanisms that underlie the high internal pressure in nematodes are unknown. Most species rapidly desiccate upon exposure to air, and transport studies using ^3H_2O indicate that the cuticle in some species, including *A. suum*, is quite permeable (102). Paradoxically, solute levels in pseudocoelomic fluid are generally iso-osmotic or even hypo-osmotic to the fluid in the digestive tract of the host. There is no biophysical model that explains how nematodes maintain a high internal pressure in the absence of a measurable external driving force (i.e., higher solute levels inside than outside). There is evidence that osmoregulation is a property of the body wall, but the mechanisms have not been characterized (117, 118).

The high internal pressure in nematodes is also needed for ingestion. Nematodes lack splanchnic muscles and the gut normally collapses under the hydrostatic pressure of the pseudocoelom. Ingested food is pumped caudally against this pressure by the pharynx, which is the only muscle in the digestive tract except for those in the rectum that regulate defecation (119). The internal pressure, by preventing the rapid passage of digesta, may contribute to the efficiency of nutrient absorption across the intestine.

In addition to its role in providing a scaffold against which the hydrostatic skeleton can be levered, the elasticity of the cuticle is another essential aspect of its role in locomotion. Although the cuticle collagen fibers are inelastic, their unusual arrangement in the composite basal layer confers elasticity. The fibers are organized into cylindrical helices which spiral around the body in opposite directions, such that fibers in the middle layer run at a 75° angle to the longitudinal axis of the parasite (116). The longitudinal muscles are attached to the cuticle at its inner surface. As these muscles contract, the helix is drawn apart. This changes the shape of the body, which becomes longer and thinner as localized volume reductions occur within regions of the pseudocoel beneath bundles of contracting muscle fibers. The control of the crosslinking reactions that provide the necessary organization of proteins in the cuticle for these mechanical roles remains to be fully described.

4.1.2.2. Nutrient absorption.   Transcuticular absorption of glucose, amino acids and other nutrients has been demonstrated in all filarial nematodes examined (91). In most filariae, the cuticle is the principal site of nutrient absorption *in vitro*, with very little occurring across the intestine. Several observations support this conclusion: (i) microfilariae of most species possess a non-functional gut; (ii) the hypodermis contains enzymes associated with amino acid transport whereas the intestine often does not; and (iii) nutrient transport across the cuticle/hypodermis complex demonstrates saturability, stereoselectivity and competitive as well as non-competitive inhibition, indicating the presence of specific carrier systems (91, 120).

Evidence for transcuticular transport of physiologically relevant quantities of nutrients in intestinal nematodes is less compelling. Most studies have used *A. suum*, which does not absorb or utilize significant quantities of glucose *in vitro* (121). The fact that some transcuticular absorption of glucose occurs *in vitro* (122), suggests that the process may be important *in vivo*. Other species of intestinal nematodes absorb and utilize large quantities of exogenous glucose *in vitro*. The importance of the cuticle in glucose absorption is suggested by studies that demonstrate that metabolism of exogenous glucose by *H. contortus* is not affected by exposure to high levels of invermectin, which paralyzes the pharynx and inhibits ingestion (123).

Amino acid transport is equally poorly understood in gastrointestinal nematodes. It has been difficult to demonstrate significant transport of amino acids across the gut. Histochemical and enzymological data demonstrate that γ-glutamyl transpeptidase is present in the cuticle–hypodermis of *A. suum* in considerably greater abundance than in the intestinal epithelium (124). This enzyme is part of an amino acid transport system found in many organisms.

4.1.2.3. Transport of inorganic ions. *Ascaris suum* has been used to study transcuticular transport of inorganic ions. These parasites maintain an electrochemical gradient across the body wall that may be driven by active ion transport (125). Na^+ and K^+ concentrations in the pseudocoelomic fluid of freshly collected parasites are approximately equal to those found in the intestinal fluid of swine, and the body wall is quite permeable to these cations (126). Conversely, the concentrations of Ca^{2+} and Mg^{2+} in the pseudocoel are maintained at constant levels, and are not affected by the concentrations of these cations in the external medium.

Active transport of Cl^- occurs across the cuticle–hypodermis complex of *A. suum* (127). Control of internal Cl^- levels appears to play an important role in maintaining the membrane potential of the somatic muscle cells, which are much more permeable to Cl^- than to other ions. A large Cl^- channel that can also conduct organic acids has been described in *A. suum* muscle membranes (128); the same channel has been found to be expressed on both the inward and outward facing membranes of the hypodermis (K. L. Blair *et al.*, unpublished results). It is not known if this channel plays a role in organic acid excretion across the cuticle (129, 130). In vertebrates, muscle membrane potential is dependent primarily on K^+.

4.1.2.4. Water balance. The cuticle is relatively permeable to water. However, most evidence suggests that nematodes are relatively insensitive to osmotic changes in the external medium (117 and 118). *A. suum*, for example, tolerates incubations for up to 4 days in dilutions of artificial sea water that range in strength from 20 to 40% (131). During extended *in vitro* incubations, the osmotic pressure of the pseudocoelomic fluid conforms roughly to that of the external medium, providing the osmolarity of the external medium is within 20–40% of that of the pseudocoelomic fluid of freshly collected worms (131).

The ability of parasitic nematodes to regulate their volume and maintain internal ionic stability probably depends on several mechanisms associated with the cuticle/hypodermis complex. Under hypo-osmotic conditions, swelling of the pseudocoel due to water accumulation is limited by the relatively inelastic cuticle. Evidence supporting the concept of aniso-osmotic extracellular regulation by a cuticle-associated process comes from studies using *A. suum* that demonstrate active transport of inorganic ions across isolated segments of body wall (127). Nematodes, like cestodes, may also adjust the osmotic pressure within their pseudocoel by regulating the rate of transmural organic acid excretion (129).

4.1.2.5. Transport of non-nutrient organic solutes. The cuticle is an important site for the absorption of anthelmintics and other small organic molecules. Accumulation of levamisole by *A. suum* can be accounted for solely by transcuticular diffusion (132). For the filariae, *Brugia pahangi* and *Dipetalonema viteae*, absorption indices for a wide range of non-electrolytes show no obvious relationship with compound lipophilicity (133). This contrasts sharply with *S. mansoni*, for which absorption can be predicted by the single variable log K (134). When other physicochemical parameters, such as molecular weight, dipole moment and total energy are considered along with log K, it is possible to make quantitative predictions about the absorption of non-electrolytes by these nematodes. This indicates that absorption by filarial nematodes is influenced by non-lipid components in the cuticle (133).

The intestinal nematodes *A. suum* and *H. contortus* have also been used in quantitative absorption kinetic studies to determine if physicochemical properties can be used to predict the absorption rates of non-nutrient organic molecules (100, 135). As in filariae, $\log K$ alone does not accurately predict how rapidly the molecules are absorbed. Studies using adult *A. suum* mechanically ligated to eliminate intestinal transport, or dissected into segments of body wall that contain no anatomical openings, have shown that the collagen and lipid components of the cuticle each present a distinct barrier to diffusion of non-nutrient organic molecules. Penetration of the collagen barrier is highly dependent on the molecular size and electrical charge of the permeant. The permeability of neutral solutes decreases with size, and positively charged molecules penetrate faster whereas anions penetrate slower than neutral solutes of comparable size. These results indicate that the cuticle of *A. suum* contains functional aqueous pores which are negatively charged and about 15 Å in radius. The functional pores that mediate permeability through the cuticle of gastrointestinal parasites must be tortuous paths through the crosslinked fibers of the cuticle.

4.1.2.6. Excretion.   Although direct evidence exists for outward transport of inorganic ions across the cuticle of *A. suum* (127), little is known about the relative importance of the cuticle, intestine or putative excretory canal system in nematodes for outward transport of metabolic end-products or other organic molecules. Organic acid end-products of carbohydrate metabolism are excreted across the cuticle in *A. suum* (129) and *H. contortus* (123). This process generates a microclimate pH within the aqueous pores of the cuticle which is maintained at pH ~ 5.0, near the pK_a of the excreted organic acids (130).

4.2. Internal Surfaces

4.2.1. Structural considerations

4.2.1.1. Gross and microscopic anatomy.   The alimentary canal of nematodes extends the full length of the organism, opening only at the mouth and anus. It consists of three regions: the stomodaeum (mouth, buccal capsule and pharynx or esophagus), the intestine and the proctodaeum (rectum and anus, includes the reproductive opening in males). The stomodaeum and proctodaeum are lined by cuticle. The cuticle in the buccal capsule is often modified to form teeth or plates which are used for attachment to host membranes or abrasion of ingested food particles. The intestinal wall consists of a single layer of columnar epithelial cells which contain microvilli on the luminal side. In *A. suum*, these microvilli increase the surface area of the intestine by a factor of ~ 90. They are coated by a unit membrane and a glycocalyx similar in appearance to that of vertebrate intestinal epithelium. The basal membrane of intestinal cells in nematodes may be smooth or folded. In most species, it appears to form a continuous layer which arises as a secretory product of the epithelial cells (136). In some species, mitochondria are concentrated near the basal lamellae (101). Some columnar cells, particularly in the anterior regions of the intestine, appear to serve glandular functions. In *A. suum*, parts of the cells containing secretory material detach from the cell and disintegrate in the lumen of the intestine. The intestine in most nematodes has no muscular layer and so does not show peristalsis. It is normally collapsed due to the high internal pressure of the pseudocoelom. Against this pressure, food particles are propelled backwards only by the pumping action of the pharynx.

4.2.1.2. Biochemistry and molecular biology of structural components.   Enzymes found in the nematode gut include the usual array of hydrolyses typical of digestive tissues. Most progress has been made in the characterization of proteases. Carboxy and thiol proteases are found in the gut of many parasitic nematodes (137). A cysteine protease has been purified and the gene encoding it cloned from *H. contortus* (138); its presence in the gastrointestinal tract is evidence of its digestive role. Since it also degrades fibrinogen, it may play a role in preventing clotting of ingested blood. The gene for a similar enzyme has been cloned from *C. elegans* (139). It is also related to the *S. mansoni* hemoglobinase and is expressed only in the gut. An esterase and vitellogenin are also expressed only in the gut of *C. elegans*, and factors that regulate this tissue-specific expression have been tentatively identified (139). Similar results have been reported for an acid phosphatase (140). The experiments with *C. elegans* could provide a number of tools for characterizing gut function in parasitic species.

The alimentary canal in nematodes also contains proteins that inhibit digestive enzymes of the host (primarily proteases; see ref. 141) or prevent coagulation of host blood (138). The structures of some of these protease inhibitors have been identified (142, 143).

Though little progress has been reported on the structural biochemistry of the gastrointestinal tract, some information is available about the collagens found in the intestinal basement membrane of *A. suum* (144). The structural organization of collagen aggregates in this tissue clearly distinguishes it from cuticular collagens, though the biochemical explanation for the differences remains unresolved.

4.2.1.3. Developmental biology.   Very little is known about the biochemistry of the gut through development of parasitic nematodes. However, considerable insight is being gained in *C. elegans* (145, 146). The identification of genes that are only expressed in the gut (see above) permits a detailed analysis of the development of this specialized tissue.

4.2.2. Functional biology

The importance of the intestine in nutrient absorption, excretion and ionic- and osmoregulation varies extensively among species and between stages of development within a species. In general, the intestine is more important in gastrointestinal than filarial nematodes. This generalization may be challenged as better culture systems for nematodes are developed or when transport studies can be extended to *in vivo* conditions (see ref. 62 for review). Since adult stages of parasitic nematodes often do not survive extended *in vitro* incubations, it is probably safe to assume that essential components are absent from the culture systems used, and that these deficiencies may directly or indirectly affect gut function of the parasites.

4.2.2.1. Digestion.   In intestinal nematodes, digestion usually begins as food particles enter the pharynx, or even earlier in species that feed by extracorporeal digestion of host tissues. *Nippostrongylus brasiliensis*, for instance, releases histolytic enzymes from the pharyngeal and subventral glands that partially digest host mucosal cells which are subsequently ingested by the pumping action of the pharynx (101). The release of digestive enzymes onto food particles is greatly reduced in saprophagous species.

The buccal cavity of some intestinal nematodes contains teeth or cutting plates which are modifications of the cuticle. These are most prevalent in species that feed on the mucosa of the alimentary or respiratory tracts of the host. Nematodes that have a small buccal cavity and rely on extracorporeal digestion of host tissues depend less heavily on the grinding action of teeth, and cuticle modifications in the buccal cavity are less extensive.

In most animal parasitic nematodes, pharyngeal glands secrete digestive enzymes onto ingested food particles as they enter the pharynx. In *A. suum*, an esterase, amylase, maltase, protease, peptidase and lipase are secreted into the pharynx during feeding. The pharynx of species that feed on host mucosal cells may also contain hyaluronidase and additional enzymes specialized for digestion of host tissues (141).

The intestine is also an important site for secretion of digestive enzymes. In *A. suum* several disaccharidases are abundant on the microvilli, including sucrase, palatinase, maltase and trehalase, with maltase activity predominating. Endopeptidases that break down hemoglobin and other host proteins are also present. Lipases have been detected in intestinal cells of several species, and their activity depends extensively on the diet of the parasite. In *Strongylus edentatus*, which feeds on host mucosal tissue, lipase activity in the intestine is 12-fold greater than that for *A. suum*, a saprophagous feeder (141).

4.2.2.2. Nutrient absorption. The intestine of intestinal parasites provides an important surface for nutrient absorption. In some species, such as *A. suum*, the intestine may be the only site where physiologically significant amounts of glucose and amino acids are absorbed (13,61). Unfortunately, almost everything we know about intestinal transport in nematodes comes from studies using *A. suum*. The atypical size of this organism suggests that it may not be a good model for the study of this process, a smaller species may be less reliant on the bulk transfer capacity of their intestines for the delivery of nutrients, other solutes and water to internal tissues. Nevertheless, among commonly available nematodes, only *A. suum* is large enough to dissect and isolate segments of intestine to study by physiological techniques.

In isolated segments of *A. suum* intestine, glucose and fructose are absorbed much more rapidly than galactose (147). The absorption of glucose occurs against a steep concentration gradient, is saturable and Na^+-sensitive (147). The non-metabolizable analog of glucose, 3-O-methylglucose, is absorbed unidirectionally (lumen to pseudocoelom) across the intestine. Its absorption against a concentration gradient depends on the presence of authentic glucose in the incubation medium (148). Together, these findings suggest that glucose absorption across the intestine of *A. suum* is energy-dependent and involves an Na^+/glucose symporter (13).

Little is known about the transport of amino acids across the intestine of nematodes. *In vitro* studies using isolated segments of intestine from *A. suum* indicate that uptake of methionine, glycine, histidine and valine is stereospecific and non-linear with respect to concentration, indicative of a mediated transport process (12). Most nematodes excrete a wide range of amino acids (149), and though some of these, such as alanine and proline, are true metabolic end-products, others must be derived from ingested materials.

Most evidence indicates that the absorption of fatty acids, mono- and triglycerides and cholesterol occurs mainly across the intestine of intestinal nematodes. Absorption of albumin-complexed palmitic acid by cells from isolated segments of *A. suum* intestine is a passive process that occurs in the absence of glucose or the presence of iodoacetate. Passage of this fatty acid from the intestinal cells into the pseudocoelom, however, requires glucose and is sensitive to metabolic inhibitors, indicative of mediated transport (150). The process of fatty acid absorption by *A. suum* intestine is stimulated by bile salts (150). Triglycerides are absorbed very slowly, whereas monoglycerides are absorbed rapidly by intestinal cells, then converted into free fatty acids before being actively transported to the pseudocoelomic fluid for dissemination to the body tissues (151).

4.2.2.3. Transport of inorganic ions and water. Direct evidence for transintestinal transport of inorganic ions is available only for *A. suum*. Concentration gradients for Na^+, K^+ and Cl^- across the intestine contribute to a 10–15 mV electrical potential (inside positive) which is abolished below 20°C and inhibited by agents that interfere with glycolysis (152, 153). Flux measurements indicate that, among the major inorganic ions, only K^+ diffuses outward from the lumen more rapidly than it diffuses inward; other ions must contribute to the electrical potential. The electrical potential is also dependent, in part, on the presence of glucose on the luminal side, and is consistent with the concept that Na^+ flux from the intestine is coupled to the absorption of glucose into the epithelial cells of the intestine.

The existence of an Na^+ gradient is essential for Na^+-coupled glucose transport in other systems. This gradient is usually maintained by an Na^+/K^+-ATPase that pumps more Na^+ ions out of the cell than K^+ ions in. However, Na^+ and glucose transport across the *A. suum* intestine is not affected by high levels of ouabain (152), indicating that, if an Na^+/K^+-ATPase regulates the concentrations of these ions, the enzyme is distinct from that characterized in vertebrates and most other invertebrates.

The extent to which the intestine of *A. suum* or other gastrointestinal nematodes is involved with osmoregulation is unknown. *A. suum* does pass water out through its anus (154), but the cuticle of all nematodes examined is permeable to water (101), and the relative importance of the cuticle, intestine or the putative excretory canaliculi located in the hypodermis for water transport has not been determined.

4.2.2.4. Excretion. The intestine, putative excretory system and the cuticle–hypodermis complex have all been suggested as possible surfaces for excretion in nematodes. Direct evidence for excretion at any of these sites is lacking. Indirect evidence against the importance of the intestine for excretion comes from studies that demonstrate that the rate of decline in the viability of some species is unaffected by treatments which prevent intestinal passage of solutes. In *A. suum*, mechanical ligation of the head and tail does not affect viability over 3-day incubations. The intrinsic excretion rates of each of the organic acid end-products of carbohydrate metabolism are also unaffected by this process (129). Similarly, pharmacological ligation of *H. contortus* using 1 nM ivermectin, which paralyzes the pharynx and inhibits bulk transport of solutes along the intestine, does not affect the rates of excretion of propionate or lactate, or the viability of these parasites estimated from ATP levels (123).

REFERENCES

1. Lumsden, R. D. and Murphy, W. (1980) Morphological and functional aspects of the cestode surface. In: *Cellular Interactions in Symbiosis and Parasitism* (eds Cook, C., Pappas, P. and Rudolph, E.), Ohio State University Press, Columbus, OH, pp. 95–130.
2. Smyth, J. D. and McManus, D. P. (1989) *The Physiology and Biochemistry of Cestodes.* Cambridge University Press, Cambridge.
3. Threadgold, L. T. (1984) Parasitic platyhelminths. In: *Biology of the Integument*, Vol. 1 (eds Berieter-Hahn, J., Matolsky, A. G. and Richards, K. S.), Springer-Verlag, Berlin, pp. 132–191.
4. Barrett, J. (1981) *Biochemistry of Parasitic Helminths*, Macmillan, London.
5. Knowles, W. J. and Oaks, J. A. (1979) Isolation and partial biochemical characterization of the brush border plasma membrane from the cestode *Hymenolepis diminuta*. *J. Parasitol.* **65**: 715–731.
6. Cain, G. D., Johnson, W. J. and Oaks, J. A. (1977) Lipids from subcellular fractions of the tegument of *Hymenolepis diminuta*. *J. Parasitol.* **63**: 486–491.
7. Frayha, G. J. and Smyth, J. D. (1983) Lipid metabolism in parasitic helminths. *Adv. Parasitol.* **22**: 309–387.
8. Barrett, J. (1983) Lipid metabolism. In: *Biology of the Eucestoda*, Vol. 2 (eds. Arme, C. and Pappas, P. W.), Academic Press, London, pp. 391–419.
9. Dunn, J. and Threadgold, L. T. (1984) *Taenia crassiceps*: temperature, polycations, and cysticercal endocytosis. *Exp. Parasitol.* **58**: 110–124.
10. Gustafsson, M. K. S., Wikgren, M. C., Karhi, T. J. and Schot, L. P. C. (1985) Immunocytochemical demonstration of neuropeptides and serotonin in the tapeworm *Diphyllobothrium dendritium*. *Cell Tissue Res.* **240**: 255–260.
11. Walker, J. and Barrett, J. (1993) Evidence for a G protein system in the tegumental brush border plasma membrane of *Hymenolepis diminuta*. *Int. J. Parasitol.* **23**: 281–284.
12. Read, C. P., Rothman, A. H. and Simmons, J. E., Jr (1963) Studies on membrane transport, with special reference to parasite-host integration. *Ann. NY Acad. Sci.* **113**: 154–205.
13. Pappas, P. W. and Read, C. P. (1975) Membrane transport in helminth parasites: a review. *Exp. Parasitol.* **37**: 469–530.
14. Phifer, K. O. (1960) Permeation and membrane transport in animal parasites: the absorption of glucose by *Hymenolepis diminuta*. *J. Parasitol.* **46**: 51–62.
15. Arme, C. and Coates, A. (1973) *Hymenolepis diminuta*: active transport of α-aminoisobutyric acid by cysticercoid larvae. *Int. J. Parasitol.* **3**: 553–560.
16. Pappas, P. W., Uglem, G. L. and Read, C. P. (1973) Mechanisms of amino acid transport in *Taenia crassiceps* larvae (Cestoda). *Int. J. Parasitol.* **3**: 641–651.
17. MacInnis, A. J., Fisher, F. M., Jr and Read, C. P. (1965) Membrane transport of purines and pyrimidines in a cestode. *J. Parasitol.* **51**: 260–267.
18. Threadgold, L. T. and Hopkins, C. A. (1981) *Schistocephalus solidus* and *Lingula intestinalis*: pinocytosis by the tegument. *Exp. Parasitol.* **51**: 444–445.
19. Hopkins, C. A., Law, L. M. and Threadgold, L. T. (1978) *Schistocephalus solidus*: pinocytosis by the plerocercoid tegument. *Exp. Parastol.* **44**: 161–172.
20. Pappas, P. W. (1984) Kinetic analysis of the membrane-bound alkaline phosphatase activity of *Hymenolepis diminuta* (Cestoda: Cyclophyllidea) in relation to development of the tapeworm in the definitive host. *J. Cell. Biochem.* **25**: 131–137.
21. Read, C. P. (1973) Contact digestion in tapeworms. *J. Parasitol.* **59**: 672–677.
22. Thomas, J. N. and Turner, S. G. (1980 A reinterpretation of the evidence for contact digestion in the tapeworm, *Hymenolepis diminuta*. *J. Physiol.* **301**: 79P–80P.
23. Uglem, G. L. (1992) Water balance and its relation to fermentation acid production in the intestinal parasites *Hymenolepis diminuta* (Cestoda) and *Moniliformis moniliformis* (Acanthocephala). *Parasitology* **77**: 874–883.
24. Pappas, P. W. and Uglem, G. L. (1990) *Hymenolepis diminuta* (Cestoda) liberates an inhibitor of proteolytic enzymes during *in vitro* incubation. *Parasitology* **101**: 455–464.
25. Branford-White, C. J., Hipkiss, J. B. and Peters, T. J. (1984) Evidence for a Ca^{2+}-dependent activator in the rat tapeworm *Hymenolepis diminuta*. *Mol. Biochem. Parasitol.* **13**: 201–211.

26. Uglem, G. L. and Just, J. J. (1983) Trypsin inhibition by tapeworms: antienzyme secretion or pH adjustment? *Science* **220**: 79–81.
27. Smyth, J. D. and Halton, D. W. (1983) *The Physiology of Trematodes*. Cambridge University Press, Cambridge.
28. Hockley, D. J. (1973) Ultrastructure of the tegument of *Schistosoma*. *Adv. Parasitol.* **11**: 233–305.
29. Proudfoot, L., Kusel, J. R., Smith, H. V. and Kennedy, M. W. (1991) Biophysical properties of the nematode surface. In: *Parasitic Nematodes — Antigens, Membranes and Genes* (ed. Kennedy, M. W.), Taylor and Francis, London, pp. 1–26.
30. MacGregor, A. N., Kusel, J. R. and Wilson, R. A. (1988) Isolation and characterization of discoid granules from the tegument of adult *Schistosoma mansoni*. *Parasitol. Res.* **74**: 250–254.
31. Wilson, R. A. and Barnes, P. E. (1974) An *in vitro* investigation of dynamic processes occurring in the schistosome tegument, using compounds known to disrupt secretory processes. *Parasitology* **68**: 259–270.
32. Hockley, D. J. and McLaren, D. J. (1973) *Schistosoma mansoni*: changes in the outer membrane of the tegument during development from cercaria to adult worm. *Int. J. Parasitol.* **3**: 13–25.
33. Wilson, R. A. and Barnes, P. E. (1979) Synthesis of macromolecules by the epithelial surfaces of *Schistosoma mansoni*: an autoradiography study. *Parasitology* **78**: 295–310.
34. McDiarmid, S. S., Podesta, R. B. and Rahman, S. M. (1982) Preparation and partial characterization of a multilamellar body fraction from *Schistosoma mansoni*. *Mol. Biochem. Parasitol.* **5**: 93–105.
35. Allan, D., Payares, G. and Evans, W. H. (1987) The phospholipid and fatty acid composition of *Schistosoma mansoni* and its purified tegumental membranes. *Mol. Biochem. Parasitol.* **23**: 123–128.
36. Meyer, F., Meyer, H. and Bueding, E. (1970) Lipid metabolism in the parasitic and free-living flatworms *Schistosoma mansoni* and *Dugesia dorotocephala*. *Biochim. Biophys. Acta* **210**: 257–266.
37. Smith, J. H., Reynolds, E. S. and vonLichtenberg, F. (1969) The integument of *Schistosoma mansoni*. *Am. J. Trop. Med. Hyg.* **18**: 28–49.
38. Cesari, I. M., Simpson, A. J. G. and Evans, W. H. (1991) Properties of a series of tegumental membrane-bound phosphohydrolase activities of *Schistosoma mansoni*. *Biochem. J.* **198**: 467–473.
39. Podesta, R. B. and McDiarmid, S. S. (1982) Enrichment and partial enzyme characterization of ATPase activity associated with the outward-facing membrane of the surface epithelial syncytium of *Schistosoma mansoni*. *Mol. Biochem. Parasitol.* **6**: 225–235.
40. Simpson, A. J. G., Schryer, M. D., Cesari, I. M., Evans, W. H. and Smithers, S. R. (1981) Isolation and partial characterization of the tegumental outer membrane of adult *Schistosoma mansoni*. *Parasitology* **83**: 163–177.
41. Fetterer, R. H. and Vande Waa, J. and Bennett, J. L. (1980) Characterization and localization of ouabain receptors in *Schistosoma mansoni*. *Mol. Biochem. Parasitol.* **1**: 209–219.
42. Pearce, E. J., Magee, A. I., Smithers, S. R. and Simpson, A. J. G. (1991) Sm25, a major schistosome tegumental glycoprotein, is dependent on palmitic acid for membrane attachment. *EMBO J.* **10**: 2741–2746.
43. Matsumoto, Y., Perry, G., Levine, R. J. C., Blanton, R., Mahmond, A. A. F. and Aikawa, M. (1988) Paramyosin and actin in schistosomal teguments. *Nature* **333**: 76–78.
44. Holy, J. M., O'Leary, K. A., Oaks, J. A. and Tracy, J. W. (1989) Immunocytochemical localization of the major glutathione *S*-transferases in the adult *Schistosoma mansoni*. *J. Parasitol.* **75**: 181–190.
45. Wijffels, G. L., Sexton, J. L., Salvatore, L. *et al.* (1992) Primary sequence heterogenecity and tissue expression of glutathione *S*-transferases of *Fasciola hepatica*. *Exp. Parasitol.* **74**: 87–99.
46. Bennett, C. E. (1975) *Fasciola hepatica*: development of caecal epithelium during migration in the mouse. *Exp. Parasitol.* **37**: 426–441.

47. Pearce, E. J. and Sher, A. (1989) Three major surface antigens of *Schistosoma mansoni* are linked to the membrane by glycosylphosphatidylinositol. *J. Immunol.* **142**: 979–984.
48. Stirewalt, M. A. (1974) *Schistosoma mansoni*: cercaria to schistosomule. In: *Advances in Parasitology* (ed. Dawes, B.), Academic Press, London, pp. 115–182.
49. McKerrow, J. H. and Doenhoff, M. J. (1988) *Schistosome* proteases. *Parasitol. Today* **4**: 334–340.
50. McKerrow, J. H., Newport, G. and Fishetson, Z. (1991) Recent insights into the structure and function of a larval proteinase involved in host infection by a multicellular parasite. *Proc. Soc. Exp. Biol. Med.* 119–124.
51. McLaren, D. J. and Hockley, D. J. (1976) *Schistosoma mansoni*: the occurrence of microvilli on the surface of the tegument during transformation from cercaria to schistosomulum. *Parasitology* **73**: 169–187.
52. Furlong, S. T. and Caulfield, J. P. (1988) *Schistosoma mansoni*: sterol and phospholipid composition of cercariae, schistosomula and adults. *Exp. Parasitol.* **65**: 222–231.
53. Snary, D., Smith, M. A. and Clegg, J. A. (1980) Surface proteins of *Schistosoma mansoni* and their expression during morphogenesis. *Eur. J. Immunol.* **10**: 573–575.
54. Ram, D., Romano, B. and Schechter, I. (1994) Immunochemical studies on the cercarial-specific calcium binding protein of *Schistosoma mansoni*. *Parasitology* **108**: 289–300.
55. Bruce, J. I., Pezzlo, F., Yajima, Y. and McCarty, J. F. (1971) An electron microscope study of *Schistosoma mansoni* migration through mouse tissue: ultrastructure of the gut during the hepatoportal phase of migration. *Exp. Parasitol.* **30**: 165–173.
56. McLaren, D. J. and Hockley, D. J. (1977) Blood flukes have a double outer membrane. *Nature* **269**: 147–149.
57. Kemp, W. M., Merritt, S. C., Borgucki, M. S., Rosier, J. G. and Seed, J. R. (1977) Evidence for adsorption of heterospecific host immunoglobulin on the tegument of *Schistosoma mansoni*. *J. Immunmol.* **119**: 1849–1854.
58. Remold, H. G., Mednis, A., Hein, A. and Caulfield, J. P. (1988) Human monocyte-derived macrophages are lysed by schistosomula of *Schistosoma mansoni* and fail to kill the parasite after activation with interferon gamma. *Am. J. Pathol.* **131**: 146–155.
59. Silk, M. H., Spence, I. M. and Gear, J. H. S. (1969) Ultrastructural studies of the blood fluke-*Schistosoma mansoni*. I. The integument. *S. Afr. J. Med. Sci.* **34**: 1–10.
60. Thompson, D. P., Pax, R. A. and Bennnett, J. L. (1982) Microelectrode studies of the tegument and subtegumental compartments of male *Schistosoma mansoni*: an analysis of electrophysiological properties. *Parasitology* **85**: 163–178.
61. Espinoza, B., Tarrab-Hazdai, R., Silman, I. and Arnon, R. (1988) Acetylcholinesterase in *Schistosoma mansoni* is anchored to the membrane via covalently attached phosphatidyl-inositol. *Mol. Biochem. Parasitol.* **29**: 171–179.
62. Pappas, P. W. (1988) The relative roles of the intestines and external surfaces in the nutrition of monogeneans, digeneans and nematodes. *Parasitology* **96**: S105–S121.
63. Isseroff, H. and Read, C. P. (1974) Studies on membrane transport. VIII. Absorption of monosaccharides by *Fasciola hepatica*. *Comp. Biochem. Physiol.* **47**A: 141–152.
64. Popiel, I. and Basch, P. F. (1984) Reproductive development of female *Schistosoma mansoni* (Digenea: Schistosomatidae) following bisexual pairing of worms and worm segments. *J. Exp. Zool.* **232**: 141–150.
65. Uglem, G. L. and Lee, K. J. (1985) *Proterometra macrostoma* (Trematoda: Azygiidae): functional morphology of the tegument. *Int. J. Parasitol.* **15**: 61–64.
66. Rogers, S. H. and Bueding, E. (1975) Anatomical localization of glucose uptake by *Schistosoma mansoni* adults. *Int. J. Parasitol.* **5**: 369–371.
67. Uglem, G. L. and Read, C. P. (1975) Sugar transport and metabolism in *Schistosoma mansoni*. *J. Parasitol.* **61**: 390–397.
68. Skelly, P. J., Kim, J. W., Cunningham, J. and Shoemaker, C. B. (1994) Cloning, characterization, and functional expression of cDNAs encoding glucose transporter proteins from the human parasite *Schistosoma mansoni*. *J. Biol. Chem.* **269**: 4247–4253.

69. Fripp, P. J. (1967) The sites of [^{14}C]-glucose assimilation in *Schistosoma haematobium*. *Comp. Biochem. Physiol.* **23**: 893–898.
70. Cornford, E. M. and Fitzpatrick, A. M. (1985) The mechanism and rate of glucose transfer from male to female schistosomes. *Mol. Biochem. Parasitol.* **17**: 131–141.
71. Isseroff, H. and Read, C. P. (1969) Studies in membrane transport. VI. Absorption of amino acids by fascioliid trematodes. *Comp. Biochem. Physiol.* **30**: 1153–1159.
72. Chappell, L. H. (1974) Methionine uptake by larval and adult *Schistosoma mansoni*. *Int. J. Parasitol.* **4**: 361–369.
73. Rumjanek, F. D. and Simpson, A. J. G. (1980) The incorporation and utilization of radiolabeled lipids by adult *Schistosoma mansoni in vitro*. *Mol. Biochem. Parasitol.* **1**: 31–44.
74. Moffat, D. and Kusel, J. R. (1992) Fluorescent lipid uptake and transport in adult *Schistosoma mansoni*. *Parasitology* **105**: 81–89.
75. Haseeb, M. A., Eveland, L. K. and Fried, B. (1985) The uptake, localization and transfer of [^{14}C]-cholesterol in *Schistosoma mansoni* males and females maintained *in vitro*. *Comp. Biochem. Physiol.* **82A**: 421–423.
76. Pax, R. A., Guo-Zhong, C. and Bennett, J. L. (1987) *Schistosoma mansoni*: measurement of Na$^+$ ion activity in the tegument and the extracellular spaces using ion-selective microelectrodes. *Exp. Parasitol.* **64**: 219–227.
77. Brodie, D. A. and Podesta, R. B. (1981) ^{3}HOH-Osmotic water fluxes and ultrastructure of an epithelial syncytium. *J. Membr. Biol.* **61**: 107–114.
78. Wilson, R. A. and Webster, L. A. (1974) Protonephridia. *Biol. Rev.* **49**: 127–160.
79. Bricker, C. S., Pax, R. A. and Bennett, J. L. (1982) Microelectrode studies of the tegument and subtegumental compartments of male *Schistosoma manson*. I. Anatomical location of sources of electrical potentials. *Parasitology* **85**: 149–161.
80. Day, T. A., Bennett, J. L. and Pax, R. A. (1992) *Schistosoma mansoni*: patch-clamp study of a nonselective cation channel in the outer tegumental membrane of females. *Exp. Parasitol.* **74**: 348–356.
81. Pax, R. A. and Bennett, J. L. (1990) Studies on intrategumental pH and its regulation in adult male *Schistosoma mansoni*. *Parasitology* **101**: 219–226.
82. Halton, D. W. (1967) Observations on the nutrition of digenetic trematodes. *Parasitology* **57**: 639–660.
83. Smith, A. M., Dowd, A. J., McGonigle, S. *et al.* (1993) Purification of a cathepsin L-like proteinase secreted by adult *Fasciola hepatica*. *Mol. Biochem. Parasitol.* **62**: 1–8.
84. Erasmus, D. A. (1977) The host–parasite interface of trematodes. *Adv. Parasitol.* **15**: 201–242.
85. Clegg, J. A. and Smithers, R. (1972) The effects of immune Rhesus monkey serum on schistosomula of *Schistosoma mansoni* during cultivation *in vitro*. *Int. J. Parasitol.* **2**: 79–98.
86. Bogitsh, B. J. and Carter, O. S. (1977) *Schistosoma mansoni*: ultrastructural studies on the esophageal secretory granules. *J. Parasitol.* **63**: 681–686.
87. Davis, A. H., Nanduri, J. and Watson, D. C. (1987) Cloning and gene expression of *Schistosoma mansoni* protease. *J. Biol. Chem.* **262**: 12851–12855.
88a. Cheng, T. C. (1977) The control of parasite: the role of the parasite. Uptake mechanisms and metabolic interference in parasites as related to chemotherapy. *Proc. Helminth. Soc. Wash.* **44**: 2–17.
88b. Bird, A. F. (1991) *The Structure of Nematodes*, 2nd edn, Academic Press, New York.
89. Blaxter, M. L., Page, A. P., Rudin, W. and Maizels, R. M. (1992) Nematode surface coats: actively evading immunity. *Parasitol. Today* **8**: 243–247.
90. Cox, G. N. (1992) Molecular and biochemical aspects of nematode collagens. *J. Parasitol.* **78**: 1–15.
91. Howells, R. E. (1980) Filariae: dynamics of the surface. In: *The Host Invader Interplay* (ed. Van den Bossche, H.), Elsevier/North Holland, Amsterdam, pp. 69–84.
92. Johnstone, I. L. (1994) The cuticle of the nematode *Caenorhabditis elegans*: a complex collagen structure. *Bioessays* **16**: 171–178.

93. Maizels, R. M., Blaxter, M. L. and Selkirk, M. E. (1993) Forms and functions of nematode surfaces. *Exp. Parasitol.* **77**: 380–384.
94. Politz, S. M. and Philipp, M. (1992) *Caenorhabditis elegans* as a model for parasitic nematodes: a focus on the cuticle. *Parasitol. Today* **8**: 6–12.
95. Preston-Meek, C. M. and Pritchard, D. I. (1991) Synthesis and replacement of nematode cuticle components. In: *Parasitic Nematodes—Antigens, Membranes and Genes* (ed. Kennedy, M. W.), Taylor and Francis, London, pp. 84–94.
96. Lee, D. L. (1970) Moulting in nematodes: the formation of the adult cuticle during the final moult of *Nippostrongylus brasiliensis*. *Tissue Cell* **2**: 139–153.
97. Bird, A. F. (1980) The nematode cuticle and its surface. In: *Nematodes as Biological Models*, Vol. 2 (ed. Zuckerman, B. M.), Academic Press, New York, pp. 213–236.
98. Wright, K. A. (1987) The nematode's cuticle, its surface, and the epidermis: function, analogy, and homology—A current consensus. *J. Parasitol.* **73**: 1077–1083.
99. Murrell, K. D., Graham, C. E. and McGreevy, M. (1983) *Strongyloides ratti* and *Trichinella spiralis*: net charge of epicuticle. *Exp. Parasitol.* **55**: 331–339.
100a. Kennedy, M. W., Foley, M., Kuo, Y. M., Kusel, J. R. and Garland, P. B. (1987) Biophysical properties of the surface lipid of parasitic nematodes. *Mol. Biochem. Parasitol.* **22**: 233–240.
100b. McLaren, D. J. (1980) *Schistosoma mansoni*: the parasite surface in relation to host immunity. In: *Tropical Medicine Research Studies* (ed. Brown, K. N.), John Wiley, Chichester.
101. Lee, D. L. and Atkinson, H. J. (1977) *Physiology of Nematodes*. Columbia University Press, New York.
102. Ho, N. F. H., Geary, T. G., Raub, T. J., Barsuhn, C. L. and Thompson, D. P. (1990) Biophysical transport properties of the cuticle of *Ascaris suum*. *Mol. Biochem. Parasitol.* **41**: 153–166.
103. Kramer, J. M. (1994) Structures and functions of collagens in *Caenorhabditis elegans*. *FASEB J.* **8**: 329–336.
104. Fujimoto, D. and Kenaya, S. (1973) Cuticulin: a non-collagen structural protein from *Ascaris* cuticle. *Arch. Biochem. Biophys.* **157**: 1–6.
105. Fetterer, R. H., Rhoads, M. L. and Urban, J. F., Jr (1993) Synthesis of tyrosine-derived cross-links in *Ascaris suum* cuticular proteins. *J. Parasitol.* **79**: 160–166.
106. Fujimoto, D., Horiuchi, K. and Hirama, M. (1981) Isotrityrosine, a new cross-linking amino acid isolated from Ascaris cuticle collagen. *Biochem. Biophys. Res. Commun.* **99**: 637–643.
107. Fetterer, R. H. (1989) The cuticular proteins from free-living and parasitic stages of *Haemonchus contortus*. I. Isolation and partial characterization. *Comp. Biochem. Physiol.* **94B**: 383–388.
108. Cox, G. N. (1990) Molecular biology of the cuticle collagen gene families of *Caenorhabditis elegans* and *Haemonchus contortus*. *Acta Trop.* **47**: 269–281.
109. Betschart, B. and Wyss, K. (1990) Analysis of the cuticular collagens of *Ascaris suum*. *Acta Trop.* **47**: 297–305.
110. Kingston, I. B. (1991) Collagen genes in *Ascaris*. In: *Parasitic Nematodes—Antigens, Membranes and Genes* (ed. Kennedy, M. W.), Taylor and Francis, London, pp. 66–83.
111. Cain, G. D. and Lutz, T. W. (1990) Extraction and characterization of neutral lipids from the cuticle of *Ascaris suum*. Abstracts, 65th Annual Meeting, American Society of Parasitologists, June 26–30, 1990, E. Lansing, MI.
112. Cox, G. N., Shamansky, L. M. and Boisvenue, R. J. (1990) *Haemonchus contortus*: evidence that the 3A3 collagen gene is a member of an evolutionarily conserved family of nematode cuticle collagens. *Exp. Parasitol.* **70**: 175–185.
113. Sinhgh, R. N. and Sulston, J. E. (1978) Some observations on moulting in *Caenorhabditis elegans*. *Nematologica* **24**: 63–71.
114. Chvapil, M., Boucek, M. and Erlich, E. (1970) Differences in the protocollagen hydroxylase activities from *Ascaris* muscle and hypodermis. *Arch. Biochem. Biophys.* **140**: 11–18.
115. Fetterer, R. H. and Wasiuta, M. (1987) *Ascaris suum*: partial isolation and characterization of hypodermis from adult female. *Exp. Parasitol.* **63**: 312–318.
116. Harris, J. E. and Crofton, H. D. (1957) Structure and function of the nematodes: internal pressure and the cuticular structure in *Ascaris*. *J. Exp. Biol.* **34**: 116–130.

117. Fusé, M., Davey, K. G. and Sommerville, R. I. (1993) Osmoregulation in the parasitic nematode *Pseudoterranova decipiens*. *J. Exp. Biol.* **175**: 127–142.

118. Fusé, M., Davey, K. G. and Sommerville, R. I. (1993) Water compartments and osmoregulation in the parasitic nematode *Pseudoterranova decipiens*. *J. Exp. Biol.* **175**: 143–152.

119. Raizen, D. M. and Avery, L. (1994) Electrical activity and behavior in the pharynx of *Caenorhabditis elegans*. *Neuron* **12**: 483–495.

120. Hayes, D. J. and Selwood, D. L. (1990) Glucose transport in *Acanthocheilonema viteae*. *Parasitology* **101**: 249–255.

121. Castro, G. A. and Fairbairn, D. (1969) Comparison of cuticular and intestinal absorption of glucose by adult *Ascaris lumbricoides*. *J. Parasitol.* **55**: 13–16.

122. Fleming, M. W. and Fetterer, R. H. (1984) *Ascaris suum*: partial isolation and characterization of hypodermis from the adult female. *Exp. Parasitol.* **63**: 312–318.

123. Geary, T. G., Sims, S. M., Thomas, E. M. *et al.* (1993) *Haemonchus contortus*: ivermectin-induced paralysis of the pharynx. *Exp. Parasitol.* **77**: 88–96.

124. Dass, P. D. and Donahue, M. J. (1986) γ-Glutamyl transpeptidase activity in *Ascaris suum*. *Mol. Biochem. Parasitol.* **20**: 233–236.

125. Pax, R. A., Geary, T. G., Bennett, J. L. and Thompson, D. P. (1995) *Ascaris suum*: characterization of transmural and hypodermal potentials. *Exp. Parasitol.* **80**: 85–97.

126. Hobson, A. D., Stephenson, W. and Eden, A. (1952) Studies on the physiology of *Ascaris lumbricoides*. II. The inorganic composition of the body fluid in relation to that of the environment. *J. Exp. Biol.* **29**: 22–29.

127. DeMello, W. C. and Tercafs, R. R. (1966) Ionic permeability of the *Ascaris* body wall. *Acta Physiol. Lat. Am.* **16**: 121–127.

128. Valkanov, M., Martin, R. J. and Dixon, D. M. (1994) The Ca-activated chloride channel of *Ascaris suum* conducts volatile fatty acids produced by anaerobic respiration: a patch clamp study. *J. Membr. Biol.* **138**: 133–141.

129. Sims, S. M., Magas, L. T., Barsuhn, C. L., Ho, N. F. H., Geary, T. G. and Thompson, D. P. (1992) Mechanisms of microenvironmental pH regulation in the cuticle of *Ascaris suum*. *Mol. Biochem. Parasitol.* **53**: 135–148.

130. Sims, S. M., Ho, N. F. H., Magas, L. T., Geary, T. G., Barsuhn, C. L. and Thompson, D. P. (1994) Biophysical model of the transcuticular excretion of organic acids, cuticle pH and buffer capacity in gastrointestinal nematodes. *J. Drug Target.* **2**: 1–8.

131. Hobson, A. D. (1948) The physiology and cultivation in artificial media of nematodes parasitic in the alimentary tract of animals. *Parasitology* **38**: 183–227.

132. Verhoeven, H. L. E., Willemsens, G. and Van den Bossche, H. (1980) Uptake and distribution of levamisole in *Ascaris suum*. In: *Biochemistry of Parasites and Host–Parasite Relationships* (ed. Van den Bossche, H.), Elsevier/North-Holland Biomedical Press, Amsterdam, pp. 573–579.

133. Court, J. P., Murgatroyd, R. C., Livingstone, D. and Rahr, E. (1988) Physicochemical characteristics of non-electrolytes and their uptake by *Brugia pahangi* and *Dipetalonema viteae*. *Mol. Biochem. Parasitol.* **27**: 101–108.

134. Bocasch, W. D., Cornford, E. M. and Oldendorf, W. H. (1981) *Schistosoma mansoni*: correlation between lipid partition coefficient and transintegumental uptake of non-electrolytes. *Exp. Parasitol.* **52**: 396–403.

135. Fetterer, R. H. (1986) Transcuticular solute movement in parasitic nematodes: relationship between non-polar solute transport and partition coefficient. *Comp. Biochem. Physiol.* **84A**: 461–466.

136. Ross, M. H. and Grant, L. (1968) On the structural integrity of basement membrane. *Exp. Cell Res.* **50**: 277–285.

137. Maki, J. and Yanagisawa, T. (1986) Demonstration of carboxyl and thiol protease activities in adult *Schistosoma mansoni*, *Dirofilaria immitis*, *Angiostrongylus cantonensis* and *Ascaris suum*. *J. Helminthol.* **60**: 31–37.

138. Pratt, D., Cox, G. N., Milhausen, M. J. and Boisvenue, R. J. (1990) A developmentally regulated cysteine protease gene family in *Haemonchus contortus*. *Mol. Biochem. Parasitol.* **43**: 181–192.

139. Ray, C. and McKerrow, J. H. (1992) Gut-specific and developmental expression of a *Caenorhabditis elegans* cysteine protease gene. *Mol. Biochem. Parasitol.* **51**: 239–250.
140. Beh, C. T., Ferrari, D. C., Chung, M. A. and McGhee, J. D. (1991) An acid phosphatase as a biochemical marker for intestinal development in the nematode *Caenorhabditis elegans*. *Dev. Biol.* **147**: 133–143.
141. von Brand, T. (1973) *Biochemistry of Parasites*, 2nd edn, Academic Press, New York.
142. Homandberg, G. A., Litwiller, R. D. and Peanasky, R. J. (1989) Carboxypeptidase inhibitors from *Ascaris suum*: the primary structure. *Arch. Biochem. Biophys.* **270**: 153–161.
143. Martzen, M. R., McMullen, B. A., Smith, N. E., Fujikawa, K. and Peanasky, R. J. (1990) Primary structure of the major pepsin inhibitor from the intestinal parasitic nematode *Ascaris suum. Biochemistry* **29**: 7366–7372.
144. Noelken, M. E., Wisdom, B. J., Jr, Dean, D. C., Hung, C.-H. and Hudson, B. G. (1986) Intestinal basement membrane of *Ascaris suum*. Molecular organization and properties of the collagen molecules. *J. Biol. Chem.* **261**: 4706–4714.
145. Aamodt, E. J., Chung, M. A. and McGhee, J. D. (1991) Spatial control of gut-specific gene expression during *Caenorhabditis elegans* development. *Science* **252**: 579–582.
146. Goldstein, B. (1992) Induction of gut in *Caenorhabditis elegans* embryos. *Nature* **357**: 255–257.
147. Sanhueza, P. (1968) Absorption of carbohydrates by intestine of *Ascaris lumbricoides in vitro. Nature* **219**: 1062–1063.
148. Beames, C. G. (1971) Movement of hexoses across the midgut of *Ascaris. J. Parasitol.* **57**: 97–102.
149. Barrett, J. (1981) Amino acid metabolism in helminths. *Adv. Parasitol.* **30**: 39–105.
150. Beames, C. G. and King, G. A. (1972) Factors influencing the movement of materials across the intestine of *Ascaris*. In: *Comparative Biochemistry of Parasites* (ed. Van den Bossche, H.), Academic Press, New York, pp. 275–282.
151. Beames, C. G., Bailey, H. H., Rock, C. O. and Schanbacher, C. M. (1974) Movement of cholesterol and β-sitosterol across the intestine of *Ascaris suum. Comp. Biochem. Physiol.* **47**A: 880–888.
152. Harpur, R. P. and Popkin, J. S. (1973) Intestinal fluid transport: studies with the gut of *Ascaris lumbricoides. Can. J. Physiol. Pharmacol.* **51**: 79–90.
153. Merz, J. M. (1977) Short-circuit current and solute fluxes across the gut epithelium of *Ascaris suum*. Unpublished PhD. dissertation, Oklahoma State University, Stillwater, OK.
154. Wright, D. J. and Newall, D. R. (1976) Osmotic and ionic regulation in nematodes. In: *Nematodes as Biological Models* (ed. Zuckerman, B. M.), Academic Press, New York.

13 Protozoan Cell Organelles

MARILYN PARSONS

Seattle Biomedical Research Institute and Department of Pathobiology, University of Washington, Seattle, WA, USA

SUMMARY

Although all eukaryotes compartmentalize functions within the cell, parasitic protozoa have evolved unique features with respect to organellar structure and function. Most highly conserved appear to be the organelles involved in protein trafficking, with the features and functions of parasite vesicular organelles generally paralleling those of host cells. Much less conserved are cytoskeletal structures: not only do parasite cytoskeletons differ from those of higher eukaryotes, but they also differ widely from one another. Organellar compartmentalization of energy metabolism is seen in most, but not all, protozoa. Mitochondria, when present, are usually at the level of one per cell, and often show stage-regulated respiratory activity. Glycosomes, the site of glycolysis in kinetoplastids, appear to be related to the peroxisomes of higher eukaryotes based on the presence of certain metabolic pathways and similarites in signals directing protein import. The hydrogenosome of trichomonads shows some features similar to mitochondria with respect to protein import, but differs dramatically in its intrinsic metabolic pathways. The study of parasite organelles has revealed that the problem of compartmentalizing function within the eukaryotic cell has been solved in a variety of ways and has suggested that the distinctive, yet essential, properties and components of these organelles may make them useful targets for anti-parasitic agents.

1. PROTOZOAL PARASITES SHOW MANY VARIATIONS FROM THE STANDARD EUKARYOTIC CELL ARCHITECTURE

Eukaryotes differ from prokaryotes in their subcellular organization. Eukaryotes all possess a nucleus, cytoskeleton and endomembrane system. Some organelles ubiquitous

Biochemistry and Molecular Biology of Parasites
ISBN 0-12-473345-X

233

in higher taxa, such as peroxisomes and even mitochondria, surprisingly, are not present in all eukaryotes. Most cytologically simpler organisms probably diverged early in the eukaryotic lineage. The most divergent eukaryotes examined thus far are Diplomonads (*Giardia*) and Microspora, both of which have relatively little subcellular organization and lack mitochondria and hydrogenosomes (1). More recently diverged protozoans such as the Kinetoplastida and Apicomplexa have a distinct organization as elaborate as that of the metazoan cell. In contrast to protozoans, parasitic worms evolved more recently and have the basic cellular body plan common to animal cells, including the mitochondrion/cytosol compartmentalization. Thus, they are more similar to their hosts than to most parasitic protozoa. This chapter will focus on differences between host and parasite organelles, and therefore will concentrate on the protozoa. Characteristics of organelles of higher eukaryotes will not be discussed.

Information about protozoan organelles is relatively limited, and is restricted to a few organisms studied either because of their importance as parasites or their utility as model systems. Fewer specific biological reagents, such as monoclonal antibodies, inhibitors, and cloned genes, are available for the study of subcellular compartmentalization in parasites than in higher eukaryotes. However, there are many outstanding ultrastructural analyses of protozoan parasites. Metabolic pathways contained within membrane-bounded organelles have been studied in some detail, and molecular cloning studies have provided additional information. This chapter focuses on selected organisms and organelle systems which have been studied in some detail and which point out unique aspects of parasite organelles; it is not intended to be encyclopedic.

2. THE CYTOSKELETON

All eukaryotes possess a cytoskeleton, and parasitic protozoa are no exception. Cytoskeletal structures, particularly microtubular elements, classically have been among the major features used in defining the taxonomic relationships of protozoa. The cytoskeleton of higher eukaryotes contains three major types of protein filaments: intermediate filaments, actin filaments and microtubules. Intermediate filaments are composed of fibrous, overlapping proteins; those found in mammalian cells (e.g. keratins and lamins) have not yet been studied in protozoa.

Actin is the most abundant protein in most mammalian cells, where it underlies the plasma membrane to form the cortex that helps to define the shape of the cell. Actin is not an abundant protein in many parasitic protozoa, although there are low levels of proteins which react with heterologous anti-actin antibodies or phalloidin, a drug which binds actin filaments (2). In *Trypanosoma brucei* an actin gene has been identified, but the level of protein product is apparently low (3). On the other hand, ameboid protists possess high levels of actin (4,5). Actin is thought to be important in phagocytosis by *Entamoeba histolytica*, and in concert with myosin II, play a role in parasite movement (5). *Plasmodium falciparum* has two distinct genes which encode actins significantly divergent from vertebrate actins. One of these is expressed only in sexual stages and appears to be more closely related to muscle than to cytoskeletal actin (actins of other unicellular eukaryotes show the reverse relationship) (6).

Microtubules form some of the most prominent structures of protozoa. They are composed of copolymerized α and β tubulin subunits, and are decorated with an array of microtubule-associated proteins (MAPs). Tubulin genes have been cloned from several parasitic protozoa and are highly conserved, even among these diverse organisms. For example, *T. brucei* and mammalian tubulins show about 80% sequence identity at the amino acid level (7).

2.1. The Mitotic Spindles of Protozoa Shows Diverse Organization

The organization of most protozoan mitotic spindles differs from that of mammalian cells (8). In kinetoplastids, mitosis is closed (i.e., the nuclear envelope never breaks down) and the microtubular spindle forms within the nucleus. No spindle pole body, such as a centriole-based organizing center, has been identified in kinetoplastids (8–10). The centrioles in trypanosomes, i.e. the basal bodies, do not participate in partitioning of the trypanosome nuclear genome—but rather in the segregation of the mitochondrial genome (11) (see section 4.1).

In most other parasitic protozoa mitosis is also closed. As shown in Fig. 13.1, the trichomonad spindle forms extranuclearly and remains so throughout the closed mitosis (12). The spindle pole bodies, termed attractophores, are also extranuclear, and appear to derive from the basal body. The Apicomplexa have a cone-shaped microtubular hemi-spindle, termed the centrocone, which is just outside of the nuclear envelope (13). The centrioles are located on top of the centrocones. Interestingly the coccidia and gregarines have only singlet microtubules in their centrioles, as opposed to the standard triplets, whereas in hematozoa the centrioles are replaced by dense bodies. The

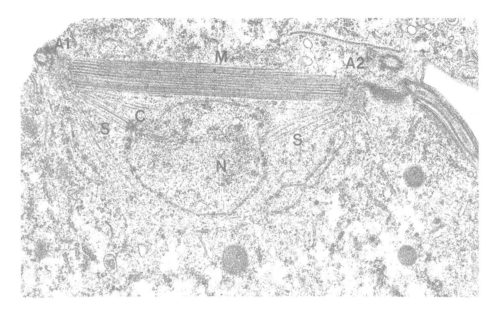

FIG. 13.1   The organization of the mitotic spindle of *T. vaginalis*. Two attractophores (A1, A2) are connected to the centriolar regions (C) through a microtubular spindle (Fch). Microtubules (Fc) between the two attractophores elongate, separating the attractophores and attached structures. (Reprinted with permission from ref. 12.)

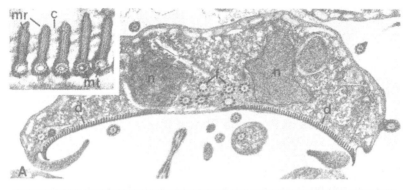

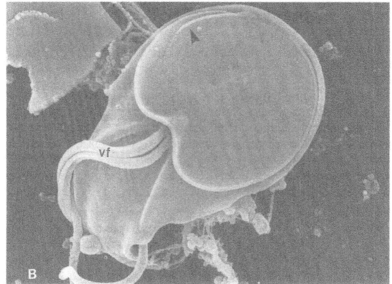

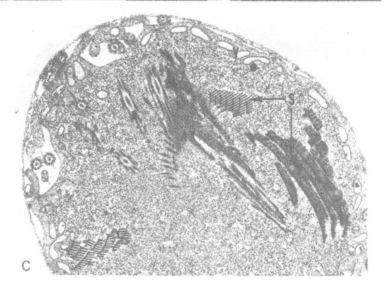

centrocone microtubules elongate and pierce the nuclear envelope, and then attach to the chromosomes. As analysis of mitotic structures in parasites is carried on the molecular level, features both conserved and distinct from those of the host will undoubtedly be revealed.

2.2. Giardia Possesses a Unique Structure, the Ventral Disk

The microtubular cytoskeleton of *Giardia* has several components, including four pairs of flagella, two specific bands of microtubules (the funis and the median body), and the ventral disk (14) (Fig. 13.2B). The ventral disk is unique to the genus *Giardia* and is important in attachment of the trophozoite to the intestinal wall. The disk is composed of a sheet of microtubules underlying the plasma membrane (Fig. 13.2A). A microribbon extends from each microtubule dorsally into the cytoplasm. At the circumference of the disk is a lateral crest (Fig. 13.2B), which probably contains contractile proteins, as revealed by staining with heterologous antibodies specific for actin, myosin, α-actinin and tropomyosin (15). Attachment to the host may involve these contractile proteins or the beating of the ventral flagella to create a suction force, in addition to possible protein–protein interactions (16, 17).

A group of disk proteins, the giardins, has been the subject of extensive studies (reviewed in ref 16). Two-dimensional gel analysis reveals at least 23 isoforms. Although their interrelationships are unclear, some giardins share antigenic determinants. Genes encoding the α, β, and γ giardin subfamilies have been cloned; these subfamilies show no obvious sequence similarities (18). The giardins are predicted to have a high α-helix content, and the β-giardins are predicted to have the coiled-coil structure of α-type fibrous proteins. The giardins have been localized to the microribbons of the ventral disk and at least some of them assemble into filaments *in vitro*. As shown in Fig. 13.2C, the disk becomes fragmented in *Giardia* cysts (19). It is not clear whether these fragments are degraded and the disk synthesized *de novo*, or whether the fragments are reassembled or used as a template for assembly upon excystation.

2.3. Trichomonads Possess Several Unusual Cytoskeletal Features

In addition to their novel mitotic spindle, trichomonads possess several unusual cytoskeletal structures. Trichomonads have a rod-like bundle of microtubules, the axostyle, which extends from the apical end (where the flagella emerge) through and past the posterior of the cell body. The apical end of the axostyle is associated with a crescent-shaped microtubular structure called the pelta. The cross-striated non-microtubular structures that lie beneath the plasma membrane are called costae. Costae are

---

FIG. 13.2 The ventral disk of *Giardia*. A. Transmission electron micrograph of a trophozoite in cross-section, showing the ventral disk (d), flagella (f) and two nuclei (n). The inset at higher magnification shows the microtubules (mt) and microribbons (m) of the ventral disk, the latter extending into the cytoplasm (c). (Reprinted with permission from ref. 95.) B. Scanning electron micrograph of the ventral surface of *Giardia* trophozoite. The lateral crest almost completely circumscribes the disk, except where layers of microtubules overlap (arrow). vf, ventral flagella. Bar = 1 μM. (Reprinted with permission from ref. 15.) C. Electron micrographs of a *Giardia lamblia* cyst. Fragments of the ventral disk (S) are visible, in cross- and longitudinal section. V, vacuoles; A, axonemes; AX, axostyle. (Reprinted with permission from ref. 96.)

thought to be modified flagellar rootlets and are connected to basal bodies. Costae are composed of multiple proteins; at least five different species, ranging from 47 to 135 kDa, have been shown to be present in *T. foetus*. The characteristics of costae, including the presence of a preponderance of high-molecular-weight proteins, resistance to chaotropic reagents, and solubility, indicated that the costal proteins are quite different from other fine filament proteins such as the giardins. The costa and axostyle may play a role in motility.

2.4. The Cytoskeleton of Kinetoplastids Reveals Fundamental Differences in its Organization from Mammalian Cytoskeletons

The cytoskeleton of *T. brucei* has been more thoroughly studied than that of other parasitic protozoa. The major structural features include the axoneme and paraflagellar rod of the flagellum and a corset of microtubules underlying the plasma membrane. These subpellicular microtubules form an ordered, cross-linked array wrapping around the entire cell except in the area of the flagellar pocket (where the flagellum emerges). Since the flagellar pocket is the site of endo- and exocytosis, this gap in the microtubular array may allow vesicles access to the cell membrane. The cytoplasmic face of the array is smooth, but the membrane face is studded with microtubule associated proteins (MAPs) that presumably crosslink microtubules to each other and to the plasma membrane. Several distinct MAPs have been identified in *T. brucei*; these localize to different positions on the subpellicular array (e.g. the membrane face (20), the intermicrotubule cross-bridges (21), and the ends of microtubules (22)), suggesting diverse functions. Although these proteins may function analogously to MAPs of mammalian cells, mammalian homologs have not been identified.

As in most other organisms, the flagellar axoneme of kinetoplastids has a typical 9 + 2 structure and arises from a centriolar basal body. Trypanosomes and other unflagellated kinetoplastids contain a second basal body; in related organisms a second flagellum arises from this basal body (23). Along the shaft of the flagellum, parallel to the axoneme, is a structure unique to the Kinetoplastida, the paraflagellar rod. In *T. brucei*, its major constituents are two almost identical fibrillar proteins that are unrelated to microtubules. The flagellum of trypanosomes remains associated with the cell body (i.e., it is recurrent) until it emerges at the anterior end of the cell (the cells swim with the tip of the flagellum forward). A protein composed of 68 amino acid repeats has been localized to fibers connecting the cell body and the paraflagellar rod (20, 24). Another protein associated with the paraflagellar rod cross-reacts with antibodies to spectrin, a cytoskeletal protein of red blood cells (20). The role of the paraflagellar rod is not known. Apart from its role in motility, the flagellum may also function as an attachment organelle (23).

During differentiation in *T. brucei*, all of the cytoskeletal structures are retained, although the relative position of the flagellum changes. In *Leishmania* species, however, the amastigote stage possesses only a very short, non-motile flagellum lacking the paraflagellar rod (25) and has an abbreviated subpellicular array. At the molecular level this is reflected by a decreased level of tubulin biosynthesis in amastigotes, which may be mediated by alterations in mRNA abundance or isoforms and/or translational controls (26, 27).

FIG. 13.3 A sheet of microtubules from *T. brucei*, labeled with antibody to tyrosinylated-tubulin to detect newly assembled microtubules. Flags (or arrowheads) indicate short, labeled microtubules inserted between unlabeled, older tubules. (Reprinted with permission from ref. 97.)

The morphogenesis and organization of the trypanosome microtubular array has many unusual features. In mammalian interphase cells single microtubules radiate out through the cytoplasm from an organizing center containing centrioles that is near the nucleus. This contrasts with the ordered array of trypanosome microtubules, shown in Fig. 13.3. During mitosis, the cytoplasmic microtubules of mammalian cells are disassembled and a mitotic spindle is formed. The astral fibers eventually elongate to reform the daughter cytoplasmic microtubules. Trypanosomes handle the replication and partitioning of their cytoskeleton quite differently from mammalian cells (28). Using antibodies to tyrosinylated-tubulin, a marker for newly assembled microtubules, it has been possible to study the incorporation of new microtubules into the array (28) (Fig. 13.3). New microtubules are inserted between existing microtubules, and do not emanate from a centriolar microtubular organizing center. What organizes the growth of pellicular microtubules in trypanosomes is not clear, but MAPs of nearby tubules may be important in this process. The fact that these studies are even possible indicates that trypanosome microtubules are more stable than their rapidly polymerizing and depolymerizing mammalian counterparts. Indeed, the subpellicular array is highly stable, and is not disassembled even during mitosis. Many drugs, such as colchicine, which disrupt polymerization of mammalian microtubules are ineffective against trypanosome microtubules and vice versa (7, 29).

2.5 *Naegleria* Species: Developmental Control of the Flagellar Apparatus

The ameboid flagellate, *Naegleria*, transform from a dividing ameboid stage to a flagellated stage when starved (30). In the ameboid stage the cells have no evidence of microtubular structures related to motility, but they actively synthesize actin. Since transformation can be induced by culture conditions, it has been possible to establish the sequence of events in the transformation to the flagellated stage. Interestingly, the tubulin used to form the flagellar structures is not derived from the pre-existing pool, but rather is transcribed and translated *de novo*. Specific α-tubulin and β-tubulin genes appear to be coordinately induced along with a flagellar calmodulin gene (31, 32). The developmental sequence also includes repression of actin synthesis (30).

3. ORGANELLES INVOLVED IN MACROMOLECULAR TRAFFICKING

As other eukaryotic cells, protozoan parasites possess organelles and vesicles involved in transport of macromolecules to and from the cell surface. Most organisms contain structures analogous to the endoplasmic reticulum, Golgi, endosomes, lysosomes and endolysosomes. Ultrastructural, cytochemical, and cell fractionation studies suggest that in most aspects vesicular functions resemble those found in higher eukaryotes. However, studies are far less complete than those performed on mammalian cell models and significant differences may yet be discovered. In addition to the pathways discussed here, *Plasmodia* and many other intracellular parasites possess another protein sorting pathway followed by parasite-encoded proteins which become localized in the host cell membrane or cytoskeleton (33). This pathway is discussed, along with the parasite plasma membranes, in Chapters 8, 11 and 16. *E. histolytica* possesses intracellular organelles morphologically and biochemically similar to mammalian lyosomes and endosomes. Lysosomal function in *E. histolytica* has been reviewed (34).

3.1. Many Parasites Possess an Inconspicuous Golgi Apparatus

The Golgi apparatus has been difficult to detect in cultured, non-encysting *Giardia* trophozoites. This led to the description of *Giardia* and related species as lacking a Golgi apparatus. However, the application of improved microscopic techniques led to the observation of perinuclear flattened stacks of membrane (35). During encystation, cyst antigens can be observed in these Golgi-like structures and also in encystation-specific vesicles. At the same time acid phosphatase (a lysosomal marker in mammalian cells) can also be seen in the Golgi and in the small lysosome-like vacuoles underlying the plasma membrane, but not in the encystation-specific vesicles. Hydrolytic enzymes within these lysosome-like vacuoles have been hypothesized to be involved in excystation (36). The major plasma membrane antigen, which is constitutively expressed, is not transported to the cell membrane by the encystation specific vesicles but by a distinct, uncharacterized pathway (35). Thus even this divergent eukaryote has the capability of sorting molecules to at least three destinations: lysosomes, secretory vesicles and plasma membrane.

   The Golgi apparatus has been difficult to detect in *Entamoeba*, but recent studies indicate its presence (34). Similarly, the Golgi of *Plasmodium* species has been elusive, however, Golgi-like vesicular elements have been observed, apparently budding off the endoplasmic reticulum (37). Furthermore, treatment of *P. falciparum* with brefeldin A, an inhibitor of secretory effort to the post-Golgi compartment blocks protein secretion and development (38). Conversely, in trichomonads and related organisms, the Golgi apparatus is highly developed. Located adjacent to the basal body and associated filaments (39), this modified Golgi has been called the parabasal body.

3.2. The Vesicular Transport System of Kinetoplastida
Resembles that of Mammalian Cells

Smooth and rough endoplasmic reticulum have been identified in kinetoplastids, where their functions are likely to be similar to that in other organisms. Trypanosomes possess a homolog of the molecular chaperone BiP which has been used as a specific marker for the endoplasmic reticulum (40). Proteins are retained in the endoplasmic reticulum by virtue of a C-terminal tetrapeptide, typically KDEL in mammalian cells. Trypano-some BiP ends in MDDL, whereas another putative endoplasmic reticulum protein ends in KQDL. Thus it appears that the retention signals in these organisms have diverged significantly from those of higher eukaryotes.

The Golgi is located between the nucleus and the flagellar pocket, the site of endo- and exocytosis (10, 25, 41). Analysis of the major surface antigen of *T. brucei* blood-forms, variant surface glycoprotein, has demonstrated that it apparently follows the typical pathway used by integral membrane proteins in other cells (41) (see Chapter 11). Immunogold localization reveals its presence in rough endoplasmic reticulum, through all the cisternae of the Golgi, in tubulovesicular elements adjacent to the trans-Golgi, as well as on the surface of the cell. Variant surface glycoprotein appears to be deposited on the plasma membrane at the flagellar pocket, where the subpellicular microtubules are absent. Evidence for active recycling of the protein was obtained (42). Cell fractionation studies indicate that, as in mammalian cells, various protein process-ing activities, such as glycosyltransferases are located in the Golgi and endoplasmic reticulum (43). The ionophore monensin dilates the Golgi of kinetoplastids (as it does in mammalian cells) and probably alters its function, but does not arrest the transport of variant surface glycoprotein (41, 44).

Microscopic analysis of cells incubated with proteins, such as albumin or transferrin, coupled to colloidal gold or horseradish peroxidase suggests that endocytotic processes are probably similar to those in mammalian cells. The flagellar pocket can be thought of as a large, permanent coated pit (45). Endocytotic vesicles bud off the flagellar pocket; the ingested proteins are found in these as well as in tubular network similar to mammalian endosomes, and in round lysosome-like vesicles (46). The coated vesicles are somewhat larger than mammalian coated vesicles, and lack proteins cross-reacting with anti-clathrin antibodies (47). Receptor-mediated endocytosis has been observed (48). Several hydrolytic enzymes are located in the acidic lysosome-like organelles of *T. brucei*, including a cysteine protease and acid phosphatase (49, 50). Bloodform *T. brucei* possess a set of glycoproteins broadly distributed among lysosomes and endosomes as are mammalian LAMPs (lysosome-associated proteins) (45). In the case of the African trypanosome *T. congolense*, bloodforms and metacyclic forms appear to be much more actively endocytosing than procyclic forms and epimastigotes (51).

Amastigotes of *L. mexicana* and *L. amazonensis* possess an abundance of large lysosome-like organelles termed megasomes (Fig. 13.4) (52). Few megasomes, if any, are present in *L. donovani* and *L. major*. Reconstruction from serial sections indicates that megasomes may take up as much as 15% of the volume of *L. amazonensis* amastigotes (53). Cytochemical and antibody staining indicates that these megasomes contain hydrolytic enzymes such as cysteine protease and arylsulfatase. The activities of the cysteine protease and arylsulfatase are highly stage-regulated, being as much as 40 times higher in amastigote than promastigote forms (52). RNAase and DNAase are also much more abundant in amastigotes. These data should be remembered when inter-preting studies concerning the stage-regulation of various proteins and transcripts.

3.3. *Plasmodia* Digest Host Cytoplasm Within Vesicles

Various, apparently membrane-bounded, structures which function in digestion are observed in electron micrographs of *Plasmodium* species. Among these are the food vacuoles, residual bodies and endocytotic vesicles of various sizes.

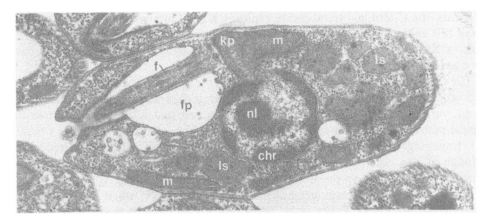

FIG. 13.4 Longitudinal section of an *L. mexicana* amastigote. Note the abundant megasomes (ls) and the kinetoplast (kp) at the base of the short flagellum (f). Also shown and identified are the flagellar pocket (fp), the nucleolus (nl), chromatin (chr), and two sections through the single mitochondrion (m). (Reprinted with permission from ref. 53.)

Vesicles containing host cell cytosol can arise by pinching off distal portions of the cytostome (a long, sometimes branched invagination of the plasma membrane), or by micropinocytosis from the cytostome or the cell membrane (the latter particularly in early trophozoites) (54, 55). Such vesicles are surrounded by two membranes, the outer derived from the parasite plasma membrane and the inner from the parasitophorous vacuole membrane that surrounds the parasite in the red blood cell. The inner membrane is eventually lost. The vesicles are the location of digestion of hemoglobin and, presumably, other host-derived nutrients. Hydrolytic enzymes including aminopeptidase, acid phosphatase and aspartic protease have been localized within the vesicles (56, 57). The intravesicular pH is low, approximately 4.8 (58). Thus these vesicles appear to be homologous to the endosome–lysosome system of higher organisms. As the contents of the vesicles is digested, the density of the matrix decreases concurrent with the growth of a crystal of hemozoin (malaria pigment, the residue left after digestion of hemoglobin) (54, 55) (Fig. 13.5). Vesicles fuse to form the residual body, which is left behind when the merozoites are released.

Digestion of host cytosol within the food vacuoles appears to be essential for parasite growth. Treatment of the parasite cultures with inhibitors of the major cysteine protease blocks parasite proliferation (59). This appears to be the result of action within the food vacuole, since the organisms show abnormal vacuole morphology and undigested hemoglobin. The antimalarial drug cloroquine also probably acts at the level of the food vacuole.

4. MANY, BUT NOT ALL, PARASITIC PROTOZOA COMPARTMENTALIZE PATHWAYS OF ENERGY METABOLISM WITHIN ORGANELLES

The more cytologically simple protozoan parasites perform most or all steps of energy metabolism within the cytosol. These parasites include *Giardia* and *Entamoeba* and the Microspora such as *Enterozoon*. More complex parasites possess one or more organelles involved in the generation of energy. Glycosomes are restricted to the Kineto-

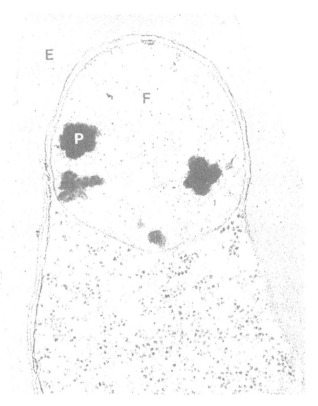

FIG. 13.5 The food vacuole of *P. gallinaceum*. Note the decreased density of the food vacuole (F) as compared to the erythrocyte cytoplasm. Also indicated are malaria pigment particles (P) (hemozoin). (Reprinted with permission from ref. 54.)

plastida, hydrogenosomes to trichomonad flagellates and certain ciliates, whereas mitochondria are found in kinetoplastids, apicomplexans and cytologically complex amebae. Pathways of energy metabolism in aerobic and anaerobic protists are detailed in Chapters 2 and 3 respectively.

Unlike mammalian cells which possess hundreds of mitochondria, protozoans with mitochondria often possess only one. The cristae are usually tubular rather than discoid. Cytochrome-mediated respiration in protozoan mitochondria appears to be fairly typical, and involves gene products specified by the nuclear and mitochondrial genomes. Only a few nuclear-encoded mitochondrial proteins have been studied in any parasitic protozoa. Thus far there is no reason to think that the pathway for mitochondrial import of nuclear-encoded proteins differs substantially in these organisms from those described in higher eukaryotes.

4.1. Kinetoplastids Have Evolved Mechanisms to Replicate and Segregate Their Unusual Mitochondrial Genome

The class Kinetoplastida is named after its unusual mitochondrial feature, the kineto-plast (Fig. 13.4). The kinetoplast is the organized network of mitochondrial DNA (or kDNA) which is localized to one end of the single mitochondrion, near the flagellar pocket. This array, which is visible microscopically, is composed of a two classes of molecules: the maxicircles and the minicircles (60, 61). In *T. brucei* there are approximately 50 maxicircles and 5000–10000 minicircles in the catenated network. The maxicircles are homogeneous, and are the homologs of other mitochondrial genomes. Many maxicircle transcripts are post-transcriptionally altered through a process known as RNA editing (see Chapter 11). Much of the information required for RNA editing is specified by small transcripts called guide RNAs arising from the minicircles. Minicircles show a high degree of sequence diversity in many species, but in others are more homogeneous. The rapid evolution of minicircle sequences, coupled with their high abundance, has led several workers to use minicircle sequences for detection and identification of parasites (62).

The presence of a single mitochondrion, as opposed to the hundreds in mammalian cells, means that the replication and partitioning of the kDNA must be carefully controlled. In mammalian cells mitochondrial DNA synthesis occurs throughout the cell cycle. In *T. brucei*, mitochondrial DNA synthesis commences about the time of nuclear DNA synthesis, just after duplication of the basal bodies (63). Minicircles are removed from the network prior to replication and then reattached (60). Newly replicated circles are punctuated by nicks or gaps, possibly informing the replication machinery that they have already been copied. The mitochondrion divides before the nucleus, so there is a period when the cell has two kinetoplasts, but only one nucleus (although the relative timing of nuclear and kinetoplast division may vary in different species). Evidence has been obtained indicating that basal bodies are important in segregation of the mitochondrial genome (64). The kinetoplast lies adjacent to the basal bodies; inhibition of the movement of basal bodies with the antimicrotubular drug ansamitocin blocks the segregation of the kDNA. Furthermore, the basal bodies appear to be physically linked to kDNA since complexes containing both can be isolated (Fig. 13.6).

4.2. *Plasmodia* Possess Two Organellar Genomes

Recent studies have shown that *P. falciparum* also possesses a single mitochondrion during its erythrocytic stages (65, 66) (Fig. 13.7). This mitochondrion enlarges and becomes multilobed as development proceeds. Eventually each daughter merozoite receives a lobe of the mitochrondrion, presumably complete with its own copy of the mitochondrial genome.

Two organellar genomes have been described in *P. falciparum* (67). One is a 6 kb linear DNA which is tandemly arrayed, and found in many other species of *Plasmodia*. A similarly sized extrachromosomal element is also present in *Theileria* species and probably in other Apicomplexa too (67). The second is a 35 kb circuit DNA, which has been studied in detail only in *P. falciparum*, but is probably present in other *Plasmodia* and *Toxoplasma* as well. Cross-hybridization studies reveal only very limited homology between the 6 kb and 35 kb molecules. The 6 kb elements contains two genes which are found in all mitochondrial genomes examined thus far and which are components of the respiratory chain: cytochrome *b* and cytochrome oxidase subunit I (67). It also encodes cytochrome oxidase subunit III and possesses sequences which appear to correspond to fragments of rRNAs. Some of the rRNA fragments are predicted to be transcribed from one strand and some from the other; they are jumbled and intersper-

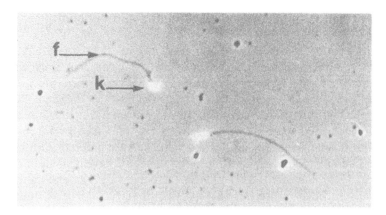

FIG. 13.6 Kinetoplasts of *T. brucei* connected to basal bodies. Flagella were prepared by detergent lysis and cytoskeletal depolymerization. The flagellar are viewed by phase contrast and the kDNA by DAPI fluorescence. DAPI is a fluorescent dye which binds to DNA. (Reprinted with permission from ref. 64.)

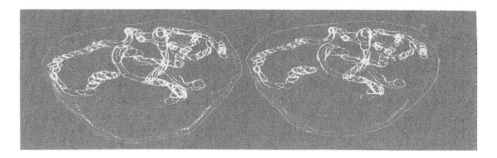

FIG. 13.7 Stereographic presentation of the three-dimensional reconstruction of the mitochondrion of a *P. falciparum* trophozoite. (Reprinted with permission from ref. 66.)

sed among the protein coding genes and each other. The predicted secondary structures of the fragments correspond to those of the highly conserved core sequences of *E. coli* rRNA (67), suggesting the possibility that these fragments participate in the formation of a functional ribosome.

Mitochondial genomes are usually circular and this, in part, led to the assumption that the 35 kb circle was the mitochondrial genome. Much of the circle has been sequenced and no genes encoding components of the mitochondrial electron transport chain have been found. The circle does encode certain subunits of a bacteria-like RNA polymerase, some ribosomal proteins, tRNAs, and ribosomal RNAs (the last in the form of an inverted repeat). These characteristics are not typical of mitochondrial genomes, although they are not uncommon in chloroplast genomes (67).

These initial studies indicate that the 6 kb molecule is the mitochondrial genome. If borne out, it would be the most compact mitochondrial genome described to date. The subcellular location of the 35 kb molecule remains a mystery. One possibility is that it is within the 'spherical body', a double-membrane enclosed organelle which is closely apposed to the mitochondrion in electron micrographs (68). The function of the spherical body is unknown.

4.3. Stage-regulation of Mitochondrial Structure and Function

Mitochondrial activity is developmentally regulated in *T. brucei* (9, 69). In slender bloodforms the mitochondrion is achristate and the citric acid cycle and cytochrome-mediated respiratory activity are absent. At this stage there are no detectable cytochromes, although an electrochemical potential can be detected (70). ATP is generated through glycolysis. As the bloodforms transform to intermediate and stumpy forms the mitochondrion appears to be in transition as well, since several characteristics associated with the developed procyclic mitochondrion are induced. Although cytochromes are still absent, the mitochondrion shows increased development of an electron-motive force (71), increased H^+-ATPase activity (72, 73), and alterations in mitochondrially encoded transcripts (61). At the ultrastructural level, tubular cristae appear (9, 23). In procyclic forms, the mitochondrion is highly ramified and has discoid cristae, a complete citric acid cycle, and a functional respiratory chain. At the molecular level, increased H^+-ATPase activity, and alterations in mitochondrially encoded transcript abundance, polyadenylation and RNA editing are observed.

Despite the apparent lack of classical mitochondrial activity in bloodforms, the mitochondrial genome is not completely silent at this stage. Mitochondrial transcripts encoding components of the trypanosome homolog of NADH dehydrogenase complex I are more abundant during this stage of the life cycle than in procyclic forms (61). Nevertheless, activity of the mitochondrial genome is not required for viability during the bloodstream stage. Mutants lacking kDNA have been generated by treating cells with DNA-intercalating agents. These cells continue to grow aggressively within the mammalian host, but they cannot transform to procyclic forms (61). Some species of trypanosomes closely related (morphologically indistinguishable) to *T. brucei* have lost most or all of their kDNA. These species, *T. equiperdum*, *T. equinum* and *T. evansi*, are restricted to the bloodform stage and are not cyclically but sexually or passively transmitted (23).

Although parasites lacking kDNA cannot develop the respiratory chain, presumably other mitochrondrial functions continue. Although the conventional view of mitochondria is as the 'powerhouse of the cell', it is clear that they perform other functions as well. The kinetoplastid mitochondrion probably functions in Ca^{2+} homeostasis as do vertebrate mitochondria (74). In bloodforms an unusual interrelationship between the glycosome and the mitochondrion exists to regenerate the NAD needed for glycolysis (see Chapter 3). This coupling requires mitochondrial glycerol phosphate dehydrogenase, ubiquinol and the alternate oxidase. Since interference with the coupling of the glycosomal and mitochondrial systems is lethal to bloodforms (75), some mitochondrial function is required even in kDNA mutants. The required proteins are presumably nuclearly encoded. Despite this metabolic partnership, glycosomes and the mitochondrion do not appear to be closely apposed within the cell (76).

In *P. falciparum* there also appears to be stage-specific regulation of mitochondrial generation of ATP. In insect stages, such as oocysts, the Krebs cycle is functional and much energy is provided by cytochrome-mediated processes. In asexual erythrocytic forms, the citric acid cycle is not functional, a rotenone-sensitive NADH dehydrogenase complex is absent (77), and the majority of ATP is produced by glycolysis (78, 79). At the ultrastructural level, mitochondria in the erythrocytic stages of mammalian malaria parasites possess few cristae, whereas gametocyte, oocyst, and sporozoite mitochondria contain well-developed cristae. Interestingly, the mitochondria of erythrocytic stages of avian malaria parasites are cristate. Unlike *T. brucei* bloodforms, erythrocytic stages of *P. falciparum* possess *a*-, *b*- and *c*-type cytochromes, as well as cytochrome oxidase (77).

Do mitochondria have any role in erythrocytic stages? The answer appears to be yes. Inhibitors of mitochondrial protein synthesis, such as tetracycline, inhibit the growth of malarial parasites, suggesting that mitochondrially encoded proteins are important for parasite growth. The cationic fluorescent dye rhodamine 123 concentrates in negatively charged cell compartments. When asexual stages are incubated with low concentrations of the dye, it specifically stains the mitochondrion (65). Inhibitors of electron transport block this effect (80). This demonstrates that a transmembrane potential is generated, probably due to a functioning electron transport chain. Potentially connecting these two observations is the finding that a particulate (probably mitochondrial) enzyme functioning in pyrimidine biosynthesis, dihydro-orotate dehydrogenase, is sensitive to inhibitors of electron transport. Since malaria parasites synthesize their pyrimidines *de novo* (Chapter 6), this pathway should be required for growth. Thus mitochondria are active in asexual erythrocytic stages, but their main role is probably not ATP synthesis.

4.4. Some Anaerobic Protozoa Contain a Unique Organelle of Energy Metabolism, the Hydrogenosome

The hydrogenosome is so named because H_2 is a major product of organellar metabolism. Hydrogenosomes have been found only in anaerobic protozoa, including the Trichomonadidae and certain anaerobic ciliates (81–83). The pathways contained in this organelle are discussed in Chapter 3. Strains lacking hydrogenosomal function can be selected by growth on metronidazole, an antiprotozoal chemotherapeutic (81). Since these strains grow more slowly than parental strains, it appears that hydrogenosomes confer an advantage to the cell but are not indispensable.

Certain characteristics of the hydrogenosome are reminiscent of mitochondria, whereas others are clearly distinct (81). As shown in Fig. 13.8, the trichomonad hydrogenosome is bounded by two closely apposed membranes, but no cristae are observed. Although both organelles decarboxylate pyruvate to acetate, the enzymes which do so are not related. The Krebs cycle and cytochromes are absent in hydrogenosomes and ATP is formed only at the level of substrate phosphorylation (see Chapter 3). Hydrogenosomes appear to lack DNA. As is the case for mitochondria, enzymes of other pathways are also associated with hydrogenosomes. These include adenylate kinase, glycerol-3-phosphate dehydrogenase and superoxide dismutase (81). Calcium is present in hydrogenosomes, suggesting that similarly to mitochondria, they may function in the regulation of intracellular Ca^{2+} (84). Whether hydrogenosomes and mitochondria arose from a common progenitor endosymbiont or resulted from distinct endosymbiontic events is an unresolved matter (81).

The apparent lack of an intrahydrogenosomal genome suggests that genes encoding hydrogenosomal proteins are nuclear. The first gene cloned from a trichomonad was that encoding hydrogenosomal ferredoxin of *T. vaginalis* (85). This [2Fe-2S] ferredoxin

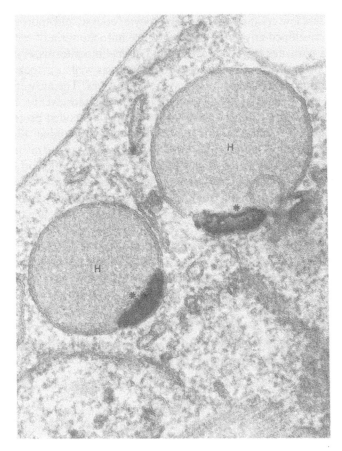

FIG. 13.8 The hydrogenosome of *T. foetus*. A thin section was stained with glutaraldehyde–osmium tetroxide–potassium ferrocyanide in the presence of 5 mM $CaCl_2$. Note the two closely apposed membranes of the hydrogenosome (H), and the electron-dense product which localizes between the two membranes (asterisk), suggesting the presence of intrahydrogenosomal calcium. (Reprinted with permission from ref. 84.)

is more closely related to the ferredoxins of the aerobic bacterium *Pseudomonas putida* and mitochondria than to the 2[4Fe-4S] ferredoxins of hydrogen producing bacteria and *E. histolytica*. Studies have revealed that this protein, as well as another hydro-genosomal protein, the β subunit of succinyl CoA synthetase, are translated on free polysomes (i.e. not membrane-bound) (86). This pathway of biogenesis is similar to that observed for mitochondria and glycosomes, but unlike that for organelles derived by budding from the endoplasmic reticulum. Mitochondrial proteins have presequences which contain targeting information and which are removed upon entry into the organelle. Similarly, the open reading frames encoded by the hydrogenosomal protein genes begin 8–10 amino acids before the amino terminus of the mature hydro-genosomal protein, suggesting that these proteins contain a short presequence (82). The hydrogenosomal presequences are similar to one another in amino acid composition, but differ from mitochondrial presequences in length and composition.

4.5. Kinetoplastids Are Unique Among Eukaryotes in Their Compartmentalization of Glycolysis

Kinetoplastida possess an unusual organelle, the glycosome. Many glycolytic enzymes are contained within the glycosome, hence its name. In all other organisms, glyolytic enzymes are cytoplasmic. Organisms may possess 10–60 or more glycosomes depending on the species examined (53, 76). As discussed in Chapter 2, this compartmentalization of glycolysis may facilitate the extremely high rate of glycolysis by these organisms, most notably *T. brucei* bloodforms. In *T. brucei*, the protein composition of the glycosomes depends on the stage of the life cycle (87). All major glycosomal proteins of bloodforms, with their absolute dependence on glycolysis, are glycolytic enzymes in the pathway from hexokinase to phosphoglycerate kinase. In procyclic forms, which are actively respiring, levels of many of the glycolytic enzymes are reduced in the glycosomes, reflecting the reduced steady-state level of their corresponding mRNAs. In the case of phosphoglycerate kinase, its glycosomal isozyme is highly expressed in bloodforms but reduced in procyclic forms, whereas the cytosolic isozyme shows the inverse pattern. Since these two proteins are transcribed as a part of a polycistronic complex, the regulation appears to be post-transcriptional (88). In procyclic forms, distinct glycosomal proteins are induced. Among those induced are malate dehydrogenase, phosphoenolypyruvate carboxykinase, and several unidentified proteins (87). The regulation of the synthesis of procyclic-specific glycosomal proteins has not been studied in detail.

The glycosome is bounded by a single membrane, and contains a dense protein core (89) (Fig. 13.9). It is not known if the components of the glycosome are ordered in a

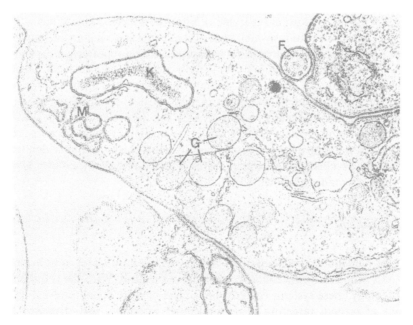

FIG. 13.9 Cross-section of *T. brucei* bloodform showing a cluster of glycosomes (G). Also visible are portions of the inactive mictochondrion (M), the kinetoplast (K), and a flagellum (F). (Photograph kindly provided by Drs Isabelle Coppens and Fred Opperdoes.)

particular way within the organelle. There is no organellar genome. Proteins destined for the glycosome are encoded by nuclear genes, translated on free ribosomes, and imported post-translationally. These characteristics are the same as those of peroxisomes of animals cells and glyoxysomes of plants (89). Glycosomes, peroxisomes and glyoxysomes belong to the class of organelles called microbodies. Microbodies typically contain several enzymes of a given pathway. Other microbodies do not contain the glycolytic enzymes found in glycosomes, but rather enzymes of other pathways such the glyoxylate cycle (glyoxysomes), or peroxide metabolism (peroxisomes). However, some enzyme systems may be found in several types of microbodies, although they may not be major constituents or the pathway may not be complete (89). For example, when peroxide metabolism is present, those enzymes are found in microbodies. Similarly most microbodies contain enzymes for the β-oxidation of fatty acids and both peroxisomes and glycosomes contain enzymes for ether–lipid biosynthesis. The evolutionary origins of glycosomes and other microbodies remain obscure, although they may represent the remnants of ancient endosymbiotic events (90).

4.6. Signals Which Route Proteins to Glycosomes Can be Studied Through Introduction of Genes into *T. brucei*

Recently sequences which can route proteins to peroxisomes have been identified (91). Firefly luciferase, a peroxisomal protein, contains a C-terminal serine-lysine-leucine which was shown to be critical for proper targeting. Indeed, most peroxisomal proteins analyzed thus far show similar sequences at their C-terminus; a minority of proteins apparently utilize amino-terminal or internal signal sequences (92, 93).

Transfection studies have allowed preliminary identification of glycosomal signal sequences. The major glycosomal isozyme of phosphoglycerate kinase has a 22-amino acid extension at the C-terminus which ends in the sequence serine-serine-leucine. When the sequence encoding this extension was appended to the reporter genes, the reporter protein was localized to glycosomes (94). Firefly luciferase is also localized in glycosomes when expressed in *T. brucei* (94). Saturation mutagenesis of the last three codons of luciferase has been used to demonstrate that most C-terminal tripeptides which function as peroxisomal targeting signals, as well as other related tripeptides, function as glycosomal targeting signals (94). However, several glycosomal proteins, including the minor glycosomal phosphoglycerate kinase (which has a C-terminus identical to the cytosolic isozyme) lack such C-terminal motifs. These additional classes of glycosomal targeting signals remain to be defined.

5. PERSPECTIVES

The study of parasite organelle systems has revealed many novel approaches to the problems and profits of intracellular compartmentalization. With the tools of modern molecular biology, these systems can now be studied in more detail to reveal underlying mechanisms of protein targeting, macromolecular assembly, and structure–function relationships. Although the volume of information concerning subcellular organization in vertebrates provides a useful model for understanding protozoan organelle biogenesis, it is clear that many aspects can and do differ in protozoa — sometimes in detail, sometimes more fundamentally. Further studies of these systems may provide opportunities for control of parasitic diseases as well as increased understanding of the diverse strageties for compartmentalization of structure and function in the eukaryotic cell.

6. ACKNOWLEDGEMENTS

The author gratefully acknowledges the excellent assistance of Teresa Hill in assembling this chapter and the support of NIH AI22635.

7. REFERENCES

1. Sogin, M. L. (1989) Evolution of eukaryotic microorganisms and their small subunit ribosomal RNAs. *Am. Zool.* **29**: 487–499.
2. Mortara, R. A. (1989) Studies on trypanosomatic actin. I. Immunochemical and biochemical identification. *J. Protozool.* **36**: 8–13.
3. Ben Amar, M. F., Pays, A., Tebabi, P. *et al.* (1988) Structure and transcription of the actin gene of *Trypanosoma brucei. Mol. Cell. Biol.* **8**: 2166–2176.
4. Fulton, C., Lai, E. Y., Lamoyi, E. and Sussman, D. J. (1986) *Naegleria* actin elicits species-specific antibodies. *J. Protozool.* **33**: 322–327.
5. Guillén, N. (1993) Cell signalling and motility in *Entamoeba histolytica. Parasitol. Today* **9**: 364–369.
6. Wesseling, J. G., Snijders, P. J. F., van Smoeren, P., Jansen, J., Smits, M. A. and Schoenmakers, J. G. G. (1989) Stage-specific expression and genomic organization of the actin genes of the malaria parasite *Plasmodium falciparum. Mol. Biochem. Parasitol.* **35**: 167–176.
7. Seebeck, T., Hemphill, A. and Lawson, D. (1990) The cytoskeleton of trypanosomes. *Parasitol. Today* **6**: 49–52.
8. Heath, I. B. (1980) Variant mitoses in lower eukaryotes: indicators of the evolution of mitosis?. *Int. Rev. Cytol.* **64**: 1–77.
9. Steiger, R. F. (1973) On the ultrastructure of *Trypanosoma (Trypanozoon) brucei* in the course of its life cycle and some related aspects. *Acta Trop.* **30**: 66–168.
10. De Souza, W. (1984) Cell biology of *Trypanosoma cruzi. Int. Rev. Cytol.* **86**: 197–283.
11. Robinson, D., Beattie, P., Sherwin, T. and Gull, K. (1991) Microtubules, tubulin, and microtubule-associated proteins of trypanosomes. *Methods Enzymol.* **196**: 285–302.
12. Brugerolle, G. (1975) Etude de la cryptopleuromitose et de la morphogenese de division chez *Trichomonas vaginalis* et chez plusieurs genres de Trichomoadines primitives. *Protistologica* **11**: 457–468.
13. Vivier, E. and Desportes, I. (1990) Phylum Apicomplexa. In: *Handbook of Protoctista* (eds Margulis, L., Corliss, J. O., Melkonian, M. and Chapman, D. J.), Jones and Bartlett Publishers, Boston, pp. 549–573.
14. Vickerman, K. (1990) Phylum Zoomastigina, Class Diplomonadida. In: *Handbook of Protoctista* (eds Margulis, L., Corliss, J. O., Melkonian, M. and Chapman, D. J.), Jones and Bartlett Publishers, Boston, pp. 200–210.
15. Feely, D. E., Schollmeyer, J. V. and Erlandsen, S. L. (1982) *Giardia* spp.: distribution of contractile proteins in the attachment organelle. *Exp. Parasitol.* **53**: 145–154.
16. Peattie, D. A. (1990) the giardins of *Giardia lamblia:* genes and proteins with a promise. *Parasitol. Today* **6**: 52–56.
17. Gillin, F. D., Boucher, S. E., Rossi, S. S. and Reiner, D. S. (1989) *Giardia lamblia:* the roles of bile, lactic acid, and pH in the completion of the life cycle *in vitro. Exp. Parasitol.* **69**: 164–174.
18. Nohria, A., Alonso, R. A. and Peattie, D. A. (1992) Identification and characterization of gamma-giardin and the gamma-giardin gene from *Giardia lamblia. Mol. Biochem. Parasitol.* **56**: 27–38.

19. Feely, D. E. (1988) Morphology of the cyst of *Giardia microti* by light and electron microscopy. *J. Protozool.* **35**: 52–54.
20. Hemphill, A., Seeback, T. and Lawson, D. (1991) The *Trypanosoma brucei* cytoskeleton: ultrastructure and localization of microtubule-associated and spectrin-like proteins using quick-free, deep-etch, immunogold electron microscopy. *J. Struct. Biol.* **107**: 211–220.
21. Woods, A., Baines, A. J. and Gull, K. (1992) A high molecular mass phosphoprotein defined by a novel monoclonal antibody is closely associated with the intermicrotubule cross bridges in the *Trypanosoma brucei* cytoskeleton. *J. Cell Sci.* **103**: 665–675.
22. Rindisbacher, L., Hemphill, A. and Seebeck, T. (1993) A repetitive protein from *Trypanosoma brucei* which caps the microtubules at the posterior end of the cytoskeleton. *Mol. Biochem. Parasitol.* **58**: 83–96.
23. Vickerman, K. (1990) Phylum Zoomastigina, Class Kinetoplastida. In: *Handbook of Protoctista* (eds Margulis, L., Corliss, J. O., Melkonian, M. and Chapman, D. J.), Jones and Bartlett Publishers, Boston, pp. 215–238.
24. Muller, N., Hemphill, A., Imboden, M., Duvalet, G., Dwinger, R. H. and Seebeck, T. (1992) Identification and characterization of two repetitive non-variable antigens from African trypanosomes which are recognized early during infection. *Parasitology* **104**: 111–120.
25. Pan, A. A. and Pan, S. C. (1986) *Leishmania mexicana*: comparative fine structure of amastigotes and promastigotes *in vitro* and *in vivo*. *Exp. Parasitol.* **62**: 254–265.
26. Fong, D., Wallach, M., Keithly, J., Melera, P. W. and Chang, K. P. (1984) Differential expression of mRNAs for alpha- and beta-tubulin during differentiation of the parasitic protozoan *Leishmania mexicana*. *Proc. Natl. Acad. Sci. USA* **81**: 5782–5786.
27. Landfear, S. M. and Wirth, D. F. (1984) Control of tubulin gene expression in the parasitic protozoan *Leishmania enriettii*. *Nature* **309**: 716–717.
28. Gull, K., Birkett, C., Gerke-Bonet, R. *et al.* (1990) The cell cycle and cytoskeletal morphogenesis in *Trypanosoma brucei*. *Biochem. Soc. Trans.* **18**: 720–722.
29. Fong, D. (1993) Effect of the anti-microtubule compound tubulozole on *Leishmania* protozoan parasites *in vitro*. *FEMS Microbiol. Lett.* **107**: 95–100.
30. Dyer, B. D. (1990) Class Zoomastigina, Class Amebomastigota. In: *Handbook of Protoctista* (eds Margulis, L. Corliss, J. O., Melkonian, M. and Chapman, D. J.) Jones and Bartlett Publishers, Boston, pp. 186–190.
31. Lai, E. Y., Remillard, S. P. and Fulton, C. (1988) The alpha-tubulin gene family expressed during cell differentiation in *Naegleria gruberi*. *J. Cell. Biol.* **106**: 2035–2046.
32. Shea, D. K. and Walsh, C. J. (1987) mRNAs for alpha- and beta-tubulin and flagellar calmodulin are among those coordinately regulated when *Naegleria gruberi* amebae differentiate into flagellates. *J. Cell. Biol.* **105**: 1303–1309.
33. Wiser, M. F. (1991) Malarial proteins that interact with the erythrocyte membrane and cytoskeleton. *Exp. Parasitol.* **73**: 515–523.
34. Schlesinger, P. (1988) Lysosomes and *Entamoeba histolytica*. In: *Amebiasis: Human infection with Entamoeba histolytica* (ed. Ravdin, J. I.), Wiley, New York, pp. 297–307.
35. Gillin, F. D., Reiner, D. S. and McCaffery, M. (1991) Organelles of protein transport in *Giardia lamblia*. *Parasitol. Today* **7**: 113–116.
36. Lindmark, D. J. (1988) *Giardia lamblia*: localization of hydrolase activities in lysosome-like organelles of trophozoites. *Exp. Parasitol.* **65**: 141–147.
37. Slomianny, C. and Prensier, G. (1990) A cytochemical ultrastructural study of the lysosomal system of different species of malaria parasites. *J. Protozool.* **37**: 465–470.
38. Crary, J. L. and Haldar, K. (1992) Brefeldin A inhibits protein secretion and parasite maturation in the ring stage of *Plasmodium falciparum*. *Mol. Biochem. Parasitol.* **53**: 185–192.
39. Dyer, B. D. (1990) Phylum Zoomastigina, Class Parabasalia. In: *Handbook of the Protoctista* (eds Margulis, L., Corliss, J. O., Melkonian, M. and Chapman, D. J.), Jones and Bartlett Publishers, Boston, pp. 252–258.
40. Bangs, J. D., Uyetake, L., Brickman, M. J., Balber, A. E. and Boothroyd, J. C. (1993) Molecular cloning and cellular localization of a BiP homologue in *Trypanosoma brucei*. Divergent ER retention signals in a lower eukaryote. *J. Cell Sci.* **105**: 1101–1113.
41. Duszenko, M., Ivanov, I. E., Ferguson, M. A., Plesken, H. and Cross, G. A. (1988) Intracellular transport of a variant surface glycoprotein in *Trypanosoma brucei*. *J. Cell Biol.* **106**: 77–86.

42. Seyfang, A., Mecke, D. and Duszenko, M. (1990) Degradation, recycling, and shedding of *Trypanosoma brucei* variant surface glycoprotein. *J. Protozool.* **37**: 546–552.

43. Grab, D. J., Ito, S., Kara, U. A. K. and Rovis, L. (1984) Glycosyltransferase activities in golgi complex and endoplasmic reticulum fractions isolated from African trypanosomes. *J. Cell Biol.* **99**: 569–577.

44. Bates, P. A., Hermes, I. and Dwyer, D. (1990) Golgi-mediated post-translational processing of secretory acid phosphatase by *Leichmania donovani* promastigotes. *Mol. Biochem. Parasitol.* **39**: 247–256.

45. Brickman, M. J. and Balber, A. E. (1993) *Trypanosoma brucei rhodesiense:* membrane glycoproteins localized primarily in endosomes and lysosomes of bloodstream forms. *Exp. Parasitol.* **76**: 329–344.

46. Webster, P. (1989) Endocytosis by African trypanosomes. I. Three-dimensional structure of the endocytic organelles in *Trypanosoma brucei* and *T. congolense*. *Eur. J. Cell Biol.* **49**: 295–302.

47. Shapiro, S. Z. and Webster, P. (1989) Coated vesicles from the protozoan parasite *Trypanosoma brucei:* purification and characterization. *J. Protozool.* **36**: 344–349.

48. Coppens, I., Opperdoes, F. R., Courtoy, P. J. and Baudhuin, P. (1987) Receptor-mediated endocytosis in the bloodstream form of *Trypanosoma brucei*. *J. Protozool.* **34**: 465–473.

49. Londsdale-Eccles, J. D. and Grab, D. J. (1987) Lysosomal and non-lysosomal peptidyl hydrolases of the bloodstream forms of *Trypanosoma brucei brucei*. *Eur. J. Biochem.* **169**: 467–475.

50. Coppens, I., Baudhuin, P., Opperdoes, F. R. and Courtoy, P. J. (1993) Role of acidic compartments in *Trypanosoma brucei*, with special reference to low-density lipoprotein processing. *Mol. Biochem. Parasitol.* **58**: 223–232.

51. Webster, P. and Fish, W. R. (1989) Endocytosis by African trypanosomes. II. Occurrence in different life-cycle stages and intracellular sorting. *Eur. J. Cell Biol.* **49**: 303–310.

52. Pupkis, M. F., Tetley, L. and Coombs, G. H. (1986) *Leishmania mexicana:* amastigote hydrolases in unusual lysosomes. *Exp. Parasitol.* **62:** 29–39.

53. Coombs, G. H., Tetley, L., Moss, V. A. and Vickerman, K. (1986) Three dimensional structure of the leishmania amastigote as revealed by computer-aided reconstruction from serial sections. *Parasitology* **92**: 13–23.

54. Aikawa, M. (1988) Fine structure of malaria parasites in the various stages of development. In: *Malaria. Principles and Practice of Malariology* (eds Wernsdorfer, W. H. and McGregor, I.), Vol. I. Churchill Livingstone, New York, pp. 97–129.

55. Slomianny, C. (1990) Three-dimensional reconstruction of the feeding process of the malaria parasite. *Blood Cells* **16**: 369–378.

56. Slomianny, C., Charet, P. and Prensier, G. (1983) Ultrastructural localization of enzymes involved in the feeding process in *Plasmodium chabaudi* and *Babesia hylomysci*. *J. Protozool.* **30**: 376–382.

57. Vander Jagt, D. L., Hunsaker, L. A., Campos, N. M. and Scaletti, J. V. (1992) Localization and characterization of hemoglobin-degrading aspartic proteinases from the malarial parasite *Plasmodium falciparum*. *Biochim. Biophys. Acta Protein Struct. Mol. Enzymol.* **1122**: 256–264.

58. Ginsburg, H. (1990) Antimalarial drugs: is the lysosomotropic hypothesis still valid?. *Parasitol. Today* **6**: 334–337.

59. Rosenthal, P. J., Wollish, W. S., Palmer, J. T. and Rasnick, D. (1991) Antimalarial effects of peptide inhibitors of a *Plasmodium falciparum* cysteine proteinase. *J. Clin. Invest.* **88**: 1467–1472.

60. Ryan, K. A., Shapiro, T. A., Rauch, C. A. and Englund, P. T. (1988) Replication of kinetoplast DNA in trypanosomes. *Annu. Rev. Microbiol.* **42**: 339–358.

61. Stuart, K. and Feagin, J. E. (1992) Mitochondrial DNA of kinetoplastids. *Int. Rev. Cytol.* **141**: 65–88.

62. Barker, R. H., Jr (1990) DNA probe diagnosis of parasitic infections. *Exp. Parasitol.* **70**: 494–499.

63. Woodward, R. and Gull, K. (1990) Timing of nuclear and kinetoplast DNA replication and early morphological events in the cell cycle of *Trypanosoma brucei*. *J. Cell Sci.* **95**: 49–57.

64. Robinson, D. R. and Gull, K. (1991) Basal body movements as a mechanism for mitochondrial genome segregation in the trypanosome cell cycle. *Nature* **352**: 731–733.
65. Divo, A. A., Geary, T. G., Jensen, J. B. and Ginsburg, H. (1985) The mitochondrion of *Plasmodium faciparum* visualized by rhodamine 123 fluorescence. *J. Protozool.* **32**: 442–446.
66. Slomianny, C. and Prensier, G. (1986) Application of the serial sectioning and tridimensional reconstruction techniques to the morphological study of the *Plasmodium falciparum* mitochondrion. *J. Parasitol.* **72**: 595–598.
67. Feagin, J. E. (1994) The extrachromosomal DNAs of Apicomplexan parasites. *Annu. Rev. Microbiol.* **48**: 81–104.
68. Kilejian, A. (1991) Spherical bodies. *Parasitol. Today* **7**: 309.
69. Vickerman, K. (1985) Developmental cycles and biology of pathogenic trypanosomes. *Br. Med. Bull.* **41**: 105–114.
70. Nolan, D. P. and Voorheis, H. P. (1992) The mitochondrion in bloodstream forms of *Trypanosoma brucei* is energized by the electrogenic pumping of protons catalysed by the F_1F_0-ATPase. *Eur. J. Biochem.* **209**: 207–216.
71. Bienen, E. J., Saric, M., Pollakis, G., Grady, R. W. and Clarkson, Jr A. B. (1991) Mitochondrial development in *Trypanosoma brucei brucei* transitional bloodstream forms. *Mol. Biochem. Parasitol.* **45**: 185–192.
72. Williams, N., Choi, S. Y.-W., Ruyechan, W. T. and Frank, P. H. (1991) The mitochondrial ATP synthase of *Trypanosoma brucei:* developmental regulation through the life cycle. *Arch. Biochem. Biophys.* **288**: 509–515.
73. Vercesi, A. E., Bernardes, C. F., Hoffman, M. E., Gadelha, F. R. and Docampo, R. (1991) Digitonin permeabilization does not affect mitochondrial function and allows the determination of the mitochondrial membrane potential of *Trypanosoma cruzi in situ. J. Biol. Chem.* **266**: 14431–14434.
74. Docampo, R. and Vercesi, A. E. (1989) Ca^{2+} transport by coupled *Trypanosoma cruzi* mitochondria *in situ. J. Biol. Chem.* **264**: 108–111.
75. Fairlamb, A. H. and Opperdoes, F. R. (1986) Carbohydrate metabolism in African trypanosomes, with special reference to the glycosome. In: *Carbohydrate Metabolism in Cultured Cells* (ed. Morgan, M. J.), Plenum, New York, pp. 183–224.
76. Tetley, L. and Vickerman, K. (1991) The glycosomes of trypanosomes: number and distribution as revealed by electron spectroscopic imaging and 3-D reconstruction. *J. Microsc.* **162**: 83–90.
77. Fry. M. and Beesley, J. E. (1991) Mitochondria of mammalian *Plasmodium* species. *Parasitology* **102**: 17–26.
78. Scheibel, L. W. (1988) Plasmodial metabolism and related organellar function during various stages of the life-cycle: carbohydrates. In: *Malaria. Principles and Practice of Malariology* (eds Wernsdorfer, W. H. and McGregor, I.), Churchill Livingstone, New York, pp. 171–217.
79. Fry, M., Webb, E. and Pudney, M. (1990) Effect of mitochondrial inhibitors on adenosinetriphosphate levels in *Plasmodium falciparum. Comp. Biochem. Physiol.* **96B**: 775–782.
80. Gutteridge, W. E. (1979) Conversion of dihydroorotate to orotate in parasitic protozoa. *Biochim. Biophys. Acta* **582**: 390–401.
81. Muller, M. (1993) The hydrogenosome. *J. Gen. Microbiol.* **139**: 2879–2889.
82. Johnson, P. J., Lahti, C. J. and Bradley, P. J. (1993) Biogenesis of the hydrogenosome in the anaerobic protist *Trichomonas vaginalis. J. Parasitol.* **79**: 664–670.
83. Lloyd, D., Hillman, K., Yarlett, N. and Williams, A. G. (1989) Hydrogen production by rumen holotrich protozoa: effects of oxygen and implications for metabolic control by in situ conditions. *J. Protozool.* **36**: 205–213.
84. De Souza, W. and Benchimol, M. (1988) Electron spectroscopic imaging of calcium in the hydrogenosomes of *Tritrichomonas foetus. J. Submicrosc. Cytol. Pathol.* **20**: 619–621.
85. Johnson, P. J., d'Oliveira, C., Gorrell, T. E. and Müller, M. (1990) Molecular analysis of the hydrogenosomal ferredoxin of the anaerobic protist *Trichomonas vaginalis. Proc. Natl. Acad. Sci. USA* **87**: 6097–6101.
86. Lahti, C. J. and Johnson, P. J. (1991) *Trichomonas vaginalis* hydrogenosomal proteins are synthesized on free polyribosomes and may undergo processing upon maturation. *Mol. Biochem. Parasitol.* **46**: 307–310.

87. Parsons, M. and Nielson, B. (1990) *Trypanosoma brucei:* two-dimensional gel analysis of the major glycosomal proteins during the life cycle. *Exp. Parasitol.* **70**: 276–285.

88. Gibson, W. C., Swinkels, B. W. and Borst, P. (1988) Post-transcriptional control of the differential expression of phosphoglycerate kinase genes in *Trypanosoma brucei. J. Mol. Biol.* **201**: 315–325.

89. Opperdoes, F. R. (1988) Glycosomes may provide clues to the import of peroxisomal proteins. *Trends Biochem. Sci.* **13**: 255–260.

90. Michels, P. A. M. and Opperdoes, F. R. (1991) The evolutionary origin of glycosomes. *Parasitol. Today* **7**: 105–109.

91. Gould, S. J., Keller, G. A., Hosken, N., Wilkinson, J. and Subramani, S. (1989) Conserved tripeptide sorts proteins to peroxisomes. *J. Cell Biol.* **108**: 1657–1664.

92. Swinkels, B. W., Gould, S. J., Bodnar, A. G., Rachubinski, R. A. and Subramani, S. (1991) A novel, cleavable peroxisomal targeting signal at the amino-terminus of the rat 3-ketoacyl-CoA thiolase. *EMBO J.* **10**: 3255–3262.

93. Kragler, F., Langeder, A., Raupachova, J., Binder, M. and Hartig, A. (1993) Two independent peroxisomal targeting signals in catalase A of *Saccharomyces cerevisiae. J. Cell Biol.* **120**: 665–673.

94. Sommer, J. M. and Wang, C. C. (1994) Targeting proteins to the glycosomes of African trypanosomes. *Annu. Rev. Microbiol.* **48**: 105–138.

95. Peattie, D. A., Alonso, R. A., Hein, A. and Caulfield, J. P. (1989) Ultrastructural localization of giardins to the edges of disk microribbons of *Giardia lamblia* and the nucleotide and deduced protein sequence of alpha giardin. *J. Cell Biol.* **109**: 2323–2335.

96. Sheffield, H. and Bjorvatn, B. (1977) Ultrastructure of the cyst of *Giardia lamblia. Am. J. Trop. Med. Hyg.* **26**: 23–30.

97. Sherwin, T. and Gull, K. (1989) Visualization of detyrosination along single microtubules reveals novel mechanisms of assembly during cytoskeletal duplication in trypanosomes. *Cell* **57**: 211–221.

14 Neurotransmitters of Helminths

RALPH E. DAVIS and ANTONY O. W. STRETTON

Department of Zoology, University of Wisconsin-Madison, Madison, WI 53706, USA

SUMMARY

Among the parasitic helminths (flatworms and nematodes), no biologically active substance has as yet been shown to meet all the criteria for neurotransmitter/neuromodulator status. There are several substances that are reasonable candidates.

For the parasitic flatworms these include acetylcholine (ACh) (inhibitory), serotonin (5-HT) (excitatory) and glutamate (excitatory in cestodes). Other active substances with weaker candidacies are dopamine (excitatory in trematodes), noradrenaline (mixed action depending on trematode species) and histamine (excitatory in cestodes). Based on the circumstantial evidence of immunocytochemical studies and by analogy with other organisms, it is reasonable to hypothesize that certain neuropeptides (e.g., NPF-related peptides, among others) may also have effects on the neuromuscular system of parasitic flatworms, but the functional studies to demonstrate this have yet to be done.

For the parasitic nematodes putative neurotransmitters and/or neuromodulators include ACh (excitatory), γ-aminobutyric acid (inhibitory), and glutamate (excitatory). Though 5-HT and the catecholamines are indicated by immunocytochemical, biochemical and anatomical techniques, their precise functions as neurosignals remains to be determined. The study of neuropeptides in nematodes has revealed the strong candidacy of the family of FMRFamide-like peptides and in particular the AF peptides which have been shown to have specific physiological effects.

Biochemistry and Molecular Biology of Parasites
ISBN 0-12-473345-X

1. INTRODUCTION

The ability of a parasite to move to different sites in its host and to maintain its position in host body fluids, where appropriate, is crucial to the parasite's survival. Understanding the mechanisms underlying its motile capabilities, particularly at the level of its neuromuscular system, is of great importance both for an understanding of basic parasite behavior and biology and for the development of anthelmintic compounds. This chapter will focus on the chemical signals affecting the neuromuscular systems of adult parasitic helminths with emphasis on putative transmitters and modulators in the trematodes, cestodes and nematodes.

Parasitic helminths exhibit a variety of putative intercellular signalling molecules. These include classical small molecule transmitters, modulators and neuropeptides. In many cases there is close spatial association of cell bodies or neurites containing certain signaling molecules with muscle tissue or with motoneurons. This provides circumstantial evidence that these chemicals represent putative transmitters or modulators involved in neuromuscular activity. In other cases physiological experiments have directly demonstrated biological activity on muscle or motoneurons.

We are taking a rather catholic view in defining the neuromuscular/motonervous system and its chemical interactions, since 'action at a distance' (as yet another level of complexity) increases the number of potential partners in intercellular signaling far beyond those that are connected by classical synapses. This broad view is especially appropriate in the parasitic platyhelminths where the acoelomate body construction makes isolated cell or individual tissue studies difficult. The pseudocoelomate construction of nematodes allows studies at select sites (e.g., neuron or muscle) and makes it possible to be more definitive in interpreting functional studies. If one employs the most rigorous criteria identifying a substance as a neurotransmitter or neuromodulator, there are currently no ligands that unequivocally meet the criteria in any parasitic helminth; it is essential that this caveat be borne in mind. A further recurring theme seems to be that parasitic helminth receptors do not fall into neat vertebrate pharmacological classifications.

In the parasitic platyhelminths, functional studies are in their infancy (1, 2) the data at the present time coming almost entirely from studies of the effects of putative bioactive signals on whole worm motility or cut worm/muscle strip preparations. The nature of these preparations (which contain a complexity of tissue or cell types usually including transverse, circular and longitudinal muscles as well as sensory, inter- and motoneurons) has made it difficult to definitively localize putative transmitter or modulator effects specifically to the neuromuscular junctions. Establishment of neurotransmitter or neuromodulator status for a given ligand at a neuromuscular junction depends on pharmacological studies using simultaneous intracellular recordings with microelectrodes in an identified presynaptic motoneuron and a postsynaptic muscle cell. For technical reasons largely related to acoelomate construction, traditional intracellular recording techniques have been very difficult to apply to the parasitic platyhelminths and hence such studies in these organisms are rare. This inability has prevented comprehensive electrophysiological analysis of platyhelminth nervous and neuromuscular systems and results in lack of information crucial to defining and assessing the neural circuits underlying locomotion and other behaviors. The recent use of patch clamp and intracellular recording technology, detegumented preparations, and isolated muscle cell preparations in *Schistosoma mansoni* (3–6) offer the prospect that identified cell types can be directly assessed pharmacologically.

In the nematode parasites, *Ascaris* in particular, the application of intracellular recording technology to elucidate neuromuscular function has a history going back through the 1960s. This history includes intracellular and patch clamp studies of muscle as well as extensive anatomical and electrophysiological studies of the motonervous system including intracellular and extracellular recordings from identified neurons. Based on recent immunocytochemical studies and the subsequent characterization of a variety of nematode neuropeptides, there are many new putative signals to be tested at specific neuromuscular sites. Rather than technical difficulties, it is the sheer number of compounds to be studied and the number of sites to be analyzed that presents the daunting challenge, though an exciting start has been made.

This chapter contrasts the parasitic platyhelminths and parasitic nematodes, and emphasizes selected well-studied species as representative of each phylum (trematodes: *Fasciola hepatica* and *Schistosoma mansoni*; cestodes: *Hymenolepis diminuta*; nematodes: *Ascaris suum*). Supplementary information about the free-living nematode *Caenorhabditis elegans* and other parasitic helminths is also presented.

2. PARASITIC PLATYHELMINTHS

The major anatomical components of the nervous system of parasitic platyhelminths are shown in the generalized schematic of Fig. 14.1. Pax and Bennett (1) have reviewed information regarding putative signaling substances in parasitic helminths; see Table 14.1 (modified from Tables 1 and 2 in their review) for a summary.

2.1. Acetylcholine (ACh)

ACh is a putative inhibitory transmitter in flatworm parasites. Pharmacological studies demonstrate that it and cholinomimetic drugs inhibit muscle, suppressing rhythmical movement and ultimately producing a flaccid paralysis. This occurs both in the trematodes (*F. hepatica* (7–10), *S. mansoni* (11–13), *Haplometra cylindracea* (14), *Diclidophora merlangi* (15)) and in the cestodes (*H. diminuta* (16, 17)).

Biochemical and histochemical studies have shown the presence of acetylcholinesterase (AChE) in these and other parasitic flatworms (8, 10, 18–22). There is also evidence for choline, the precursor to ACh (10), and choline acetyltransferase (ChAT), the synthetic enzyme for ACh (8, 10, 18). The cholinesterase of the trematodes, *F. hepatica* (8, 23) and *S. mansoni* (18, 24, 25), and the cestodes, *H. diminuta* and *H. nana* (21), is predominantly an acetylcholinesterase with lesser or no non-specific esterase activity depending on the species. The most pronounced histochemical staining occurs in central nervous system structures (central ganglia and associated commissures/nerve cords), the pharynx and suckers. Staining is also found in fine ramifications throughout the parenchyma/subintegumental muscle regions presumably associated with neuro-muscular innervation, and in muscularized reproductive structures.

In the guinea-pig ileum bioassay, extracts and homogenates of *S. mansoni* produce spasmogenic activity; this activity is abolished with cholinesterase treatment or blocked with the muscarinic antagonist atropine. This suggests the involvement of endogenous ACh or a closely related substance (8, 11). In *H. diminuta* (26), ACh-like immunoreactivity has been reported in neuron-like cells in and around the longitudinal nerve cords, in fine processes that pass close to or surround deep longitudinal muscle blocks possibly representing cholinergic innervation, and in fine processes to the sucker-associated musculature.

Gaining a sense of the neuromuscular organization in parasitic flatworms has been particularly daunting and perhaps some of this difficulty stems from the complex morphology of the nerve-muscle interface itself. This is best illustrated in the cestodes. Although there is evidence that cestode neurons send out processes to innervate discrete muscles in the classical neuron-to-muscle style of vertebrates (27, 28), there is also evidence of the complex sarconeural construction seen in other invertebrates (e.g., nematodes) where muscle cells send many fine processes to the nerves, synapses being made *en passant* in a polyneuronal, multiterminal fashion (29). When this pattern is combined with the diffuse nexus of small nerves/neurons in the peripheral nervous system as is typically seen in parasitic flatworms, the organization is extremely challenging to interpret (cf. Fig. 14.1).

Because of the relatively small size of most parasitic flatworms, pharmacological studies have largely depended on external application of drugs in behavioral assays which assess worm motility (30, 31). The primary tool for these studies has been the muscle strip preparation. In lieu of a pure muscle strip preparation (with no neuronal elements present), attempts have been made to block neuronal input using high magnesium salines (13), an approach which, once neuronal blockade has been confirmed, will permit more definitive statements about muscle membrane receptors. In most studies a force transducer is used to characterize drug efficacy relative to on-going spontaneous or evoked muscle contraction, i.e., effects on frequency and amplitude are assessed.

TABLE 14.1   Anatomical distribution and functional effects of major putative transmitters at neuromuscular sites in parasitic platyhelminths (modified from ref. 1).

Neurotransmitter	CNS	PNS	Holdfast structures	Body muscle	Reproductive structures	Gut	Putative function
Cestodes							
Acetylcholine	+	+	+	+	+	−	Inhibitory
Serotonin	+	+	+	+	+	−	Excitatory
Catecholamines	+	+	+	+	−	−	?
Glutamate	−	+	−	−	−	−	Excitatory
Neuropeptides	+	+	+	+	+	+	?
Trematodes							
Acetylcholine	+	+	+	+	+	+	Inhibitory
Serotonin	+	+	+	+	+	+	Excitatory
Catecholamines	+	+	+	?	+	−	Mixed responses
Neuropeptides	+	+	+	+	+	−	?

CNS, central nervous system; PNS, peripheral nervous system; +, present; −, absent; ?, unknown.

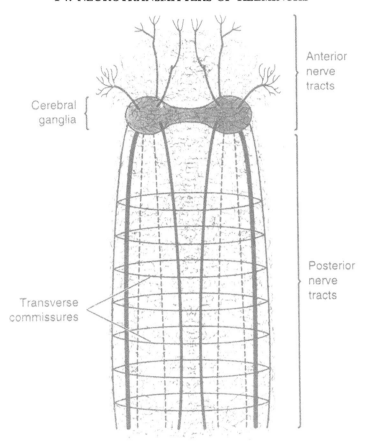

FIG. 14.1 Generalized schematic of the parasitic platyhelminth nervous system. The central nervous system is composed of a pair of cerebral ganglia connected by a commissure and a variable number of anterior and posterior nerve tracts arising from these ganglia. The anterior tracts innervate (where present) anterior suckers and other attachment structures, pharynx and portions of anterior body wall musculature. The anterior central nervous system of cestodes shows additional structures, e.g. the presence of anterior nerves to small rostellar ganglia and their associated commissure and nerves. The major posterior nerve cords run longitudinally in a submuscular position and are variable in number: typically three pairs in trematodes and five pairs in cestodes; numerous smaller longitudinal nerve bundles mark more posterior regions in some species. The peripheral nervous system consists of transverse commissures which interconnect these cords. The transverse commissures typically occur at regular intervals yielding an orthogonal arrangement of circular rings along the body length. An intricate meshwork of fine neural processes gives the peripheral nervous system a nerve net-like structure. Discrete nerve plexuses occur in association with specific structures that require innervation, e.g. suckers, pharynx, reproductive structures and muscle associated with ducts.

In *F. hepatica* and schistosomes, the nicotinic agonist carbachol is a more potent agonist of muscle relaxation/inhibition than is ACh (perhaps because of its insensitivity to esterase). Muscarine is ineffective. Application of the muscarinic antagonist atropine stimulates increased muscle tone, contraction and/or motility and reverses the paralysis induced by carbachol and ACh (9–11, 13, 32). The stimulatory effect of atropine and the ability of eserine (a cholinesterase inhibitor) to relax muscle have led several investigators to suggest that a tonic release of transmitter from cholinergic cells may be occurring (9, 10, 13). Nicotine is a more effective agonist than ACh in *F. hepatica* (also perhaps because of its esterase insensitivity (9, 10)) but is ineffective in schistosomes (11, 30). The nicotinic antagonist, *d*-tubocurarine, is ineffective in schistosomes (11, 33); in *F. hepatica*, the results with *d*-tubocurarine are inconsistent (8–10). In contrast, in the trematode *H. cylindracea*, *d*-tubocurarine is an effective antagonist and atropine is largely ineffective in reversing the ACh effect. ACh, carbachol and nicotine all inhibit the amplitude and frequency of contraction (14).

In the cestodes, *H. diminuta*, *H. microstoma* (17, 34) and *Grillotia erinaceus* (35), ACh also relaxes longitudinal muscle strips or cut whole worm preparations. This suggests an inhibitory function. In *G. erinaceus*, carbachol mimics ACh, though nicotine is ineffective; atropine is an ineffective antagonist of the ACh effect.

Thus, flatworm cholinergic receptors display a mixture of nicotinic and muscarinic properties (10, 11, 17, 30, 32) and are therefore pharmacologically different from the classical vertebrate cholinergic receptors. (Nicotine and muscarine have been used classically in vertebrate pharmacology to define two different classes of cholinergic receptors.) These pharmacological differences may be due to a number of factors: lack of specificity of 'vertebrate' blockers for flatworm receptors, the presence of a variety of cholinergic receptors on the same or different cells (neuron and/or muscle), activation of multicomponent pathways (e.g., activation of excitatory neuronal input to inhibitory fibers presynaptic to muscle), etc.

2.2. Serotonin (5-HT)

In parasitic flatworms, 5-HT serves a variety of functions from control of carbohydrate metabolism to intra- and interorganism (host-parasite) communication (36). Rhythmical movement is stimulated by 5-HT and related indoleamines (e.g., lysergic acid diethylamide) (7, 8, 37). This stimulatory effect, like the ACh inhibitory effect, can be obtained in the absence of the central ganglia suggesting a peripheral site of action (37). The stimulation of motility (increased contraction amplitude and frequency) *in vitro* suggests that 5-HT or a closely related indoleamine is an excitatory transmitter in both the trematodes *F. hepatica* (9, 37, 38), *S. mansoni* (11, 12, 39, 40), *Clonorchis sinensis* (39), *D. merlangi* (15) and the cestodes *Taenia pisiformis* (39), *Mesocestoides corti* (41), *H. diminuta* (16, 42, 43), *Dipylidium caninum* (44).

In histochemical, biochemical or immunocytochemical studies, 5-HT has been reported in several species of trematodes (45–51) and cestodes (22, 27, 41, 42, 46, 52–57). There has been difficulty demonstrating 5-HT *per se* in *F. hepatica* in histochemical and biochemical studies (46, 58); there is, however, immunocytochemical evidence for wide distribution of 5-HT-like immunoreactivity in the central and peripheral nervous systems of this organism (47, 51).

In flatworms, 5-HT-like immunoreactivity is most pronounced in CNS structures (anterior ganglia, longitudinal and transverse nerve cords), but it is also seen in a peripheral nexus of small frequently varicose nerve fibers associated with subtegumental muscles, the musculature of the suckers, rostellum (where present and functional), and the muscular lining of various reproductive ducts/structures. In general, the distribution of putative 5-HT staining roughly parallels the distribution of cholinergic elements as suggested by cholinesterase localization (21, 23, 42, 47, 59, 60). In the trematode *D.*

merlangi and the cestodes *Proteocephalus pollanicola* and *Moniezia expansa*, the overall cholinesterase and peptidergic staining seem to parallel one another more and dominate the CNS; the serotonergic staining pattern is somewhat different, being most prominent in the PNS (22, 57, 61).

Though there is evidence for the synthetic enzyme aromatic amino acid decarboxylase (45,62) and the degradative enzyme monoamine oxidase (MAO) in *S. mansoni* (63), it has been suggested in *S. mansoni* (30, 62, 64) that the predominant source of 5-HT in the parasite might be the host. In *H. diminuta* (65,66) the same suggestion has been made, though there is evidence of endogenous synthesis as well (67,68). Host 5-HT and circadian variations in the levels of 5-HT in the worm have been proposed as playing a role in the migratory behavior of *H. diminuta* (65,69,70).

Pharmacologically, the antagonists methysergide, cyproheptidine, dihydro-ergotamine, bromolysergic acid diethylamide and metergoline block the effects of 5-HT in *S. mansoni* (13, 32, 33, 71, 72). Metergoline in longitudinal muscle of *S. mansoni* (13) and methysergide in *H. cylindracea* (14) not only block 5-HT-induced stimulation but each antagonist alone can be inhibitory. These results suggest that, like ACh, 5-HT might be released tonically. In contrast, neither methysergide nor ketanserin is effective in the monogenean, *D. merlangi* (15) or the cestode *G. erinaceus* (35). In *F. hepatica*, MAO inhibitors mimic 5-HT in stimulating motility. The amine-depleting agents chloroamphetamine and reserpine inhibit motility (9). Similar inhibitory results with reserpine have been obtained in *S. mansoni* (33) and *H. cylindracea* (14).

Attempts have been made to characterize the pharmacology of the fluke receptors with a battery of 5-HT antagonists. Bearing in mind that it is not possible to evaluate the putative neuromuscular junction receptors specifically, the antagonist profiles (order of potency) show that the 5-HT receptors of *F. hepatica* and *S. mansoni* are in most respects similar to each other but quite different from those described in vertebrates (6, 72–76). In a comparison of 5-HT agonists on a *F. heptica* muscle strip preparation, there was some similarity to the vertebrate 5-HT$_1$ receptor, assuming agonist action was mediated by one receptor type (77). As with the cholinergic receptors, it is possible that a variety of 5-HT receptors occur in parasitic helminths, and that these receptors do not fall precisely into the pharmacological receptor classification of vertebrates (36, 72, 78). As with cholinergic agents, 5-HT effectively stimulates contractions in both longitudinal and circular muscles of *S. mansoni* even in high magnesium media, suggesting that 5-HT receptors are present on muscle membranes (13). Studies on isolated muscle fibers in *S. mansoni* have yet to show the increase in contraction expected in light of the 5-HT-induced contractions seen in whole worm studies, but there may be a number of extenuating reasons for this (6). In isolated muscle fibers, contractions induced by elevated potassium are dependent on extracellular calcium and require 5-HT for maintenance (6). Forskolin, an adenylate cyclase activator mimics 5-HT, and H-89, a protein kinase inhibitor, blocks the 5-HT effect suggesting mediation by a cAMP-dependent pathway (4,6). Phorbol ester induces muscle contracture that is calcium-dependent, but does not involve muscle depolarization. These contractions seem to involve protein kinase C activation and enhancement of sarcolemmal calcium channel activity (5). It remains to be seen precisely how 5-HT may be involved in these effects and in general whether it serves as a transmitter, modulator, or both.

2.3. Catecholamines

Less work has been done on the catecholamines than on ACh and 5-HT and the conclusions regarding function are more tentative. Biochemical and histochemical studies indicate the presence of the catecholamines dopamine (DA) and noradrenaline (NA) in the trematodes *S. mansoni* (NA, but little or no DA) and *F. hepatica* (DA, but little or no NA) (32, 33, 40, 45, 46, 79, 80).

In *S. mansoni* NA and DA are inhibitory and cause a lengthening of the worm, although there is no obvious effect on motility (12, 13, 32, 33, 72). The pharmacology of the receptors is more DA-like than NA-like (e.g., they are haloperidol- and spiroperidol-sensitive and relatively insensitive to adrenergic blockers). NA has been detected in *S. mansoni* but DA has not; it has been suggested that NA is the endogenous ligand which binds to a DA-type of receptor (32, 33). This receptor is believed to be located primarily on the circular muscles and not on the longitudinal muscles, the latter being responsible for locomotion (13). There are likely to be DA-like receptors on cells presynaptic to muscle as indicated by the consistent relaxation of muscle in the presence of DA in normal magnesium media (i.e., when neurotransmission is presumed to be intact) (13).

In *F. hepatica*, DA (37) and NA (9) have been reported to be inhibitory (although a subsequent report has shown that DA stimulates motility (9)). In contrast, in *D. merlangi*, DA (like 5-HT) increased the frequency, amplitude and tone of spontaneous contractions and seems to have an excitatory function; NA produced similar though weaker effects (15). The failure of dopaminergic and adrenergic blockers to reverse the effects of DA and NA in *D. merlangi* suggests that the invertebrate receptors differ from their vertebrate counterparts (15). Even between species there seem to be differences in the catecholaminergic systems.

In the cestode *Diphyllobothrium dendriticum*, catecholamines are found primarily in the CNS but also peripherally in stained fibers traversing through or along longitudinal and transverse muscle layers (81). The extent and distribution of catecholaminergic fibers parallels the serotonergic and, to a lesser extent, the peptidergic cells and fibers (53, 81, 82). DA dominates the scolex and neck region whereas NA and adrenaline (A) were more evenly distributed, with A predominating (56, 83).

Small amounts of DA, NA and octopamine have been demonstrated in *H. diminuta* (46, 83). Although the adult cestode *Spirometra mansonoides* lacked catecholamines, the larvae did express them (52); in contrast, the plerocercoid larvae of *D. dendriticum* did not have significant catecholamine amounts whereas the adult did (56). These studies emphasize the importance of examining different developmental stages in order to assess the presence/role of a given neuroactive substance. The precise physiological functions of the catecholamines in cestodes have been difficult to determine.

2.4. Glutamate

Glutamate is a putative neurotransmitter at the neuromuscular interface in cestodes (84). Extracellular recordings from the longitudinal nerve cord of the cestodarian, *Gyrocotyle fimbriata*, showed increased spike activity in response to applied L-glutamate or L-aspartate which was blocked by the antagonist 2-amino-4-phosphonobutyrate (85).

In *H. diminuta*, glutamate produced strong longitudinal body strip contractions and a high affinity glutamate transport mechanism was detected in tissue slices (86). Release of glutamate stimulated by high potassium was partially calcium-dependent and antagonized by magnesium; furthermore, tetrodotoxin and zero sodium blocked the release of glutamate that had been exogenously supplied (43, 87). A polyclonal antiserum gave consistently intense staining in cell bodies and neurites associated with the longitudinal nerve cords (84). Glutamate, kainate and *N*-methyl-D-aspartate have no effect on muscle tone in detegumented *S. mansoni* (5) and there is no other evidence that glutamate is a neurotransmitter in trematodes (32).

2.5. γ-Aminobutyric Acid (GABA)

There is no evidence that GABA or glycine function as neuromuscular transmitters in parasitic flatworms (32). The absence of GABA is particularly surprising given its wide occurrence in both vertebrates and invertebrates.

2.6. Histamine

Significant numbers of histaminergic neurons have been reported in larval *D. dendriticum* (88). Though histaminergic neurons have not been identified in *H. diminuta*, histamine selectively affects motility in the posterior region of the strobila in a concentration-dependent manner: 10^{-9} M significantly increased motility in this region, whereas 10^{-6} M and 10^{-3} M inhibited it (16). As with 5-HT and some of the catecholamines, it has been suggested that histamine may be a host-generated substance.

2.7. Neuropeptides

A variety of neuropeptide immunoreactivities have been described in trematodes (48, 61, 89–97) and cestodes (22, 53, 55, 57, 82, 98–102). In the nervous system, neuropeptide immunoreactive staining is widespread in the CNS and PNS though it is typically most prominent in the CNS.

Of the antisera tested so far, those with the widest immunoreactivity across species and most commonly associated with neuromuscular locations (proximity of neurites to muscle) include pancreatic polypeptide-like peptide(s) (PP) of the neuropeptide Y (NPY) superfamily and Phe-Met-Arg-Phe-amide (FMRFamide)-like peptide(s), an invertebrate family of neuropeptides (22, 53, 55, 57, 61, 82, 89–92, 95, 103). In parasitic flatworms immunoreactivities to members of the NPY superfamily (neuropeptide Y, peptide YY and pancreatic polypeptide) and FMRFamide are likely to be largely or completely due to the native neuropeptide F (NPF) (22, 93, 100, 101). This 39 amino acid neuropeptide was first isolated and sequenced from the cestode *Moniezia expansa* and is viewed as a closely related member of the NPY superfamily; it represents the first invertebrate member of this family to be isolated and sequenced. Similar staining patterns for the PP and PYY members of the NPY superfamily, FMRFamide and NPF have been obtained. Ultrastructural studies have co-localized NPF, PP and FMRFamide to a specific population of large dense-core vesicles in neurons of some species. NPF quenches PP and FMRFamide immunoreactivities but not vice versa. These results obtained in species of trematodes and cestodes suggest that most, if not all, of the

PP/FMRFamide immunoreactivity may be due to NPF-like peptide(s) (22, 93, 96, 101, 102). The common C-terminal motifs of NPF, other members of the NPY superfamily and FMRFamide mean that great caution must be exercised in the interpretation of immunocytochemical experiments using antisera that recognize this region of the peptides.

In *F. hepatica* (90) and *S. mansoni* (48,95), antisera to members of the NPY superfamily of peptides, FMRFamide, and, to a lesser extent, substance P (in select muscular sites) give immunoreactive staining in the neurons innervating the subtegumental muscle, oral and ventral suckers, and the muscular walls of various reproductive organs and ducts (in addition to a pronounced occurrence centrally). Antisera to FMRFamide-like peptides prominently stain the neurons innervating longitudinal and transverse muscle bundles of *D. dendriticum* (53,82); neurons innervating the bothridial musculature are also a major site of PP-, FMRFamide-, and small cardioactive peptide B-like immunoreactivity (53, 98, 104). Gastrin/cholecystokinin-like immunoreactivity is associated with neurons leaving the peduncle ganglia innervating the clamp musculature in *D. merlangi* (61). In *S. mansoni* there is evidence for the presence of pro-opiomelanocortin-related peptides (adrenocorticotropin (ACTH), melanocyte-stimulating hormone (MSH), beta-endorphin (BEnd)) (105); a high affinity delta-like opioid receptor and a ligand very similar to methionine-enkephalin (met-enkephalin) have also been reported (106). This enkephalinergic system may participate in both internal and host–parasite signaling.

Limited and/or weak staining is seen at select sites and in select species for a variety of other peptide antisera (e.g., peptide histidine-isoleucine, calcitonin gene-related peptide, the tachykinin family members substance P and neurokinin A, etc. (22, 55, 91, 99)). These various immunoreactivities, though not necessarily prominent in extent or amount, nevertheless suggest the presence of a wide diversity of potentially bioactive peptides in the nervous system of parasitic flatworms. This neurochemical complexity is illustrated in cestodes where 20 different peptide immunoreactivities outnumber the seven putative 'classical' (small molecule) transmitters reported to date for this group; in the trematodes there are approximately 24 different peptide immunoreactivities in a wide range of species. These numbers will no doubt continue to increase.

Although a variety of peptide-like immunoreactivities have been described, many of which occur in neuronal-like cells in the peripheral nerve nexus, great caution must be exercised in interpreting their function; for example, some of these are associated with putative sensory receptors instead of serving as neurotransmitters or neuromodulators of neuromuscular function (48, 82, 91, 92); peptidergic innervation of reproductive ducts and structures is also commonly seen, e.g., the muscular wall of the ootype (22, 55, 57, 90, 107, 108).

Data demonstrating a neuropeptide effect on motility are scarce. In *S. mansoni* neither FMRFamide nor substance P have any effect on muscle tone or activity, however it is not clear that they are able to reach their putative target sites intact (5). Endogenous FMRFamide-like and substance P-like peptides have not yet been isolated and tested.

Somatostatin significantly increases motility in anterior and mid-regions of the strobila of *H. diminuta* (16). Somatostatin-like immunoreactivity, however, has not been shown in *H. diminuta*, though it is present in lateral longitudinal nerve cords of *Trilocularia acanthiaevulgaris* (104). As with the other putative neuroactive substances, its precise site of action is unknown. Though NPF was first isolated in a parasitic flatworm and is believed to be widely distributed in invertebrate phyla (all sequences to data contain C-terminal -RXRFa, where X is either P or T) (109), its function in flatworms has yet to be identified. In the mollusc *Aplysia*, NPF has been identified, cloned and physiologically characterized; it produces a prolonged neuronal inhibition (110). Current efforts are aimed at identifying and synthesizing NPF and other neuropeptide(s) from a variety of parasitic flatworms, isolating the receptor(s) and characterizing function with an ultimate goal of exploiting this knowledge in the development of new anthelminthics.

3. PARASITIC NEMATODES

One of the strengths of working with nematodes is that they come in different sizes. This allows different nematodes to be exploited for their distinctive experimental advantages. The small free-living nematode *Caenorhabditis elegans* has been developed into a system where genetic analysis, the molecular isolation and sequencing of genes, the description of the developmental lineage of each adult cell and of the morphology and synaptic connectivity of each neuron, is extremely sophisticated. Large nematodes such as *Ascaris suum* have large cells that can be penetrated with microelectrodes for electrophysiological experiments and can be dissected for the biochemical analysis of single cells. Figure 14.2 is a tracing of a transverse section of *Ascaris* that shows the relationship between nerve and muscle. At the morphological level, the neurons of *C. elegans* and *A. suum* are remarkably similar (111, 112, 121). This similarity may also extend to the physiological level. For example, the cholinergic pharmacology and the cellular localization of GABA are very similar in these two nematodes. At least in some respects, *C. elegans* can be used as a model organism for the exploration of fundamental biological mechanisms also present in parasitic nematodes.

3.1. Acetylcholine

ACh is the best established transmitter candidate in nematodes. It was first isolated from *A. suum* (113). Microiontophoresis of ACh onto the ends of muscle arms at the nerve cords of *A. suum* produces depolarizations that can be blocked by *d*-tubocurarine and potentiated by AChE inhibitors (114). In addition to the ACh receptors at the nerve cords, muscles also have extrasynaptic receptors (115–117). Biochemical assays show that the excitatory motoneurons contain ChAT whereas the inhibitory moto-neurons do not (118). Stimulation of the excitors leads to muscle cell depolarization which is blocked by *d*-tubocurarine (119).

In the absence of applied ACh, *d*-tubocurarine hyperpolarizes and AChE inhibitors depolarize muscle cells. This suggest that ACh is tonically released onto muscle, as in the cestodes (114). In intracellular recordings it has been shown that motoneurons continuously release transmitter at their normal resting potentials; this can be reduced by hyperpolarizing the neuron and increased by depolarization (120). Thus, moto-neurons are a source of tonically released transmitter. Some of the interneurons that synapse onto the excitatory motoneurons are also cholinergic (121); however, the total number of interneurons that are cholinergic is presently unknown.

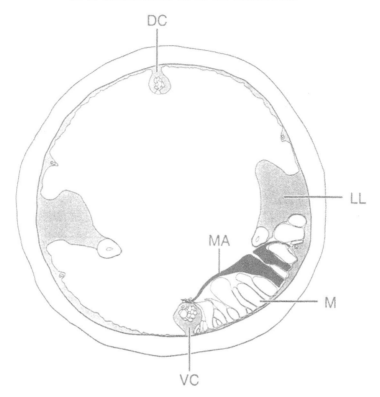

FIG. 14.2 Tracing of a transverse section of the parasitic nematode *Ascaris* showing the relationship between nerve and muscle. The hypodermis (gray) surrounding the pseudocoelom is raised into four longitudinal bands: the larger lateral lines (LL) which also include the excretory duct, and the two major nerve cords, dorsal (DC) and ventral (VC). Muscle cells (M) are indicated in the right ventral quadrant, but have been omitted from the left ventral and both dorsal quadrants. The pharynx has also been omitted. The two filled muscle cells (black) send arms (MA) to the nerve cords, one to the ventral cord and the other to one of the four minor nerve cords. This section was taken a few millimeters behind the tip of the head. Further posteriorly, the number of muscle cells increases dramatically and the density of the muscle arms at the nerve cord increases.

The cholinergic system is not confined to neurons. In *A. suum* there is ChAT activity in hypodermal tissue, and this accounts for over 95% of the total ChAT activity (122). Thus it is likely that a second, non-neural, cholinergic signaling system exists in *A. suum*. The physiological role of this system is not known. However, it is clear that one cannot assume that the effects of cholinergic drugs on behavior are solely due to their action on the neuromuscular system.

The pharmacology of ACh in *A. suum* has been explored extensively (117, 123). The muscle ACh receptor most resembles a vertebrate nicotinic receptor, and has properties of both neuromuscular and ganglionic receptors. Nicotinic receptors have also been reported in *C. elegans*, *H. contortus* and *D. vitae* (124–127). The anthelminthics morantel, pyrantel and levamisole are cholinergic agonists in *A. suum* (128) and *C. elegans* (124, 129).

A search for muscarinic cholinergic function has shown the presence of high affinity binding sites for *N*-methylscopolamine (NMS) in homogenates of *C. elegans* (130). A similar binding site has been found in *A. suum* L2 larvae (131). The cellular location and function of this binding site is not known. A second distinct action of NMS has been described — it opens channels in some classes of motoneuron (132). However, since a relatively high concentration is required for this effect, it is not mediated by the high affinity receptor. Furthermore, although the action of ACh on motoneurons is blocked by *d*-tubocurarine, the action of NMS is not, so the natural ligand of the low affinity NMS receptor is unlikely to be ACh, and remains to be discovered.

Several classes of mutants affecting cholinergic function have been isolated in *C. elegans*. Alleles of *cha-1*, a structural gene (probably the only structural gene) for ChAT, were isolated as mutations resistant to AChE inhibitors (133). The *cha-1* gene has now been cloned (134). *C. elegans* contains at least three forms of AChE, encoded by three genes, *ace-1*, *ace-2* and *ace-3* (135–136). There appears to be functional redundancy, since single *ace* mutants are indistinguishable behaviorally from wild-type worms. Double mutants are uncoordinated and triple mutants (in all three genes) are lethal. Analysis of genetic mosaics has shown that *ace-1* is expressed in muscle cells (137).

Genes encoding ACh receptors have been identified by isolating mutants resistant to the cholinergic agonist levamisole (124, 129). Two of them have been cloned and sequence analysis shows that *unc-38* encodes a polypeptide with sequence homology to α-subunits of the ACh receptor from other organisms, and *unc-29* encodes a subunit resembling non-alpha chains (138).

A further cholinergic function has been identified in *C. elegans*. The gene *unc-17*, which is very closely linked to *cha-1*, encodes a protein that is analogous to the vertebrate transporter that concentrates acetylcholine in synaptic vesicles (139).

3.2. γ-aminobutyric Acid (GABA)

GABA was first proposed as an inhibitory neurotransmitter in *A. suum* by del Castillo *et al.* (140), who also showed that the anthelmintic piperazine was a GABA agonist (141). Both GABA and piperazine open Cl^- channels in muscle cells (116, 142). Muscle relaxes when GABA is applied to dissected preparations of *A. suum* (143), *C. elegans* (124), or *Dipetalonema vitae* (127). The selective localization of GABA-like immunoreactivity in inhibitory motoneurons, but not in excitatory motoneurons strongly supports the suggestion that GABA is an inhibitory neuromuscular transmitter (144).

There is a total of 30 GABA-immunoreactive neurons in *A. suum*, about 10% of the total neurons (145). Of these, 19 are the inhibitory motoneurons that act on somatic muscle, four (the RME neurons in the nerve ring) are probably head motoneurons innervating head muscles, and one is probably a tail motoneuron (DVB) that innervates anal muscles. Thus 24 of the 30 GABA-containing neurons appear to be motoneurons. This leaves only six GABA-containing neurons in the head ganglia, the only interneurons that contain GABA. It is surprising that this transmitter system is so sparsely represented among the interneurons. A similar distribution is found in identified neurons of *C. elegans* (146, 147). GABA-immunoreactivity has also been found in neurons of *G. ulmi* (148).

The cellular localization of a sodium-dependent GABA uptake system has been described in *A. suum* (149). [³H]GABA accumulates in muscle cells and a small subset of neurons. These neurons show only partial overlap with the neurons that express GABA. The dorsal and ventral inhibitory motoneurons show intense GABA staining, but do not take it up. Certain identified head neurons take up GABA, but do not have GABA-immunoreactivity. Another subset of head neurons has both GABA uptake and immunoreactivity. These observations raise functional questions. The first concerns the site of synthesis of the GABA that is detected immunocytochemically. Since they cannot take it up, presumably the dorsal and ventral motoneurons synthesize the GABA they contain; indeed, the dorsal and ventral nerve cords contain glutamate decarboxylase (GAD), the GABA biosynthetic enzyme (C. D. Johnson and A. O. W. Stretton, unpublished). The neurons that can transport GABA might derive it entirely from an external source and might lack GAD. The localization of GAD must be determined at the single cell level, but no satisfactory antibody for nematode GAD exists. A second question is whether all neurons that contain GABA-immunoreactivity use GABA as a transmitter. GABA content and uptake are only two of several properties linked to GABAergic transmission. To identify the putative signaling dyads, it is important to know which cells can release the GABA they contain and which cells carry GABA receptors. At present, the most direct method for determining these properties is by electrophysiology. Among the GABA-containing neurons this has been applied only to the dorsal and ventral GABA-containing motoneurons, which do indeed make inhibitory synapses to muscle (119).

The pharmacology of GABA receptors on *A. suum* muscle has been explored. Tests with agonists suggest similarities with vertebrate GABA$_A$ receptors (150), but several GABA$_A$ antagonists, including bicucculine and picrotoxin, are inactive. Certain arylaminopyridazine-GABA derivatives, however, act as competitive GABA antagonists in *A. suum* (117, 151), a discovery that has filled an important gap in nematode pharmacology. Both GABA and piperazine open Cl$^-$ channels with a conductance of 22 pS (142). By substituting a series of anions for Cl$^-$, the diameter of the GABA-gated pore is estimated to be between 0.29 and 0.33 nm (152).

The usual degradative enzyme for GABA, a GABA α-ketoglutarate transaminase, has been identified in *C. elegans* (153) and *Nippostrongylus brasiliensis* (154).

GABA appears to play an important role in the control of locomotory movements. In *A. suum* it is known that the GABA-containing inhibitory motoneurons receive their sole synaptic input from the excitatory motoneurons innervating the opposite muscle, thus ensuring reciprocity between dorsal and ventral musculature (155). Similar synaptic connections have been identified morphologically in *C. elegans* (111). There are several uncoordinated mutants of *C. elegans* that affect the GABA-containing neurons. These are being used to analyze the role of GABA in locomotion and other functions (147, 156). In many GABA-defective mutants, the contraction of dorsal and ventral muscle is simultaneous instead of the alternating contractions seen during normal locomotion.

3.3. Glutamate

Glutamate may serve as a neuroactive substance in nematodes, though no satisfactory demonstrations of cellular localization in subsets of neurons have been made. Membrane-associated binding sites for glutamate have been found in *C. elegans* and *H. contortus* (157, 158). Glutamate binding is inhibited by aspartate and, to a lesser extent, by quisqualate, but not by kainate or *N*-methyl-D-aspartate (NMDA).

MK-801, a potent non-competitive antagonist of the vertebrate NMDA receptor (159) has a lethal effect on *C. elegans* (160). Unlike at the MK-801 binding site in rat brains, however, at the *C. elegans* binding site no glutamate agonists that were tested (including kainate, quisqualate, NMDA and glutamate itself) had any effect on glutamate binding, so there is no indication that the MK-801 receptor in nematodes is related to glutamate action. Polyamines compete with MK-801 binding, as does argiopine, a polyamine-containing spider toxin which is lethal to *C. elegans*.

Two observations strengthen the idea that a glutamatergic system exists in nematodes. First, when *C. elegans* mRNA is injected into *Xenopus laevis* oocytes, a glutamate-sensitive Cl^- channel is expressed (161). This channel is sensitive to the anthelmintic ivermectin. Recently, this channel has been cloned and has been shown to be related to glycine and $GABA_A$ receptors (162). Second, in *A. suum* one of the classes of dorsal excitatory motoneurons, class DE2, is depolarized by glutamate, as well as kainate and aspartate, but not by NMDA (Davis and Stretton, unpublished). It is likely that glutamate or a related compound is the endogenous neurotransmitter that activates these channels. A second, slower, depolarizing effect of glutamate is found in motoneurons and in hypodermis; this appears to be due to a glutamate pump (Davis and Stretton, unpublished). As with ACh, it seems that the actions of classical 'neuro'-transmitters are not confined to the nervous system.

3.4. Serotonin (5-HT)

Early work on 5-HT in nematodes was summarized by Willett (163). 5-HT has been reported in *Goodyus ulmi* (148), *C. elegans* (164), *A. suum* (165), *Ascaridia galli* (166) and *Trichostrongylus colubriformis* (167) but was not detected in *H. contortus* (168) or *Phocanema decipiens* (169). The earlier indication that 5-HT is only found in nematodes which are animal intestinal parasites (163) is clearly no longer tenable since it is present in *C. elegans* and *G. ulmi*, two free-living nematodes. However, this suggestion did lead to the concern that the source of 5-HT in intestinal parasitic nematodes might be exogenous, and provoked a search for an endogenous biosynthetic pathway. The two biosynthetic enzymes, tryptophan hydroxylase, which converts tryptophan into 5-hydroxytryptophan (5-HTP), and aromatic amino acid decarboxylase, which converts 5-HTP into 5-HT, have been characterized in female *A. suum* (170). Both activities are present in muscle and intestinal tissues, the intestine having the higher activity. Neither enzyme was detectable in cuticle, female gonadal tissue or hemolymph. Both tryptophan and 5-HT are taken up from the bathing solution into muscle and intestine, and can also apparently cross the cuticle (171).

Two reports raise concern over the authenticity of the identification of 5-HT in some nematodes. Willett (163) summarized a series of experiments on *Panagrellus redivivus*, showing that 5-HT itself could not be detected in this nematode; instead there is a compound resembling 5-HT, but heavier by 58 mass units. In *A. galli*, 5-HTP, 5-HIAA and 5-HT were identified chromatographically in extracts of whole worms, but attempts to produce 5-HT from radiolabeled tryptophan resulted in a compound clearly distinct from 5-HT (172). Complete chemical characterization will reveal whether these compounds are important signaling molecules or are intermediates of the degradative pathways.

The degradative pathway for 5-HT in nematodes has been controversial. Reports of hydroxyindole acetic acid (HIAA) in *A. galli* and *A. suum* imply that oxidative deamination by an MAO-like enzyme occurs (166). No MAO activity was detected in *Prionchulus punctatus, Panagrellus redivivus, Aphelencus avenae* and *C. elegans* (163). 5-HT is *N*-acetylated in *A. galli, A. suum, Brugia pahangi* and *N. brasiliensis* (173, 174). Cut pieces of *B. pahangi* or *A. galli* exposed to radiolabeled 5-HT produced *N*-acetyl-5-HT, but no HIAA was detected. Perhaps MAO is restricted to regions inaccessible to the labeled 5-HT (173).

The cellular localization of 5-HT has been studied more clearly in *C. elegans* and *A. suum*. In both species a pair of neurosecretory neurons in the pharynx (the NSM neurons) show prominent staining with an anti-5-HT antibody (164, 165, 175). In *A. suum* these neurons are highly atypical in that they are extensively branched with many varicosities on the outside surface of the pharynx. In *C. elegans*, but not in *A. suum*, the HSN neurons, a pair of motoneurons that control egg-laying at the vulval muscles, contain 5-HT-immunoreactivity as do six additional neurons in the head. In males of both *A. suum* and *C. elegans* there are five additional 5-HT-immunoreactive cells in the tail; they project to a plexus surrounding the pre-anal ganglion (164; Johnson and Stretton, unpublished). It is not yet known whether these neurons contain the 5-HT synthetic enzymes or whether they accumulate it from the hemolymph. The functional relationship between the 5-HT of the neurons, muscle cells and intestine is not yet clear.

Selective uptake of ^{3}H-labeled 5-HT into neurons occurs in *P. decipiens* (169). Uptake is into neuronal processes in the pharyngeal nerve cords and into processes extending to two large lateral ganglion cells. These cells do not correspond to the NSM cells of *C. elegans* and *A. suum* which are confined to the pharynx itself. It would be very interesting to compare these results with anti-5-HT staining of neurons in *P. decipiens*. 5-HT receptors occur in membrane-enriched preparations from muscle and intestine of *A. suum* (176). These receptors are pharmacologically related, but not identical, to the 5-HT_{1A} and 5-HT_2 receptors of mammals (177).

What does 5-HT do? In *A. suum*, 5-HT stimulates glycogenolysis in muscle (178) and increases cyclic AMP levels and activation of phosphorylase while inactivating glycogen synthase (179). Injection of 5-HT into *A. suum* produces a reversible paralysis of the worm (180), suggesting a possible role in locomotion.

In *C. elegans* 5-HT stimulates egg-laying and inhibits pharyngeal pumping (181). Imipramine, a 5-HT uptake inhibitor, also stimulates egg-laying, but in worms in which HSN neurons are missing, 5-HT (but not imipramine) produces egg-laying, suggesting that 5-HT acts directly on the egg-laying muscles, and that the source of the 5-HT is the HSN neurons (182). A stimulatory effect of 5-HT on egg-laying has also been reported in *G. ulmi* (148).

3.5. Catecholamines and Related Amines

Willett (163) has reviewed the literature prior to 1980. DA, octopamine, and NA as well as their precursors and metabolites have been reported in *A. suum, C. elegans, A. galli, T. colubriformis, N. brasiliensis* and *Litomosoides carinii* (167). In *C. elegans*, formaldehyde-induced fluorescence (FIF) revealed eight dopaminergic sensory neurons; in *A. suum*, only four of these sensory neurons show unequivocal DA-like FIF (183). Tyrosine hydroxylase and aromatic amino acid decarboxylase activity are found in *C. elegans*, so a complete biochemical synthetic pathway for dopamine is present. In the presence of reserpine, FIF is lost from the processes but retained in the cell bodies. Since the classical action of reserpine is to inhibit the pump that concentrates dopamine in synaptic vesicles, this is taken as evidence that these cells concentrate DA in synaptic vesicles. It is therefore surprising that the reserpine-treated worms showed no behavioral deficit. One possible explanation is that DA is only one of several transmitters in these cells. The deirid sensory cells (ADE neurons) of *A. suum* have been shown to contain at least four peptide-like immunoreactivities plus DA (FMRFamide, cholecystokinin, small cardioactive peptide B, and neuropeptide Y (184)).

In *C. elegans*, mutants in four genes (*cat-1, cat-2, cat-3* and *cat-4*) showed reductions in both DA content and FIF intensity; in the most severe case (*cat-2*), DA and FIF were undetectable. None of these mutants have detectable changes in level of either biosynthetic enzyme nor any obvious behavioral defects.

Uptake of ^{3}H-labeled NA and ^{3}H-labeled DOPA into a small subset of non-pharyngeal head neurons was seen in *P. decipiens* (169). As with 5-HT, biosynthetic enzymes were detected in non-neural tissue (185), raising the question of whether the DA in neurons is synthesized in those neurons, or is synthesized elsewhere and concentrated by selective uptake.

Octopamine is present in *C. elegans* (181), but its cellular distribution is not yet known. It acts antagonistically to 5-HT, to produce suppression of egg-laying and stimulation of pharyngeal pumping.

The degradative pathways for catecholamines are also controversial. There are reports of MAO activity in *A. galli* (166), and others showing that *N*-acetylation is the major (and probably the exclusive) metabolic reaction for both DA and octopamine in *A. galli, B. pahangi, N. brasiliensis, A. suum* and *C. elegans* (173, 174). *O*-methylated products of octopamine, DA, epinephrine and NA were described in *T. colubriformis*, suggesting that at least some nematodes have catechol *O*-methyl transferase activity (167).

3.6. Neuropeptides

Considering the powerful roles that neuropeptides play as neurotransmitters and/or neuromodulators in both vertebrates and invertebrates, it is surprising that their presence in nematodes has only been explored relatively recently. However, it is now clear that neuropeptides are present in nematodes and have potent effects on behavior. They may mediate yet more intercellular signaling systems where chemical intervention may lead to effective anthelminthic action. Because of the increasing genetic drug resistance to existing anthelminthics, it is important to discover new families of drugs that affect different targets, and the neuropeptides offer attractive leads for the discovery of such drugs.

A wide range of peptide-like immunoreactivities has been detected in nematodes. In *G. ulmi* ACTH- and FMRFamide-immunoreactivities were found in neurons, but vasoactive intestinal peptide (VIP) and somatostatin-immunoreactivities were absent (148). FMRFamide-immunoreactivity has been found in neurons in *P. redivivus, C. elegans, Heterodera glycines* and *A. suum* (184, 186–188). Bombesin-immunoreactivity has been reported in nematodes of the genera *Haemonchus, Onchocerca, Ascaris* and *Dirofilaria*. The reactivity was highest in the hypodermal–cuticular interface, so it may be non-neural (189). Adipokinetic-hormone-immunoreactivity is present in *P. redivivus* neurons; a peptide fraction obtained from *P. redivivus* evokes an adipokinetic response in locusts (190, 191).

In *A. suum*, a survey of 45 different antisera raised against 42 peptides showed 12 different peptide-like immunoreactivities (184). Each antiserum stained a different subset of neurons, thus formally defining at least one distinct epitope that is selectively localized in those neurons. The 12 positive immunoreactivities were seen with anti-bodies raised against luteinizing hormone releasing hormone (LHRH), small cardioac-tive peptide B, neuropeptide Y (NPY), FMRFamide, cholecystokinin (CCK), gastrin, α-melanocyte stimulating hormone (αMSH), corticotropin releasing factor (CRF), calcitonin-gene related peptide (CGRP), vasoactive intestinal peptide (VIP) and two peptides from *Aplysia*, named L11 and 12_B. The number of neurons stained by different antisera varied widely, ranging from just one pair of neurons, in the case of the LHRH-like immunoreactivity, to over half of the neurons in the worm, for FMRFam-ide-like immunoreactivity. There was no obvious relationship between the occurrence of staining and the phylogenetic relationships to the animal in which the peptide had originally been discovered. No neuropeptide-specific staining was obtained with anti-bodies to ACTH, bombesin, calcitonin, β-endorphin, FSH, GIP, glucagon, growth hormone, GRF, insulin, leu- and met-enkephalin, LH, motilin, neurophysin, neuro-tensin, oxytocin, prolactin, proctolin, somatostatin, substance P, TSH, vasopressin, bag cell peptide, egg-laying hormone, peptides A, 9, and 16A from *Aplysia* and red pigment concentrating hormone.

A second study of *A. suum*, in which 31 antisera were tested, showed positive immunoreactivity with antisera to 11 peptides: pancreatic polypeptide, peptide YY, neuropeptide Y, gastrin, cholecystokinin, substance P, atrial natriuretic peptide, salmon gonadotropin-releasing hormone, mammalian gonadotropin-releasing hormone, chromogranin A, and FMRFamide (192).

Comparison of these two studies showed that in almost all cases there were differences in the reported cellular staining patterns; some differences were minor and others were large. Almost certainly these differences arose because antibodies with different specificities were used. This emphasizes the caution that is appropriate when interpreting the results of immunocytochemical experiments which use antibodies raised against peptides isolated from other organisms (the antibodies used were raised against mammalian, molluscan or arthropod peptides). A positive result merely indicates the presence of a cellular epitope (not necessarily a peptide) that cross reacts with the antibody, and a negative result shows its absence. Whether or not the epitope is present on a peptide that is functionally or evolutionarily related is a matter for further experiments, ideally leading to the isolation and sequence determination of the epitope-bearing molecule. Since there is no necessary coincidence between an epitope and the part of the peptide responsible for biological activity, it is not surprising that different antibodies against a non-endogenous peptide should recognize different molecules. For example, four different monoclonal antibodies raised against mam-malian cholecystokinin recognized four different subsets of neurons in *A. suum* (Sithigorngul, Cowden, and Stretton, unpublished).

Immunocytochemical studies are valuable, however, because they identify classes of neurons, or other cells, that contain a particular immunoreactivity, and allow the peptide chemist to make a strategic choice of which peptide-like immunoreactivity to purify for sequence determination. Furthermore, such studies identify an assay (immunoassay) that can be used to monitor the fractionation. Ultimately, it will be necessary to check the cross reactivity of each of the antisera used for immunocytochemistry with each of the endogenous peptides that are isolated. It will then be possible to interpret more rigorously the present immunocytochemical results. Ideally, monospecific antibodies against each peptide will be raised, as they have for the *Ascaris* peptides AF1 and AF2 (193).

Among the cells in *A. suum* that were specifically stained, there are many examples of neurons showing co-localization of different peptide-immunoreactivity in the same cell (184). For example, a pair of identified neurons called the RIG neurons contained six different immunoreactivities. Whether some, or perhaps all, of the peptide-immunoreactivities are due to coexistence of multiple epitopes on the same molecule, or whether they represent the expression of the products of several neuropeptide-encoding genes in the same cell, will not be clear until the peptide sequences are known so that their cross reactivity with these antisera can be directly tested.

The comparative cellular distribution of the related peptides in different nematodes is extremely interesting, because of its implication for the universality (or not) of signaling mechanisms, and their relevance to fundamental issues such as the basis of speciation, and practical issues such as the cross-species effectiveness of prospective anthelminthics. Unfortunately the data available at present are sparse, the most extensive being the comparisons of the FRMFamide-immunoreactivity in *A. suum* (184), *C. elegans* (188) and *P. redivivus* (199, 203). Although some similar neurons in *A. suum* and *C. elegans* contain FMRFamide-immunoreactivity, there are many others that differ in their content. About 10% of the neurons in *C. elegans* contain FMRFamide-immunoreactivity, compared to over 50% of the neurons in *A. suum*. There are subsets of neurons that are positive in *C. elegans* but negative in *A. suum*, and the converse is also true. In other species, it is harder to make the comparisons since individual neurons have not yet been identified.

Anatomical results obtained with the positive antisera were helpful in making a decision on which of the neuropeptides to attempt to purify from *A. suum*. FMRFamide-immunoreactivity occurs in neurons that innervate somatic muscles. These results suggested that putative FMRFamide-like peptides might be involved in motor control.

HPLC fractionation of extracts of *A. suum* revealed considerable molecular heterogeneity among the FMRFamide-like peptides; at least 10 peaks were found (194). After extensive purification 18 peptides have been isolated and sequenced (194–197 and unpublished). The first two to be sequenced were named AF1 and AF2 (AF = *Ascaris* FMRFamide-like peptide). Their sequences were Lys-Asn-Glu-Phe-Ile-Arg-Phe-amide and Lys-His-Glu-Tyr-Leu-Arg-Phe-amide respectively (194, 196).

Synthetic AF1 and AF2 have potent biological activity on the neuromuscular system of *A. suum*. Injection of AF1 or AF2 produced a local paralysis near the injection site. In a muscle strip preparation, both AF1 and AF2 produced multiple effects on muscle tension, including relaxation, contraction, and induction of rhythmic activity (196, 197). Intracellular recording techniques show that AF1, at 10^{-9}–10^{-7} M, short-circuits electrical activity in inhibitory, but not excitatory motoneurons by opening channels in their membranes (194). The input resistance of these cells is drastically reduced by AF1; this effect is due, at least in large part, to the presence of receptors for AF1 on the inhibitory motoneuron since, when synaptic transmission is blocked with Co^{2+}, the

TABLE 14.2 Subfamilies of AF peptides based on chemical and functional relationships.

Group I	AF1	KNEFIRFa
	AF2	KHEYLRFa
	AF8	KSAYMRFa
	AF5	SGKPTFIRFa
Group II	AF6	FIRFa
	AF7	AGPRFIRFa
Group III	AF3	AVPGVLRFa
	AF4	GDVPGVLRFa
	AF10	GFGDEMSMPGVLRFa
Group IV	AF11	SDIGISEPNFLRFa
		...PNFLRFa*
Group V	AF9	GLGPRPLRFa

*Known only from partial sequences with a C-terminal PNFLRFamide.

AF1-induced effect still occurs. Since these motoneurons depend on their unusually high membrane resistance for long-distance signaling (they do not produce classical action potentials) (198), AF1-induced reductions in membrane resistance would lead to their being effectively removed from the locomotory circuit. This is a possible explanation for the paralysis which occurs in AF1-injected worms, although there may be additional sites of action not yet studied.

Besides AF1 and AF2, 16 other *Ascaris* FMRFamide-like peptides have been isolated and sequenced (12 sequences are shown in Table 14.2) (194–197; Cowden and Stretton, unpublished). These sequences are unique and not related by post-translational modification or, in most cases, by proteolysis; they are also different from FMRFamide-like sequences reported in other organisms, whether obtained from isolated and sequenced peptides, or by deduction from the DNA sequences of genes. It is clear that there is a family of FMRFamide-like peptides in *A. suum*. They can be divided into several subfamilies related by sequence. Initial experiments on their bioactivity suggest that these may also be functional subfamilies. Measurements of effects on muscle tension, and on input resistance of four types of motoneurons, show that there are at least four classes of biological activity controlled by these peptides; AF1 and AF2 form one class that causes contraction and the generation of rhythmic activity in muscle (194, 197), AF3 and AF4 cause muscle contraction (195), AF5 and AF7 reduce the input resistance of both excitatory and inhibitory motoneurons, and AF11 relaxes muscle and increases the input resistance of inhibitory motoneurons (Davis and Stretton, unpublished).

To determine whether the AF peptides are members of an intragenic family or are the products of multiple genes will require cloning and sequencing of the *A. suum* gene or genes that encode them. Recently, six peptides of the subgroup sharing C-terminal PGVLRFamide (AF3, AF4, AF10 and three new peptides) have been shown to be encoded by a complete transcript that includes none of the other AF peptides (Edison, Messinger and Stretton, unpublished). It seems likely that there are multiple genes for the AF peptides.

Specific antibodies to AF1 and AF2 show that they are present in different subsets of neurons (193). Their expression in different cells may be possible because they may be encoded by different genes or may be translated from alternatively spliced transcripts of a single gene.

Two peptides, SDPNFLRFamide and SADPNFLRFamide have been isolated from *P. redivivus* (199). The first shows bioactivity on *A. suum* that differs from that of AF1 and AF2 — it causes relaxation and hyperpolarization of muscle membranes (200, 201). Specific antibodies to AF1 and AF2 failed to detect immunoreactivity in *C. elegans* (202), suggesting that this family is not present. Recently however, AF2 and AF8 have been isolated from *P. redivivus*; AF8 produces a contraction of *Ascaris* muscle (203).

The approach for peptide discovery described above depends on the availability of antibodies against previously characterized peptides, and will only detect peptides that share epitopes with known peptides. This approach is unacceptably limiting: there may well be *A. suum* peptides which would not be detected by existing antibodies, either because they belong to so far undiscovered 'universal' families or because they are structurally unique, and are found only in nematodes, or even only in *A. suum*. There are several existing methods for detecting such peptides. These include purification by the classical method of bioassay (204) and searching for putative peptides by screening cDNA libraries (205, 206, 207). Li (205, 206) has isolated and sequenced a cDNA that encodes eight different putative peptides, seven of which have C-terminal-PNFLRFamide sequences. Six of these peptides have been isolated from extracts of *C. elegans* (202). Two of the sequences are the same as *Panagrellus* peptides PF1 and PF2. Another cDNA, an alternatively spliced version of the genomic sequence, results in a sequence encoding only seven of the peptides (206). With *C. elegans* selected as a model organism for sequencing the entire genome, the sequences of all the putative peptide-encoding genes will be available; this offers an exciting new potential source of sequence information.

In another alternative method, a crude peptide extract (from *A. suum* or *C. elegans*) is conjugated to carrier protein, and used as an immunogen for generating monoclonal antibodies (208). Screening of hybridoma-conditioned culture fluids is carried out by immunocytochemistry on *A. suum* whole mounts. Clones are selected if they produce antibodies staining subsets of cells different from those stained by any previously tested antibody or antiserum. This criterion thus defines novel epitopes. This technology has general applicability and has recently been extended so that the monoclonal antibodies can be used to detect with high sensitivity the unknown peptide during fractionation (193). Compared to radioimmunoassay, there is the considerable advantage that a radiolabeled peptide is not required, an important consideration when the chemical nature of the peptide is still unknown! Several new immunoreactivities have been found by this method. The world of neuropeptides in nematodes seems to be very complex indeed.

REFERENCES

1. Pax, R. A. and Bennett, J. L. (1991) Neurobiology of parasitic platyhelminths: possible solutions to the problems of correlating structure with function. *Parasitology* **102**: S31–S39.
2. Pax, R. A. and Bennett, J. L. (1992) Neurobiology of parasitic flatworms: how much 'neuro' in the biology? *J. Parasitol.* **78**: 194–205.

3. Blair, K. L., Day, T. A., Lewis, M. C., Bennett, J. L. and Pax, R. A. (1991) Studies on muscle cells isolated from *Schistosoma mansoni*: a Ca^{2+}-dependent K^+ channel. *Parasitology* 102: 251–258.

4. Blair, K. L., Bennett, J. L. and Pax, R. A. (1993) Serotonin and acetylcholine: further analysis of praziquantel-induced contraction of magnesium paralyzed *Schistosoma mansoni*. *Parasitology* 107: 387–395.

5. Blair, K. L., Bennett, J. L. and Pax, R. A. (1994) *Schistosoma mansoni*: myogenic characteristics of phorbol-ester induced muscle contraction. *Exp. Parasitol.* 78: 302–316.

6. Day, T. A., Bennett, J. L. and Pax, R. A. (1994) Serotonin and its requirement for maintenance of contractility in muscle fibers isolated from *Schistosoma mansoni*. *Parasitology* 108: 425–432.

7. Chance, M. R. A. and Mansour, T. E. (1949) A kymographic study of the action of drugs on the liver fluke (*Fasciola hepatica*). *Br. J. Pharmacol.* 4: 7–13.

8. Chance M. R. A. and Mansour, T. E. (1953) A contribution to the pharmacology of movement in the liver fluke. *Br. J. Pharmacol.* 8: 134–138.

9. Holmes S. D. and Fairweather, I. (1984) *Fasciola hepatica*: the effects of neuropharmacological agents upon *in vitro* motility. *Exp. Parasitol.* 58: 194–208.

10. Sukhdeo, M. V. K., Sangster, N. C. and Mettrick, D. F. (1986) Effects of cholinergic drugs on longitudinal muscle contractions of *Fasciola hepatica*. *J. Parasitol.* 72: 858–864.

11. Barker, R. L., Bueding, E. and Timms, A. R. (1966) The possible role of acetylcholine in *Schistosoma mansoni*. *Br. J. Pharmacol.* 26: 656–665.

12. Hillman, G. R. and Senft, A. W. (1973) Schistosome motility measurements: response to drugs. *J. Pharmacol. Exp. Ther.* 185: 177–184.

13. Pax, R. A., Siefker, C. and Bennett, J. L. (1984) *Schistosoma mansoni*: differences in acetylcholine, dopamine, and serotonin control of circular and longitudinal parasite muscles. *Exp. Parasitol.* 58: 314–324.

14. McKay, D. M., Halton, D. W., Allen, J. M. and Fairweather, I. (1989) The effects of cholinergic and serotonergic drugs on motility *in vitro* of *Haplometra cylindracea* (Trematoda: Digenea). *Parasitology* 99: 241–252.

15. Maule, A. G., Halton, D. W., Allen, J. M. and Fairweather, I. (1989) Studies in motility *in vitro* of an ectoparasitic monogenean, *Diclidophora merlangi*. *Parasitology* 98: 85–93.

16. Sukhdeo, M. V. K., Hsu, S. C., Thompson, C. S. and Mettrick, D. F. (1984) *Hymenolepis diminuta*: behavioral effects of 5-hydroxytryptamine acetylcholine, histamine and somatostatin. *J. Parasitol.* 70: 682–688.

17. Thompson, C. S., Sangster, N. C. and Mettrick, D. F. (1986) Cholinergic inhibition of muscle contraction in *Hymenolepis diminuta* (Cestoda). *Can. J. Zool.* 64: 2111–2115.

18. Bueding, E. (1952) Acetylcholinesterase activity of *Schistosoma mansoni*. *Br. J. Pharmacol.* 7: 563–566.

19. Graff, D. J. and Read, C. P. (1967) Specific acetylcholinesterase in *Hymenolepis diminuta*. *J. Parasitol.* 53: 1030–1031.

20. Halton, D. W. and Morris, G. P. (1969) Occurrence of cholinesterase and ciliated sensory structures in a fish gill fluke, *Diclidophora merlangi* (Trematoda, monogenea). *Z. Parasitenkd.* 33: 21–30.

21. Wilson, V. C. L. C. and Schiller, E. L. (1969) The neuroanatomy of *Hymenolepsis diminuta* and *H. nana*. *J. Parasitol.* 55: 261–270.

22. Maule, A. G., Halton, D. W., Shaw, C. and Johnston, C. F. (1993) The cholinergic, serotoninergic and peptidergic components of the nervous system of *Moniezia expansa* (Cestoda: Cyclophyllidea). *Parasitology* 106: 429–440.

23. Halton, D. W. (1967) Histochemical studies of carboxylic esterase activity in *Fasciola hepatica*. *J. Parasitol.* 53: 1210–1216.

24. Bueding, E., Schiller, E. L. and Bourgeois, J. G. (1967) Some physiological, biochemical and morphologic effects of Tris (*p*-aminophenyl) carbonium salts (TAC) on *Schistosoma mansoni*. *Am. J. Trop. Med. Hyg.* 16: 500–515.

25. Fripp, P. J. (1967) Biochemical localization of esterase activity in schistosomes. *Exp. Parasitol.* 21: 380–390.

26. Samii, S. I. and Webb, R. A. (1990) Acetylcholine-like immunoreactivity in the cestode *Hymenolepsis diminuta*. *Brain Res.* **513**: 161–165.
27. Webb, R. A. and Mizukawa, K. (1985) Serotonin-like immunoreactivity in the cestode *Hymenolepsis diminuta*. *J. Comp. Neurol.* **234**: 431–440.
28. Lumsden, R. D. and Hildreth, M. B. (1983) The fine structure of adult tapeworms. In: *Biology of the Eucestoda*. Vol. 1 (eds Arme, C. and Pappas, P. W.), Academic Press, New York, pp. 177–233.
29. Webb, R. A. (1987) Innervation of muscle in the cestode *Hymenolepsis microstoma*. *Can. J. Zool.* **65**: 928–935.
30. Hillman, G. R. (1983) The neuropharmacology of schistosomes. *Pharmacol. Ther.* **22**: 103–115.
31. Sukhdeo, M. V. K. and Mettrick, D. F. (1987) Parasite behaviour: understanding platyhelminth responses. *Adv. Parasitol.* **26**: 73–144.
32. Mellin, T. N., Busch, R. D., Wang, C. C. and Kath, G. (1983) Neuropharmacology of the parasitic trematode, *Schistosoma mansoni*. *Am. J. Trop. Med. Hyg.* **32**: 83–93.
33. Tomosky, T. K., Bennett, J. L. and Bueding, E. (1974) Tryptaminergic and dopaminergic responses of *Schistosoma mansoni*. *J. Pharmacol. Exp. Ther.* **190**: 260–271.
34. Thompson, C. S. and Mettrick, D. F. (1984) Neuromuscular physiology of *Hymenolepsis diminuta* and *H. microstoma* (Cestoda). *Parasitology* **89**: 567–578.
35. Ward, S. M., Allen, J. M. and McKerr, G. (1986) Neuromuscular physiology of *Grillotia erinaceus* metacestodes (Cestoda: Trypanorhyncha) *in vitro*. *Parasitology* **93**: 121–132.
36. Mansour, T. E. (1984) Serotonin receptors in parasitic worms. *Adv. Parasitol.* **23**: 1–36.
37. Mansour, T. E. (1957) The effect of lysergic acid diethylamide, 5-hydroxytryptamine and related compounds on the liver fluke, *Fasciola hepatica*. *Br. J. Pharmacol.* **12**: 406–409.
38. Beernink, K. D., Nelson, S. D. and Mansour, T. E. (1963) Effect of lysergic acid derivatives on the liver fluke *Fasciola hepatica*. *Int. J. Neuropharmacol.* **2**: 105–112.
39. Mansour, T. E. (1964) The pharmacology and biochemistry of parasitic helminths. *Adv. Pharmacol.* **3**: 129–165.
40. Bennett, J., Bueding, E., Timms, A. R. and Engstrom, R. G. (1969) Occurrence and levels of 5-hydroxytryptamine in *Schistosoma mansoni*. *Mol. Pharmacol.* **5**: 542–545.
41. Hariri, M. (1974) Occurrence and concentration of biogenic amines in *Mesocestoides corti* (Cestoda). *J. Parasitol.* **60**: 737–743.
42. Lee, M. B., Bueding, E. and Schiller, E. L. (1978) The occurrence and distribution of 5-hydroxytryptamine in *Hymenolepsis diminuta* and *H. nana*. *J. Parasitol.* **64**: 257–264.
43. Thompson, C. S. and Mettrick, D. F. (1989) The effects of 5-hydroxytryptamine and glutamate on muscle contraction in *Hymenolepsis diminuta* (Cestoda). *Can. J. Zool.* **67**: 1257–1262.
44. Terada, M., Ishii, A. I., Kino, H. and Sano, M. (1982) Studies on chemotherapy of parasitic helminths (VI) Effects of various neuropharmacological agents on the motility of *Dipylidium caninum*. *Jap. J. Pharmacol.* **32**: 479–488.
45. Bennett, J. and Bueding, E. (1971) Localization of biogenic amines in *Schistosoma mansoni*. *Comp. Biochem. Physiol.* **39**: 859–867.
46. Chou, T.-C. T., Bennett, J. and Bueding, E. (1972) Occurrence and concentrations of biogenic amines in trematodes. *J. Parasitol.* **58**: 1098–1102.
47. Fairweather, I., Maule, A. G., Mitchell, S. H., Johnston, C. F. and Halton, D. W. (1987) Immunocytochemical demonstration of 5-hydroxytryptamine (serotonin) in the nervous system of the liver fluke, *Fasciola hepatica* (Trematoda, Digenea). ,on *Parasitol. Res.* **73**: 255–258.
48. Gustafsson, M. K. S. (1987) Immunocytochemical demonstration of neuropeptides and serotonin in the nervous system of adult *Schistosoma mansoni*. *Parasitol. Res.* **74**: 168–174.
49. Halton, D. W., Maule, A. G., Johnston, C. F. and Fairweather, I. (1987) Occurrence of 5-hydroxytryptamine (serotonin) in the nervous system of a monogenean, *Diclidophora merlangi*. *Parasitol. Res.* **74**: 151–154.
50. Reuter, M. (1987) Immunocytochemical demonstration of serotonin and neuropeptides in the nervous system of *Gyrodactylus salaris* (Monogenea). *Acta Zool (Stockh)* **68**: 187–193.

51. Sukhdeo, S. C. and Sukhdeo, M. V. K. (1988) Immunohistochemical and electrochemical detection of serotonin in the nervous system of *Fasciola hepatica*, a parasitic flatworm. *Brain Res.* **463**: 57–62.

52. Tomosky-Sykes, T. K., Mueller, J. F. and Bueding E. (1977) Effects of putative neurotransmitters on the motor activity of *Spirometra mansonoides*. *J. Parasitol.* **63**: 492–494.

53. Gustafsson, M. K. S., Wikgren, M. C., Karhi, T. J. and Schot, L. P. C. (1985) Immunocytochemical demonstration of neuropeptides and serotonin in the tapeworm *Diphyllobothrium dendriticum*. *Cell Tissue Res.* **240**: 255–260.

54. Gustafsson, M. K. S. (1991) Skin the tapeworms before you stain their nervous system! *Parasitol. Res.* **77**: 509–516.

55. McKay, D. M., Fairweather, I., Johnston, C. F., Shaw, C. and Halton, D. W. (1991) Immunocytochemical and radioimmunometrical demonstration of serotonin- and neuropeptide-immunoreactivities in the adult rat tapeworm, *Hymenolepis diminuta* (Cestoda: Cyclophyllidea). *Parasitology* **103**: 275–289.

56. Eriksson, K., Gustafsson, M. and Akerlind, G. (1993) High-performance liquid chromatographic analysis of monoamines in the cestode *Diphyllobothrium dendriticum*. *Parasitol. Res.* **79**: 699–702.

57. Marks, N. J., Halton, D. W., Shaw, C. and Johnston, C. F. (1993) A cytochemical study of the nervous system of the proteocephalidean cestode, *Proteocephalus pollanicola*. *Int. J. Parasitol.* **23**: 617–625.

58. Tomosky-Sykes, T. K., Jardine, I., Mueller, J. F. and Bueding, E. (1977) Sources of error in neurotransmitter analysis. *Anal. Biochem.* **83**: 99–108.

59. Shield, J. M. (1971) Histochemical localization of monoamines in the nervous system of *Dipylidium caninum* (Cestoda) by the formaldehyde fluorescence technique. *Int. J. Parasitol.* **1**: 135–138.

60. Shield, J. M. (1969) *Dipylidium caninum, Echinococcus granulosus* and *Hydatigera taeniaeformis*: histochemical identificataion of cholinesterases. *Exp. Parasitol.* **25**: 217–231.

61. Maule, A. G., Halton, D. W., Johnston, C. F., Shaw, C. and Fairweather, I. (1990a) The serotonergic, cholinergic and peptidergic components of the nervous system in the monogenean parasite, *Diclidophora merlangi*: a cytochemical study. *Parasitology* **100**: 255–273.

62. Bennett, J. and Bueding, E. (1973) Uptake of 5-hydroxytryptamine in *Schistosoma mansoni*. *Mol. Pharmacol.* **9**: 311–319.

63. Nimmo-Smith R. H. and Raison, C. G. (1968) Monoamine oxidase activity of *S. mansoni*. *Comp. Biochem. Physiol.* **62C**: 403–416.

64. Mansour, T. E. (1979) Chemotherapy of parasitic worms: new biochemical strategies. *Science* **205**: 462–469.

65. Cho, C. H. and Mettrick, D. F. (1982) Circadian variation in the distribution of *Hymenolepsis diminuta* (Cestoda) and 5-hydroxytryptamine levels in the gastro-intestinal tract of the laboratory rat. *Parasitology* **84**: 431–441.

66. Cyr, D., Gruner, S. and Mettrick, D. F. (1983) *Hymenolepsis diminuta*: Uptake of 5-hydroxytryptamine (serotonin), glucose and changes in worm glycogen levels. *Can. J. Zool.* **61**: 1469–1474.

67. Ribeiro, P. and Webb, R. A. (1983) The synthesis of 5-hydroxytryptamine from tryptophan and 5-hydroxytryptophan in the cestode *Hymenolepsis diminuta*. *Int. J. Parasitol.* **13**: 101–106.

68. Ribeiro, P. and Webb, R. A. (1984) The occurrence, synthesis and metabolism of 5-hydroxytryptamine and 5-hydroxytryptophan in the cestode *Hymenolepsis diminuta*: a high performance liquid chromatographic study. *Comp. Biochem. Physiol.* **79C**: 159–164.

69. Mettrick, D. F. and Cho, C. H. (1982) Changes in tissue and intestinal serotonin (5-HT) levels in the laboratory rat following feeding and the effect of 5-HT inhibitors on the migratory response of *Hymenolepsis diminuta* (Cestoda). *Can. J. Zool.* **60**: 790–797.

70. Mettrick, D. F. and Podesta, R. B. (1982) Effect of gastrointestinal hormones and amines on intestinal motility and the migration of *Hymenolepsis diminuta* in the rat small intestine. *Int. J. Parasitol.* **12**: 151–154.

71. Hillman, G. R., Olsen, N. J. and Senft, A. W. (1974) Effect of methysergide and dihydroergotamine on *Schistosoma mansoni*. *J. Pharmacol. Exp. Ther.* **188**: 529–535.

72. Willcockson, W. S. and Hillman, G. R. (1984) Drug effects on the 5-HT response of *Schistosoma mansoni*. *Comp. Biochem. Physiol.* **77C**: 199–203.
73. McNall, S. J. and Mansour, T. E. (1983) A novel serotonin receptor in the liver fluke *Fasciola hepatica*. *Fed. Proc.* **42**: 1876.
74. McNall, S. J. and Mansour, T. E. (1984) Novel serotonin receptors in *Fasciola*: characterization by studies on adenylate cyclase activation and [³H]LSD binding. *Biochem. Pharmacol.* **33**: 2789–2797.
75. McNall, S. J. and Mansour, T. E. (1984) Desensitization of serotonin-stimulated adenylate cyclase in the liver fluke *Fasciola hepatica*. *Biochem. Pharmacol.* **33**: 2799–2805.
76. Estey, S. J. and Mansour, T. E. (1987) Nature of serotonin-activated adenylate cyclase during development of *Schistosoma mansoni*. *Mol. Biochem. Parasitol.* **26**: 47–60.
77. Tembe, E. A., Holden-Dye, L., Smith, S. W., Jacques, P. A. and Walker, R. J. (1993) Pharmacological profile of the 5-hydroxytryptamine receptor of *Fasciola hepatica* body wall muscle. *Parasitology* **106**: 67–73.
78. Ribeiro, P. and Webb, R. A. (1987) Characterization of a serotonin transporter and an adenylate cyclase-linked serotonin receptor in the cestode *Hymenolepis diminuta*. *Life Sci.* **40**: 755–768.
79. Bennett, J. L. and Gianutsos, G. (1977) Distribution of catecholamines in immature *Fasciola hepatica*: a histochemical and biochemical study. *Int. J. Parasitol.* **7**: 221–225.
80. Gianutsos, G. and Bennett, J. L. (1977) The regional distribution of dopamine and norepinephrine in *Schistosoma mansoni* and *Fasciola hepatica*. *Comp. Biochem. Physiol.* **58C**: 157–159.
81. Gustafsson, M. K. S. and Eriksson, K. (1991) Localization and identification of catecholamines in the nervous system of *Diphyllobothrium dendriticum* (Cestoda). *Parasitol. Res.* **77**: 498–502.
82. Gustafsson M. K. S., Lehtonen, M. A. I. and Sundler, F. (1986) Immunocytochemical evidence for the presence of 'mammalian' neurohormonal peptides in neurons of the tapeworm *Diphyllobothrium dendriticum*. *Cell Tissue Res.* **243**: 41–49.
83. Ribeiro, P. and Webb, R. A. (1983a) The occurrence and synthesis of octopamine and catecholamines in the cestode *Hymenolepis diminuta*. *Mol. Biochem. Parasitol.* **7**: 53–62.
84. Webb, R. A. and Eklove, H. (1989) Demonstration of intense glutamate-like immunoreactivity in the longitudinal nerve cords of the cestode *Hymenolepis diminuta*. *Parasitol. Res.* **75**: 545–548.
85. Keenan, L. and Koopowitz, H. (1982) Physiology and *in situ* identification of putative aminergic neurotransmitters in the nervous system of *Gyrocotyle fimbriata*, a parasitic flatworm. *J. Neurobiol.* **13**: 9–21.
86. Webb, R. A. (1986) The uptake and metabolism of L-glutamate by tissue slices of the cestode *Hymenolepis diminuta*. *Comp. Biochem. Physiol.* **85C**: 151–162.
87. Webb, R. A. (1988) Release of exogenously supplied [³H] glutamate and endogenous glutamate from tissue slices of the cestode *Hymenolepis diminuta*. *Can. J. Physiol. Pharmacol.* **66**: 889–894.
88. Wikgren, M., Reuter, M., Gustafsson, M. K. S. and Lindroos, P. (1990) Immunocytochemical localization of histamine in flatworms. *Cell Tissue Res.* **260**: 479–484.
89. Basch, P. F. and Gupta, B. C. (1988) Immunocytochemical localization of regulatory peptides in six species of trematode parasites. *Comp. Biochem. Physiol.* **91C**: 565–570.
90. Magee, R. M., Fairweather, I., Johnston, C. F., Halton, D. W. and Shaw, C. (1989) Immunocytochemical demonstration of neuropeptides in the nervous system of the liver fluke, *Fasciola hepatica* (Trematoda, Digenea). *Parasitology* **98**: 227–238.
91. McKay, D. M., Halton, D. W., Johnston, C. F., Fairweather, I. and Shaw, C. (1991a) Cytochemical demonstration of cholinergic, serotonergic and peptidergic nerve elements in *Gorgoderina vitelliloba* (Trematoda: Digenea). *Int. J. Parasitol.* **21**: 71–80.
92. Maule, A. G., Halton, D. W., Johnston, C. F., Fairweather, I. and Shaw, C. (1989b) Immunocytochemical demonstration of neuropeptides in the fish-gill parasite, *Diclidophora merlangi* (Monogenoidea). *Int. J. Parasitol.* **19**: 307–316.
93. Maule, A. G., Brennan, G. P., Halton, D. W., Shaw, C., Johnston, C. F. and Moore, S. (1992b) Neuropeptide F-immunoreactivity in the monogenean parasite *Diclidophora merlangi*. *Parasitol. Res.* **78**: 655–660.

94. Skuce, P. J., Johnston, C. F., Fairweather, I., Halton, D. W. and Shaw, C. (1990a) A confocal scanning microscope study of the peptidergic and serotonergic components of the nervous system in larval *Schistosoma mansoni*. *Parasitology* **101**: 227–234.

95. Skuce, P. J., Johnston, C. F., Fairweather, I., Halton, D. W., Shaw, C. and Buchanan, K. D. (1990b) Immunoreactivity to the pancreatic polypeptide family in the nervous system of the adult human blood fluke, *Schistosoma mansoni*. *Cell Tissue Res.* **261**: 573–581.

96. Brennan, G. P., Halton, D. W., Maule, A. G. *et al.* (1993a) Immunoelectron microscopical studies of regulatory peptides in the nervous system of the monogenean parasite, *Diclidophora merlangi*. *Parasitology* **106**: 171–176.

97. Brownlee, D. J. A., Brennan, G. P., Halton, D. W., Fairweather, I. and Shaw, C. (1994) Ultrastructural localization of FMRFamide- and pancreatic polypeptide immunoreactivities within the central nervous system of the liver fluke *Fasciola hepatica* (Trematoda: Digenea). *Parasitol. Res.* **80**: 117–124.

98. Gustafsson, M. K. S. and Wikgren, M. C. (1989) Development of immunoreactivity to the invertebrate neuropeptide small cardiac peptide B in the tapeworm *Diphyllobothrium dendriticum*. *Parasitol. Res.* **75**: 396–400.

99. Fairweather, I. and Halton, D. W. (1991) Neuropeptides in platyhelminths. *Parasitology* **102**: S77–S92.

100. Maule, A. G., Shaw, C., Halton, D. W. *et al.* (1991) Neuropeptide F: a novel parasitic flatworm regulatory peptide from *Moniezia expansa* (Cestoda: Cyclophyllidea). *Parasitology* **102**: 309–316.

101. Maule, A. G., Shaw, C., Halton, D. W., Brennan, G. P., Johnston, C. F. and Moore, S. (1992a) Neuropeptide F (*Moniezia expansa*): localization and characterization using specific antisera. *Parasitology* **105**: 505–512.

102. Brennan, G. P., Halton, D. W., Maule, A. G. and Shaw, C. (1993b) Electron immunogold labeling of regulatory peptide immunoreactivity in the nervous system of *Moniezia expansa* (Cestoda: Cyclophyllidea). *Parasitol. Res.* **79**: 409–415.

103. Fairweather, I., Macartney, G. A., Johnston, C. F., Halton, D. W. and Buchanan, K. D. (1988) Immunocytochemical demonstration of 5-hydroxytryptamine (serotonin) and vertebrate neuropeptides in the nervous system of excysted cysticercoid larvae of the rat tapeworm, *Hymenolepis diminuta* (Cestoda, Cycloophyllidea). *Parasitol. Res.* **74**: 371–379.

104. Fairweather, I., Mahendrasingam, S., Johnston, C. F., Halton, D. W., McCullough, J. S. and Shaw, C. (1990) An ontogenetic study of the cholinergic and serotonergic nervous systems in *Trilocularia acanthiaevulgaris* (Cestoda, Tetraphyllidea). *Parasitol. Res.* **76**: 487–496.

105. Duvaux-Miret, O. and Capron, A. (1992) Proopiomelanocortin in the helminth *Schistosoma mansoni*. Synthesis of beta-endorphin, ACTH, and alpha-MSH: existence of POMC-related sequences. *Ann. NY Acad. Sci.* **650**: 245–250.

106. Duvaux-Miret, O., Leung, M. K., Capron, A. and Stefano, G. B. (1993) *Schistosoma mansoni*: an enkephalinergic system that may participate in internal and host-parasite signaling. *Exp. Parasitol.* **76**: 76–84.

107. Halton, D. W., Brennan, G. P., Maule, A. G., Shaw, C., Johnston, C. F. and Fairweather, I. (1991) The ultrastructure and immunogold labelling of pancreatic polypeptide-immunoreactive cells associated with the egg-forming apparatus of a monogenean parasite, *Diclidophora merlangi*. *Parasitology* **102**: 429–436.

108. Maule, A. G., Halton, D. W., Johnston, C. F., Shaw, C. and Fairweather, I. (1990b) A cytochemical study of the serotonergic, cholinergic and peptidergic components of the reproductive system in the monogenean parasite, *Diclidophora merlangi*. *Parasitol. Res.* **76**: 409–419.

109. Halton, D. W., Shaw, C., Maule, A. G., Johnston, C. F. and Fairweather, I. (1992) Peptidergic messengers: a new perspective of the nervous system of parasitic platyhelminthes. *J. Parasitol.* **78**: 179–193.

110. Rajpara, S. M., Garcia, P. D., Roberts, R. *et al.* (1992) Identification and molecular cloning of a neuropeptide Y homolog that produces prolonged inhibition in *Aplysia* neurons. *Neuron* **9**: 505–513.

111. White, J. G., Southgate, E., Thomson, J. N. and Brenner, S. (1976) The structure of the ventral nerve cord of *Caenorhabditis elegans*. *Phil. Trans. R. Soc. B* **275**: 298–327.

112. Stretton, A. O. W., Fishpool, R. M., Southgate, E. *et al.* (1978) Structure and physiological activity of the motorneurons of the nematode *Ascaris*. *Proc. Natl. Acad. Sci. USA* **75**: 3493–3497.
113. Mellanby, H. (1955) The identification and estimation of acetylcholine in three parasitic nematodes, *Ascaris lumbricoides*, *Litomosoides carinii*, and the microfilariae of *Dirofilaria repens*. *Parasitology* **45**: 287–294.
114. del Castillo, J., de Mello, W. C. and Morales, T. (1963) The physiological role of acetylcholine in the neuromuscular system of *Ascaris lumbricoides*. *Arch. Int. Physiol. Biochim.* **71**: 741–757.
115. Brading, A. F. and Caldwell, P. C. (1971) The resting membrane potential of the somatic muscle cells of *Ascaris lumbricoides*. *J. Physiol.* **217**: 605–624.
116. Martin, R. J. (1982) Electrophysiological effects of piperazine and diethylcarbazine on *Ascaris suum* somatic muscle. *Br. J. Pharmacol.* **77**: 255–265.
117. Martin, R. J., Pennington, A. J., Duittoz, A. H., Robertson, S. and Kusel, J. R. (1991) The physiology and pharmacology of neuromuscular transmission in the nematode parasite *Ascaris suum*. *Parasitology* **102**: S41–S58.
118. Johnson, C. D. and Stretton, A. O. W. (1985) Localization of choline acetyltransferase within identified motorneurons of the nematode *Ascaris*. *J. Neurosci.* **5**: 1984–1992.
119. Walrond, J. P., Kass, I. S., Stretton, A. O. W. and Donmoyer, J. E. (1985) Identification of excitatory and inhibitory motorneurons in the nematode *Ascaris* by electrophysiological techniques. *J. Neurosci.* **5**: 1–8.
120. Davis, R. E. and Stretton, A. O. W. (1989) Signaling properties of *Ascaris* motorneurons: graded active responses, graded synaptic transmission, and tonic transmitter release. *J. Neurosci.* **9**: 415–425.
121. Angstadt, J. D. and Stretton, A. O. W. (1983) Intracellular recordings and Lucifer Yellow fills of interneurons in the nematode *Ascaris*. *Soc. Neurosci. Abstr.* **9**: 302.
122. Johnson, C. D. and Stretton, A. O. W. (1980) Neural control of locomotion in *Ascaris*: anatomy, electrophysiology, and biochemistry. In: *Nematodes as Biological Models*, vol. 1 (ed. Zuckerman, B. M.), Academic Press, New York, pp. 159–195.
123. Colquhoun, L., Holden-Dye, L. and Walker, R. J. (1991) The pharmacology of cholinoceptors on the somatic muscle cells of the parasitic nematode *Ascaris suum*. *J. Exp. Biol.* **158**: 509–530.
124. Lewis, J. A., Wu, C.-H., Berg, H. and Levine, J. H. (1980) The genetics of levamisole resistance in the nematode *Caenorhabditis elegans*. *Genetics* **95**: 905–928.
125. Lewis, J. A., Fleming, J. T., McLafferty, S., Murphy, H. and Wu, C. (1987) The levamisole receptor, a cholinergic receptor of the nematode *Caenorhabditis elegans*. *Mol. Pharmacol.* **31**: 185–193.
126. Sangster, N. C., Davis, C. W. and Collins, G. H. (1991) Effects of cholinergic drugs on longitudinal contractions in levamisole-susceptible and -resistant *Haemonchus contortus*. *Int. J. Parasitol.* **21**: 689–695.
127. Christ, D., Goebel, M. and Saz, H. J. (1990) Actions of acetylcholine and GABA on spontaneous contractions of the filariid *Dipetalonema viteae*. *Br. J. Pharmacol.* **101**: 971–977.
128. Harrow, I. D. and Gration, K. A. F. (1985) Mode of action of the anthelmintics morantel, pyrantel and levamisole on muscle cell membrane of the nematode *Ascaris suum*. *Pestic. Sci.* **16**: 662–672.
129. Brenner, S. (1974) The genetics of *Caenorhabditis elegans*. *Genetics* **77**: 71–94.
130. Culotti, J. G. and Klein, W. L. (1983) Occurrence of muscarinic acetylcholine receptors in wild-type and cholinergic mutants of the nematode *Caenorhabditis elegans*. *J. Neurosci.* **3**: 359–368.
131. Segerberg, M. A. (1989) Nematode cholinergic pharmacology. PhD thesis. University of Wisconsin-Madison.
132. Segerberg, M. A. and Stretton, A. O. W. (1993) Actions of cholinergic drugs in the nematode *Ascaris suum*: complex pharmacology of muscle and motorneurons. *J. Gen. Physiol.* **101**: 1–26.
133. Rand, J. B. and Russell, R. L. (1984) Choline acetyltransferase-deficient mutants of the nematode *Caenorhabditis elegans*. *Genetics* **106**: 227–248.

134. Alfonso-Pizzaro, A., Grundahl, K. and Rand, J. B. (1990) Molecular analysis of the choline acetyltransferase structural gene from the nematode *Caenorhabditis elegans. Soc. Neurosci. Abstr.* **16**: 199.
135. Culotti, J. G., von Ehrenstein, G., Culotti, M. R. and Russell, R. L. (1981) A second class of acetylcholinesterase mutants of *Caenorhabditis elegans. Genetics* **97**: 281–305.
136. Johnson, C. D., Duckett, J. G., Culotti, J. G., Herman, R. K., Meneely, P. M. and Russell, R. L. (1981) An acetylcholinesterase mutant of the nematode *Caenorhabditis elegans. Genetics* **97**: 261–279.
137. Herman, R. K. and Kari, C. K. (1985) Muscle-specific expression of a gene affecting acetylcholinesterase in the nematode *Caenorhabditis elegans. Cell.* **40**: 509–514.
138. Lewis, J. A. and Fleming, J. T. (1991) Cloning nematode acetylcholine receptor genes. *Abstr. Neurotox '91*, p. 8.
139. Alfonso, A., Grundahl, K., Duerr, J. S., Han, H. P. and Rand, J. B. (1993) The *Caenorhabditis elegans* unc-17 gene: a putative vesicular acetylcholine transporter. *Science* **261**: 617–619.
140. del Castillo, J., de Mello, W. C. and Morales, T. (1964) Inhibitory action of gamma aminobutyric acid (GABA) on *Ascaris* muscle. *Experientia* **20**: 141–143.
141. del Castillo, J., de Mello, W. C. and Morales, T. (1964) Mechanism of the paralysing action of piperazine on *Ascaris* muscle. *Br. J. Pharmacol.* **22**: 463–477.
142. Martin, R. J. (1985) Gamma-aminobutyric acid- and piperazine-activated single channel currents from *Ascaris suum* body muscle. *Br. J. Pharmacol.* **84**: 445–461.
143. Ash, A. S. F. and Tucker, J. F. (1967) The bioassay of gamma aminobutyric acid using a muscle preparation from *Ascaris lumbricoides. J. Pharm. Pharmacol.* **19**: 240–245.
144. Johnson, C. D. and Stretton, A. O. W. (1987) GABA-immunoreactivity in inhibitory motor neurons of the nematode *Ascaris. J. Neurosci.* **7**: 223–235.
145. Guastella, J., Johnson, C. D. and Stretton, A. O. W. (1991) GABA-immunoreactive neurons in the nematode *Ascaris. J. Comp. Neurol.* **307**: 584–597.
146. McIntire, S. L., Kaplan, E. J. and Horvitz, H. R. (1993) The GABAergic nervous system of *Caenorhabditis elegans. Nature* **364**: 337–341.
147. McIntire, S. L., Jorgensen, E. and Horvitz, H. R. (1993) Genes required for GABA function in *Caenorhabditis elegans. Nature* **364**: 334–337.
148. Leach, L., Trudgill, D. L. and Gahan, P. B. (1987) Immunocytochemical localization of neurosecretory amines and peptides in the free-living nematode *Gooeyus ulmi. Histochemistry* **19**: 471–475.
149. Guastella, J. and Stretton, A. O. W. (1991) Distribution of ^{3}H-GABA uptake sites in the nematode *Ascaris. J. Comp. Neurol.* **307**: 598–608.
150. Holden-Dye, L., Krogsgaard-Larsen, P., Neilsen, L. and Walker, R. J. (1989) GABA receptors on the somatic muscle cells of the parasitic nematode, *Ascaris suum*: stereoselectivity indicates similarity to a GABA$_a$-type agonist recognition site. *Br. J. Pharmacol.* **98**: 841–850.
151. Duittoz, A. H. and Martin, R. J. (1991) Effects of the arylaminopyrazine-GABA derivatives SR95103 and SR95531 on the *Ascaris* muscle GABA receptor: the relative potency of the antagonists in *Ascaris* is different to that at vertebrate GABA$_a$ receptors. *Comp. Biochem. Physiol.* **98C**: 417–422.
152. Parri, H. R., Holden-Dye, L. and Walker, R. J. (1991) Studies on the ionic selectivity of the GABA-operated chloride channel on the somatic muscle bag cells of the parasitic nematode *Ascaris suum. Exp. Physiol.* **76**: 597–606.
153. Hedgecock, E. (1976) GABA metabolism in *Caenorhabditis elegans.* PhD thesis. University of California, Santa Cruz.
154. Watts, S. D. M. and Atkins, A. M. (1984) Kinetics of 4-aminobutyrate: 2-oxyglutarate aminotransferase from *Nippostrongylus brasiliensis. Mol. Biochem. Parasitol.* **12**: 207–216.
155. Walrond, J. P. and Stretton, A. O. W. (1985) Reciprocal inhibition in the motor nervous system of the nematode *Ascaris*: direct control of ventral inhibitory motorneurons by dorsal excitatory motorneurons. *J. Neurosci.* **5**: 9–15.
156. Thomas, J. H. (1990) Genetic analysis of defecation in *Caenorhabditis elegans. Genetics* **124**: 855–872.

157. Schaeffer, J. M., White, T., Bergstrom, A. R., Wilson, K. E. and Turner, M. (1990) Identification of glutamate-binding sites in *Caenorhabditis elegans*. *Pesticide Biochem. Physiol.* **36**: 220–228.
158. Rohrer, S. P., Evans, W. D. and Bergstrom, A. (1990) A membrane associated glutamate binding protein from *Caenorhabditis elegans* and *Haemonchus contortus*. *Comp. Biochem. Physiol.* **95C**: 223–228.
159. Wong, E. H. F., Kemp, J. A., Priestly, A., Knight, A. R., Woodruff, G. N. and Iversen, L. L. (1986) The anticonvulsant MK-801 is a potent N-methyl-D-aspartate antagonist. *Proc. Natl. Acad. Sci. USA* **83**: 7104–7108.
160. Schaeffer, J. M., Bergstrom, A. R. and Turner, M. J. (1989) MK-801 is a potent nematocidal agent. Characterization of MK-801 binding sites in *Caenorhabditis elegans*. *Biochem. J.* **260**: 923–926.
161. Arena, J. P., Liu, K., Paress, P. S. and Cully, D. F. (1991) Avermectin-sensitive chloride current induced by *Caenorhabditis elegans* RNA in *Xenopus* oocytes. *Mol. Pharmacol.* **40**: 368–374.
162. Cully, D. F., Vassilatis, D. K., Liu, K. K. *et al.* (1994) Cloning of an avermectin-sensitive glutamate-gated chloride channel from *Caenorhabditis elegans*. *Nature* **371**: 707–711.
163. Willett, J. D. (1980) Control mechanisms in nematodes. In: *Nematodes as Biological Models*, vol. 1 (ed. Zuckerman, B. M.), Academic Press, New York, pp. 197–225.
164. Desai, C., Garriga, G., McIntire, S. L. and Horvitz, H. R. (1988) A genetic pathway for the development of the *Caenorhabditis elegans* HSN motor neurons. *Nature* **336**: 638–646.
165. Stretton, A. O. W. and Johnson, C. D. (1985) GABA and 5HT immunoreactive neurons in *Ascaris*. *Soc. Neurosci. Abstr.* **11**: 626.
166. Mishra, S. K., Sen, R. and Ghatak, S. (1984) *Ascaris lumbricoides* and *Ascaridia galli*: biogenic amines in adults and developmental stages. *Exp. Parasitol.* **57**: 34–39.
167. Fransden, J. C. and Bone, L. W. (1987) Biogenic amines and their metabolites in *Trichostrongylus colubriformis*, a nematode parasite of ruminants. *Comp. Biochem. Physiol.* **87C**: 75–77.
168. Rogers, W. P. and Head, R. (1972) The effect of the stimulus for infection on hormones in *Haemonchus contortus*. *Comp. Gen. Pharmacol.* **3**: 6–10.
169. Goh, S. L. and Davey, K. G. (1976) Selective uptake of adrenaline, DOPA, and 5-hydroxy-tryptamine by the nervous system of *Phocanema decipiens* (Nematoda): a light autoradio-graphic and ultrastructural study. *Tissue Cell* **8**: 421–435.
170. Chaudhuri, J., Martin, R. E. and Donahue, M. J. (1988) Tryptophan hydroxylase and aromatic L-amino acid decarboxylase activities in the tissues of adult *Ascaris suum*. *Int. J. Parasitol.* **18**: 341–346.
171. Chaudhuri, J., Martin, R. E. and Donahue, M. J. (1988) Evidence for the absorption and synthesis of 5-hydroxytryptophan in perfused muscle and intestinal tissue and whole worms of adult *Ascaris suum*. *Parasitology* **96**: 157–170.
172. Smart, D. (1988) Investigations of the synthesis and metabolism of 5-hydroxytryptamine in *Ascaridia galli* (Nematoda). *Int. J. Parasitol.* **18**: 747–752.
173. Isaac, R. E., Muimo, R. and Macgregor, A. N. (1990) N-acetylation of serotonin, octopamine and dopamine by adult *Brugia pahangi*. *Mol. Biochem. Parasitol.* **43**: 193–198.
174. Isaac, R. E., Eaves, L., Muimo, R. and Lamango, N. (1991) N-acetylation of biogenic amines in *Ascaridia galli*. *Parasitology* **102**: 445–450.
175. Brownlee, D. J. A., Fairweather, I., Johnston, C. F. and Shaw, C. (1994) Immunocytochemical demonstration of peptidergic and serotoninergic components of the enteric nervous system of the roundworm, *Ascaris suum* (Nematoda, Ascaroidea). *Parasitology* **108**: 89–103.
176. Chaudhuri, J. and Donahue, M. J. (1989) Serotonin receptors in the tissues of adult *Ascaris suum*. *Mol. Biochem. Pharmacol.* **35**: 191–198.
177. Williams, J. A., Shahkolahi, A. M., Abbassi, M. and Donahue, M. J. (1992) Identification of a novel 5-HT$_N$ (Nematoda) receptor from *Ascaris suum* muscle. *Comp. Biochem. Physiol.* **101C**: 469–474.
178. Donahue, M. J., Yacoub, N. J. and Harris, B. J. (1982) Correlation of muscle activity with glycogen metabolism in muscle of *Ascaris suum*. *Am. J. Physiol.* **242**: R514–R521.

179. Donahue, M. J., Yacoub, N. J., Michinoff, C. A., Massaracchia, R. A. and Harris, B. J. (1981) Serotonin (5-hydroxytryptamine): a possible regulator of glycogenolysis in perfused muscle segments of *Ascaris suum. Biophys. Biochem. Res. Commum.* **101**: 112–117.

180. Buchanan, C. A. and Stretton, A. O. W. (1991) The effects of biogenic amines on *Ascaris* locomotion. *Soc. Neurosci. Abstr.* **17**: 279.

181. Horvitz, R. H., Chalfie, M., Trent, C., Sulston, J. E. and Evans, P. D. (1982) Serotonin and octopamine in the nematode *Caenorhabditis elegans. Science* **206**: 1012–1014.

182. Trent, C., Tsung, N. and Horvitz, H. R. (1983) Egg-laying defective mutants of the nematode *Caenorhabditis elegans. Genetics* **104**: 619–647.

183. Sulston, J., Dew, M. and Brenner, S. (1975) Dopamine neurons in the nematode *Caenorhabditis elegans. J. Comp. Neurol.* **163**: 215–244.

184. Sithigorngul, P., Stretton, A. O. W. and Cowden, C. (1990) Neuropeptide diversity in *Ascaris*: an immunocytochemical study. *J. Comp. Neurol.* **294**: 362–376.

185. Smart, D. (1988) Catecholamine synthesis in *Ascaridia galli* (Nematoda). *Int. J. Parasitol.* **18**: 485–492.

186. Atkinson, H. J., Isaac, R. E., Harris, P. D. and Sharpe, C. M. (1988) FMRFamide-like immunoreactivity within the nervous system of the nematodes *Panagrellus redivivus, Caenorhabditis elegans*, and *Heterodera glycines. J. Zool. (Lond)* **216**: 663–671.

187. Davenport, T. R. B., Lee, D. L. and Isaac, R. E. (1988) Immunocytochemical demonstration of a neuropeptide in *Ascaris suum* (Nematoda) using an antiserum to FMRFamide. *Parasitology* **97**: 81–88.

188. Schinkmann, K. and Li, C. (1992) Localization of FMRFamide-like peptides in *Caenorhabditis elegans. J. Comp. Neurol.* **316**: 251–260.

189. Huntingdon, M. K., Mackenzie, C. D., Geary, T. G. and Williams, J. F. (1991) Bombesin-like immunoreactivity in parasitic helminths. *Abstr. Am. Soc. Parasitol.* p. 64.

190. Davenport, T. R. B., Isaac, R. E. and Lee, D. L. (1991) The presence of peptides related to the adipokinetic hormone/red pigment concentrating hormone family in the nematode, *Panagrellus redivivus. Gen. Comp. Endocrinol.* **81**: 419–425.

191. Davenport, T. R. B., Eaves, L. A., Hayes, T. K., Lee, D. L. and Isaac, R. E. (1994) The detection of AKH/HrTH-like peptides in *Ascaridia galli* and *Ascaris suum* using an insect hyperglycaemic bioassay. *Parasitology* **108**: 479–485.

192. Brownlee, D. J. A., Fairweather, I., Johnston, C. F., Smart, D., Shaw, C. and Halton, D. W. (1993) Immunocytochemical demonstration of neuropeptides in the central nervous system of the roundworm, *Ascaris suum* (Nematoda: Ascaroidea). *Parasitology* **106**: 305–316.

193. Sithigorngul, P. and Stretton, A. O. W. (1991) Differential distribution of AF1, a FMRFamide-like peptide in the *Ascaris* nervous system, revealed by a specific monoclonal antibody. *Abstr. Soc. Neurosci.* **17**: 279.

194. Cowden, C., Stretton, A. O. W. and Davis, R. E. (1989) AF1, a sequenced bioactive neuropeptide isolated from the nematode *Ascaris. Neuron* **2**: 1465–1473.

195. Cowden, C. and Stretton, A. O. W. (1995) Eight novel FMRFamide-like peptides isolated from the nematode *Ascaris suum. Peptides* **16**: 491–500.

196. Cowden, C. and Stretton, A. O. W. (1990) AF2, a nematode neuropeptide. *Soc. Neurosci. Abstr.* **16**: 305.

197. Cowden, C. and Stretton, A. O. W. (1993) AF2, an *Ascaris* neuropeptide: isolation, sequencing and bioactivity. *Peptides* **14**: 423–430.

198. Davis, R. E. and Stretton, A. O. W. (1989) Passive membrane properties of motorneurons and their role in long distance signaling in the nematode *Ascaris. J. Neurosci.* **9**: 403–414.

199. Geary, T. G., Price, D. A., Bowman, J. W. *et al.* (1992) Two FMRFamide-like peptides from the free-living nematode *Panagrellus redivivus. Peptides* **13**: 209–214.

200. Bowman, J. W., Geary, T. G. and Thompson, D. P. (1991) Electrophysiological characterization of the effects of nematode FMRFamide-like neuropeptides on *Ascaris suum* muscles. *Abstr. Neurotox '91*, 129.

201. Franks, C. J., Holden-Dye, L., Williams, R. G., Pang, F. Y. and Walker, R. J. (1994) A nematode FMRFamide-like peptide, SDPNFLRFamide (PF1), relaxes the dorsal muscle strip preparation of *Ascaris suum. Parasitology* **108**: 229–236.

202. Rosoff, M. L., Doble, K. E., Price, D. A. and Li, C. (1993) The *flp-1* propeptide is processed into multiple, highly similar FMRFamide-like peptides in *Caenorhabditis elegans*. *Peptides* **14**: 331–338.
203. Maule, A. G., Shaw, C., Bowman, J. W. *et al.* (1994) KSAYMRFamide: a novel FMRF-amide-related heptapeptide from the free-living nematode, *Panagrellus redivivus*, which is myoactive in the parasitic nematode, *Ascaris suum*. *Biochem. Biophys. Res. Commun.* **200**: 973–980.
204. Smart, D., Preston, C. M. and Lloyd, D. (1990) The use of a novel motility assay for the isolation of bioactive peptides from the free-living nematode *Panagrellus redivivus*. *Comp. Biochem. Physiol.* **95B**: 335–339.
205. Li, C. (1990) FMRFamide-like peptides in *C. elegans*: developmental expression and cloning and sequencing of the gene. *Soc. Neurosci. Abstr.* **16**: 305.
206. Rosoff, M. L., Burglin, T. R. and Li, C. (1992) Alternatively spliced transcripts of the *flp-1* gene encode distinct FMRFamide-like peptides in *Caenorhabditis elegans*. *J. Neurosci.* **12**: 2356–2361.
207. Nambu, J. R., Taussig, R., Mahon, A. C. and Scheller, R. H. (1983) Gene isolation with cDNA probes from identified *Aplysia* neurons: neuropeptide modulators of cardiovascular physiology. *Cell* **35**: 47–56.
208. Sithigorngul, P., Cowden, C., Guastella, J. and Stretton, A. O. W. (1989) Generation of monoclonal antibodies against nematode peptide extract: another approach for identifying unknown peptides. *J. Comp. Neurol.* **284**: 389–397.

15 Reproduction and Development in Helminths

H. RAY GAMBLE, RAYMOND H. FETTERER
and JOSEPH F. URBAN, JR

*United States Department of Agriculture, Agricultural Research Service,
Livestock and Poultry Sciences Institute, Parasite Biology and Epidemiology
Laboratory, Beltsville, MD, USA*

SUMMARY

Evolution of successful mechanisms of transmission in helminth parasites has
resulted in an array of reproductive and developmental strategies. Despite
considerable descriptive literature, little knowledge exists on molecular aspects of
reproduction. One exception is the study of a conserved polypeptide, the major
sperm protein, associated with motility in nematode sperm. Eggshell composition
and assembly have been studied in some detail for helminths. Shells are generally
multilayered, with some type of cross linking for stabilization, and are designed
to be resistant to environmental challenges. Egg hatching is triggered by alterations in the external environment and effected changes in eggshell permeability
and by intrinsic and extrinsic enzymes. Hatched helminth larvae transform
through various larval stages, in the trematodes and cestodes, or mature through
a series of four molts in the nematodes. Where dormant stages occur, mechanisms
of encystment and excystment rely on environmental cues. Nematode molting is
also dependent on environmental stimuli, and stages outside the host often
become quiescent until they contact or are ingested by a suitable host. Molting
and exsheathment are separate processes in many nematodes with distinct stimuli
and effector mechanisms. Helminth reproduction and subsequent development
is regulated by numerous secreted and internal factors, including pheromones
for mate finding and pairing, and ecdysteroids in regulation of growth. Some

Biochemistry and Molecular Biology of Parasites
ISBN 0-12-473345-X

associations of ecdysteroids and molting have also been demonstrated. Much remains to be understood about the biochemical bases and molecular aspects of reproduction and development in helminths. Utilization of new methods for *in vitro* cultivation of these parasites will assist in future studies.

1. INTRODUCTION

Helminths have evolved a variety of mechanisms for reproduction, dispersal and development. These adaptations are required because opportunities for transmission of parasites, in contrast to many other infectious organisms, are generally infrequent. Thus the perpetuation of a species relies on the production of large numbers of next generation organisms. Similarly, parasite development takes many forms in order to survive, encounter and successfully infect a new host. To accomplish these goals, parasites pass through a number of different stages, in the external environment and in one or more intermediate hosts, prior to arriving in the definitive host, where sexual reproduction takes place.

Although there is a vast amount of literature detailing the morphological events surrounding parasite reproduction and development, the underlying biochemical mechanisms controlling these complex processes are largely unknown. Specific cues for initiation of new developmental processes clearly exist, and are of importance in understanding, and perhaps controlling, parasites. The basis for major shifts in body form, internal and external organs, growth rates and asexual division processes is undoubtedly under tight genetic control, yet studies of developmental processes at the molecular level are scarce in the field of parasitology. This chapter, will highlight several events in parasite development that have been studied at the biochemical and molecular level and point to other areas where such studies might aid in understanding the organism and also in developing new control strategies for helminth parasites.

2. STRATEGIES FOR REPRODUCTION

Reproductive processes of helminths are diverse. The adults are hermaphroditic (monoecious) or dioecious, requiring, in some cases, no fertilization, in others, self- or cross-fertilization in hermaphrodites or insemination of females by males. Reproduction by monoecious species is simplified since no mate finding is necessary. In those species where separated sexes exist, and in hermaphroditic species where cross-fertilization is preferred, various adaptations have increased the likelihood of worm pairing and mate finding in the host.

There is little biochemical information on spermiogenesis, gametogenesis and fertilization. One area which has received some attention is the unique mechanism of motility in nematode sperm. Nematode sperm are non-flagellated cells, lacking an acrosome and flagella, and exhibiting pseudopodial locomotion. Studies on *Caenorhabditis*, *Ascaris* and several other parasitic nematodes have demonstrated that a highly conserved polypeptide, designated the major sperm protein (MSP) is responsible for sperm motility (1). The MSP is found only in nematode testes, and comprises more than 15% of total protein in the sperm cell. During spermiogenesis, MSP is synthesized and assembles into filamentous fibrous bodies which transport MSP and other sperm proteins to the spermatid. MSP then dissociates in the cytoplasm until later in spermio-genesis when it reassembles into 2–3 nm filaments in the pseudopod to facilitate ameboid movement (2, 3). Several genes encoding MSPs have been sequenced and reveal considerable conservation at the amino acid level in different species of nematodes (1).

3. BIOCHEMICAL ASPECTS OF DEVELOPMENTAL PROCESSES

3.1. Eggshell Formation

For oviparous and ovoviviparous reproductive strategies, the eggshell is of considerable importance in protecting the developing embryo from adverse environmental influen-ces. Shell formation is initiated immediately after fertilization by processes which are quite different in the platyhelminths and the nematodes. The chemical composition of eggshells and the sources of their origin and assembly are only partially understood.

In the platyhelminths, the fertilized oocyte is surrounded by molecules produced by two specialized structures, the vitellarium and the Mehlis's gland. In schistosomes, vitelline cells, which are deposited around the fertilized ova, exocytose hydrophobic droplets in the space between the oocyte and an outer boundary formed by Mehlis' gland secretions (4). Eggshell precursor polypeptides, originating from vitelline cells, have been cloned from several species of schistosomes (5, 6) and are shown to be rich in glycine (40–50%) and tyrosine (10%) residues. Several mechanisms for assembly of the eggshell have been proposed. Although the Mehlis' gland was originally thought to contribute crosslinking agents to shell formation, this has not been confirmed. It now appears that the viscous secretions of the Mehlis' gland, suggested to be lipoprotein in nature, serve as a buffer region between vitelline cells and the ootype during shell deposition. Eggshell precursor polypeptides, liberated from vitelline cells by exocytosis, remain as hydropho-bic aggregates until they contact the external shell boundary. Hydrophobicity is thought to be due to a histidine-rich region of these molecules (4). Low pH in the exocytotic vesicles might serve to inhibit crosslinking mechanisms (phenol oxidase activity) of shell precursors. Chemical changes, perhaps in response to higher pH at the lipoprotein boundary membrane, result in physical changes to the vitelline cell aggregates which fuse into an evenly distributed layer around the shell surface.

The eggshell is stabilized by crosslinking of its components. A number of studies have suggested the presence of sclerotin or quinone-tanned proteins in eggshells of trematodes. Sclerotinization or tanning occurs by enzymatic oxidation of phenols, either free phenols or phenolic amino acid constituents of proteins, to quinones. The highly reactive quinones form covalent bonds with free amino groups in adjacent peptide chains resulting in a highly stable, insoluble and rigid protein (7). Sclerotiniz-ation is most often associated with the presence of the copper-containing enzyme phenol oxidase. Histochemical studies of eggshells and eggshell precursors from a variety of trematode species have demonstrated cytochemical reactions characteristic of phenols and phenol oxidase (reviewed in ref. 8). Studies of *Schistosoma mansoni* suggest a definite role for phenol oxidase in tanning of the eggshell. Diethyldithiocarbamate and disulfiram, copper chelators and potent inhibitors of phenol oxidase and related enzymes, inhibit *S. mansoni* phenol oxidase; disulfiram also prevents egg laying *in vivo* (9). In addition, phenol oxidase is exclusively associated with the female schistosome (10).

Recent studies have identified shell precursors from *S. mansoni* and *Fasciola hepatica* with molecular masses of 14–48 kDa (5, 6) further suggesting a role for sclerotinization in eggshell formation. A 31 kDa precursor from *F. hepatica* is particularly unique in that it is rich in dihydroxyphenylalanine (DOPA), a potential precursor of quinone crosslinking (11). A 35 kDa eggshell precursor, identified by pulse labeling *S. mansoni* with [^{14}C]tyrosine, is converted into an approximately 100 kDa protein in shells of newly laid eggs most likely due to a quinone-mediated crosslinking mechanism (12).

In addition to sclerotinization there is some older evidence suggesting the existence of other mechanisms of protein stabilization. Keratin or keratin-like proteins characterized by high sulfur (presumably cysteine) content have been identified by histochemical methods in several trematode species and may be characteristic of paramphistomes (8, 13). Crosslinking by oxidative coupling of tyrosine-residues to form dityrosine may also occur in trematode eggshells (14). Dityrosine crosslinks are well known in nematode cuticles (15) but detailed analyses of tyrosine derived crosslinks in eggshells has not been performed.

In contrast to the contributions of other organs to the platyhelminth eggshell, the nematode eggshell is formed from elements present in the fertilized oocyte (16). Initiation of eggshell formation is a direct result of fertilization. Typically, the shell consists of three layers, an external lipoprotein vitelline layer bounded by a trilaminate membrane, a medial layer consisting of varying amounts of chitin and protein, and a basal lipoprotein layer (Fig. 15.1). One or more additional external layers may be contributed by uterine elements. Eggshell formation in nematodes occurs typically as follows. Upon fertilization, a secondary oolemma is formed and the outer membrane separates from the oolemma and transforms into the vitelline layer. *N*-acetyl-D-glucosamine, derived from stored glycogen, amino acids and acetate (from ascarosides) are crosslinked by chitin synthetase and deposited into the intermembrane space. Varying amounts of chitin spindles are mixed into a protein matrix to make up the most prominent layer of the eggshell. This protein/chitin layer likely accounts for the rigid structure of the shell. Lipids, also produced by the oocyte, are deposited beneath the chitin layer by extrusion of refringent granules from the egg cytoplasm. The lipid layer in the *Ascaris* eggshell has been studied in some detail. During oogenesis, ascarosides are deposited from the ovarian tissue. Ascarosides are high-molecular-weight glycolipids, derived from the glycone ascarylose (3,6-dideoxy-L-mannose) and long chain (C_{25}–C_{34}) branched or unbranched secondary alcohols (monol ascarosides) or unbranched diols (diol ascarosides). Prior to fertilization, ascarosides exist as acetate or propionate esters. Following fertilization, ascaroside esters in the refringent granules are converted into free ascarosides which form the basal lipid layer of the eggshell. Coincident with processing of ascaroside esters is the liberation of acetate which is thought to be used in the subsequent synthesis of *N*-acetylglucosamine, through an acetyl-CoA intermediate, and incorporation into chitin.

In the Oxyurida and the Ascaridida, in addition to the three layers of eggshell originating from the oocyte, additional layers are deposited by the uterine cells over the vitelline layer. Uterine cell contributions to the eggshell have not been studied in depth, although these deposits are thought to be lipoprotein in nature and have been shown to assume a complex architecture (16).

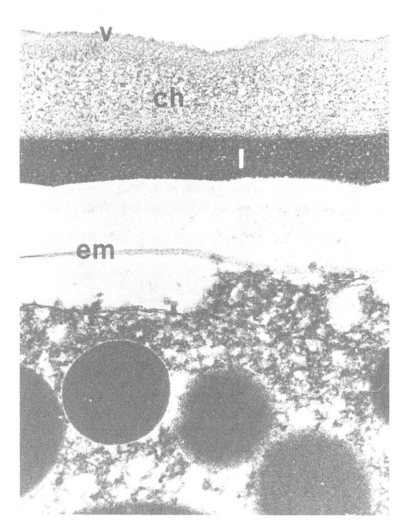

FIG. 15.1 Electron micrograph of the eggshell of the nematode *Haemonchus contortus* (from sheep) showing the vitelline (v), chitin (ch) and lipid (l) layers and the embryo bounded by the embryonic membrane (em).

The mechanisms for the stabilization of nematode eggshells are known with less certainty than in trematodes. In *A. suum*, eggshells recovered from the host feces are brown and relatively insoluble compared to shells of eggs recovered from the uterus or laid *in vitro*. Phenols have been extracted from the eggshells of *A. suum*, but phenol oxidase has not been demonstrated (16). Phenols have also been demonstrated in the eggshell of *Trichuris ovis* (17) and the presence of phenol oxidase in female *Trichuris suis* has been demonstrated (18) and localized to the eggshell (Fetterer, unpublished observation). The role of sclerotinization or of other stabilization mechanisms for nematode eggshell proteins remains undefined.

Helminth eggshells have limited permeability and thus are protected from adverse environmental conditions. The permeability of nematode eggshells is limited to passage of respiratory gases and lipid solvents, although water may move slowly through the shell (16). The lipid layer is thought to be primarily responsible for regulating permeability. Changes in the permeability of the shell are characteristic of hatching in nematodes.

3.2. Mechanisms of Egg Hatching

Escape from the eggshell is critical to parasite larvae, as defects in hatching preclude the organism from further development. Hatching mechanisms include mechanical activity of the larvae inside the eggshell, enzymatic secretions of the larvae, or a combination of the two.

In those trematodes that have operculate eggs, hatching involves several mechanisms including permeability changes in the eggshell, an increase in internal pressure due to the influx of water and the physical opening of the operculum solely by force or in conjunction with the release of enzymes which attack the edges of the operculum. These series of events are stimulated by factors in the environment, including light, osmotic pressure and carbon dioxide tension. Hatching of non-operculate, schistosome eggs has received considerable attention although the precise mechnanism remains unknown (19). Schistosome eggs hatch when eggs are placed in fresh water; this alteration of the osmotic gradient across the eggshell results in diffusion of water into the egg and perhaps in the rupture of the shell. The possibility of enzymes participating in this process is suggested by the finding that a leucine aminopeptidase (LAP) is associated with the vitelline membrane of schistosome eggs and that its activity is regulated by salt concentration in a manner similar to hatching stimuli.

Stimulation of hatching of cestode eggs is also elicited by factors related to their environment. Pseudophyllidan species such as *Diphyllobothrium* respond to light and hatch very rapidly. It is not clear, however, whether enzymes are involved in this process. In some Cyclophyllidean species, mechanical distruption of the eggshell and underlying embryonic layers followed by action of host digestive enzymes is responsible for liberating the oncosphere (20). In the Taeniidae the eggshell is a very thin membrane and is generally not seen in fecal specimens; host digestive enzymes are thought to act upon the thickened embryophore region, which consists of keratin blocks (21).

Hatching in nematodes is initiated by various stimuli resulting in events including an alteration of the permeability of the eggshell, an influx of water, an increase in larval activity, and, in some cases, release of enzymes. Several studies (22, 23) have demonstrated alterations in permeability of the nematode eggshell prior to hatching by examining influx of compounds of known molecular weight. In general, permeability changes result from a breakdown of the lipid layer of the eggshell, as a result of either mechanical disruption or perhaps the action of enzymes. Permeability is characterized by an efflux of trehalose and an influx of water, often increasing the volume of the shell. It has been suggested that, in *Ascaris suum*, the penetration of the shell by carbon dioxide causes hydration and poration of the ascaroside layer. The loss of trehalose, a compound which presumably functions in cold tolerance and protection from desiccation (24), is accompanied by influx of water allowing the larva more room to move. Subsequent changes in the eggshell are probably elicited by enzymes released from the larva. Enzymes reported from hatching eggs of nematodes include a chitinase, a lipase (esterase), α- and β-glycosidases, a trypsin-like protease, and LAP from *A. lumbricoides* (23,25), chitinase, collagenase, lipase and actylcholinesterase from *Trichostrongylus colubriformis* (26), and lipase and LAP from *Haemonchus contortus* (27). The action of certain enzymes likely softens the eggshell allowing more flexibility and movement by the larva, making it easier for the larva to mechanically disrupt the shell. However, the role of specific enzymes in hatching remains to be established.

3.3. Transformation of Larval Stages

3.3.1. Platyhelminths

The transition of parasites through larval stages to the adult is characterized by a number of changes in body form and growth. These transformations are regulated by factors both intrinsic and extrinsic to the parasite. Biochemical events which occur during these changes are, for the most part, not well understood. However, some examples of parasite development at the biochemical level have been documented.

Metacercariae of a number of species of trematodes go through a process of encystment prior to ingestion by the definitive host. Excystment is initiated in response to environmental cues and is affected by host parasite factors. In *Fasciola* sp., the metacercarial cyst is produced by several types of cystogenous glands (28). Secretions of these glands produce four layers of cyst including an outer, tanned protein or sclerotin layer, a fibrous layer consisting of mucoprotein and acid mucopolysaccharides, a mucopolysaccharide layer consisting of glycoproteins and acid and basic muco-polysaccharides, and an inner layer of keratin rods in a lipoprotein matrix. Quinone tanning of the outer layer of the cyst is indicated by the presence of basic proteins and phenols in the cyst of *F. hepatica* and of phenolase in cercariae of *Fascioloides magna*.

Excystment of metacercariae is initiated by changes in the external environment, indicating contact with the definitive host. These factors generally include digestion of the cyst by host enzymes, but may also include changes in temperature, gases and osmolality (29). In *F. hepatica*, the outer layers of the cyst are partially digested by host pepsin and trypsin. However, excystment can not be completed in the absence of additional factors including an increase in temperature, elevated carbon dioxide, a low redox potential, and host bile salts. It is also speculated that parasite enzymes contribute to digestion of the cyst, although such enzymes have not yet been identified.

Cestodes apparently rely entirely on host enzymes for digestion of intermediate cystic stages in the definitive host. Requirements for various species include host pepsin, trypsin, pancreatin and bile salts (21).

3.3.2. Nematodes

In contrast to the platyhelminths, whose life history involves major morphogenic events, changes in gross body form which occur in nematodes developing from hatched larvae to adults are predominantly increases in size. To accommodate growth, it is necessary for nematodes periodically to cast the cuticle and this pattern is established as four molts in the life cycle. The structural biochemistry of the nematode cuticle has been discussed in Chapter 12. The initiation of the molting or ecdysis process in nematodes begins with the formation of a new cuticle underlying the old, probably synthesized by the hypodermis. Cuticle development is characterized by bursts of collagen synthesis (30); other components of the cuticle are also synthesized during molting periods and secretory products of multivesicular structures in the hypodermis are thought to contribute protein precursors to the new cuticle (31).

Separate stimuli are responsible for the synthesis of a new cuticle and the casting or ecdysing of the old cuticle (32). Multiple cuticles can develop under appropriate conditions and in the absence of stimuli for ecdysis. Thus, in at least some nematodes, cuticle development may be a function of a normal growth process dictated by constraints such as size and elasticity of each cuticle stage, whereas ecdysis or exsheathment is specifically triggered by environmental stimuli.

For ecdysis to occur, the old and new cuticles have to separate, then the old cuticle is cast. These events might also be elicited by different processes as they often do not occur in immediate sequence. Thus, synthesis of a new cuticle and cuticular separation might be coupled but be separate from ecdysis. The infective larvae of *Haemonchus contortus* secrete glycoprotein into the intercuticular space between the second molt (2M) and third molt (3M) cuticle (33); this material was suggested to act as a lubricant and to increase internal osmotic pressure, allowing movement of the larvae until environmental cues trigger the ecdysis. The process by which the old and new cuticles separate is not known. It is possible that once the cuticles are completely formed, charge differences at the surface interfaces (the old basal membrane and the trilaminate membrane of the new epicuticle) simply allow separation. Alternatively, worm secretions into the intercuticular space might passively, by increased turgor pressure, or actively, by enzymatic processes, lead to separation.

The final event in molting is the actual casting of the old cuticle, often referred to as exsheathment (Fig. 15.2). Exsheathment is caused by mechanisms including mechanical disruption of the cuticle, enzymatic digestion of the cuticle, or a combination of enzymatic and mechanical methods. In tissue-migrating worms it is easy to visualize exsheathment by strictly mechanical means as migration through tissue places considerable stress on the surface of the worm. Nevertheless, the cuticle to be cast must be more susceptible to abrasion and loss than the underlying cuticle and worm tissue.

In several nematodes, specific enzymes have been implicated in exsheathment. In *H. contortus*, a zinc metalloprotease is released by third stage larvae upon ingestion by the host or following appropriate stimuli *in vitro* (34). This protease digests a specific region of the second molt cuticle, providing an opening for the rapid exsheathment of the L3. The specificity of this protease, which has a similar activity against second molt cuticles of other trichostrongyles, is related to the composition of the cuticle. A chemically unique region of the second molt cuticle is found at the 20th annulus where the protease-mediated digestion occurs (35,36). At this location, a 160 kDa polypeptide is incorporated into the cuticle, and makes this restricted region sensitive to the parasite protease; this cuticle polypeptide contains limited collagen-like domains that likely serve to anchor it in the cuticle. Although this is the first description of a specific enzyme which has a definitive role in the ecdysis process, it is likely that other such enzymes will be identified in the future. Several enzymes including chitinase, collagenase, lipase and leucine aminopeptidase have been reported in exsheathing fluids (27), but, the function of these enzymes remains to be demonstrated. As discussed by Hotez *et al.* (37) the complex of parasite enzymes mediating tissue invasion and internal digestive processes likely have multiple roles and might well function in some aspects of cuticle degradation and ecdysis.

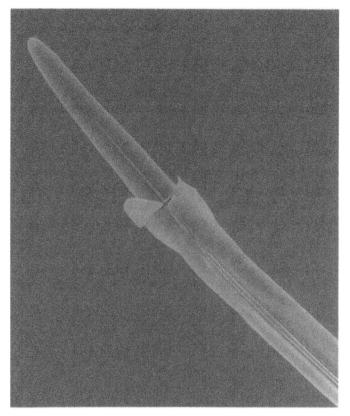

FIG. 15.2 Scanning electron micrograph showing exsheathment of an infective larvae of the nematode *Haemonchus contortus*. Exsheathment occurs following digestion of a specialized region of the cuticle near the anterior end by a parasite-secreted zinc metalloprotease.

Specific triggers are responsible for the induction of ecdysis. Although the proper stimuli for ecdysis can be applied *in vitro* (for example, CO_2 and increased temperature in the case of *H. contortus*), it is unclear how worms receive and act on these signals. Several authors have proposed that neurotransmitters induce the release of exsheathing enzymes (38, 39). For example, development of the free-living nematode *Caenorhabditis elegans* has been linked to chemosensory neuron stimulation (40). Environmental sensory input, in the form of increased food supply (usually bacteria) or a decrease in the concentration of a density-dependent pheromone will trigger chemosensory neurons to initiate growth of arrested dauer larvae to mature adults.

4. BIOCHEMICAL CUES AFFECTING DEVELOPMENT

4.1. Pheromones and Other Reproductive Cues

Chemical communication between sexes or hermaphrodites for the purpose of pairing and mating has been reported in a number of species of trematodes and nematodes (41).

Compounds involved in chemoattraction (pheromones) have been reported to be free sterols (cholesterol), sterol esters and peptides (42, 43). In some cases, pheromones have been localized to reproductive tissues of female worms (44).

Communication between sexes of schistosomes apparently extends beyond attractants, as survival and reproductive development of female worms is dependent on pairing with a male (45). Observations regarding pairing and development include a stimulation of reproductive morphogenesis in females by lipid extracts of male worms (46), increased tyrosine uptake by females stimulated with male extract (47), transfer of glucose from male to female worms (48) and a concomitant depletion of male glycogen reserves (49), and an increase in DNA synthesis in paired versus unpaired female worms (50). Some evidence has been presented which suggests that male worms aid in the mechanics of the feeding process of female worms (51); these authors suggest that distinct stimuli might operate to stimulate feeding and growth with undetermined factors from the male worm inducing reproductive maturation in female schistosomes.

It has been reported that tumor necrosis factor-α, a host-derived cytokine, stimulates egg-laying in S. mansoni worm pairs in vitro (52). This finding suggests that products of the host immune response to the parasite or products characteristic of the local environment within the host, can influence parasite reproduction. If parasites have receptors for host-derived cytokines, then the phylogenetic and functional relationship between the host cytokine network and that reported in invertebrates (53) becomes intriguing to explore in terms of parasite development.

4.2. Hormones Influencing Development

Although our knowledge of the regulation of helminth growth and development is at an elementary stage compared with some other invertebrates, an increasing body of research suggests the existence of hormonal regulation of various developmental processes in helminths.

Although originally identified in insects, ecdysteroids have been reported to occur in both platyhelminth and nematode parasites, as well as in species from other invertebrate phyla. Within the platyhelminths, ecdysteroids have been identified in the cestodes, *Echinococcus granulosus*, *Hymenolepis diminuta* and *Monezia expansa* and the trematodes, *Schistosoma mansoni* and *Trichobilharzia ocellata*; ecdysone, 20-hydroxyecdysone and 20,26-dihydroxyecdysone are the most common compounds identified (54–58). Active secretion of 20-hydroxyecdysone by *H. diminuta in vitro* suggests that ecdysteroids are of parasite origin (59). However, steroid precursors are most likely obtained from the host since helminths are considered to be incapable of *de novo* synthesis of steroids (60). Ecdysteroids have been circumstantially implicated in regulation of growth and development in cestodes and trematodes. Exogenous 20-hydroxyecdysone stimulates growth and asexual reproduction in *Mesocestoides corti* larvae *in vivo* and *in vitro* (61). Increased ecdysteroid levels in anterior segments of *M. expansa* further implicate ecdysteroids in the regulation of growth (55). Ecdysteroids have also been detected in cestode and trematode eggs and immunocytochemical observations demonstrated ecdysteroid immunoreactive material in the oviduct of *S. mansoni* (62).

Because of the apparent similarity between molting in nematodes and arthropods, it is assumed that ecdysone regulates molting and development in nematodes. Support for this hypothesis has come from the modulation of developmental processes by exogenously applied ecdysteroids and the identification of 20-hydroxyecdysone and related compounds in nematodes. Ecdysone was first isolated from adult *Ascaris suum* (63). This finding has been confirmed by a number of studies utilizing HPLC, radioimmunoassay and mass spectroscopy reporting the occurrence of ecdysone and 20-hydroxyecdysone in several nematode species including *A. suum*, *Parascaris equorum*, *Dirofilaria immitis*, *Brugia pahangi* and *Nippostrongylus braziliensis* (64–66). Polar conjugated ecdysteroids have been detected in *A. suum* eggs with relatively high levels during the early stages of embryogenesis which decrease dramatically on completion of embryonation (67). Although the occurrence of ecdysteroids has been documented, attempts to demostrate synthesis of 20-hydroxyecdysone from radiolabeled precursors has been unsuccessful (60).

Exogenously applied ecdysteroids affect the molting of nematodes. Ecdysone stimulated molting *in vitro* in *A. suum*, *H. polygyrus* and *D. immitis* (68, 69) and a correlation between the occurrence of both the third and fourth molts in *A. suum* with 20-hydroxyecdysone levels has been reported (69). In addition, azosteroids, synthetic steroid antagonists, prevent the *in vitro* molting and growth and development of *O. ostertagi* and *N. braziliensis* larvae (70).

The occurrence and secretion of ecdysteroids by helminths has led to the proposal to use ecdysteroids secreted by the parasite into the host serum as a means of diagnosis. However, the correlation between occurrence of infection and ecdysteroid level seems too variable and unreliable to be of diagnostic value (68).

Although, ecdysteroids occur in helminth tissue and affect a number of aspects of development, the true hormonal role of ecdysone in helminths remains to be documented. Insect juvenile hormones (JH) have also been implicated as regulators of nematode development, particularly in relationship to molting and exsheathment, but there is little analytical evidence supporting the occurrence of JH in nematodes. However, exogenously applied JH and synthetic JH analogs have been reported to alter developmental processes of nematodes (71–73). A concerted effort to apply modern analytical methods could resolve the question of the presence of insect JH in nematodes.

4.3. Density-dependent Development

Several studies have shown that establishment and growth of parasites in the host is regulated by factors including the size of the parasite population (74, 75). Considerable work has been conducted in the tapeworm *Hymenolepis diminuta*, on the 'crowding effect' (76), whereby worm size is directly dependent on worm density. Variation in worm size has been shown to result from a reduced level of DNA synthesis in germinative regions of the cestode (77). By chemically analyzing worm excretions and their effects on DNA synthesis, it was determined that succinate, acetate, D-glucosaminic acid, and cGMP were the products responsible for limiting DNA synthesis in crowded worms (78) and this effect was confirmed using an *in vivo* system (79).

5. DEVELOPMENT *IN VITRO*

Advances in cultivation technology (80) through the recognition of specific growth factors, nutrients and hormones, an awareness of appropriate pH, Po_2, Pco_2, Eh, osmotic pressure conditions and the concentration of important physiological ions, the development of systems for waste removal and spatial arrangements for organism pairing and copulation *in vitro*, have led to the successful cultivation of most of the major helminth parasites of medical, veterinary and biological importance. These techniques have been broadly applied to evaluate host immune responses to the parasite *in vitro*, to generate parasite excretory and secretory products with various, biological activities, to describe metabolic pathways and biochemical processes, related to growth and differentiation, and to test compounds as essential nutrients, growth factors, and for anthelmintic activity. The application of cultivation techniques to the study of helminth growth, development and differentiation can be illustrated by way of a synopsis of procedures used for the cultivation of *Ascaris suum* from egg to egg-laying adults.

With the first description of the cultivation of *Nippostrongylus braziliensis* from egg to adult (81), it became clear that the complexity of the *in vitro* culture system must reflect the complexity of the life cycle of the parasite. Properties of the ambient environment appropriate for each stage of development have to be reproduced *in vitro* to successfully support the development of the parasite. Fertilized *A. suum* eggs provide a homogeneous population of cells that advance synchronously to first-stage larvae in distilled water while only temperature and aeration (82) need to be controlled. Cleavinger *et al.* (83) demonstrated transcription of the actin gene in the early 4- to 8-cell stage of embryogenesis accounting for early cytoplasmic actin synthesis. Actin genes and genes for the major sperm protein (MSP), which are only found in worm testicular tissues have been compared in a variety of free-living and parasitic nematode species (1). The number of MSP genes varies from one in *Ascaris* to 60 in *Caenorhabditis*, where they comprise a large multigene family. In contrast, there are over 20 germ line genes for actin in *Ascaris* where active muscle synthesis is required during various phases of growth, especially during the adult stage. Extracts of 32–64-cell *Ascaris* embryos provide a unique cell-free system for the catalysis of a number of macromolecular processes including transcription by RNA polymerase II and III, *cis*- and *trans*-splicing, and translation (84). Free and conjugated ecdysone and 20-hydroxyecdysone in embryonating eggs between 3 and 12 days of incubation have been postulated to regulate larval morphogenesis and/or eggshell tanning (67), and have been considered as activators of oocyte development. The nature of ecdysteroids as regulators of growth and specific gene activation during oocyte development, spermatogenesis and embryogenesis remains to be definitively explored (60).

The mature egg, containing infective first-molt and second-stage larvae, can be easily surface sterilized chemically and an aseptic hatching process can yield large numbers of larvae suitable for cultivation. Development of second-stage larvae to mature egg-laying adults *in vitro* has been accomplished in a three-step roller culture system (85). Each step requires complex media, with specific supplements and changes in gas phases to trigger development through each larval stage to the mature adult. The first culture step is particularly suitable for studies of early larval development because a large and synchronous population of second-stage larvae advances to late-third-stage (97%) in a basal medium supplemented with L-cysteine. Development to fourth-stage, in step two, is stimulated by an increase in the concentration of amino acids in the basal medium, whereas subsequent development to young and mature adults, in step three, is triggered when bovine hemin is added to the medium. The morphogenesis of larval and adult stages *in vitro* is identical to that observed *in vivo*, however, development from late-third-stage is generally slower, growth of late larval and adult stages retarded, and spermatogenesis, oogenesis and egg-laying are delayed. Development of second-stage larvae to mid-third-stage can also be accomplished in defined media using a stationary culture system, but the yields are generally < 85% with no further stage advancement.

A common tactic used to obtain high yields of synchronous developing larvae is to isolate larvae that have advanced to the stage of interest in the host. Rabbits infected with *A. suum* eggs harbor a high percentage of late-third-stage (97%) in the lungs at 7 days after inoculation. These can be isolated aseptically to provide larvae for cultivation and for stage-specific infection of a host. Application of a two-step roller culture system consisting of basal medium for 7 days followed by a medium supplemented with bovine hemin, in a gas phase of 85% N/5% O_2/10% CO_2, can produce optimally 100% fourth-stage larvae at a time comparable to that *in vivo*, and good yields of late-fourth-stage, fourth-molt and young adult worms within 28 days of culture of larvae derived from rabbit lung (86). *A. suum* mature under these conditions, copulate *in vitro* and produce fertilized eggs. Other significant observations made in these systems include feeding of adult worms on dead larval stages; the loss of the genital girdle or body constriction at the level of the vulva in females releasing fertilized eggs; and the application of a one-step culture system for development of *A. suum* third-stage larvae to mature adults similar to one used to support comparable development of *Oesophagostomum radiatum* (87). This suggests that common molecular triggers stimulate the development of phylogenetically distant nematode parasite species of the mammalian intestine. Culture systems starting with *A. suum* third-stage larvae, have been used as a model for anaerobic mitochondrial metabolism and the aerobic–anaerobic transition of third-stage larvae developing to fourth-stage (88) (Chapter 4), whereas a stationary culture system using defined media has been used to assay the effects of biogenic amines and invertebrate hormones on development (89) and changes in cuticular proteins from different stages of *A. suum* (90). These and other systems of parasitic nematode development *in vitro* are excellent models for gene expression in helminths associated with stage development (91). *In vitro* cultivation, in general, is a useful way to obtain stages of parasites not easily available *in vivo*, but suitable for growth under controlled conditions.

6. REFERENCES

1. Scott, A. L., Dinman, J., Sussman, D. J. and Ward, S. (1988) Major sperm protein and actin genes in free-living and parasitic nematodes. *Parasitology* **98**: 471–478.
2. Roberts, T. M., Pavalko, F. M. and Ward, S. (1986) Membrane and cytoplasmic proteins are transported in the same organelle complex during nematode spermiogenesis. *J. Cell Biol.* **102**: 1787–1796.
3. Sepsenwol, S., Ris, H. and Roberts, T. M. (1989) A unique cytoskeleton associated with crawling in the amoeboid sperm of the nematode *Ascaris suum*. *J. Cell Biol.* **108**: 55–56.
4. Wells, K. E. and Cordingley, J. S. (1991) *Schistosoma mansoni*: eggshell formation is regulated by pH and calcium. *Exp. Parasitol.* **73**: 295–310.
5. Bobeck, L. A., LoVerde, P. T. and Rekosh, D. M. (1989) *Schistosoma haematobium*: analysis of eggshell protein genes and their expression. *Exp. Parasitol.* **68**: 17–30.
6. Johnson, K. S., Taylor, D. W. and Cordingley, J. S. (1987) A possible eggshell protein gene from *Schistosoma mansoni*. *Mol. Biochem. Parasitol.* **22**: 89–100.

7. Anderson, S. O. (1985) Sclerotization and tanning of the cuticle. In: *Comprehensive Insect Physiology, Biochemistry and Pharmacology* (eds Kerkut, G. and Gilbert, L.), Pergamon Press, Oxford.

8. Smyth, J. D. and Halton, D. W. (1983) *The Physiology of Trematodes.* Cambridge University Press, Cambridge.

9. Bennett J. L. and Gianutos, G. (1978) Disulfiram: a compound that selectively induces abnormal egg production and lowers norepinephrine levels in *Schistosoma mansoni. Biochem. Pharmacol.,* **27**: 817–820.

10. Seed, J. L. and Bennett, J. L. (1980) *Schistosoma mansoni:* phenol oxidase's role in egg-shell formation. *Exp. Parasitol.* **49**: 430–441.

11. Wait, J. H. and Rice-Ficht, A. C. (1987) Presclerotized eggshell protein from the liver fluke *Fasciola hepatica. Biochemistry* **26**: 7819–7825.

12. Kawanaka, M. (1991) Identification of putative eggshell precursor protein of the female *Schistosoma japonicum. Int. J. Parasitol.* **21**: 225–231.

13. Madhavi, R. (1966) Egg-shells in Paramphistomatidae (Trematoda:Digena). *Experientia* **22**: 93–94.

14. Ramalingham, K. (1973) The chemical nature of the egg-shell of helminths. 1. Absence of quinone tanning in egg-shell of the liver fluke *Fasciola hepatica. Int. J. Parasitol.* **3**: 67–75.

15. Fetterer, R. H. and Rhoads, M. L. (1991) Tyrosine-derived cross-linking amino acids in the sheath of *Haemonchus contortus* infective larvae. *J. Parasitol.* **76**: 619–624.

16. Wharton, D. A. (1983) The production and functional morphology of helminth eggshells. *Parasitology* **86**: 85–97.

17. Monne, L. and Honig, G. (1954) On properties of egg envelopes of various parasitic nematodes. *Arkv. Zool.* **7**: 261–272.

18. Fetterer, R. H. and Hill, D. E. (1993) The occurrence of phenol oxidase activity in female *Trichuris suis. J. Parasitol.* **79**: 155–159.

19. Wells, K. E. and Cordingley, J. S. (1991) *Schistosoma mansoni:* eggshell formation is regulated by pH and calcium. *Exp. Parasitol.* **73**: 295–310.

20. Holmes, S. D. and Fairweather, I. (1982) *Hymenolepis diminuta:* the mechanism of egg hatching. *Parasitology* **85**: 237–250.

21. Smyth, J. D. and McManus, D. P. (1989) *The Physiology and Biochemistry of Cestodes.* Cambridge University Press, Cambridge.

22. Barrett, J. (1976) Studies on the induction of permeability in *Ascaris lumbricoides* eggs. *Parasitology* **73**: 109–121.

23. Fairbairn, D. (1961) The in vitro hatching of *Ascaris lumbricoides* eggs. *J. Zool.* **39**: 153–162.

24. Perry, R. N. (1989) Dormancy and hatching of nematode eggs. *Parasitol. Today* **5**: 377–383.

25. Rogers, W. P. (1960) The physiology of the infective process of nematodes: the stimulus from the animal host. *Proc. R. Soc. London* **B152**: 367–386.

26. Bone, L. W. and Parish, E. J. (1988) Egg enzymes of the ruminant nematode *Trichostrongylus colubriformis. Int. J. Invent. Repro. Dev.* **14**: 299–302.

27. Rogers, W. P. and Brooks, F. (1977) The mechanism of hatching of eggs of *Haemonchus contortus. Int. J. Parasitol.* **7**: 61–65.

28. Dixon, K. E. (1965) The structure and histochemistry of the cyst wall of the metacercariae of *Fasciola hepatica. Parasitology* **55**: 215–226.

29. Lackie, A. M. (1975) The activation of infective stages of endoparasites of vertebrates. *Biol. Rev.* **50**: 285–323.

30. Selkirk, M. E. (1991) Structure and biosynthesis of cuticular proteins of lymphatic filarial parasites. In: *Parasitic Nematodes: Antigens, Membranes, and Genes* (ed. Kennedy, M. W.). Taylor amd Francis, Bristol, PA.

31. Lee, D. L. (1970) Moulting in nematodes: the formation of the adult cuticle during final moult in *Nippostrongylus braziliensis. Tissue Cell* **2**: 139–153.

32. Sommerville, R. I. and Murphy, C. R. (1983) Reversal of order of ecdysis in *Haemonchus contortus. J. Parasitol.,* **69**: 368–371.

33. Bird, A. F. (1990) Vital staining of glycoprotein secreted by infective third stage larvae of *Haemonchus contortus* prior to exsheathing. *Int. J. Parasitol.* **20**: 619–623.

34. Gamble, H. R., Lichtenfels, J. R. and Purcell, J. P. (1989) Scanning electron microscopy of the ecdysis of *Haemonchus contortus* infective larvae. *J. Parasitol.*, **75**: 303–307.
35. Gamble, H. R., Purcell, J. P. and Fetterer, R. H. (1989) Purification of a 44 kDa protease which mediates the ecdysis of infective *Haemonchus contortus* larvae. *Mol. Biochem. Parasitol.* **33**: 49–58.
36. Gamble, H. R., Purcell, J. P. and Fetterer, R. H. (1990) Biochemical characterization of cuticle polypeptides from the infective larvae of *Haemonchus contortus. Comp. Biochem. Physiol.* **96B**: 421–429.
37. Hotez, P., Haggerty, J., Hawson, J. *et al.* (1990) Infective *Ancylostoma* hookworm larval metalloproteases and their functions in tissue invasion and ecdysis. *Infect. Immun.* **58**: 3883–3892.
38. Davey, K. G. and Kan, S. P. (1968) Molting in a parasitic nematode, *Phocanema decipiens.* IV. Ecdysis and its control. *Can. J. Zool.* **46**: 893–898.
39. Rogers, W. P. and Petronijevic, T. (1982) The infective stage and the development of nematodes. In: *Biology and Control of Endoparasites* (eds Symons, L. E. A., Donald, A. D. and Dineen, J. K.), Academic Press, New York.
40. Bargmann, C. I. and Horvitz, H. R. (1991) Control of larval development by chemosensory neurons in *Caenorhabditis elegans. Science* **251**: 1243–1246.
41. Fried, B. (1986) Chemical communication in hermaphroditic digenetic trematodes. *J. Chem. Ecol.* **12**: 1659–1677.
42. Fried, B. and Robinson, G. A. (1981) Pairing and aggregation of *Amblosoma suwaense* metacercariae in vitro and partial characterization of lipids involved in chemoattraction. *Parasitology* **82**: 225–229.
43. Fried, B., Tancer, R. B. and Fleming, S. J. (1980) In vitro pairing of *Echinostoma revolutum* metacercariae and adults, and characterization of worm products involved in chemoattraction. *J. Parasitol.* **66**: 1014–1018.
44. Riga, E. and MacKinnon, B. M. (1987) Chemical communication in *Heligosomoides polygyrus*: the effect of age and sexual status of attracting and responding worms and localization of the sites of pheromone production in the female. *Can. J. Zool.* **66**: 1943–1947.
45. Armstrong, J. C. (1965) Mating behavior and development of schistosomes in the mouse. *J. Parasitol.* **51**: 605–616.
46. Shaw, J. R., Marshall, I. and Erasmus, D. A. (1977) *Schistosoma mansoni*: in vitro stimulation of vitelline cell development by extracts of male worms. *Exp. Parasitol.* **42**: 12–20.
47. Popiel, I. and Erasmus, D. A. (1981) *Schistosoma mansoni*: changes in the rate of tyrosine uptake by unisexual females after stimulation by males and male extracts. *J. Helminthology.* **55**: 33–37.
48. Cornford, E. M. and Fitzpatrick, A. M. (1985) Glucose transfer from male to female schistosomes: definition of the mechanism and rate constant. *Mol. Biochem. Parasitol.* **17**: 131–141.
49. Cornford, E. M. and Huot, M. E. (1981) Glucose transfer from male to female schistosomes. *Science* **213**: 1269–1271.
50. den Hollander, J. E. and Erasmus, D. A. (1985) *Schistosoma mansoni*: male stimulation and DNA synthesis by the female. *Parasitology* **91**: 449–457.
51. Gupta, B. C. and Basch, P. F. (1987) The role of *Schistosoma mansoni* males in feeding and development of female worms. *J. Parasitol.* **73**: 481–486.
52. Amiri, P., Locksley, R. M., Parslow, T. G. *et al.* (1992) Tumour necrosis factor-alpha restores granulomas and induces parasite egg-laying in schistosome-infected SCID mice. *Nature* **356**: 604–607.
53. Beck, G. and Habicht, G. S. (1991) Primitive cytokines: harbingers of vertebrate defence. *Immunol. Today* **12**: 180–183.
54. Henk, D. F. G. Schallig, Young, N. J., Magee, R. M., deJong-Brink, M. and Rees, H. H. (1991) Identification of free and conjugated ecdysteroids in cercariae of the schistosome *Trichobilharzia ocellata. Mol. Biochem. Parasitol.* **49**: 169–176.
55. Mendis, A. W., Rees, H. H. and Goodwin, T. W. (1984) The occurrence of ecdysteroids in the cestode *Moniezia expansa. Mol. Biochem. Parasitol.* **10**: 123–128.

56. Mercer, J. G., Munn, A. E. and Rees, H. H. (1987) *Echinococcus granulosus*: occurrence of edysteroids in protoscoleces and lygated cyst fluid. *Mol. Biochem. Parasitol.* **24**: 203–214.
57. Mercer, J. G., Munn, A. E., Arme, C. and Rees, H. H. (1987) Analysis of ecdysteroids in different developmental stages of *Hymenolepis diminuta*. *Mol. Biochem. Parasitol.* **25**: 61–71.
58. Torpier, G., Hin, M., Nirde, P., DeReggi, M. and Capron, A. (1982) Detection of ecdysteroids in the human trematode *Schistosoma mansoni*. *Parasitology* **84**: 123–130.
59. Mercer, J. G., Munn, A. E., Arme, C. and Rees, H. H. (1987) Ecdysteroid excretion by adult *Hymenolepis diminuta* in vitro. *Mol. Biochem. Parasitol.* **26**: 225–234.
60. Barker, G. C., Chitwood, D. J. and Rees, H. H. (1990) Ecdysteroids in helminths and annelids. *Invert. Repro. Dev.* **18**: 1–11.
61. Kowalski, J. C. and Thorson, R. E. (1976) Effect of certain lipid compounds on growth and asexual multiplication of *Mesocestoides corti* (Cestoda) *tetrathydria*. *Int. J. Parasitol.* **6**: 327–331.
62. Basch, P. F. (1986) Immunocytochemical localization of ecdysteroids in life history stages of *Schistosoma mansoni*. *Comp. Biochem. Physiol.* **83A**: 199–202.
63. Horn, D. H. S., Wilkie, J. S. and Thompson, J. A. (1974) Isolation of β-ecdysone (20-hydroxyecdysone) from the parasitic nematode *Ascaris lumbricoides*. *Experientia* **30**: 1109.
64. Cleator, M., Delves C. J., Howells, R. E. and Rees, H. H. (1987) Identity and tissue localization of free and conjugated ecdysteroids in adults of *Dirofilaria immitis* and *Ascaris suum*. *Mol. Biochem. Parasitol.* **25**: 93–105.
65. Mercer, J. G., Barker, G., Howells, R. E. and Rees, H. H. (1990) Investigations of ecdysteroid excretion by adult *Dirofilaria immitis* and *Brugia pahangi*. *Mol. Biochem. Parasitol.* **38**: 89–96.
66. O'Hanlon, G. M., Cleator, M., Mercer, J. G., Howells, R. and Rees, H. H. (1991) Metabolism and fate of ecdysteroids in nematodes *Ascaris suum* and *Parascaris equorum*. *Mol. Biochem. Parasitol.* **47**: 179–188.
67. Fleming, M. W. (1987) Ecdysteroids during embryonation of eggs of *Ascaris suum*. *Comp. Biochem. Physiol.* **87A**: 803–805.
68. Barker, G. C. and Rees, H. H. (1990) Ecdysteroids in Nematodes. *Parasitol. Today* **12**: 384–387.
69. Fleming, M. W. (1985) *Ascaris suum*: role ecdysteroids in moulting. *Exp. Parasitol.* **60**: 207–210.
70. Bottjer, K. P., Whisenton, L. R. and Weinsten, P. (1984) Ecdysteroid-like substance in *Nippostrongylus braziliensis*. *J. Parasitol.* **70**: 986–987.
71. Boisvenue, R. J., Emmick, J. L. and Galloway, R. B. (1977) *Haemonchus contortus*: effects of compounds with juvenile hormone activity on the in vitro development of infective larvae. *Exp. Parasitol.* **42**: 67–72.
72. Rogers, W. P. (1973) Juvenile and moulting hormones from nematodes. *Parasitology* **67**: 105–113.
73. Rogers, W. P. (1980) The action of insect juvenile hormone on hatching of eggs of nematode *Haemonchus contortus* and its role in development of infective and non-infective stages. *Comp. Biochem. Physiol.* **66A** 631–635.
74. Nollen, P. M. (1983) Patterns of sexual reproduction among parasitic platyhelminths. *Parasitology* **86**: 99–120.
75. Ractliffe, L. H., LeJambre, L. F., Uhazy, L. S. and Whitlock, J. H. (1971) Density dependence of the weight of *Haemonchus contortus* adults. *Int. J. Parasitol.* **1**: 297–301.
76. Roberts, L. S. (1961) The influence of population density on patterns and physiology of growth in *Hymenolepis diminuta* in the definitive host. *Exp. Parasitol.* **18**: 332–371.
77. Roberts, L. S. and Insler, G. D. (1982) Developmental physiology of cestodes. XVII. Some biological properties of putative 'crowding factors' in *Hymenolepis diminuta*. *J. Parasitol.* **68**: 263–269.
78. Zavras, E. T. and Roberts, L. S. (1985) Developmental physiology of cestodes: cyclic nucleotides and the identity of putative crowding factors in *Hymenolepis dimunta*. *J. Parasitol.* **71**: 96–105.
79. Cook, R. L. and Roberts, L. S. (1991) In vivo effects of putative crowding factors on development of *Hymenolepis diminuta*. *J. Parasitol.* **77**: 21–25.
80. Smyth, J. D. (1990) *In Vitro Cultivation of Parasitic Helminths* CRC Press, Boca Raton.

81. Weinstein, P. P. and Jones, M. F. (1956) The in vitro cultivation of *Nippostrongylus muris* to the adult stage. *J. Parasitol.* **42**: 215–236.
82. Boisvenue, R. J. (1990) Effects of aeration and temperature in vitro and in vivo studies on developing and infective eggs of *Ascaris suum. J. Helminthol. Soc. Wash.* **57**: 51–56.
83. Cleavinger, P. J., McDowell, J. W. and Bennett, K. L. (1989) Transcription in nematodes: early *Ascaris* embryos are transcriptionally active. *Dev. Biol.* **133**: 600–604.
84. Nilsen, T. W. (1993) Trans-splicing of nematode premessenger RNA. *Annu. Rev. Microbiol.* **47**: 413–440.
85. Douvres, F. W. and Urban, J. F., Jr (1983) Factors contributing to the in vitro development of *Ascaris suum* from second-stage larvae to mature adults. *J. Parasitol.* **69**: 549–558
86. Douvres, F. W. and Urban, J. F., Jr (1986) Development of *Ascaris suum* from in vivo-derived third-stage larvae to egg-laying adults in vitro. *Proc. Helminthol. Soc. Wash.* **53**: 256–262.
87. Douvres, F. W. (1983) The in vitro cultivation of *Oesophagostomum radiatum*, the nodular worm of cattle. III. Effects of bovine heme on development of adults. *J. Parasitol.* **69**: 570–576.
88. Komuniecki, R. W. and Komuniecki, P. R. (1993) Aerobic/anaerobic transitions in energy metabolism during the development of the parasitic nematode, *Ascaris suum.* In: *Biology of Parasitism* (ed. Boothroyd, J.), Wiley Less Press, New York.
89. Fleming, M. W. (1988) Hormonal effects on the in vitro larval growth of the swine intestinal roundworm, *Ascaris suum. Invert. Repro. Dev.* **14**: 153–160.
90. Fetterer, R. H., Hill, D. E. and Urban, J. F., Jr (1990) The cuticular biology in developmental stages of *Ascaris suum. Acta Trop.* **47**: 289–295.
91. Cox, G. N., Shamansky, L. M. and Boisvenue, R. J. (1990) *Haemonchus contortus*: evidence that the 3A3 collagen gene is a member of an evolutionarily conserved family of nematode cuticle collagens. *Exp. Parasitol.* **70**: 175–185.

16 Invasion Mechanisms

JEAN FRANÇOIS DUBREMETZ[1]
and JAMES H. McKERROW[2]

[1]INSERM Unité 42–369, 59651 Villeneuve d'Ascq Cedex, France
and [2]Departments of Pathology, Medicine and Pharmaceutical Chemistry,
University of California, and Department of Veterans Affairs Medical
Center, San Francisco, CA, USA

SUMMARY

The initial interaction between parasite and host often involves invasion of a parasite into host tissue or host cells. Single cell protozoan parasites often evade the host immune response by residing within host cells. They must enter the host cell with minimal trauma to ensure that they preserve an environment suitable for their replication and not trigger a lethal host immune response. This intracellular invasion paradigm is also shared by a helminth parasite, *Trichinella spiralis*. Most other helminths reside in the extracellular space. Often an invasive larval form will penetrate skin or the mucosa of the gastrointestinal tract to gain entry into the host. Migration can be fairly extensive. To complete this invasion process, helminth parasites have evolved sensory organs for finding and navigating within the host, attachment mechanisms utilizing specialized mouth structures or glue-like secretions, and hydrolytic enzymes to digest macromolecular barriers in the extracellular space.

1. INVASION INTO CELLS

1.1. How to Enter a Cell: Phagocytosis or Invasion

Intracellular parasitism implies entry of a pathogen into the cytoplasm of a host cell. Various strategies have been evolved by microorganisms to achieve this goal. Because

Biochemistry and Molecular Biology of Parasites
ISBN 0-12-473345-X

some host cells may have the ability to internalize foreign material by phagocytosis, distinction between parasite invasion or uptake is often a challenging issue. One must closely analyze the entry event to decide whether phagocytosis is involved. Phagocytosis is a peculiar type of endocytosis that is usually performed by specialized cells. It involves binding of a particle to the cell plasmalemma through various ligand–receptor interactions (integrins, Fc receptors, hydrophobic interactions), zippering of the membrane around the particle and internalization into the resulting vacuole. The vacuole membrane is directly derived from the plasmalemma of the cell and will be enriched in some molecules (receptors) if these were involved in the binding. The vacuole enters the endocytic pathway and fuses eventually with lysosomes after acidification by a proton ATPase located in its membrane. Various markers of the successive compartments are known, especially the lysosomal membrane glycoproteins (LGP) that are markers of late endosomes and lysosomes (1). Recently, transfection of genes encoding receptors in cells normally unable to express these molecules has permitted the dissection of the respective roles of these molecules and of their domains in internalization (2). Another characteristic of phagocytosis is that it is driven by the polymerization of actin around the developing phagosome and therefore sensitive to inhibitors of microfilament formation such as cytochalasins.

Host cell invasion by many microorganisms does not utilize the phagocytosis route and the vacuole that is formed either does not enter the endocytic pathway or is destroyed before the phagolysosome stage. An artificial gradation can be described starting from *Sporozoa* that appear to have created an internalization procedure entirely distinct from phagocytosis; through *Trypanosoma cruzi* that may use the phagocytic system to get into the cell but then escape into the cytoplasm; to *Leishmania* promastigotes that are internalized and live into a mostly typical phagolysosome (Fig. 16.1).

1.2. Creating a New Compartment in the Host Cell: Sporozoa

A major feature of Sporozoa is to have evolved specific structures and organelles for host cell invasion; no other group except Microsporidia has achieved such a complex differentiation. Among Sporozoa, the *Plasmodium* genus has been the most thoroughly investigated. Most of our knowledge on host cell invasion by Sporozoa has been obtained from studies of erythrocyte invasion by malaria merozoites (3). Studies on other members of the group that invade nucleated cells have shown that the invasion process is not significantly different from that described in *Plasmodium* (4). The very conserved organization of the invasive stage (zoite) among Sporozoa is a key to the explanation of invasion in this group. Indeed, all zoites of Sporozoa share a cellular polarity and specialized organelles organized into an apical complex (from which derives the other name of the phylum: Apicomplexa). The apical complex occupies the anterior part of the cell and comprises a cytoskeleton (anterior rings, subpellicular microtubules, inner membrane complex) and vesicular electron dense organelles (rhoptries, micronemes, dense granules). All the components of this structure are considered to be involved in host cell invasion. The invasion phenomenon itself can be divided into three successive steps: recognition and attachment, internalization and vacuole development, vacuole maturation.

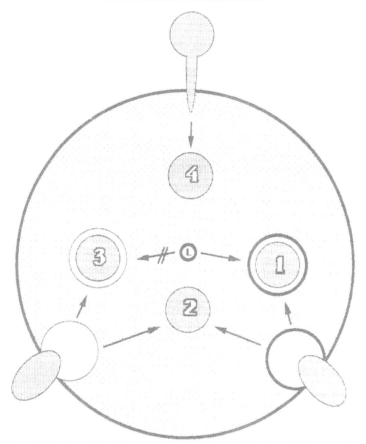

FIG. 16.1   Schematic drawing of the different ways of invasion used by intracellular parasites (L: lysosome).
1. *Leishmania* sp.; 2. *Trypanosoma cruzi* or piroplasms; 3. Sporozoa except piroplasms; 4. Microsporidia.

1.2.1. Recognition and attachment (5)

Surface to surface binding between zoite and host cell is supposed to involve not only zoite surface molecules and host cell surface, but also bridging molecules exocytosed by the zoite (erythrocyte binding antigen in *Plasmodium falciparum*, 135 kDa Duffy receptor in *P. knowlesi* or *P. vivax*, both of which are contained in micronemes) or components of the extracellular matrix (laminin in *Toxoplasma*). The red cell surface receptors used by *P. falciparum* which include glycophorins, sialic acids and band 3 are still incompletely identified. *P. vivax* and *P. knowlesi* use different receptors among which is a 45 kDa glycoprotein that carries the Duffy determinant.

A reorientation step has been observed in *Plasmodium* after attachment. It may result from a gradient of receptor distribution on merozoite surface, or from the presence of apically located higher affinity receptors. This reorientation brings the apex in contact with the cell membrane, which is necessary for internalization. Such a passive reorientation step may not be present in other Sporozoa where gliding motility or conoid flexing could bring about the apical contact.

The recognition or receptor–ligand interaction between *Plasmodium* and its host cell is highly specific as illustrated by the very narrow range of invasion capabilities of these organisms. The matter is much less clear for other Sporozoa, the prototype of which is *Toxoplasma*. These can invade any type of cell *in vitro* except the erythrocyte, and the efficiency of invasion may be modulated by receptor–ligand interaction or a ubiquitous membrane molecule such as cholesterol might serve as a receptor triggering invasion for these zoites (6).

1.2.2. Internalization, parasitophorous vacuole formation

1.2.2.1. Motility and moving junction.   Once a zoite has made an apical contact with the target cell, there is a close junction between both plasmalemma. Freeze fracture shows a rhomboidal array of intramembrane particles which is likely to be a crystalline array of lipids. As demonstrated in *P. knowlesi*, all the preliminary steps can occur in the presence of cytochalasin B that inhibits both the zoite motility and invasion. Inhibition of zoite motility by other agents also blocks invasion. This has been observed in all Sporozoa studied so far and strongly suggests that invasion operates by using the gliding motility system of the zoites. This is likely to involve an actin-based motor located in the inner membrane complex of the zoite (7). The apical junction turns into a annulus through which the zoite glides into the developing vacuole.

1.2.2.2. Vacuole membrane.   The origin of the vacuole membrane is unclear. It is in continuity with the plasmalemma of the host cell through the moving junction during the invasion process, but its structure is dramatically different. It is almost completely devoid of intramembranous particles or host cell surface proteins. Contribution of host cell to the vacuole membrane is therefore probably restricted to lipids.

1.2.2.3. Exocytosis of organelles.   The zoite contribution to the vacuole is believed to occur through the exocytosis of specialized apical organelles named rhoptries. Rhoptries are pedunculated organelles, the ducts of which extend toward the apex of the zoite and are open for exocytosis during the invasion. A complex set of proteins, some of which are grouped in families, have been identified in these organelles in both *P. falciparum* and *Toxoplasma*. Some of these are associated with the developing vacuole membrane, thus confirming the exocytic process (8, 9). Ultrastructural (*Plasmodium*) or biochemical (*Toxoplasma*) data on rhoptries suggest the presence of lipids that could also be inserted in the vacuole membrane upon exocytosis. Enzyme activities such as proteases (10) and phospholipases (4) may be involved in invasion. These activities could modify cell surface proteins or lipids or the underlying cytoskeleton during formation of the vacuolar membrane.

The invasion by Sporozoa is shown diagrammatically in Fig. 16.2.

1.2.3. Vacuole maturation

Once the vacuole is completed a maturation step occurs, during or shortly after invasion, that is characterized by the exocytosis of dense granules from the parasite. The proteins of the dense granules are targeted to either the vacuolar space (11), a vacuolar membranous network and the vacuole membrane (12) or the inner side of the

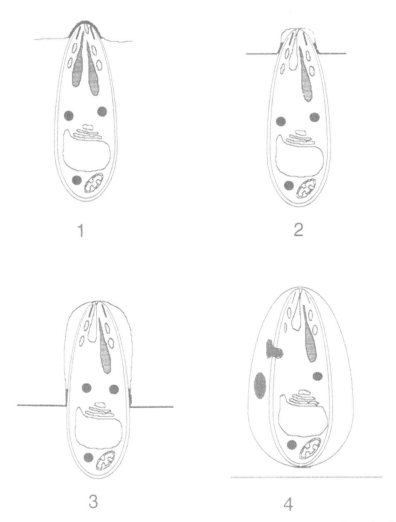

FIG. 16.2 Diagram of invasion by Sporozoa. 1, Recognition and moving junction formation (microneme exocytosis); 2,3, rhoptry exocytosis, zoite entry; 4, dense granule exocytosis, vacuole remodeling.

host cell membrane (13). Vacuole maturation reflects the development of metabolic exchanges between parasite and host cell. The vacuole is entirely isolated from the endosomal traffic of the host cell and the parasite must rely on transmembrane transport to get its nutrients. The contribution of the parasite to the vacuole includes the necessary transporters or channels in that membrane. In *P. falciparum* the existence of direct channels between the vacuole and the extracellular space has been claimed (14).

A peculiar type of vacuole maturation occurs in *Piroplasma* where the vacuole disappears after the exocytosis of dense granules (15). The organism develops directly in the cytoplasm of the host cell. The mechanism of vacuole lysis is not known. This type of intracellular behavior is morphologically shared with *Trypanosoma cruzi* (described below), but whether any molecular function is shared remains to be investigated.

1.3. Escaping from a Phagosome into the Cytoplasm: *Trypanosoma cruzi*

This organism has an intracellular phase in the vertebrate host only and invasion is performed by trypomastigotes derived from the insect gut or previous round of intracellular multiplication. Although these organisms are highly motile, their active involvement in the invasion has long been controversial. Fibronectin augments the internalization of *T. cruzi* in phagocytic and non-phagocytic cells (16). It acts as a molecular bridge between parasite and host through attachment to a beta 1 integrin.

Attachment is energy dependent on the part of the parasite and does not need host cell metabolism since fixed cells can be attached and invaded (17). Although entry of *T. cruzi* in phagocytes is believed to be mediated by phagocytosis, typical phagocytic receptors (FcR, CR3) do not seem to be involved (18). Penetration in non-phagocytic cells seems to be an active process that is not blocked by cytochalasin D, which inhibits actin-based systems (17). During the first 60 min, *T. cruzi* resides in a vacuole that acquires lysosomal glycoproteins in its membrane which means that it fuses with endocytic vesicles en route to lysosomes. The vacuole acidifies and a pore-forming protein, antigenically similar to C9, the membranolytic component of complement, is released in the vacuole (19). The vacuole membrane becomes discontinuous and disappears leaving the parasite directly in the cytoplasm where it transforms into amastigotes.

1.4. Living in a Phagosome: *Leishmania*

Infectious forms of *Leishmania* (metacyclic promastigotes) possess two types of surface molecules involved in attachment and invasion. These are the lipophosphoglycan (LPG) and the major surface protease GP63. Both of these ligands can independently mediate attachment of the parasite to macrophages. *Leishmania* do not invade non-phagocytic cells. Internalization may require synergy of both parasite ligands as has been shown with artificial systems of beads coated with both molecules separately or associated. The cellular receptors involved in binding and internalization may be multiple but the complement component CR3 seems to be a major receptor that binds both LPG and GP63 (20). The vacuole containing the parasite is acidified (21) and its membrane becomes LGP-positive but MPR (mannose-6-phosphate receptor) negative. This corresponds to a lysosomal compartment. Only some extracellular ligands appear in the vacuole via receptor-mediated endocytosis; this means that it fuses only with some of the receptor-mediated endocytic pathways (20, 22). The survival and development of *Leishmania* sp. within the hostile environment represented by the lysosomal contents is supposed to be mediated by several protective mechanisms including the inhibition of the macrophage respiratory burst by LPG (see Chapter 9) and the activity of parasitic enzymes (proteases, superoxide dismutase and acid phosphatase) that may counteract the activity of lysosomal enzymes.

1.5. A Multicellular Organism Within a Cell: *Trichinella*

The L1 larvae of this nematode are born in the lamina propria of the gut, travel in the blood and eventually enter a striated skeletal muscle cell. The mechanism of host cell penetration is not known but what has been extensively studied is the transformation of the host cell by the parasite after entry (23). The multinucleated muscle fiber loses the myofibrils that are replaced by smooth membranes and mitochondria; the nuclei enlarge and develop prominent nucleoli; the cell glycocalix is replaced by a thick collagen coat and around the nurse cell thus formed, angiogenesis is triggered and a circulatory plexus develops. This transformation reflects a dramatic alteration of the host cell. This is believed to be triggered by the worm which sends information modifying gene expression. The exact nature of the message is not known but proteins secreted in the cell by the larva are likely to play a role in this modulation. A 43 kDa glycoprotein exocytosed from granules found in specialized cells of the worm (stichocytes) is targeted to the nuclei of the host cell where it can be immunodetected and may play a role in modulation of host genomic expression; other secreted parasite molecules could also be involved in this transformation.

1.6. Injection into a Cell: *Microsporidia*

These parasites are quite original with respect to invasion since they inject themselves through the plasmalemma of the host cell. The microsporidian spore is a highly organized system of a coiled hollow polar filament that everts at excystation to open a hole in a target cell and inject the microsporidium. The parasite then develops directly in the cytoplasm of the host cell. Triggering of the spore discharge seems to be initiated by ionic changes in the medium (24).

2. INVASION BY HELMINTHS

Helminths are multicellular parasites, often a millimeter or more in length. With rare exceptions (see *Trichinella*), invasion of and seclusion within host cells is not feasible. When speaking of host invasion by helminth parasites, one considers tissue invasion or, more specifically, migration through extracellular barriers.

Different helminth parasites may have quite distinct pathways of invasion and migration in the host, and may employ different mechanisms to facilitate that invasion. The port of entry for several parasitic nematodes and trematodes is the skin. However, even within this group, the exact mode of entry may vary significantly (Fig. 16.3). Invasive larvae of *Strongyloides*, hookworm and schistosomes may enter the skin directly, without the need for an insect vector bite or accidental trauma. Larvae of *Brugia* and *Wuchereria* are deposited on skin from the salivary gland of the mosquito vector and follow the mosquito bite path to a lymphatic vessel. The L3 larvae of *Onchocerca*, are also deposited at the vector bite site, but then migrate for a considerable distance through dermal connective tissue before reaching their final site of adult residence.

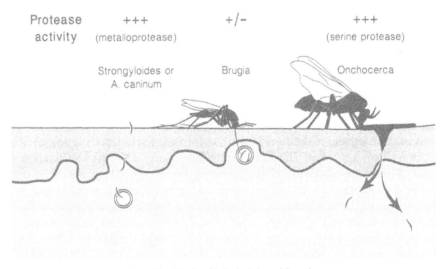

FIG. 16.3   Mode of helminth larval invasion.

The intestinal wall is another barrier for helminth parasite invasion. Here, the common endpoint of invasion is the lumen of a blood vessel, with subsequent dissemination to multiple organs. After reaching the capillary bed of a distant organ, for example, the lung, the larvae may migrate out of the capillary bed into the bronchial tree or interstitial tissue.

Finally, there are examples of the reverse pathway. Schistosome eggs must leave the lumen of mesenteric blood vessels and cross the intestinal wall in a path exactly opposite that of invasive gastrointestinal parasites. What are the macromolecular barriers that impede each of these different parasite migrations?

2.1. Pathways of Helminth Invasion

The skin is an excellent barrier to microbial and parasitic infections. The most superficial layer of the skin is composed of flattened squamous cells, which are highly keratinized. Beneath this is the epidermal layer composed of cells tightly interconnected by desmosomes and other intercellular structures. These, in turn, are attached to the basement membrane composed of covalently bound or interwoven macromolecules. Between the basement membrane and a target blood vessel is an extracellular matrix rich in type I collagen, elastin and proteoglycan. Elastin and type I collagen are both interwoven fibrillar molecules, whereas the carbohydrate-rich proteoglycan behaves like a hydrated gel. For details of these macromolecular interactions, the reader is referred to reviews on the structure of skin.

The wall of small venules and lymphatics is composed of a basement membrane not unlike that found between the epidermis and dermis, and an endothelial cell layer that provides a barrier to movement of macromolecules into and out of the lumen.

The wall of the small intestine and colon is composed of a columnar epithelial cell layer, often producing a thick mucus coat. This mucus coat can be a barrier to microbial or parasitic invasion from the gut lumen (25). As in the skin, this epithelium rests on a basement membrane below which there is an intervening extracellular matrix before the vascular bed of the lamina propria is reached.

2.2. Specific Steps in the Invasion Process

Our knowledge of host tissue invasion by helminth parasites is still at a very basic level. Nevertheless, there are several steps that are logically required. These will be briefly outlined and then discussed individually in more detail.

(1) Tactic responses. For the parasite entering the host from the external environ-ment, this is the process by which it finds the host. For the parasite in the lumen of the gut or in the bloodstream, this is the process that defines where it will go and, equally important, when it will stop.

(2) Attachment. Prior to invasion it is necessary for parasites, particularly those coming from the external environment, to attach to the host. During the actual process of invasion, cyclical attachment and release from extracellular matrix or other structures may be necessary for motility.

(3) Digestion of macromolecular barriers. Particularly in migration through skin and connective tissue, the interactive extracellular matrix must be breached before a multicellular organism can invade.

(4) Evasion of host immunity. Implicit in the ability to invade is the ability to evade host responses that would block parasite migration or be directly lethal to the organism. Mechanisms by which parasite helminths evade the host immune response have been extensively analyzed and debated, and the reader is referred to the numerous reviews on this subject, particularly for the well-studied example of schistosomiasis (26).

2.2.1. Tactic responses and migration

Some information is available for those helminth parasites which enter the host from external environments. Schistosome cercariae and larvae of *Stronglyloides* and hook-worm respond to specific signals that identify the host as such and have been the most extensively studied.

A response to light ensures that schistosome cercariae reside in the same part of a pond or lake as the host (27). Schistosome cercariae which infect humans are shed from the intermediate host snail under stimulus of light following a period of darkness. *Schistosoma douthitti*, in contrast, which infects nocturnal mammals such as voles and muskrats, leaves the snail in darkness following a period of light (28). Cercariae of *Diplostomums pathaceum*, which infect fish, are stimulated to swimming bursts by water turbulence or touch; *Trichobilharzia ocellata*, which infects ducks, is stimulated by shadows (27).

To contact host skin and invade, schistosome cercariae follow a thermal gradient and then use an elegantly adapted penetration response triggered by specific fatty acids present on the skin (29–31). Specific free fatty acids like linoleic acid will stimulate cercariae to invade *in vitro*. Whether this is receptor-mediated is not known, but cercariae can metabolize linoleic acid to eicosanoids (29), which are potential 'second messengers'. Upon stimulation cercariae lose their glyocalyx and tails and release a protease from preacetabular gland cells to facilitate invasion.

Hookworm and *Stronglyloides* larvae enhance their chance of encountering host skin by aggregating in clusters at the highest points of blades of grass (32). Hookworm larvae also recognize mammalian hosts by sensing CO_2 and are stimulated to penetrate by host skin proteins (33). The molecular details of this interaction have yet to be elucidated.

Clues to other signals for attachment and invasion of the host by helminth parasites have come from work on fish parasites such as *Acanthostomum brauni*. Cercariae of *A. brauni* do not recognize small molecules such as amino acids, monosaccharides or electrolytes, but do respond to hyaluronic acid and glycoproteins for attachment (34). Penetration of the host is triggered by free fatty acids and mucus components present on the fish skin surface (35).

Monogenean parasites of fish skin lay eggs which fall to the sea or river bottom. Hatching of eggs of *Entobdella soleae* is timed by photoperiod to occur just after dawn, when the common sole (*Solea solea*) begins resting on the sandy bottom. Hatching is also stimulated by mucus from the host fish skin. The ciliated oncomiracidium that emerges is phototactic and may also sense currents (36).

Thermosensing undoubtedly plays a role in the directional migration or invasion of some nematode parasites as it does in schistosome cercariae. Clues to mechanisms of thermosensing in nematodes have come from studies of the free-living nematode *Caenorhabditis elegans*. *C. elegans* larvae and adults migrate to the temperature at which they have been growing when they have been placed in thermal gradient (37). Mutants with thermotactic abnormalities have been isolated. Many of them have lost the ability to migrate towards a specific temperature, whereas others show reversal of the usual behavior. Many, although not all, of these thermotactic mutants are also defective in chemotaxis. Chemotaxis was one of the first sensory behaviors noted in *C. elegans*. Worms are attracted to cyclic nucleotides, both anions and cations, some amino acids, and extracts of bacteria. All of the chemotactic mutants have been morphologically mapped to specific cells with corresponding sensory abnormalities in the head of the worm.

Much less is known about the factors that determine directionality of invasion by gut-dwelling helminths or what determines how far an L3 larva of *Onchocerca volvulus* will migrate in the dermis before deciding to stop and create a subcutaneous nodule. Similarly, little is known about how microfilariae of *Onchocerca* distribute themselves in such a manner that there is a high likelihood of being picked up by the insect vector. A particularly striking example is *Onchocerca cervipedis*, which infects deer. The adult *Onchocerca* live in the hindquarters of the deer and the microfilariae migrate across the body of the deer, up the neck, and into the ears. The blackfly vectors recognize the upright ears of the deer and congregate there.

Although light microscopic observation of parasite movement, and detailed anatomic descriptions of many parasitic trematode and nematode larvae have been completed, much less is known about the physiology of movement (27). Movement in aqueous environments of the ciliated miracidia and fork-tailed cercariae of schistosomes, and oncomiracidia of monogenean fish parasites, is very rapid (38–40). In the case of nematode parasites, clues as to mechanisms of motility are available from studies of the anatomically related, free-living nematode *Caenorhabditis elegans*, and the parasite nematode *Ascaris suum*. A number of genes have been identified by mutational analysis of *C. elegans* that are directly involved in motility. Many of these mutations (called *unc*, for *unc*oordinate) were among the first isolated 20 years ago (41). As expected, the *unc* phenotype was found to result both from defects in structural components of the musculature (such as myosin and paramyosin) as well as from mutations affecting the nervous system of the worm (37). For example, the gene *unc*-15 (linkage group I) is the structural gene for paramyosin. Normal thick filaments are lacking in null mutants, and paracrystalline inclusions are seen in missense mutants. *Unc*-54 (linkage group I) is a mutant that maps to the structural gene for the major myosin heavy chain myoB. Worms with mutant alleles have only one third the wild type number of thick filaments. Mutants in the *unc*-51 gene are paralyzed and tend to curl. This appears to be due to aberrant extensions of neurons which fail to connect to the dorsal nerve core.

Because of its size and availability, *Ascaris suum* has been the parasitic nematode of choice for neuromuscular analysis. The *Ascaris* nervous system contains approximately 90 motoneurons and interneurons (42). If both the head and tail of the worm are removed, these 90 motoneurons and interneurons are sufficient to control locomotion. *Ascaris* moves by propagating a wave of motion along its body axis anteriorly or posteriorly (42). The muscle cells that are innervated are arranged in two longitudinal masses, the dorsal and the ventral. They are independently innervated by neurons from the dorsal or ventral nerve cord (42). This distinctive pattern of nerve–muscle interaction ensures that the propagated wave moves through the axis of the worm in the dorsal–ventral plane.

2.2.2. Attachment

Many nematode parasites, such as hookworm adults and *Anisakis* adults, have evolved mouth structures which allow attachment to intestinal mucosa (43). Monogenean parasites of fish skin, which must attach to a rapidly moving host and resist water shear, also have highly evolved attachment structures (38). Schistosome cercariae emit a sticky mucus substance from the posterior acetabular glands, which allow them to attach to each other as well as to skin (40). The propensity of cercariae to attach to each other by mucus secretions may be a mechanism by which acetabular gland proteases, involved in digesting tissue barriers, can be concentrated from several organisms at a specific site on the host. Cercariae may be induced to cluster together on skin in response to L-arginine emitted from their secretory glands during the initial stages of skin exploration (27).

2.2.3. Digestion of Macromolecular Barriers

Although it is plausible that enzymes such as hyaluronidases, glycosidases or lipases may digest certain barriers that face invading helminths, evidence to date suggests that these functions are largely performed by proteolytic enzymes (44, 45). The best evidence is from studies of *Strongyloides* L3 larvae and schistosome cercariae. In both of these cases, secreted proteases with the capacity to degrade connective tissue or basement membrane molecules have been identified. Specific inhibitors of these proteases will inhibit invasion of skin by the larvae (46, 47). Related proteases have been identified in many other parasites, including hookworm (48, 49), *Anisakis* (43), *Ascaris* (50), *Toxocara* (51) and *Onchocerca* (52). To date, the nematode proteases implicated in invasion of helminths are serine or metalloproteases. The enzymes discovered in excretory/secretory products of invasive stages of helminth parasites also might be important in anticoagulation, digestion, immune evasion, or morphologic transformation. To assign an invasive function to a parasite protease, it must be shown that the protease activity actually functions outside the parasite through secretion or by localization on the external surface. The most direct test of the function of these enzymes is to show that blocking the action of the protease by irreversible inhibitors, or active-site-directed antibodies, inhibits invasion.

2.3. *Entamoeba histolytica*—a Protozoan Parasite that Invades Extracellularly

In contrast to the other protozoa discussed in this chapter, *Entamoeba histolytica* is not at any stage in its life cycle an intracellular parasite. Rather, it is more like the helminth parasites in that it invades tissue extracellular matrix, producing destructive lesions in the wall of the colon, and has the capacity to metastasize to other organs (53). A second distinctive aspect of infection by *E. histolytica* is that invasion of the host is by no means necessary for parasite replication or transmission. Only 10% or less of all cases of amebiasis result in invasive infections. For decades there has been controversy as to whether the invasive phenotype is due to host or bacterial factors acting on genetically monomorphic amebae, or whether it is an intrinsic characteristic of a specific strain or subspecies of *E. histolytica* (54). Recent analysis of *E. histolytica* isolates by restriction fragment-length polymorphism, and Southern blot analysis of specific virulence factor genes suggests a genetic basis for invasive disease (54, 55). It is now argued that pathogenic and non-pathogenic amebae are actually two species, *E. dispar* and *E. histolytica*. More than a dozen genes have been isolated from *Entamoeba* laboratory strains, and each of these shows at least some sequence divergence between the two proposed species. This includes both ribosomal genes as well as genes coding for specific enzymes like superoxide dismutase and cysteine proteinases. One caveat in interpreting these data is that the number of clinical isolates of amebae examined for each of these genes is still relatively small. Some of the differences in the sequences between genes from *E. histolytica* versus *E. dispar* is of the order of 12% or less. Sequence variation in the same genes among individual isolates or clones of the same 'species' can be as large. Orozco (56) has argued that 'pathogenic genes' may be present in non-pathogenic organisms, but in low copy number. The issue of whether gene amplification can alter the phenotype of amebae remains an open question. Because of the clonal evolution of protozoa, the term 'pseudospecies', used for viruses, may be the more accurate term in light of the genomic plasticity of protozoa.

The molecular and biochemical steps in invasion of the bowel wall by pathogenic amebae are becoming clear. First, trophozoites of *E. histolytica* have specific surface lectins and adhesion-mediating glycoproteins which allow attachment to intestinal mucus, target cells or host extracellular matrix (57, 58). Attachment appears to be necessary for target cell lysis, which is mediated at least in part by a cytolytic ion channel-forming protein secreted by the trophozoites (59, 60). Isolation of a gene coding for the cytolytic factor suggests it may be related to mellitin of bee venom (61).

Pathogenic, invasive trophozoites of *E. histolytica* also release proteinases (62–65). The major enzyme is a cysteine proteinase with cathepsin B-like substrate specificity and structural homology to members of the papain superfamily (55, 66). Release of this enzyme correlates with pathogenicity of clinical isolates and induces an immune reponse in infected individuals (67). Direct inhibition of this enzyme by specific, irreversible cysteine protease inhibitors blocks the cytopathic effect of trophozoites (68). Cytopathic effect refers to the ability of trophozoites to destroy a monolayer of tissue culture cells *in vitro*. Direct assays with purified cysteine protease have confirmed its ability to degrade a number of extracellular matrix macromolecules, including type I collagen, fibronectin and laminin. It also produces a specific cleavage in complement factor C3, producing biologically active C3 cleavage products (69).

Other potential virulence factors associated with invasion by *E. histolytica* include a metallocollagenase and phospholipases (58, 70). As a group, the virulence factors identified in extracts or secretions of pathogenic *E. histolytica* can explain in large part the cytolysis and tissue destruction that characterize *E. histolytica* invasion. Host neutrophils also contribute to the tissue destruction seen in amebic liver abscesses (71).

If host invasion is a dead-end for trophozoites of *E. histolytica*, what has maintained the invasive phenotype in a fairly competitive population of intestinal microorganisms? One possibility is that the invasive phenotype confers other adaptive advantages. The release of proteolytic enzymes and cytolytic factors may enhance degradation and uptake of nutritional molecules from host or bacteria. Pathogenicity and invasiveness not only correlate with the release of specific virulence factors but also enhanced phagocytosis. With controversial exceptions, only pathogenic (potentially invasive) isolates have been successfully weaned to long-term and stable axenic culture.

3. REFERENCES

1. Kornfeld, S. and Mellman, I. (1989) The biogenesis of lysosmes. *Annu. Rev. Cell Biol.* **5**: 483–525.
2. Joiner, K. A., Furhman, S. A., Miettinen, H. M., Kasper, L. H. and Mellman, I. (1990) *Toxoplasma gondii*—fusion competence of parasitophorous vacuoles in Fc receptor transfected fibroblasts. *Science* **249**: 641–646.
3. Bannister, L. H. and Dluzewski, A. R. (1990) The ultrastructure of red cell invasion in malaria infections—a review. *Blood Cells* **16**: 257–292.
4. Schwartzman, J. D. and Saffer, L. S. (1992) How *Toxoplasma* gets in and out of cells. In: *Intracellular Parasites* (eds. Harris, J. R. and Avila, J. L.), Plenum, New York, pp. 333–364.
5. Hadley, T. J. and Miller L. H. (1988) Invasion of erythrocytes by malaria parasites: erythrocyte ligands and parasite receptors. *Prog. Allergy* **41**: 49–71.
6. Pfefferkorn, E. (1990) Cell biology of *Toxoplasma gondii*. In: *Modern Parasite Bology: Cellular, Immunological and Molecular Aspects* (ed. Wyler, D. J.), Freeman, New York.
7. King, C. A. (1988) Cell motility of sporozoan protozoa. *Parasitol. Today* **4**: 315–319.
8. Perkins, M. E. (1991) Rhoptry organelles of apicomplexan parasites. *Parasitol. Today* **8**: 28–32.
9. Saffer, L. D., Mercereau-Puijalon, O., Dubremetz, J. F. and Schwartzman, J. D. (1992) Localization of a *Toxoplasma gondii* rhoptry protein by immunoelectron microscopy during and after host cell penetration. *J. Protozool.* **39**: 526–530.
10. Schrevel, J., Deguercy, A., Mayer, R. and Monsigny, M. (1990) Proteases in malaria-infected red blood cells. *Blood Cells* **16**: 563–584.

11. Entzeroth, R., Dubremetz, J. F., Hodick, D. and Ferreira, E. (1986) Immunoelectromicro-scopic demonstration of the exocytosis of dense granules contents into the secondary parasitophorous vacuole of *Sarcocystis muris* (Protozoa, Apicomplexa). *Eur. J. Cell Biol.* **41**: 182–188.

12. Achbarou, A., Mercereau-Puijalon, O., Sadak, A. *et al.* (1991) Differential targeting of dense granules proteins in the parasitophorous vacuole of *Toxoplasma gondii. Parasitology* **103**: 321–329.

13. Aikawa, M., Torii, M., Sjölander, A., Berzins, K., Perlman, P. and Miller, L. H. (1990) Pf155/RESA antigen is localized in dense granules of Plasmodium *falciparum* merozoites. *Exp. Parasitol.* **71**: 326–329.

14. Pouvelle, B., Spiegel, R., Hsiao, L. *et al.* (1991) Direct access to serum macromolecules by intraerythrocytic malaria parasites. *Nature* **353**: 73–75.

15. Shaw, M. K., Tilney, L. W. and Musoke, A. J. (1991) The entry of *Theileria parva* sporozoites into bovine lymphocytes: evidence for MHC class I involvement. *J. Cell Biol.* **113**: 87–101.

16. Ouaissi, M. A. and Capron, A. (1989) Some aspects of protozoan parasite–host cell interactions with special reference to RGD-mediated recognition process. *Microb. Pathol.* **6**: 1–5.

17. Schenkman, S., Robbins E. S. and Nussenzweig, V. (1991) Attachment of *Trypanosoma cruzi* to mammalian cells requires parasite energy, and invasion can be independent of the target cell cytoskeleton. *Infect. Immun.* **59**: 645–654.

18. Hall, B. F. and Joiner, K. A. (1991) Strategies of obligate intracellular parasites for evading host defences. *Immunoparasitol. Today* **12**: 124–129.

19. Andrews, N. W., Abrams, C. K., Slatin, S. L. and Griffiths, G. (1990) A *T. cruzi* secreted protein immunologically related to the complement component C9: evidence for membrane pore-forming activity at low pH. *Cell* **61**: 1277–1287.

20. Russell, D. G., Medina-Acosta E. and Golubev, A. (1991) The interface between the *Leishmania*-infected macrophage and the host's immune system. *Behring Inst. Mitt.* **88**: 68–79.

21. Antonine, J. C., Prina, E., Jouanne, C. and Bongrand, P. (1990) Parasitophorous vacuoles of *Leishmania amazonensis*-infected macrophages maintain an acidic pH. *Infect. Immun.* **58**: 779–787.

22. Rabinovitch, M., Topper, G., Cristello, P. and Rich, A. (1985) Receptor-mediated entry of peroxidases into the parasitophorous vacuole of macrophages infected with *Leishmania mexicana amazonensis. J. Leuk. Biol.* **37**: 247–261.

23. Despommiers, D. D. (1990) *Trichinella spiralis*: the worm that would be a virus. *Parasitol. Today* **6**: 193–196.

24. Pleshinger, J. and Weidner, E. (1986) The microsporidian spore invasion tube. IV. Discharge activation begins with pH-triggered Ca^{2+} influx. *J. Cell Biol.* **100**: 1834–1838.

25. Tse, S.-K. and Chadee, K. (1991) The interaction between intestinal mucus glycoproteins and enteric infections. *Parasitol. Today* **7**: 163–172.

26. Pearce, E. J., Basch, P. F. and Sher, A. (1986) Evidence that the reduced surface antigenicity of developing *Schistosoma mansoni* schistosomula is due to antigen shedding rather than host molecule acquisition. *Parasite Immunol.* **8**: 79–94.

27. Haas, W. (1992) Physiological analysis of cercarial behavior. *J. Parasitol.* **78**: 243–255.

28. Loker, E. S. (1978) Normal development of *Schistosomatium douthitti* in the snail *Lymnaea catascopium. J. Parasitol.* **64**: 977–985.

29. Fusco, A. C., Salafsky, B. and Delbrook, K. (1985) *Schistosoma mansoni*: production of cercarial eicosanoids as correlates of penetration and transformation. *J. Parasitol.* **72**: 397–404.

30. Fusco, A. C., Salafsky, B., Vanderkooi, G. and Shibuya, T. (1991) *Schistosoma mansoni*: the role of calcium in the stimulation of cercarial proteinase release. *J. Parasitol.* **77**: 649–657.

31. Haas, W. and Schmitt, R. (1982). Characterization of chemical stimuli for the penetration of *Schistosoma mansoni* cercariae. *Z. Parasitenk.* **66**: 293–307.

32. Katz, M., Despommier, D. D. and Gwadz, R. W. (1982) *Parasitic Diseases.* Springer-Verlag, New York.

33. Haas, W., Granzer, M. and Brockelman, C. R. (1990) *Opisthorchis viverrini*: finding and recognition of the fish host by the cercariae. *Exp. Parasitol.* **71**: 422–431.
34. Haas, W. and de Nuñez, M. O. (1988) Chemical signals of fish skin for the attachment response of *Acanthostomum brauni* cercariae. *Parasitol. Res.* **74**: 552–557.
35. Ostrowski de Nuñez, M. and Haas, W. (1991) Penetration of stimuli of fish skin for *Acanthostomum brauni* cercariae. *Parasitology* **102/1**: 101–104.
36. Kearn, G. C. (1986) Role of chemical substances from fish hosts in hatching and host-finding in monogeneans. *J. Chem. Ecol.* **12**: 1651.
37. Waterston, R. H., Chalfie, M. and White, J. (1988) The nematode *Caenorhabditis elegans* (ed. Wood, W. B.), Cold Spring Harbor Laboratory, pp. 281–392.
38. Kearn, G. C. (1976) Body surface of fishes. In: *Ecological Aspects of Parasitology* (ed. Kennedy, C. R.), North-Holland, Amsterdam, pp. 185–205.
39. MacInnis, A. J. (1976) How parasites find hosts: some thoughts on the inception of host–parasite integration. In: *Ecological Aspects of Parasitology* (ed. Kennedy, C. R.), North-Holland, Amsterdam, pp. 3–20.
40. Stirewalt, M. A. (1974). *Schistosoma mansoni*: Cercaria to schistosomule. *Adv. Parasitol.* **12**: 115–180.
41. Brenner, S. (1974) The genetics of *Caenorhabditis elegans*. *Genetics* **77**: 71–94.
42. Stretton, A. O. W., Davis, R. E., Angstadt, J. D., Donmoyer, J. E. and Johnson, C. D. (1985) Neural control of behaviour in *Ascaris*. *Trends Neurosci.* **8**: 294–300.
43. Sakanari, J. A. and McKerrow, J. H. (1989) Anisakiasis. *Clin. Microbiol.* **2**: 278–284.
44. Fishelson, Z., Amiri, P., Friend, D. S. *et al.* (1992) *Schistosoma mansoni*: cell-specific expression and secretion of a serine protease during development of cercariae. *Exp. Parasitol.* **75**: 87–98.
45. McKerrow, J. H., Sakanari, J. A., Brown, M. *et al.* (1989) Proteolytic enzymes as mediators of parasite invasion of skin. In: *Models of Dermatology*, Vol. 4 (eds. Maibach, H. and Lowe, N. J.), Karger, Basel, Switzerland, pp. 276–284.
46. Cohen, F. E., Gregoret, L., Amiri, M. P., Aldape, K., Railey, J. and McKerrow, J. H. (1991) Arresting tissue invasion of a parasite by protease inhibitors chosen with the aid of computer modeling. *Biochemistry* **30**: 11221–11229.
47. McKerrow, J. H., Brindley, P., Brown, M., Gam, A. A. and Neva, F. (1990) *Strongyloides stercoralis*: identification of a protease that facilitates penetration of skin by the infective larvae. *Exp. Parasitol.* **70**: 134–143.
48. Hotez, P., Trang, N., McKerrow, J. H. and Cerami, A. (1985) Isolation and characterization of a proteolytic enzyme from the adult hookworm. *J. Biol. Chem.* **260**: 7343–7348.
49. Pritchard, D. I., McKean, P. G. and Schad, G. A. (1990) An immunological and biochemical comparison of hookworm species. *Parasitol. Today* **6**: 154–156.
50. Knox, D. P. and Kennedy, M. W. (1988) Proteinases released by the parasitic larval stages of *Ascaris suum* and their inhibition by antibody. *Mol. Biochem. Paristol.* **28**: 207–216.
51. Robertson, B. D., Bianco, A. E., McKerrow, J. H. and Maizels, R. M. (1989) *Toxocara canis*: proteolytic enzymes secreted by the infective larvae. *Exp. Parasitol.* **69**: 30–36.
52. Lackey, A., James, E. R., Sakanari, J. A. *et al.* (1989) Extracellular proteases of *Onchocerca*. *Exp. Parasitol.* **68**: 176–185.
53. Ravdin, J. (ed.) (1988) Amebiasis: Human Infection by *Entamoeba histolytica*. John Wiley, New York.
54. McKerrow, J. H. (1992) Pathogenesis in amebiasis: is it genetic or acquired? *Inf. Ag. Dis.* **1**: 11–14.
55. Tannich, E., Horstmann, R. D., Knobloch, J. and Arnold, H. H. (1989) Genomic DNA differences between pathogenic and non-pathogenic *Entamoeba histolytica*. *Proc. Natl. Acad. Sci. USA* **86**: 5118–5122.
56. Orozco, E. (1992) Pathogenesis in amebiasis. *Inf. Ag. Dis.* **1**: 19–21.
57. Petri, W. A., Smith, R. D., Schlesinger, P. H., Murphy, C. F. and Ravin, J. I. (1987) Isolation of the galactose-binding lectin that mediates the in vitro adherence of *Entamoeba histolytica*. *J. Clin. Invest.* **80**: 1238–1244.

58. Ravdin, J. (1986) Pathogenesis of disease caused by *Entamoeba histolytica*: studies of adherence, secreted toxins, and contact-dependent cytolysis. *Rev. Infect. Dis.* **8**: 247–260.

59. Keller, F., Hanke, W., Trissl, D. and Bakker-Grunwald, T. (1989) Pore-forming protein from *Entamoeba histolytica* forms voltage- and pH-controlled multi-state channels with properties similar to those of the barrel-stave aggregates. *Biochim. Biophys. Acta* **982**: 89–93.

60. Young, J. D., Young, T. M., Lu, L. P., Unkeless, J. C. and Cohn, Z. A. (1982) Characterization of a membrane pore-forming protein from *Entamoeba histolytica*. *J. Exp. Med.* **156**: 1677–1690.

61. Leippe, M., Sebastian, E., Schoenberger, O. L., Horstmann, R. D. and Muller-Eberhard, H. J. (1991) Pore-forming peptide of pathogenic *Entamoeba histolytica*. *Proc. Natl. Acad. Sci. USA* **88**: 7659–7663.

62. Keene, W. E., Petitt, M. G., Allen, S. and McKerrow, J. H. (1986) The major neutral proteinase of *Entamoeba histolytica*. *J. Exp. Med.* **163**: 536–549.

63. Luaces, A. L. and Barrett, A. J. (1988) Affinity purification and biochemical characterization of histolysin, the major cysteine proteinase of *Entamoeba histolytica*. *Biochem. J.* **250**: 903–909.

64. Lushbaugh, W. B. (1988) Proteinases of *Entamoeba histolytica*. In: *Amebiasis: Human Infection by Entamoeba histolytica* (ed. Ravdin, J. J.), John Wiley, New York, pp. 219–231.

65. Scholze, H. and Werries, E. (1986) Cysteine proteinase of *Entamoeba histolytica*. I. partial purification and action on different enzymes. *Mol. Biochem. Parasitol.* **18**: 103–112.

66. McKerrow, J. H., Bouvier, J., Sikes, A. and Keene, W. E. (1991) The cysteine proteinase of *Entamoeba histolytica*: cloning of the gene and its role as a virulence factor. In: *Biochemical Protozoology as a Basis for Drug Design* (eds. North, G. C. and North, M.), Taylor & Francis, London, pp. 245–250.

67. Reed, S., Keene, W. E. and McKerrow, J. H. (1989) Thiol proteinase expression correlates with pathogenicity of *Entamoeba histolytica*. *J. Clin. Microbiol.* **27**: 2772–2777.

68. Keene, W. E., Hidalgo, M. E., Orozco, E. and McKerrow, J. H. (1990) *Entamoeba histolytica*: correlation of the cytopathic effect of virulent trophozoites with secretion of a cysteine proteinase. *Exp. Parasitol.* **71**: 199–206.

69. Reed, S. L., Keene, W. E., McKerrow, J. H. and Gigli, I. (1989) Cleavage of C3 by a neutral cysteine proteinase of *Entamoeba histolytica*. *J. Immunol.* **143**: 189–195.

70. Muñoz, M., Rojkind, M., Calderon, J., Tanimoto, M., Arias-Negrete, S. and Martínez-Palomo, A. (1984). *Entamoeba histolytica*: collagenolytic activity and virulence. *J. Protozool.* **31**: 468–470.

71. Pérez-Tamayo, R., Martinez, R. D., Montfort, I., Becker, I., Tello, E. and Pérez-Monfort, R. (1991) Pathogenesis of acute experimental amebic liver abscess in hamsters. *J. Parasitol.* **77**: 982–988.

17 Concepts of Chemotherapy

J. JOSEPH MARR[1] and BUDDY ULLMAN[2]

[1]*Ribozyme Pharmaceuticals, Inc., Boulder, CO, USA and* [2]*Department of Biochemistry and Molecular Biology, Oregon Health Sciences University, Portland, OR, USA*

SUMMARY

To achieve a good therapeutic index, an antiparasitic agent must be directed against a peculiarity in the organism's metabolism. This may manifest itself as a specific morphologic structure, such as the hydrogenosome in *Trichomonas*; or a variation in metabolism, such as the inability of hemoflagellates to synthesize purines *de novo*. Specific enzymes may be selectively targeted, such as the ornithine decarboxylase in African trypanosomes. Nevertheless, inhibition of metabolic processes is not sufficient to achieve a chemotherapeutic effect, although it is a necessary precondition.

11. BIOCHEMICAL CONSIDERATIONS

Many metabolic processes are common to most life forms, but there is a richness in the variety and variations of these processes. These similarities impose requirements for both selectivity and potency of inhibition in an antiparasitic agent. Selectivity is particularly important. The agent must inhibit the biochemistry of the parasite, yet leave the host untouched. Moreover, inhibition of an enzyme or several steps in a metabolic pathway often calls into play mechanisms to bypass the points of inhibition. These mechanisms may be alternate pathways to achieve the same product; the metabolic block may be overcome by the accumulating precursor; or the end product may be obtained from the host.

Biochemistry and Molecular Biology of Parasites
ISBN 0-12-473345-X

2. BIOLOGICAL CONSIDERATIONS

In the translation of experimental chemotherapy into clinical results, it is important that the form of the parasite in the human host is the same as the form being studied experimentally. For example, the free-living forms of many parasites, such as the hemoflagellates, are relatively easy to grow in culture; however, the important pathogenic form in the vertebrate host is an intracellular organism. To be certain that the experimental conclusions can be applied clinically, one must test them in the appropriate stage of the life cycle of the parasite. Even under circumstances where the correct morphologic form is studied, as in tissue culture, it is possible that the organism itself may have been modified by prolonged culture. In this modification it may lose its virulence.

3. PHARMACOLOGICAL CONSIDERATIONS

In addition to the parasite-related considerations described above, there are many additional issues which must be addressed before any potential chemotherapeutic agent can be considered a candidate for clinical studies. These generally come under the headings of pharmacokinetics and pharmacodynamics. The former generally deals with the rates of absorption and elimination of a compound whereas the latter is concerned with the distribution of the compound in the body, its effects on particular organ systems, and other interactions with host metabolic processes.

Pharmacological considerations may be linked with the character of the chemical compound include absorption from the gastrointestinal tract; distribution of the compound; metabolism; and elimination. This includes questions of penetration into various compartments of the body and cells; this may vary greatly and have a substantial influence on the efficacy of therapy. For example, agents directed to the treatment of cysticercosis must penetrate not only the tissue compartments, including the brain, where the parasite resides, but must also penetrate the cyst within which the parasite is located.

Since most biochemical processes are common to all organisms, the efficacy of an antiparasitic agent generally is a balance between its potency against the parasite's metabolism and its activity against the human host's metabolism. This latter is generally described as toxicity of the compound and the net effect of the two processes is termed the therapeutic index. In general, the more specific the compound for the metabolism of the parasite the greater the therapeutic index.

In addition to the above, there are other considerations of chemical and pharmaceutical development which, though divorced from the scientific considerations described, are none the less important in drug delivery to the patient. They may be important enough to cause an otherwise good chemotherapeutic agent not to be considered for clinical use. These include the stability of the compound in a wide variety of climatological conditions, its acceptability to a patient population — taste, color, size of the capsule or tablet, pain on injection, frequency of delivery — and, finally, the cost to make the compound and deliver it to the patient. For a compound to be effective against diseases of the developing world, it must be very inexpensive and usually should be given by mouth.

4. MECHANISMS

Treatment of parasitic disease has been largely empirical since relatively little was known about the biochemistry of these organisms. In the past two decades, however, there has been a remarkable expansion of our knowledge. This is only beginning to change approaches to chemotherapy but has enabled the scientific world to ask questions with far more meaning. In this section we will explore the potential relationships of the new biochemical and molecular biological knowledge as they may relate to chemotherapy rather than to review the extant knowledge on this topic (1, 2).

4.1. Intermediary Metabolism

4.1.1. Carbohydrate metabolism

Intermediary metabolism of carbohydrates, since it was the earliest studied in these organisms, has been a consideration from the point of view of chemotherapy for many years. One of the earliest rational exploitations of this was the inhibition of the glyceraldehyde-3-phosphate oxidase of these organisms by salicylhydroxamic acid. This proposal, which had excellent biochemical logic because of the reliance of African trypanosomes upon this mechanism for regenerating oxidized pyridine nucleotides, ultimately proved to be of no clinical value. Since then, many studies have demonstrated the essential peculiarity of parasitic carbohydrate metabolism, aerobic fermentation, in which intermediary products of glycolysis serve to eliminate reducing equivalents from the cell even in the presence of oxygen. Nevertheless, it has not been possible to exploit these mechanisms for chemotherapy.

More recently, glycosomes have been described (3). These structures, reviewed in Chapter 2, provide a very different packaging for the glycolytic enzymes, but have not been exploited for purposes of chemotherapy. This may be because the compartmentalization of these enzymes, which provides a very favorable biochemical environment for the capture of energy, still involves the same enzymes analogous to those found in mammalian cells. Inhibitors must be based on differences in the structures of these enzymes which differentiate them from their mammalian counterparts. This may not provide the necessary differences in specificity which are requisite for chemotherapy.

4.1.2. Anaerobic energy metabolism

Anaerobic parasites, such as *T. vaginalis* and *E. histolytica*, lack functional mitochondria. Pyruvate generated by the glycolytic pathway is further metabolized within the cytosol of *Giardia* and *Entamoeba* species and in the hydrogenosome of these parasites, a subcellular organelle unique to trichomonads (4). Under anaerobic conditions, pyruvate is converted into acetyl Co-A and CO_2 by pyruvate:ferredoxin oxidoreductase with the concomitant reduction of ferredoxin. Metronidazole and other nitro-imidazole derivatives are toxic to these anaerobes after reduction of the nitro moiety by ferredoxin. As these compounds are not reduced in mamalian cells, selectivity of their antiparasitic effects is assured (see Chapter 3).

Carbohydrate metabolism in African trypanosomes is compartmentalized within the glycosome (Chapter 2). Many structural and kinetic differences exist between the glycosomal and cytoplasmic glycolytic pathways. The high isoelectric point of the glycosomal enzymes (5) has been postulated to be the mechanism by which suramin exerts its antitrypanosomal effect. Reducing equivalents produced by glycolysis within the glycosome are transferred to a mitochondrial glycerol-3-phosphate oxidase complex. Inhibitors of this complex, such as salicylhydroxamic acid (SHAM) and esters of dihydroxybenzoate, in conjunction with glycerol, have exhibited good antitrypanosomal activity *in vitro* and in animal models (6) but not in clinical trials (see above).

4.1.3. Glycosylphosphatidylinositol anchors

Important constituents in the outer portion of the plasma membrane of many eukaryotic cells are the glycosylphosphatidylinositols (GPI). In parasitic protozoa, these materials anchor the surface membrane molecules which appear to be necessary for parasite infectivity and survival and may constitute the major glycolipids class. These include the variant surface glycoproteins in the African trypanosomes, Ssp-4 and the 90 kDA major surface glycoproteins of the amastigotes and trypomastigotes of *T. cruzi*, respectively. Similar materials constitute the major surface glycoproteins of promastigotes and amastigotes of leishmania. These glycolipids have the structure glycan-inositolphospholipid. There are important differences between the protozoan and mammalian GPI anchors in mammalian cells. The mammalian anchors differ in that they contain α-linked mannose and β-linked GalNAc in addition to ethanolamine phosphate residues (7).

The best understood unique cell surface proteins are the variant surface glycoproteins (VSG) that, through the process of antigenic variation, enable African trypanosomes to escape the host immune system (8). VSGs are anchored to the parasite cell surface by a glycosyl phosphatidylinositol (GPI) anchor that contains myristic acid as its sole fatty acid component (9). GPI biochemistry has been most extensively studied in *T. brucei* which does not synthesize fatty acids *de novo*. Salvage of myristic acid from the host is essential for the creation of the GPI anchor. One myristic acid analog, 10-(propoxy)decanoic acid, can be incorporated into the GPI anchor and is toxic to the bloodstream form of *T. brucei* (10). This compound is ineffective against mammalian cells, which use myristic acid only in the modification of the NH_2-terminus of certain proteins, or *T. brucei* procyclic forms, which lack VSGs. Related surface structures are described in Chapters 11 and 12.

4.1.4. Trypanothione

Trypanothione metabolism in the hemoflagellates may emerge as an important chemotherapeutic target. It is a compound which is essential for the survival of the parasite and is absent from its host (11).

Glutathione is the major intracellular thiol in aerobic organisms. It has many functions, and its common theme is to provide the cell with a reducing environment. In trypanosomatids, greater than 70% of the total intracellular glutathione is present as the spermidine conjugate trypanothione. The levels of this compound are maintained by the activity of trypanothione reductase. Inhibitors of this enzyme may serve as potent chemotherapeutic agents for the trypanosomatids. Biosynthesis of trypanothione is another potential target for chemotherapy since it utilizes enzymes which are not normally active in the mammalian cell (see Chapter 9).

4.1.5. Proteases

One of the newest and potentially most interesting areas for chemotherapy is the inhibition of proteases. This is of intense pharmaceutical interest for the management of high blood pressure and of viral diseases. The difficulty with this approach lies in the fact that the proteases use common mechanisms in the mammalian host and micro-organisms. Nevertheless, specificity of inhibition has been achieved for the mammalian enzyme renin for the control of high blood pressure and for the protease used for replication of the human immunodeficiency virus which causes AIDS. For this reason, there is reason to be encouraged that inhibition of proteases in parasitic organisms will be an effective means of chemotherapy.

There are four major classes of proteases for which the catalytic mechanisms have been defined. These are described according to the active component in the enzyme which assists in the binding of the substrate. These are: aspartic acid, cysteine, serine and metalloproteases. There are other proteases but their catalytic mechanisms are unclear. For a review of proteases and their mechanisms and inhibitors see (refs 12 and 13 and also Chapter 5).

The proteases have diverse functions in the various parastic organisms. Some of these functions are known, such as the use in tissue invasion by *Entamoeba histolytica* and digestion of hemoglobin by *Plasmodium* species. The major surface antigen of various leishmania species, GP63, has been extensively studied and is a zinc metallo-endopeptidase. It is proteolytically active at the surface of the promastigotes and is conserved among leishmania species. Its specific function is unknown. A protease which is receiving increasing study is the aspartic protease of *Plasmodium* which is important in the first step of digesting hemoglobin. Thereafter, other proteases assist in the degradation of the globin molecule. Inhibition of this enzyme is an attractive chemo-therapeutic target since it would require agents which are quite different from the 4-aminoquinolines and related compounds or antifolates which are currently in use. Proteases have been described in *T. cruzi*, *T. brucei*, leishmania, trichomonads and *Giardia* although their roles are unclear. In plasmodia, in addition to the proteases mentioned above, other proteases are important in initiating infection and completing the erythrocytic stage of infection (14). For a more extensive discussion of proteases see (ref. 15). This intriguing chemotherapeutic target will undoubtedly be explored in the coming years. It is relatively simple to inhibit proteinases and the challenge will be to prepare not peptide inhibitors but rather peptide mimetics which will be active against these enzymes. This is an exercise in synthetic medicinal chemistry and the results are not necessarily predicted by knowledge of the peptide involved in enzyme inhibition. Recently, screens of small molecule structural databases with molecular models of the *P. falciparum* cysteine protease, and the *S. mansoni* cercerial elastase have uncovered novel inhibitors of both enzymes (16).

4.1.6. Purine and pyrimidine metabolism

The metabolism of purines in protozoan parasites differs significantly from that in mammals (Chapter 6). Since there is no *de novo* synthesis, interdiction of the salvage pathway has far greater implications in these organisms than a similar interruption in human metabolic pathways. In addition, some of the enzymes of the salvage pathway in these parasites will accept purine analogs and metabolize them to nucleotides. These same analogs are not accepted by the corresponding human enzymes. The mucosal pathogens also lack *de novo* synthesis and have purine salvage pathways which do not cross over between adenine and guanine. This suggests that inhibition of either adenine or guanine metabolism would be sufficient to prevent growth of the parasite. In humans, there are ample opportunities for crossover. With respect to DNA metabolism, some parastic protozoans (the best studied is *Giardia*) lack ribonucleotide reductase. This means that they are totally dependent on salvage of preformed deoxynucleosides. The biochemical differences between parasitic protozoans and humans are significant enough for one to suspect that purine analogs might be effective inhibitors of these organisms. Pyrimidine metabolism, by contrast, appears to be very similar in humans and most parasites and inhibition of the enzymes of these metabolic pathways probably does not carry the same specificity as does inhibition of purine metabolism (17).

Perhaps the best examples in which the purine pathway has served as a therapeutic target are the pyrazolopyrimidines against the trypanosomatids. For instance, the hypoxanthine–guanine phosphoribosyltransferase of trypanosomatids will initiate the metabolism of the pyrazolopyrimidines as though they are purines (17–19). The best studied of these is allopurinol, an isomer of hypoxanthine. It is phosphoribosylated to the IMP analog by the parasite phosphoribosyltransferase, then aminated to the analog of AMP. After two phosphorylation steps it is converted into an analog of ATP and incorporated into RNA. This brings about inhibition of purine nucleotide biosynthesis in these organisms and halts protein synthesis. This metabolic scheme exists in all the hemoflagellates studied.

This has been translated into clinical benefits. There have been several reports of the efficacy of allopurinol in visceral leishmaniasis and one study using allopurinol riboside in cutaneous leishmaniasis. More recently, a controlled study in Colombia has documented the efficacy of allopurinol in cutaneous leishmaniasis (20). In this study the cure rate of allopurinol was 80% compared to 74% using allopurinol and glucantime together. The cure rate of glucantime alone was 36%. A similar study of the efficacy of allopurinol in chronic asymptomatic Chagas' disease showed that allopurinol in doses of 600 or 900 mg day^{-1} was equal efficacy to nifurotimox or benznidazole (21).

Another pathway of potential interest is the xanthine phosphoribosyltransferase. This enzyme is present in leishmania yet absent in humans. This raises the possibility that xanthine analogs could serve as specific chemotherapeutic agents since they would be activated by an enzyme not found in the host and incorporated into nucleotides by the parasite. In addition, xanthine analogs are likely to be oxidized by xanthine oxidase in the host and excreted, thus eliminating potential problems of toxicity.

The simple purine salvage scheme of the mucosal pathogens (*Giardia*, *Trichomonas*, and *Entamoeba* species) with the lack of crossover from adenine to guanine metabolism means that there is an absolute requirement for each of these purines. This suggests that inhibition of either pathway will inhibit the organisms' growth (Chapter 6).

4.1.7. Polyamine metabolism

Like purine metabolism, the polyamine biosynthetic pathway has served as another paradigm for rational therapeutic intervention in parasitic disease. Polyamine synthesis in *T. brucei* and *Leishmania*, like that in mammalian cells, is initiated by the enzyme ornithine decarboxylase (ODCase), although *T. cruzi* may synthesize polyamines by converting arginine into agmatine, a reaction catalyzed by arginine decarboxylase (ADCase). The growth of *T. brucei in vitro* can be inhibited by α-difluoro-methylornithine (DFMO), an irreversible inhibitor of ODCase (22). DFMO also eliminates parasitemia in both bloodstream and late-stage central nervous system infections of *T. brucei* in mice (23). Whether the antitrypanosomal effect can be attributed to putrescine, spermidine and/or trypanothione depletion has not been resolved. *L. donovani* promastigotes are also sensitive to DFMO (23), but the intracellular amastigote forms of this parasite may be refractory to the drug due to the host polyamine reservoir. α-Difluoromethylarginine, an inhibitor of ADCase activity, inhibits *T. cruzi* infectivity and multiplication in host cells and may, therefore, have therapeutic applicability (23). (See Chapter 7 for further discussion.)

As DFMO is relatively non-toxic toward humans and was active against *T. brucei* in a mouse model, it was tested and shown to be effective against *T. b. gambiense* infections in both early and late-stage disease (24). The mechanism of selective action of DFMO against *T. brucei* cannot be ascribed to a differential sensitivity of the parasite and human enzymes to DFMO inactivation, since the enzymes are equally sensitive to the drug. Rather, the selectivity can be ascribed to the differential stabilities of the parasite and mammalian ODCases. Mammalian ODCase is one of the most short-lived proteins ($t_{1/2} = <1\,h$), whereas the parasite enzyme is more stable ($t_{1/2} = \gg 6\,h$). The molecular basis of this differential lability can be attributed to a 37 amino acid COOH terminal sequence in the mammalian ODC that is rich in PEST sequences that target the enzyme for intracellular degradation. The parasite ODCase enzymes lack this COOH terminus (25). The therapeutic potential of DFMO stems from the ability of host cells to re-express catalytically active ODCase molecules after the drug has cleared.

Although DFMO has shown clinical activity toward *T.b. gambiense*, it is not particularly effective toward *T.b. rhodesiense*. DFMO requires prolonged administration and large doses so a more potent polyamine inhibitor with a broader spectrum of action will be needed. Recently, 5'-{[(Z)-4-amino-2-butenyl]methylamino}-5'-deoxyadenosine, an irreversible inhibitor of *S*-adenosylmethionine decarboxylase (SAMDCase), the enzyme that provides a substrate for spermidine synthase, has been shown to be effective against *P. falciparum in vitro* (26) and for *T. b. brucei* and *T. b. rhodesiense* in mice (27). Other polyamine analogs, such as the bis(benzyl)polyamines, have also displayed activity against *P. falciparum* both *in vitro* and *in vivo*.

4.1.8. Folic acid antagonists

Folic acid antagonists have been successfully exploited in the treatment of several parasitic diseases in humans, most prominently malaria, toxoplasmosis and pneumocystosis (28). Two classes of antifolate analogs have been used in therapy of these afflictions: the diaminopyrimidines pyrimethamine and trimethoprim; and the *p*-aminobenzoic acid (PABA) antagonists, the sulfonamides and sulfones. Pyrimethamine/sulfadoxine can be administered for malaria treatment and prophylaxis, since the two classes of drugs work synergistically. Pyrimethamine/sulfadiazine is the regimen of choice in the handling of *T. gondii* infections, while trimethoprim/sulfamethoxazole is exploited in the treatment and prophylaxis of *P. carinii* infections. The incidences of the latter two diseases have increased dramatically as a consequence of the AIDS epidemic, which has precipitated the search for improved therapy.

Pyrimethamine and trimethoprim are inhibitors of both bacterial and protozoal dihydrofolate reductase (DHFR) enzymes but do not affect the mammalian enzyme. Further specificity is achieved by the use of PABA antagonists, since they are competitive inhibitors of the protozoal dihydropteroate (DHP) synthase reaction, which condenses PABA with hydroxymethyldihydropteridine to form DHP, an intermediate in the tetrahydrofolate (THF) biosynthetic pathway. Protozoa synthesize THF *de novo* whereas humans require dietary folate. For this reason sulfur drugs are selective and virtually non-toxic to humans.

The DHFR and thymidylate synthase (TS) activities in many parasitic protozoa exist as a single bifunctional protein, termed the DHFR-TS complex (29). Since DHFR and TS are biochemically distinct enzymes in humans, DHFR-TS offers an avenue for therapeutic exploitation. Recent nutritional (30) and biochemical (31) evidence has suggested that *Leishmania*, unlike humans, have the capacity to convert pterins into folates (and perhaps vice versa) implying that these pathways might serve as therapeutic targets.

4.1.8. Quinolines

The 4-aminoquinolines (quinine, chloroquine and mefloquine) and the 8-aminoquinoline, primaquine, are the most widely used agents in the therapy and prevention of malaria. The activity of the quinoline drugs is both species- and stage-specific. Despite their prevalent usage, the precise mode of action of the aminoquinolines is still not clear. Aminoquinolines are cationic amphophilic drugs, thereby exhibiting a high degree of lipophilicity. Furthermore, they are weak bases and tend to accumulate within intracellular acidic compartments such as the parasite food vacuole of the intra-erythrocytic stage of the parasite. It has been suggested that chloroquine may act by binding to heme degradation products generating a toxic byproduct or by directly inhibiting enzymes within the malarial food vacuole (32, 33). Recently, it has been proposed that these drugs may act by inhibiting the heme polymerase that generates non-toxic insoluble hemozoin from the toxic soluble heme molecule (34).

4.2. Anthelminthic Agents

4.2.1. Avermectins

Avermectins are a family of 16-membered cyclic lactones that possess broad and potent antihelminthic activity (35). A 22,23-dihydro derivative of avermectin B_1, ivermectin, has been used for both medical and veterinary use and is effective in eradicating the microfilariae of *O. volvulus* (36). In addition, avermectins are active against ecto-parasites and are powerful insecticides (37). Avermectins obstruct postsynaptic potentials via augmentation of Cl^- ion conductance (38, 39). Avermectins are antagonists of the γ-aminobutyric acid (GABA) receptor that opens the GABA-gated Cl^- channel, and directly actuate glutamate-sensitive currents (40–42). The selectivity of avermectin action toward nematodes and ectoparasites can be imputed to a difference in the avermectin binding characteristics in nematode and mammalian membrane systems and to the failure of these drugs to cross the blood–brain barrier.

4.2.2. Other Antihelminthic Agents

These drugs are empirical in design. The most important of these agents will be discussed only briefly since they fail to illustrate rational biochemical principles. The drugs of choice in the treatment of helminthic diseases vary depending on whether the parasite is a Nematode, Cestode or Trematode and upon the location of the parasite in the host. Drugs currently employed for the treatment of intestinal roundworms include pyrantel pamoate or mebendazole for ascariasis, trichuriasis, enterobiasis and hookworm infections and thiabendazole for strongyloidiasis (28). Non-intestinal nematode infections are much more difficult to treat. If an antinematode drug is employed, it is often effective against only specific stages of the parasite. For instance, diethylcarbamazine and ivermectin are both lethal to the microfilariae of *W. bancrofti* but are ineffectual in the treatment of filariasis. Niridazole and ivermectin, however, are highly effective in the treatment of *D. medinensis* and *O. volvulus* infections, respectively (28).

Infections with mature cestodes can be readily treated with a single dose of niclosamide. The drug is equally active against *D. latum*, *T. solium* and *T. saginata*. The preferred treatment for larval tapeworm infections, such as cysticercosis, is praziquantel, a drug that is also used to treat fluke infestations, such as schistosomiasis and clonorchiasis. These agents have been in clinical use for many years, and the interested reader is referred to standard pharmacology or infectious disease texts for further information (1).

4.3. Arsenicals and Antimonials

Arsenic and antimony compounds have served as traditional empirical antiparasitic reagents. The application of such drugs is still widespread. Individuals suffering from African trypanosomiasis with central nervous system involvement are generally treated with melarsoprol, a trivalent arsenical that is toxic to the human host. The first-line drug of choice for the treatment of leishmaniasis is sodium stibogluconate, a pentavalent antimonial that also has deleterious side-effects. The precise modes of action of arsenicals and antimonials are not known, although these compounds are known to bind to sulfhydryl groups. Arsenicals and antimonials may inhibit central energy metabolizing enzymes such as those of the glycolytic pathway and tricarboxylic acid cycle.

4.4. Drug Resistance

In recent years, drug resistance has emerged as a major impediment to the successful treatment and control of diseases of parasitic origin, particularly malaria but other parasitic diseases as well. Chemotherapeutic failure as a consequence of drug resistance in malaria is a catastrophic and currently intractable worldwide public health problem. Relatively few details concerning the biochemical and molecular mechanisms by which parasites become resistant or refractory to drugs in field isolates have been discerned. Malaria resistant to chloroquine is found throughout many endemic areas, a phenomenon that has accelerated the use of less efficacious and more costly therapeutic agents.

The parasites are developing resistance to these alternative drugs as well. In fact, resistance to multiple antimalarials has been observed in both clinical isolates (43) and in drug-selected organisms *in vitro* (44). In some strains, drug resistance has been ascribed to diminished drug accumulation or augmented drug efflux (45). The resistance of *P. falciparum* strains to multiple drugs and the observation that calcium antagonists can increase drug accumulation and growth susceptibility in chloroquine-resistant parasites is analogous to the phenomenon of multidrug resistance in mammalian cells (46). Several *mdr* genes have been identified within the genome of *P. falciparum* (47,48), and higher *mdr* gene copy number and increased *mdr* gene expression were reported among several strains of drug-resistant *P. falciparum* (49). Analysis of recombinant progeny between chloroquine-sensitive and chloroquine-resistant *P. falciparum* has indicated that the drug insensitivity and rapid drug efflux phenotype of the resistant strains is not linked to the presence or amplification of either of the two malarial *mdr* genes (49). More recent evidence has suggested that overexpression of one *mdr* gene, *pfmdrl*, is linked to mefloquine resistance (50).

Since *Leishmania* can be cultured easily, this genus has served as an *in vitro* prototype for investigating drug resistance in parasitic protozoa. Two classes of drug-resistant *Leishmania* have been generated and characterized. The first is that in which a genetic lesion leads to impaired transport or metabolic function. Mutations in transport pathways (51) and in metabolic enzymes (52) have provided considerable insight into the avenues by which certain nutrients are metabolized in *Leishmania*.

A second class of drug-resistant *Leishmania* is that which has undergone gene or DNA sequence amplification in response to drug selection (53). Gene amplification is usually induced by exposing wild-type *Leishmania* to gradually increasing concentrations of drug. Strains that have amplified specific genes or DNA sequences survive selective pressure by overexpressing specific transcripts and protein. Amplification of target genes has been reported for *Leishmania* species isolated for their resistance to methotrexate (MTX), 10-propargyl-5,8-dideazafolate (CB3717), tunicamycin (TM), mycophenolic acid (MPA), and DL-α-difluoromethylornithine (DFMO), or vinblastine toxicity (54–58). Drug-induced DNA amplifications in *Leishmania* are often arrayed as extrachromosomal circular molecules (53). These circular DNA molecules include the R region found in MTX- and CB3717-selected *Leishmania* that have amplified the dihydrofolate reductase–thymidylate synthase gene, the H region that is amplified in cell lines selected for their resistance to a variety of structurally and mechanistically unrelated drugs and in several unselected laboratory strains (59), a circle of variable size in TM-selected lines. The H region contains a gene, *pgpA*, homologous to the mammalian *mdr* gene, and amplification of the *L. major lmpgpA* gene after transfection into wild-type parasites confers resistance to arsenite and certain antimonials but not to MTX or to several of the hydrophobic mammalian P-glycoprotein substrates (60). A second gene encoding a mammalian P-glycoprotein family member, *ldmdrl*, has recently been identified and established to bestow concurrent resistance to several hydrophobic drugs after amplification by either direct drug selection or subsequent to transfection (61). Recently, linear extrachromosomal DNA molecules associated with drug resistance have been identified in several drug-selected *L. donovani* strains. These include a 280 kb DNA in an MPA-resistant strain that encompasses the IMP dehydrogenase gene (62) and a 140 kb linear DNA in DFMO-generated parasites that contains the ornithine decarboxylase gene (63).

The circular and linear extrachromosomal elements in *Leishmania* can either be maintained or lost in the absence of selective pressure exerted by the corresponding drug. It has been conjectured that the formation and maintenance of extrachromosomal circles is stabilized only after prolonged exposure of cells to drug. However, after two years of maintenance under selective pressure, the linear DNA molecules still retain their inherent stability (MPA resistance) or instability (DFMO) resistance. The *cis* acting elements that provide centromeric and telomeric function to leishmanial DNAs have not yet been elucidated.

4.5. Conclusion

Classical antiparasitic agents have served medicine relatively well. It is widely recog-nized that less expensive, efficacious, and less toxic agents are needed if treatment is to be brought to the large populations. The application of biochemistry to parasitic drug research and the advent of molecular biology in the understanding of parasite biology have brought this aspect of science forward significantly. The application of purine analogs, polyamine inhibitors and the avermectins are only the beginning in our search for improved antiparasitic chemotherapy.

5. ACKNOWLEDGEMENTS

This project was supported by Grants AI 23682 and AI 32036 from the National Institutes of Allergy and Infectious Diseases and in part by a grant from the Burroughs Wellcome Fund (BU). BU is a Burroughs Wellcome Fund Scholar in Molecular Parasitology.

REFERENCES

1. Marr, J. J. (1994) Antiprotozoal and anthelminthic chemotherapy. In: *Infectious Diseases*, 5th edn (eds Hoeprich, P. D., Jordan, M. C. and Ronald, A. R.), J. B. Lippincott, Philadelphia, pp. 289–298.
2. Ginger, C. D. (1991) Possibilities for new antiprotozoa drugs. In: *Biochemical Protozoology* (eds Coombs, G. and North, N.), Taylor and Francis, London, pp. 605–621.
3. Opperdoes, F. R. (1987) Compartmentation of carbohydrate metabolism in trypanosomes. *Annu. Rev. Microbiol.* **41**: 127–151.
4. Muller, M. (1988) Energy metabolism of protozoa without mitochondria. *Annu. Rev. Microbiol.* **42**: 465–488.
5. Wierenga, R. K., Swinkels, B. W., Michels, P. A. M. *et al.* (1987) Common elements on the surface glycolytic enzymes from *Trypanosoma brucei* may serve as topogenic signals for import into glycosomes. *EMBO J* **6**: 215–221.
6. Clarkson, A. B. Jr and Brohn, G. H. (1976) Trypanosomiasis: an approach to chemotherapy by the inhibition of carbohydrate catabolism. *Science* **194**: 204–206.
7. McCombille, M. J. (1991) Glycosylated phosphatidylinositols. In: *Biochemical Protozoology* (eds Coombs, G. and North, M.), Taylor and Francis, London, pp. 286–304.
8. Boothroyd, J. C. (1985) Antigenic variation in African trypanosomes. *Annu. Rev. Microbiol.* **39**: 475–502.
9. Doering, T. L., Masterson, W. J., Hart, G. W. and Englund, P. T. (1990) Biosynthesis of glycosyl phosphatidylinositol membrane anchors. *J. Biol. Chem.* **265**: 611–614.

10. Doering, T. L., Raper, J., Buxbaum, L. U. *et al.* (1991) An analog of myristic acid with selective toxicity for African trypanosomes. *Science* **252**: 1851–1854.
11. Henderson, G. B. and Cerami, A. (1989) Trypanothione is the primary target for arsenical drugs against African trypanosomes. *Proc. Natl. Acad. Sci. USA.* **86**: 2607–2611.
12. Barrett, A. J. and McDonald, J. K. (1980) *Mammalian Proteases: A Glossary and Bibliography*, Vol. I, Academic Press, London.
13. Barrett, A. J. and Salvesen, G. (1986) Proteinase inhibitors. *Research Monographs in Cell and Tissue Physiology*, Vol. 12, Elsevier, Amsterdam.
14. Rosenthal, P. J. (1991) Proteinases of Mammalian parasites, In: *Biochemical Protozoology* (eds Coombs, G. and North, M.) Taylor and Francis, London pp. 257–269.
15. North, M. J. (1991) Proteinases of parasitic protozoa: an overview. In: *Biochemical Protozoology* (eds Coombs, G. and North, M.), Taylor and Francis, London, pp. 180–185.
16. Ring, C. S., Sun, E., McKerrow, J. H. *et al.* (1993) Structure-based inhibition design by using protein models for the development of antiparasitic agents. *Proc. Natl. Acad. Sci. USA* **90**: 3583–3587.
17. Marr, J. J. (1991) Purine metabolism in parasitic protozoa and its relationship to chemotherapy. In: *Biochemical Protozoology* (eds Coombs, G. and North, M.) Taylor and Francis, London pp. 524–536.
18. Marr, J. J. and Berens, R. L. (1983) Pyrazolopyrimidine metabolism in pathogenic trypanosomatids. *Mol. Biochem. Parasitol.* **7**: 339–356.
19. Marr, J. J. (1991) Purine analogues as chemotherapeutic agents in leishmaniasis and American trypanosomiasis. *J. Lab Clin. Med.* **118**: 111–119.
20. Martinez, S. and Marr, J. J. (1992) Efficacy of allopurinol in the treatment of American cutaneous leishmaniasis. *N. Engl. J. Med.* **326**: 741–744.
21. Gallerano, R. H., Sosa, R. R. and Marr, J. J. (1990) Therapeutic efficacy of allopurinol in patients with chronic Chagas' disease. *Am. J. Trop. Med. Hyg.* **43**: 159–166.
22. Bacchi, C. J., Nathan, H. C., Clarkson, A. B., Bienen E. J., Bitonti, A. J. and McCann P. O. (1987) Effects of the ornithine decarboxylase inhibitors DL-alpha-difluoromethylornithine and alpha-monofluromethyldehydroornithine methyl ester alone in combination with serum against *Trypanosoma brucei brucei* central nervous system models. *Am. J. Trop. Med. Hyg.* **36**: 46–52.
23. Bacchi, C. J. and McCann, P. P. (1987) Parasitic protozoa and polyamines. In: *Inhibition of Polyamine Metabolism* (eds McCann, P. P., Pegg, A. E. and Sjoerdsma, A.) Academic Press, New York, pp. 317–344.
24. Bitonti, A. J., McCann, P. P., Bacchi, C. J. and Sjoerdsma, A. (1991) Polyamine biosynthesis as a target for the chemotherapy of trypanosomatid infections. In: *Biochemical Protozoology* (eds Coombs, G. and North, M.), Taylor and Francis, London, pp. 286–304.
25. Hanson, S., Adelman, J. and Ullman, B. (1992) Amplification and molecular cloning of the ornithine decarboxylase gene of *Leishmania donovani*. *J. Biol. Chem.* **267**: 2350–2359.
26. Wright, P. S., Byers, T. L., Cross-Doersen, C. E., McCann, P. P. and Bitonti, A. J. (1991) Irreversible inhibition of S-adenosylmethionine decarboxylase in *Plasmodium falciparum*-infected erythrocytes: growth inhibition *in vitro*. *Biochem. Pharmacol.* **41**: 1731–1718.
27. Bitonti, A. J., Byers, T. L., Bush, T. L. *et al.* (1990) Cure of *Trypanosoma brucei brucei* and *Trypanosoma brucei rhodesiense* infections in mice with an irreversible inhibitor of S-adenosylmethionine decarboxylase. *Antimicrob. Agents. Chemother.* **34**: 1485–1490.
28. Katz, M., Despommier, D. D. and Gwadz, R. (1989) *Parasitic Diseases*. Springer-Verlag, New York.
29. Garrett, C. E., Coderre, J. A., Meek, T. D. *et al.* (1985) A bifunctional thymidylate synthetase-dihydrofolate reductase in protozoa. *Mol. Biochem. Parasitol.* **11**: 257–265.
30. Beck, J. T. and Ullman, B. (1990) Nutritional requirements of wild-type and folate transport-deficient *Leishmania donovani* for pterins and folates. *Mol. Biochem. Parasitol.* **43**: 221–230.
31. Beck, J. T. and Ullman, B. (1991) Biopterin conversion to reduced folates by *Leishmania donovani* promastigotes, *Mol. Biochem. Parasitol.* **49**: 21–28.
32. Fitch, C. D., Chevli, R., Banyal, H. S., Phillips, G., Pfaller, M. A. and Krogstad, D. J. (1982) Lysis of *Plasmodium falciparum* by ferriprotoporphyrin IX and a chloroquine–ferri-protoporphyrin IX complex. *Anitmicrob. Agents Chemother.* **21**: 819–822.

33. Vander Jagt, D. L., Hunsaker, L. A. and Campos, N. M. (1986) Characterization of a hemoglobin-degrading, low molecular weight protease from *Plasmodium falciparum*. *Mol. Biochem. Parasitol.* **18**: 389–400.

34. Slater, A. F. G. and Cerami, A. (1992) Inhibition by chloroquine of a novel haem polymerase enzyme activity in malaria trophozoites. *Nature* **355**: 167–169.

35. Putter, I., MacConnell, J. G., Presiser, F. A., Haidri, A. A., Ristich, S. S. and Dybas, R. A. (1981) Avermectins: novel insecticides, ascarids, and nematicides from a soil microorganism. *Experientia* **37**: 963–964.

36. Cupp, E. W., Bernardo, M. J., Kiszewski, A. E. *et al.* (1986) The effects of ivermectin on transmission of *Onchocerca volvulus*. *Science* **231**: 740–742.

37. Oslind, D. A., Cifelli, S. and Lang, R. (1979) Insecticidal activity of the antiparasitic avermectins. *Vet. Rec.* **105**: 168.

38. Fritz, L. C., Wang, C. C. and Gorio, A. (1979) Avermectin B1a irreversibly blocks postsynaptic potentials at the lobster neuromusular junction by reducing muscle membrane resistance. *Proc. Natl. Acad. Sci. USA.* **76**: 2062–2066.

39. Mellin, T. N., Busch, R. D. and Wang, C. C. (1983) Postsynaptic inhibition of invertebrate neuromuscular transmission by avermectin B1a. *Neuropharmacology* **22**: 89–96.

40. Holden-Dye, L., Hewitt, G. M., Wann, K. T., Krogsgaard-Larsen, P. and Walker, R. J. (1988) Studies involving avermectin and the 4-aminobutyric acid (GABA) receptor of *Ascaris suum* muscle. *Pestic. Sci* **24**: 231–245.

41. Cully, D. F. and Paress, P. S. (1991) Solubilization and characterization of a high affinity ivermectin binding site from *Caenorhabditis elegans*. *Mol. Pharmacol.* **40**: 326–332.

42. Arena, J. P., Liu, K. K., Paress, P. S., Schaeffer, J. M. and Cully, D. F. (1992) Expression of a gluatamate-activated chloride current in *Xenopus* oocytes injected with *Caenorhabditis elegans* RNA *Mol. Brain Res.* **15**: 339–348.

43. Webster, H. K., Thaithong, S., Pavanand, K., Yongvanitchit, K., Pinswasdi, C. and Boudreau, E. F. (1985) Cloning and characterization of mefloquine-resistant *Plasmodium falciparum* from Thailand. *Am. J. Trop. Med. Hyg.* **34**: 1022–1027.

44. Oduola, A. M., Milhous, W. K., Weatherly, N. F., Bowdre, J. H. and Desjardins, R. E. (1988) *Plasmodium falciparum*: induction of resistance to mefloquine in cloned strains by continuous drug exposure *in vitro*. *Exp. Parasitol.* **67**: 354–360.

45. Krogstad, D. J., Gluzman, I. Y., Kyle, D. E., Oduola, A. M. J., Martin, S. K., Milhous, W. K. and Schlesinger, P. H. (1987) Efflux of chloroquine from *Plasmodium falciparum*: mechanism of chloroquine resistance. *Science* **238**: 1283–1285.

46. Endicott, J. A. and Ling, V. (1989) The biochemistry of P-glycoprotein mediated multidrug resistance. *Annu. Rev. Biochem.* **58**: 137–171.

47. Wilson, C. M., Serrano, A. E., Wasley, A., Bogenschutz, M. P., Shankar, A. H. and Wirth, D. F. (1989) Amplification of a gene related to mammalian *mdr* genes in drug-resistant *Plasmodium falciparum*. *Science* **244**: 1184–1186.

48. Wellems, T. E., Panton, L. J., Gluzman, I. Y. *et al.* (1990) Chloroquine resistance not linked to *mdr*-like genes in a *Plasmodium falciparum* cross. *Nature* **345**: 253–255.

49. Foote, S. J., Thompson, J. K., Cowman, A. F. and Kemp, D. J. (1989) Amplification of the multidrug resistance gene in some chloroquine-resistant isolates of *P. falciparum*. *Cell* **57**: 921–930.

50. Wilson, C. M., Volkman, S. K., Thaithong, S. *et al.* (1993) Amplification of *pfmdr1* associated with mefloquine and halofantine resistance in *Plasmodium falciparum* from Thailand. *Mol. Biochem. Parasitol.* **57**: 151–160.

51. Kaur, K., Coons, T., Emmett, K., and Ullman, B. (1988) Methotrexate-resistant *Leishmania donovani* genetically deficient in the folate-methotrexate transporter. *J. Biol. Chem.* **263**: 7020–7028.

52. Iovannisci, D. M., Goebel, D., Kaun, K., Allen, K. and Ullman, B. (1984). Genetic analysis of adenine metabolism in *Leishmania donovani*: evidence for diploidy at the adenine phosphoribosyl transferase locus. *J. Biol. Chem.* **259**: 14617–14623.

53. Beverley, S. M. (1991) Gene amplification in *Leishmania*. *Annu. Rev. Microbiol.* **45**: 417–444.

54. Beverley, S. M., Coderre, J. A., Santi, D. V. and Schimke, R. T. (1984) Unstable DNA amplifications in methotrexate-resistant *Leishmania* consist of extrachromosomal circles which relocalize during stabilization. *Cell* **38**: 431–439.
55. Garvey, E. P., Coderre, J. A. and Santi, D. V. (1985) Selection and properties of *Leishmania tropica* resistant to 10-propargyl-5,8-dideazafolate, an inhibitor of thymidylate synthetase. *Mol. Biochem. Parasitol.* **17**: 79–91.
56. Katakura, K., Peng, Y., Pithawalla, R., Detke, S. and Chang, K. P. (1991) Tunicamycin-resistant variants from five species of *Leishmania* contain amplified DNA in extrachromosomal circles of different sizes with a transcriptionally active homologous region. *Mol. Biochem. Parasitol.* **44**: 233–244.
57. Wilson, K., Collart, F. R., Huberman, E., Stringer, J. R. and Ullman, B. (1991) Amplification and molecular cloning of the IMP dehydrogenase gene of *Leishmania donovani. J. Biol. Chem.* **266**: 1665–1671.
58. Hanson, S., Adelman, J. and Ullman, B. (1992) Amplification and molecular cloning of the ornithine decarboxylase gene from *Leishmania donovani. J. Biol. Chem.* **267**: 2350–2359.
59. Ouellette, M., Fase-Fowler, F. and Borst, P. (1990) The amplified H circle of methotrexate-resistant *Leishmania tarentolae* contains a novel P-glycoprotein gene. *EMBO J* **9**: 1027–1033.
60. Callahan, H. and Beverley, S. M. (1991) Heavy metal resistance: a new role for P-glycoproteins in *Leishmania. J. Biol. Chem.* **266**: 18427–18430.
61. Henderson, D. M., Sifri, C. D., Rodgers, M., Wirth, D. F., Hendrickson, N. and Ullman, B. (1992) Multidrug resistance in *Leishmania donovani* is conferred by amplification of a gene homologous to the mammalian *mdr*1 gene. *Mol. Cell. Biol.* **12**: 2855–2865.
62. Wilson, K., Beverly, S. M. and Ullman, B. (1992) Stable amplification of a linear extrachromosomal DNA in mycophenolic acid-resistant *Leishmania donovani. Mol. Biochem. Parasitol.* **55**: 197–206.
63. Hanson, S., Beverly, S. M., Wagner, W. and Ullman, B. (1992) Unstable amplification of two extrachromosomal elements in α-difluoromethylornithine resistant *Leishmania donovani. Mol. Cell. Biol.* **12**: 5499–5507.

Index

Note: Page references in italics refer to diagrams and tables. Textual references may also appear on those pages.

Printed and bound by CPI Group (UK) Ltd, Croydon, CR0 4YY

03/10/2024

01040323-0012